NEUROSCIENCE
Fundamentals for Rehabilitation

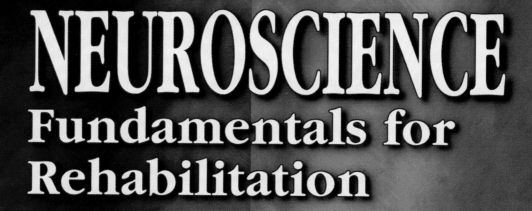

NEUROSCIENCE
Fundamentals for
Rehabilitation

LAURIE LUNDY-EKMAN, PT, PhD
Associate Professor of Physical Therapy
Pacific University
Forest Grove, Oregon

W.B. SAUNDERS COMPANY
A Division of Harcourt Brace & Company

Philadelphia   London   Toronto
Montreal   Sydney   Tokyo

W. B. SAUNDERS COMPANY

A Division of Harcourt Brace & Company

The Curtis Center
Independence Square West
Philadelphia, Pennsylvania 19106

Library of Congress Cataloging-in-Publication Data

Lundy-Ekman, Laurie.
   Neuroscience : Fundamentals for rehabilitation / Laurie Lundy-
Ekman.
     p. cm.
   ISBN 0–7218–4717–0
   1. Neurosciences. 2. Neurophysiology. 3. Nervous system—
Diseases—Patients—Rehabilitation. 4. Medical rehabilitation.
5. Physical therapy. I. Title.
   [DNLM: 1. Nervous System Physiology. 2. Nervous System—anatomy &
histology. 3. Nervous System—physiopathology. 4. Rehabilitation.
5. Physical Therapy. WL 102 L982n 1998]
RC341.L86 1998
612.8—dc21
DNLM/DLC   97–50316

NEUROSCIENCE: FUNDAMENTALS FOR REHABILITATION   ISBN 0–7216–4717–0

Printed in the United States of America

Last digit in the print number: 9   8   7   6   5   4   3

To Andy and Lisa

Contributor

Chapter 2   Physical and Electrical Properties of Cells in the Nervous System
Chapter 3   Synapses and Synaptic Transmissions
Chapter 4   Neuroplasticity

Anne Burleigh-Jacobs BS, PhD
Dasstraat 18
Hengelo OV
The Netherlands

Preface

This text is unique in specifically addressing neuroscience issues important to the practice of physical rehabilitation. For example, clinical issues including abnormal muscle tone, chronic pain, and vestibular function are stressed, while topics often covered extensively in neuroscience texts, such as the function of neurons in the retina, are omitted.

A logical progression of information—from molecular and cellular levels, to systems, then to regions—provides a framework for the student to understand neural organization. Special elements are interspersed throughout the text to enhance the learning experience: personal stories contributed by people with neurological disorders give the information an immediacy and a connection with reality that are sometimes missing from textbook presentations; clinical notes containing case examples challenge the student to apply the information to clinical situations; and disease profiles supply a quick summary of the features of neurological disorders commonly encountered in clinical practice. Brief introductions to evaluation techniques serve as guides for practicing clinical examinations. The reference and bibliographic information is highly selective, reflecting only materials that are most likely to be useful.

My primary goal in writing this text is to provide the depth of information appropriate for entry level clinicians, by avoiding excessive detail and emphasizing clinically relevant neuroscience. As Peter Greene* observed in a different context, once the essential information is organized into a framework, it is possible to know less and yet understand more. For the novice, excessive detail obscures important information and interferes with comprehension. By stressing the organization and relationships within the nervous system, this text supports future clinicians by establishing the framework essential for understanding the nervous system.

*Greene, P. H. (1972). Problems of organization of motor systems. In: Rosen, R, Snell, P. M. (Eds.). Progress in theoretical biology. San Diego, CA: Academic Press Inc., pp. 304–338.

Acknowledgments

Numerous people have contributed to this book. Many students and people with neurological disorders have served as my teachers. Anne Burleigh-Jacobs provided insightful critiques of the entire book, contributed three chapters, and provided encouragement. Renate Powell has given me unfailing support and many helpful criticisms. Mike Studer, Robert Rosenow, Nancy Heinly, Mike Hmura, Susan Hendrickson, and Daiva Banaitis also provided helpful comments on individual chapters. Lisa Million and Rodd Ambroson produced the original illustrations, applying considerable artistry and knowledge to each figure. Dr. Melvin Ball generously provided color photographs of many pathological conditions. Bob Sellin contributed the nerve conduction velocity data from his clinical practice. The authors of the personal stories willingly shared their experiences and are each credited at the end of their contribution. At W. B. Saunders Company, Margaret Biblis and Shirley Kuhn took a sincere interest in the project and expertly facilitated the entire affair. Senior editor Andrew Allen provided magical solutions to problems. My parents and sisters have been supportive and a source of inspiration. My friends Jane and John Lebens, my husband Andy, and my daughter Lisa have been remarkably understanding, good humored, and patient throughout the writing of this book.

Contents

1 **Introduction to Neuroscience** 1
What Is Neuroscience? 2
Organization of This Book 3
Introduction to Neuroanatomy 5
Clinical Application of Learning
  Neuroscience 15

2 **Physical and Electrical Properties
of Cells in the Nervous System** . . 19
Anne Burleigh-Jacobs
Supporting Cells 20
Structure of Neurons 23
Direction of Information Flow
  in Neurons 26
Transmission of Information by Neurons 27
Interactions Between Neurons 37
Diseases Affecting the Cells of the Nervous
  System 37

3 **Synapses and Synaptic
Transmissions** 43
Anne Burleigh-Jacobs
Introduction 44
Electrical Potentials at Synapses 46
Synaptic Receptors 49
Neurotransmitters 53
Toxins Affecting Neural Signaling
  at Synapses 55

4 **Neuroplasticity** 57
Anne Burleigh-Jacobs
Introduction 58
Habituation 58
Learning and Memory 58
Recovery from Injury 59
Metabolic Effects of Brain Injury 64
Effects of Rehabilitation on Plasticity 65
Neurotransplantation and Stereotaxic
  Surgical Approaches 66

5 **Development of the Nervous
System** 69
Introduction 70
Developmental Stages in Utero 70
Formation of the Nervous System 71
Cellular-Level Development 76
Developmental Disorders: In Utero
  and Perinatal Damage of the Nervous
  System 78
Nervous System Changes During
  Infancy 82

6 **Somatosensory System** 85
Introduction 86
Peripheral Somatosensory Neurons 86
Pathways to the Brain 92

7   *Somatosensation: Clinical Applications* 107
Introduction   108
Testing Somatosensation   108
Electrodiagnostic Studies   113
Sensory Syndromes   115
Clinical Perspective on Pain   118
Specific Types of Pain   120

8   *Autonomic Nervous System* 131
Sensory Receptors   132
Afferent Pathways   132
Central Regulation of Visceral
    Function   132
Efferent Pathways   135
Sympathetic Nervous System   137
Parasympathetic Nervous System   141
Clinical Correlations   143

9   *Motor System* 149
Introduction   150
Lower Motoneurons   151
Spinal Region   152
Descending Motor Pathways (Upper
    Motoneurons)   159
Control Circuits   164

10   *Clinical Disorders of the Motor System* 175
Muscle Strength and Muscle Bulk   176
Involuntary Muscle Contractions   176
Muscle Tone   176
Disorders of Lower Motoneurons   177
Upper Motoneuron Syndrome   178
Types of Upper Motoneuron
    Lesions   181
Treatment of Upper Motoneuron
    Lesions   184
Voluntary Motor System
    Degeneration   185

Pathology of the Basal Ganglia   185
Hyperkinetic Disorders   189
Signs and Symptoms of Cerebellar
    Disease   191
Three Fundamental Types
    of Movement   192
Testing the Motor System   198

11   *Peripheral Nervous System* 207
Introduction   208
Anatomy of Peripheral Nerves   208
Neuromuscular Junction   212
Dysfunction of Peripheral Nerves   212
Classification of Neuropathies   212
Dysfunctions of the Neuromuscular
    Junction   217
Myopathy   217
Electrodiagnostic Studies   217
Clinical Application   217

12   *Spinal Region* 223
Anatomy of the Spinal Region   224
Functions of the Spinal Cord   227
Spinal Cord Motor Coordination   229
Spinal Control of Pelvic Organ
    Function   231
Effects of Segmental and Tract Lesions
    in the Spinal Region   232
Spinal Region Syndromes   235
Effects of Spinal Region Dysfunction
    on Pelvic Organ Function   237
Traumatic Spinal Cord Injury   238
Specific Disorders Affecting Spinal
    Region Function   243

13   *Cranial Nerves* 249
Introduction   250
Cranial Nerve I: Olfactory   251
Cranial Nerve II: Optic   251

Cranial Nerves III, IV, and VI:
   Oculomotor, Trochlear,
   and Abducens   251
Cranial Nerve V: Trigeminal   255
Cranial Nerve VII: Facial   259
Cranial Nerve VIII:
   Vestibulocochlear   259
Cranial Nerve IX:
   Glossopharyngeal   261
Cranial Nerve X: Vagus   265
Cranial Nerve XI: Accessory   266
Cranial Nerve XII: Hypoglossal   266
Cranial Nerves Involved in Swallowing
   and Speaking   266
Systems Controlling Cranial Nerve
   Lower Motoneurons   266
Disorders of the Cranial Nerves   267
Testing Cranial Nerves   270

14   **Brain Stem Region** *277*
Introduction   278
Anatomy of the Brain Stem   278
Reticular Formation   280
Reticular Nuclei and Their
   Neurotransmitters   282
Medulla   284
Pons   286
Midbrain   288
Cerebellum   288
Disorders in the Brain Stem
   Region   289

15   **Auditory, Vestibular,
   and Visual Systems** *299*
Introduction   300
Auditory System   300
Vestibular System   301
Visual System   303
Disorders of Auditory, Vestibular,
   and Visual Systems   308

16   **Cerebrum** *315*
Introduction   316
Diencephalon   316
Subcortical Structures   319
Cerebral Cortex   320
Limbic System   327
Emotions and Behavior   330
Psychological and Somatic
   Interactions   330
Memory   333
Communication   335
Comprehension of Spatial
   Relationships   336
Use of Visual Information   336
Consciousness   337

17   **Cerebrum: Clinical
   Applications** *341*
Signs of Damage to Cerebral
   Systems   342
Diseases and Disorders Affecting
   Cerebral Function   350

18   **Support Systems: Blood Supply
   and Cerebrospinal Fluid
   System** *363*
Introduction   364
Cerebrospinal Fluid System   364
Vascular Supply   368

Appendix ■ *Neurotransmitters
   and Neuromodulators*   383
   Introduction   383
   Neurotransmitters   383
   Neuromodulators   384
Answers   387
Glossary   407
Index   425

Atlas

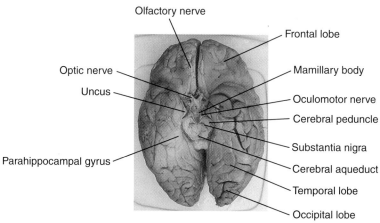

Olfactory nerve

Optic nerve

Uncus

Parahippocampal gyrus

Frontal lobe

Mamillary body

Oculomotor nerve

Cerebral peduncle

Substantia nigra

Cerebral aqueduct

Temporal lobe

Occipital lobe

A1

Inferior cerebrum and midbrain. The pons, medulla, and cerebellum have been removed.

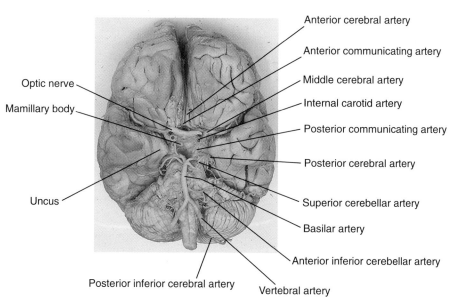

Anterior cerebral artery

Anterior communicating artery

Optic nerve

Mamillary body

Middle cerebral artery

Internal carotid artery

Posterior communicating artery

Posterior cerebral artery

Uncus

Superior cerebellar artery

Basilar artery

Anterior inferior cerebellar artery

Posterior inferior cerebral artery

Vertebral artery

A2

Inferior brain and its arterial system.

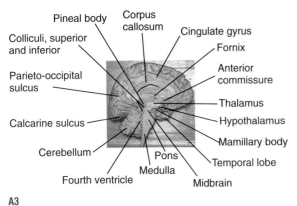

Pineal body
Corpus callosum
Cingulate gyrus
Colliculi, superior and inferior
Fornix
Anterior commissure
Parieto-occipital sulcus
Thalamus
Calcarine sulcus
Hypothalamus
Mamillary body
Cerebellum
Pons
Temporal lobe
Medulla
Fourth ventricle
Midbrain

A3
Midsagittal view of the brain.

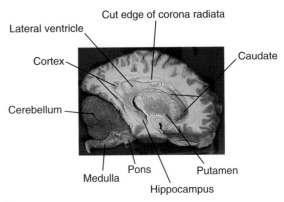

Cut edge of corona radiata
Lateral ventricle
Cortex
Caudate
Cerebellum
Medulla
Pons
Putamen
Hippocampus

A4
Lateral view of the interior right hemisphere. The lateral two-thirds of the right hemisphere have been removed to show internal structures.

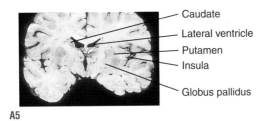

Caudate
Lateral ventricle
Putamen
Insula
Globus pallidus

A5
Coronal section through the anterior limb of the internal capsule.

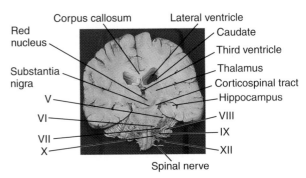

Corpus callosum
Lateral ventricle
Red nucleus
Caudate
Third ventricle
Substantia nigra
Thalamus
V
Corticospinal tract
VI
Hippocampus
VII
VIII
X
IX
XII
Spinal nerve

A6
Coronal section through the third ventricle.

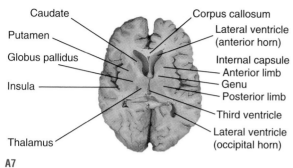

Caudate

Putamen

Globus pallidus

Insula

Thalamus

Corpus callosum

Lateral ventricle (anterior horn)

Internal capsule
Anterior limb
Genu
Posterior limb

Third ventricle

Lateral ventricle (occipital horn)

A7

Horizontal section through the cerebrum.

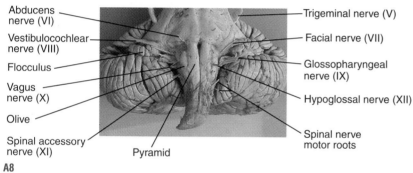

Abducens nerve (VI)

Vestibulocochlear nerve (VIII)

Flocculus

Vagus nerve (X)

Olive

Spinal accessory nerve (XI)

Pyramid

Trigeminal nerve (V)

Facial nerve (VII)

Glossopharyngeal nerve (IX)

Hypoglossal nerve (XII)

Spinal nerve motor roots

A8

Anterior view of the pons, medulla, and cerebellum.

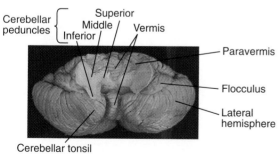

Cerebellar peduncles
Superior
Middle
Inferior
Vermis

Paravermis

Flocculus

Lateral hemisphere

Cerebellar tonsil

A9

Anterior view of the cerebellum with the brain stem removed.

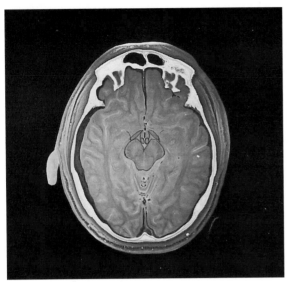

A10

Contrast-enhanced CAT (computerized axial tomography) scan of a normal brain. Horizontal section. (With permission from Bo, W. J., et al. [1990]. Basic atlas of sectional anatomy with correlated imaging. [2nd ed.]. Philadelphia: W. B. Saunders.)

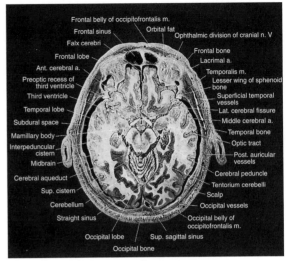

Frontal belly of occipitofrontalis m.
Frontal sinus
Orbital fat
Ophthalmic division of cranial n. V
Falx cerebri
Frontal lobe
Frontal bone
Ant. cerebral a.
Lacrimal a.
Preoptic recess of third ventricle
Temporalis m.
Lesser wing of sphenoid bone
Third ventricle
Superficial temporal vessels
Temporal lobe
Lat. cerebral fissure
Subdural space
Middle cerebral a.
Mamillary body
Temporal bone
Interpeduncular cistern
Optic tract
Midbrain
Post. auricular vessels
Cerebral aqueduct
Cerebral peduncle
Sup. cistern
Tentorium cerebelli
Cerebellum
Scalp
Straight sinus
Occipital vessels
Occipital belly of occipitofrontalis m.
Occipital lobe
Sup. sagittal sinus
Occipital bone

A11

Photograph of brain section imaged in A10. (With permission from Bo, W. J., et al. [1990]. Basic atlas of sectional anatomy with correlated imaging. [2nd ed.]. Philadelphia: W. B. Saunders.)

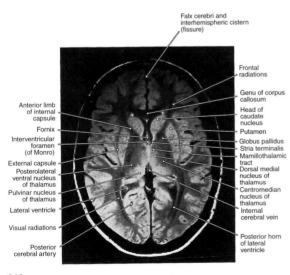

Falx cerebri and interhemispheric cistern (fissure)
Frontal radiations
Genu of corpus callosum
Anterior limb of internal capsule
Head of caudate nucleus
Fornix
Putamen
Interventricular foramen (of Monro)
Globus pallidus
Stria terminalis
External capsule
Mamillothalamic tract
Posterolateral ventral nucleus of thalamus
Dorsal medial nucleus of thalamus
Pulvinar nucleus of thalamus
Centromedian nucleus of thalamus
Lateral ventricle
Internal cerebral vein
Visual radiations
Posterior cerebral artery
Posterior horn of lateral ventricle

A12

MRI (magnetic resonance imaging) of a normal brain. Horizontal section. (With permission from Pomeranz, S. J. [1989]. Craniospinal magnetic resonance imaging. Philadelphia: W. B. Saunders.)

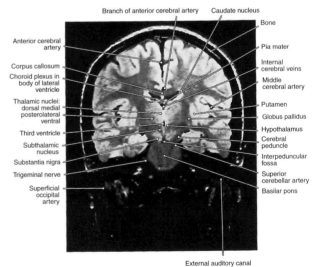

Branch of anterior cerebral artery
Caudate nucleus
Bone
Anterior cerebral artery
Pia mater
Corpus callosum
Internal cerebral veins
Choroid plexus in body of lateral ventricle
Middle cerebral artery
Thalamic nuclei: dorsal medial posterolateral ventral
Putamen
Globus pallidus
Third ventricle
Hypothalamus
Subthalamic nucleus
Cerebral peduncle
Substantia nigra
Interpeduncular fossa
Trigeminal nerve
Superior cerebellar artery
Superficial occipital artery
Basilar pons
External auditory canal

A13

MRI (magnetic resonance imaging) of a normal brain. Coronal section. (With permission from Pomeranz, S. J. [1989]. Craniospinal magnetic resonance imaging. Philadelphia: W. B. Saunders.)

Atlas

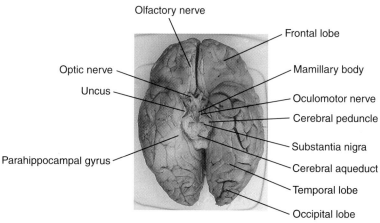

Olfactory nerve

Frontal lobe

Optic nerve

Mamillary body

Uncus

Oculomotor nerve

Cerebral peduncle

Substantia nigra

Parahippocampal gyrus

Cerebral aqueduct

Temporal lobe

Occipital lobe

A1

Inferior cerebrum and midbrain. The pons, medulla, and cerebellum have been removed.

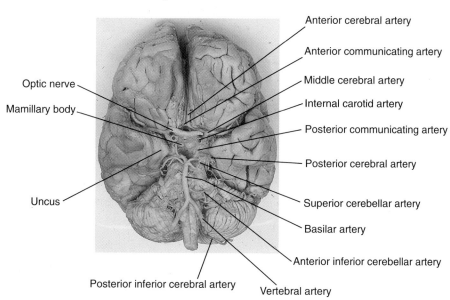

Anterior cerebral artery

Anterior communicating artery

Middle cerebral artery

Optic nerve

Internal carotid artery

Mamillary body

Posterior communicating artery

Posterior cerebral artery

Superior cerebellar artery

Uncus

Basilar artery

Anterior inferior cerebellar artery

Posterior inferior cerebral artery

Vertebral artery

A2

Inferior brain and its arterial system.

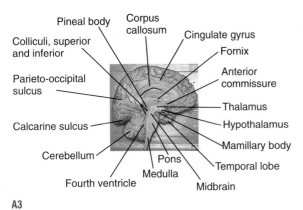

Pineal body · Corpus callosum · Cingulate gyrus
Colliculi, superior and inferior
Fornix
Parieto-occipital sulcus
Anterior commissure
Calcarine sulcus
Thalamus
Hypothalamus
Cerebellum
Mamillary body
Pons
Temporal lobe
Fourth ventricle
Medulla
Midbrain

A3

Midsagittal view of the brain.

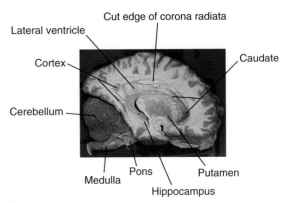

Cut edge of corona radiata
Lateral ventricle
Cortex
Caudate
Cerebellum
Medulla
Pons
Hippocampus
Putamen

A4

Lateral view of the interior right hemisphere. The lateral two-thirds of the right hemisphere have been removed to show internal structures.

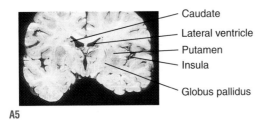

Caudate
Lateral ventricle
Putamen
Insula
Globus pallidus

A5

Coronal section through the anterior limb of the internal capsule.

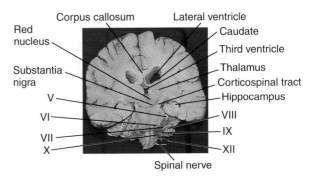

Red nucleus
Corpus callosum
Lateral ventricle
Caudate
Third ventricle
Substantia nigra
Thalamus
Corticospinal tract
V
Hippocampus
VI
VIII
VII
IX
X
XII
Spinal nerve

A6

Coronal section through the third ventricle.

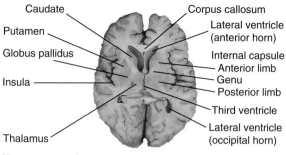

Caudate

Putamen

Globus pallidus

Insula

Thalamus

Corpus callosum

Lateral ventricle (anterior horn)

Internal capsule
Anterior limb
Genu
Posterior limb

Third ventricle

Lateral ventricle (occipital horn)

A7

Horizontal section through the cerebrum.

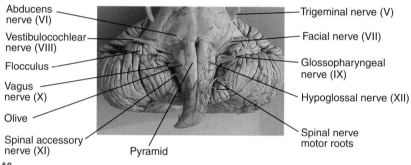

Abducens nerve (VI)

Vestibulocochlear nerve (VIII)

Flocculus

Vagus nerve (X)

Olive

Spinal accessory nerve (XI)

Pyramid

Trigeminal nerve (V)

Facial nerve (VII)

Glossopharyngeal nerve (IX)

Hypoglossal nerve (XII)

Spinal nerve motor roots

A8

Anterior view of the pons, medulla, and cerebellum.

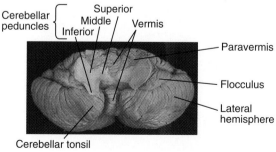

Cerebellar peduncles

Superior
Middle
Inferior

Vermis

Paravermis

Flocculus

Lateral hemisphere

Cerebellar tonsil

A9

Anterior view of the cerebellum with the brain stem removed.

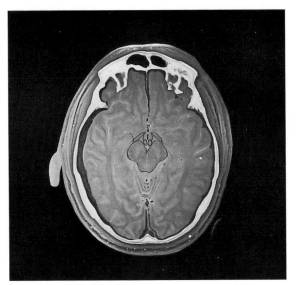

A10

Contrast-enhanced CAT (computerized axial tomography) scan of a normal brain. Horizontal section. (With permission from Bo, W. J., et al. [1990]. Basic atlas of sectional anatomy with correlated imaging. [2nd ed.]. Philadelphia: W. B. Saunders.)

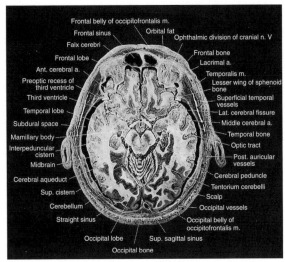

A11

Photograph of brain section imaged in A10. (With permission from Bo, W. J., et al. [1990]. Basic atlas of sectional anatomy with correlated imaging. [2nd ed.]. Philadelphia: W. B. Saunders.)

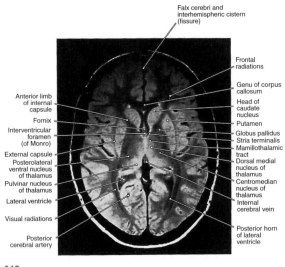

A12

MRI (magnetic resonance imaging) of a normal brain. Horizontal section. (With permission from Pomeranz, S. J. [1989]. Craniospinal magnetic resonance imaging. Philadelphia: W. B. Saunders.)

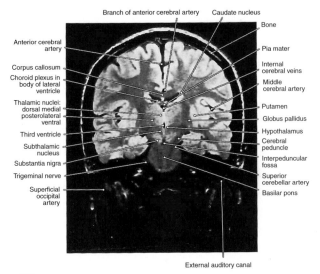

A13

MRI (magnetic resonance imaging) of a normal brain. Coronal section. (With permission from Pomeranz, S. J. [1989]. Craniospinal magnetic resonance imaging. Philadelphia: W. B. Saunders.)

1

Introduction to Neuroscience

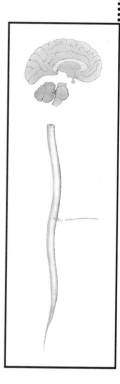

Every day, people live with disabilities related to nervous system damage or disease. People who have experienced brain damage, spinal cord injury, birth defects, and neurological diseases must cope with the effects. Tasks as seemingly simple as sitting, standing, walking, getting dressed, and remembering a name may become incredible challenges. Physical and occupational therapy play a crucial role in helping people regain the ability to function as independently as possible. The design of physical and occupational therapy treatments and management of each individual case are dependent on our understanding of the nervous system and continued research.

WHAT IS NEUROSCIENCE?

The quest to understand the nervous system is called **neuroscience.** Neuroscience is a relatively new science, concerned with the development, chemistry, structure, function, and pathology of the nervous system. Rigorous scientific research on neural function has a relatively short history, beginning in the later 1800s. At that time, physiologists Fritsch and Hitzig reported that electrical stimulation to specific areas of an animal's cerebral cortex elicits active movement, and physicians Broca and Wernicke separately confirmed, by autopsy, localized brain damage in people who had language deficits after stroke. About the same time, Hughlings Jackson proposed that multiple brain areas are essential for complex functions such as perception, action, and language.

About 1890, Cajal, a neuroanatomist, established that each nerve cell (neuron) is a distinct, individual cell, not directly continuous with other nerve cells. Sherrington, a physiologist studying involuntary reactions that occur in response to stimuli, proposed that nerve cells were linked by specialized connections he named synapses. The next major advances in understanding the nervous system did not occur until the 1950s, when both the electron microscope and the microelectrode were developed. The electron microscope allows visualization of cellular organelles, and the microelectrode can record the activity of a single nerve cell.

Beginning in the 1970s, new imaging techniques were developed that create clear images of the living spinal cord and brain, unobscured by the surrounding skull and vertebrae. These imaging techniques provide physiological and pathological information never before available. Computerized axial tomography (CAT), positron emission tomography (PET), and magnetic resonance imaging (MRI) scans all use computerized analysis to create an image of the nervous system (see Atlas and Fig. 18–19). In a CAT scan, a series of x-rays are analyzed by a computer to generate an image of the density of various areas in the nervous system. PET scans require injecting a radioactive substance into the bloodstream, then using a special camera to record signals emitted by radioactive decay. The amount of radioactive decay in an area is proportional to the local blood flow, so the computer-generated image is an indirect indicator of nerve cell activity. MRI uses a magnetic field to align naturally occurring protons in the body, and then a radio frequency pulse temporarily disrupts the alignment. The signal emitted as the nuclei return to their original position is detected, and these signals are converted into images of the nervous system. MRI can produce excellent three-dimensional images of nervous system tissues and provide information about activity-related changes in blood flow.

All of these techniques used historically to examine neural function are still used today, though with many refinements. Current approaches to understanding the nervous system include multiple levels of analysis:

- Molecular
- Cellular
- Systems
- Behavioral
- Cognitive

Molecular Neuroscience Molecular neuroscience investigates the chemistry and physics involved in neural function. Studies of the ionic exchanges required for a nerve cell to conduct information from one part of the nervous system to another and the chemical transfer of information between nerve cells are molecular-level neuroscience. Reduced to their most fundamental level, sensation, moving, understanding, planning, relating, speaking, and most other human functions depend on chemical and electrical changes in nervous system cells.

Cellular Neuroscience Cellular neuroscience considers distinctions between different types of cells in the nervous system and how each cell type functions. Inquiries into how an individual neuron processes and conveys information, how information is transferred among neurons, and the roles of non-neural cells in the nervous system are cellular-level questions.

Systems Neuroscience Systems neuroscience investigates groups of neurons that perform a common function. Systems-level analysis studies the connections, or circuitry, of the nervous system. Examples are the proprioceptive system, which conveys position and movement information from the musculoskeletal system to the central nervous system, and the motor system, which controls movement.

Behavioral Neuroscience Behavioral neuroscience looks at the interaction among systems that influence behavior. For example, studies of postural control investigate the relative influence of visual, vestibular, and proprioceptive sensations on balance under different conditions.

Cognitive Neuroscience Cognitive neuroscience covers the fields of thinking, learning, and memory. Studies of planning, using language, and the differences between memory for specific events and memory for performing motor skills are examples of cognitive-level analysis.

What Do We Learn from These Studies?

From a multitude of investigations at all levels of analysis in neuroscience, we have begun to be able to answer questions such as the following:

- How do ions influence nerve cell function?
- How does a nerve cell convey information from one location in the nervous system to another?
- How is language formed and understood?
- How does information about a hot stove encountered by a fingertip reach conscious awareness?
- How are the abilities to stand and walk developed and controlled?
- How can modern medicine contribute to the recovery of neural function?
- How can physical and occupational therapy assist a patient in regaining maximal independence after neurological injury?

The answers to these questions are explored in this text. The purpose of this text is to present the information that is essential for understanding the neuro-

logical disorders encountered by therapists. Therapists who specialize in neurological rehabilitation typically treat clients with brain and spinal cord disorders. However, clients with neurological disorders are not confined to neurological rehabilitation; therapists specializing in orthopedics frequently treat clients with chronic neck or low back pain, nerve compression syndromes, and other nervous system problems. Regardless of area of specialty, a thorough knowledge of basic neuroscience is important for every therapist.

ORGANIZATION OF THIS BOOK

The information in this text is presented in six parts:

- I. *Cellular level:* Structure and functions of the cells in the nervous system
- II. *Development:* How the nervous system forms
- III. *Somatic and autonomic systems:* Groups of neurons that perform a common function
- IV. *Regions:* Areas of the nervous system
- V. *Support systems:* Blood supply and cerebrospinal fluid systems
- VI. *Appendix:* Neurotransmitters

Cellular Level

Cells in the nervous system are neurons and glia. A **neuron** is the functional unit of the nervous system, consisting of a nerve cell body and the processes that extend outward from the cell body: dendrites and the axon.

- Neurons that convey information into the central nervous system are afferent.
- Neurons that transmit information from the central nervous system to peripheral structures are efferent.
- Neurons that connect only with other neurons are interneurons.

Glia are non-neuronal cells that do not convey information but provide services for the neurons. Some specialized glial cells form myelin sheaths, the coverings that surround and insulate axons in the nervous system and aid in the transmission of electrical signals. Other types of glia nourish, support, and protect neurons.

Development of the Human Nervous System

The development of the human nervous system in utero and through infancy is considered in this section. Common developmental disorders are also described.

Somatic and Autonomic Systems

The nervous system is composed of many smaller systems, each with distinct functions. Many systems are discussed in the context of an appropriate region of the nervous system; for example, the cognitive system is discussed in the chapter on the cerebrum. However, three systems extend through all regions of the nervous system: the somatosensory, autonomic, and somatic motor systems. The somatosensory system conveys information from the skin and musculoskeletal system to areas of the brain. The autonomic system provides bidirectional communication between the brain and smooth muscle, cardiac muscle, and gland cells. The somatic motor system transmits information from the brain to skeletal muscles. Because much of the nervous system is devoted to somatosensory, autonomic, and motor functions, being familiar with these three systems gives meaning to many of the terms encountered in studying the regions of the nervous system.

Regions of the Nervous System

The structures of the nervous system can be classified as the peripheral, spinal, brain stem and cerebellar, and cerebral regions (Fig. 1–1).

Peripheral Nervous System The peripheral nervous system consists of all parts of the nervous system that are not encased in the vertebral column or skull. Peripheral nerves, such as median, ulnar, sciatic, and cranial nerves, are groups of axons.

Spinal Region The spinal region includes all parts of the nervous system encased in the vertebral column. In addition to the spinal cord, axons attached to the cord are within the spinal region until the axons exit the intervertebral foramen.

Brain Stem and Cerebellar Region The brain stem connects the spinal cord with the cerebral region. The major divisions of the brain stem are the medulla, pons, and midbrain (see Fig. 1–1). Although the cra-

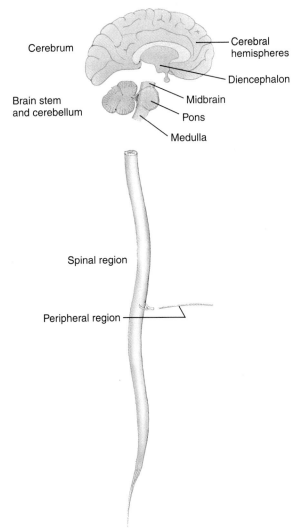

FIGURE 1–1

Regions of the nervous system. Regions are listed on the left, and subdivisions are listed on the right.

nial nerve receptors and axons are part of the peripheral nervous system, because most cranial nerves are functionally and structurally closely related to the brain stem, the cranial nerves are discussed with the brain stem region in this text. Connected to the posterior brain stem is the cerebellum.

Cerebral Region The most massive part of the brain is the cerebrum, consisting of the diencephalon and cerebral hemispheres (see Fig. 1–1). The diencephalon, in the center of the cerebrum, is almost completely surrounded by the cerebral hemispheres. The thalamus and hypothalamus are major structures of the diencephalon. Cerebral hemispheres consist of the cerebral cortex, axons connecting the cortex with other parts of the nervous system, and deep nuclei.

Support Systems

The cerebrospinal fluid and vascular systems provide essential support to the nervous system. Cerebrospinal fluid fills the ventricles, four continuous cavities within the brain, and then circulates on the surface of the central nervous system. Membranous coverings of the central nervous system, the meninges, are part of the cerebrospinal fluid system. The arterial supply of the brain is delivered by the internal carotid and vertebral arteries.

Appendix

The neurotransmitter appendix lists the actions of the most common neurotransmitters and neurological conditions associated with abnormal levels of the transmitters.

INTRODUCTION TO NEUROANATOMY

A general knowledge of basic neuroanatomy is required before proceeding in this text. As noted earlier, the nervous system is divided into four regions: peripheral, spinal, brain stem and cerebellar, and cerebral regions. The peripheral nervous system consists of all nervous system structures not encased in bone. The central nervous system, encased in the vertebral column and skull, includes the spinal cord, brain stem and cerebellar, and cerebral regions.

Parts of the nervous system are also classified according to the type of non-neural structures they innervate. Thus, the somatic nervous system connects with cutaneous and musculoskeletal structures, the autonomic nervous system connects with viscera, and the special sensory systems connect with visual, auditory, vestibular, olfactory, and gustatory (taste) structures.

Terms used to describe locations in the nervous system are listed in Table 1–1.

Planes are imaginary lines through the nervous system (Fig. 1–2). There are three planes:

- Sagittal
- Horizontal
- Coronal

A sagittal plane divides a structure into right and left portions. A midsagittal plane divides a structure into right and left halves, while a parallel cut produces parasagittal sections. A horizontal plane cuts across a structure at right angles to the long axis of the structure, creating a horizontal, or cross, section. A coronal plane divides a structure into anterior and posterior portions. The plane of an actual cut is used to name the cut surface; for example, a cut through the brain along the coronal plane is called a coronal section.

Gross Neuroanatomy and Neurophysiology

CELLULAR LEVEL

Differences in cellular constituents produce an obvious feature, the difference between gray and white

TABLE 1–1			
TERMS USED TO DESCRIBE LOCATIONS IN THE NERVOUS SYSTEM			
Term	**Definition**	**Antonym**	**Definition of Antonym**
Superior	The part is situated above another part	Inferior	The part is below another part
Rostral	Toward the head	Caudal	Toward the tail or coccyx
Anterior or ventral	Toward the front	Posterior or dorsal	Toward the back
Medial	Toward the midline	Lateral	Farther from the midline
Proximal	Nearest the point of origin	Distal	Farther from the point of origin
Ipsilateral	On the same side of the body	Contralateral	On the opposite side of the body

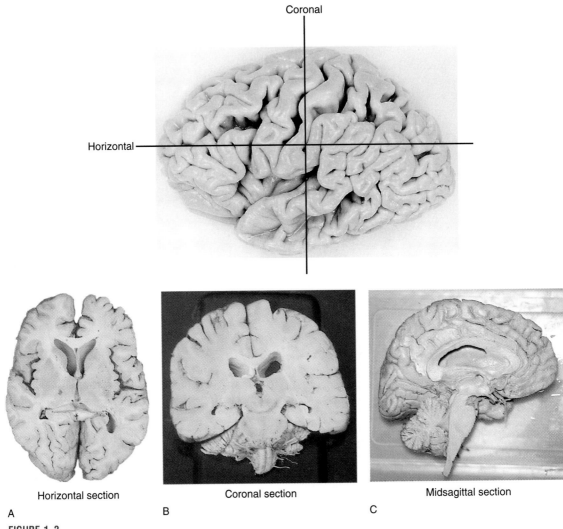

FIGURE 1-2

Planes and sections. (Lateral view courtesy of Dr. Melvin J. Ball.)

matter, in sections of the central nervous system (Fig. 1–3). White matter is composed of axons, projections of nerve cells that usually convey information away from the cell body, and myelin, an insulating layer of cells that wraps around the axons. Areas with a large proportion of myelin appear white because of the high fat content of myelin. A bundle of myelinated axons that travel together in the central nervous system is called a tract, lemniscus, fasciculus, column, peduncle, or capsule.

Areas of the central nervous system that appear gray contain primarily neuron cell bodies. These areas are called gray matter. Groups of cell bodies in the peripheral nervous system are called ganglia. In the central nervous system, groups of cell bodies are most frequently called nuclei, although gray matter on the surface of the brain is called cortex.

The axons in white matter convey information among parts of the nervous system. Information is integrated in gray matter.

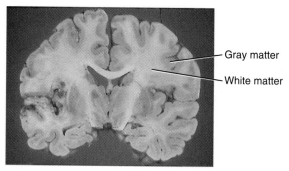

FIGURE 1–3

A coronal section of the cerebrum, revealing white and gray matter. White matter is composed of axons surrounded by large quantities of myelin. Gray matter is mainly composed of neuron cell bodies.

PERIPHERAL NERVOUS SYSTEM

Within a peripheral nerve are afferent and efferent axons. Afferent axons carry information from peripheral receptors toward the central nervous system; for example, an afferent axon transmits information about touch stimuli in the hand to the central nervous system. Efferent axons carry information away from the central nervous system; for example, efferent axons carry motor commands from the central nervous system to skeletal muscles. Peripheral components of the somatic nervous system include axons, sensory nerve endings, and glial cells. In the autonomic nervous system, entire neurons, sensory endings, synapses, ganglia, and glia are found in the periphery.

Sensory receptors respond to various stimuli in the periphery. Peripheral nerves convey information from sensory receptors into the central nervous system, and also transmit signals from the central nervous system to skeletal and smooth muscle and glands.

SPINAL REGION

Within the vertebral column, the spinal cord extends from the foramen magnum (the opening at the inferoposterior aspect of the skull) to the level of the first lumbar vertebra. Distally, the spinal cord ends in the conus medullaris. The spinal cord has 31 segments, with a pair of spinal nerves arising from each segment.

Each spinal nerve is connected to the cord by a dorsal root and a ventral root (Fig. 1–4). An enlargement of the dorsal root, the dorsal root ganglion, contains the cell bodies of sensory neurons. Cell bodies of neurons forming the ventral root are located within the spinal cord. The union of the dorsal and ventral roots forms the spinal nerve. The spinal nerve exits the vertebral column via openings between vertebra, then divides into dorsal and ventral rami that communicate with the periphery.

Cross sections of the spinal cord reveal centrally located gray matter forming a shape similar to the letter H surrounded by white matter (Fig. 1–5). Each side of the gray matter is subdivided into ventral, lateral, and dorsal horns. These horns contain cell bodies of motoneurons, interneurons, and the endings of sensory neurons. The gray matter commissure connects

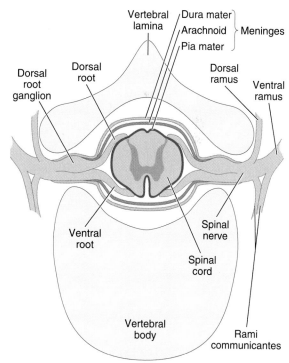

FIGURE 1–4

Spinal region. The spinal nerve is formed of axons from the dorsal and ventral roots. The bifurcation of the spinal nerve into dorsal and ventral rami marks the transition from the spinal to the peripheral region.

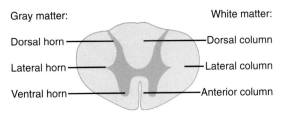

FIGURE 1–5

Cross section of the spinal cord. The central gray matter is divided into horns and a commissure. The white matter is divided into columns.

the lateral areas of gray matter. The white matter is divided into three areas (funiculi):

- Ventral column
- Lateral column
- Dorsal column

The meninges, connective tissue surrounding the spinal cord and brain, are discussed later in this chapter.

The spinal cord's two main functions are:

- To conduct information between the periphery and the brain
- To process information

The cord conveys somatosensory information to the brain and also conveys signals from the brain to neurons that directly control movement. An example of spinal cord processing of information is the reflexive movement of a limb away from a painful stimulus. Within the cord are the necessary circuits to orchestrate the movement.

BRAIN STEM AND CEREBELLAR REGION

Many fiber tracts carrying motor and sensory information travel through the brain stem, while other fiber tracts begin or end within the brain stem. In addition, the brain stem also contains important groups of neurons that control equilibrium (sensations of

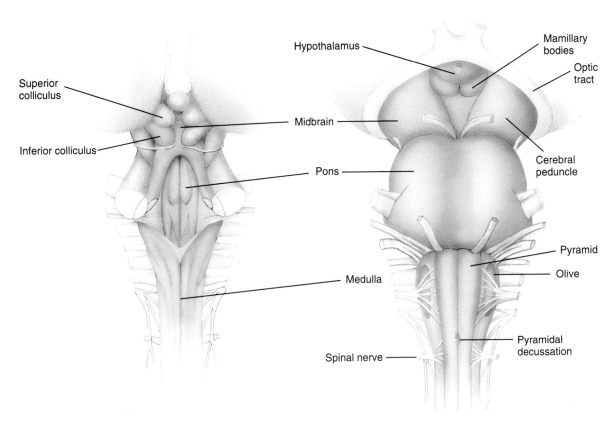

FIGURE 1–6

Brain stem: posterior and anterior views.

head movement, orienting to vertical, postural adjustments), cardiovascular activity, respiration, and other functions. External features of the brain stem are illustrated in Figure 1–6. The parts of the brain stem are the medulla, pons, and midbrain.

The medulla is continuous with the spinal cord. Features of the anterior surface of the medulla are the olive, the pyramid, and the roots of four cranial nerves. The olive is an oval bump on the superior anterolateral surface of the medulla. Axons projecting from the cerebral cortex to the spinal cord form the pyramids. As these fibers cross the midline, they form the pyramidal decussation.

Superior to the medulla is the pons. The junction of the medulla and pons is marked by a transverse line. The ventral part of the pons forms a large bulge anteriorly, containing fiber tracts and interspersed nuclei. Four cranial nerves attach to the pons.

The superior section of the brain stem is the midbrain. The anterior portion of the midbrain is formed by two cerebral peduncles, containing fibers that descend from the cerebral cortex. Dorsally, the tectum of the midbrain consists of four small rounded bodies, two superior colliculi and two inferior colliculi. The colliculi are important for orientation to stimuli. Two cranial nerves arise from the midbrain.

The brain stem conveys information between the cerebrum and the spinal cord, integrates information, and controls vital functions (i.e., respiration, heart rate).

Cranial Nerves Twelve pairs of cranial nerves emerge from the ventral surface of the brain (Fig. 1–7). Each cranial nerve is designated by a name and by a Roman numeral. The numbering is according to the site of attachment to the brain, from anterior to posterior. In general, cranial nerves innervate structures in the head, face, and neck. The exception is the vagus nerve, which innervates thoracic and abdominal viscera in addition to structures in the head and neck.

Some cranial nerves are purely sensory. The purely sensory cranial nerves are the olfactory (I), optic (II),

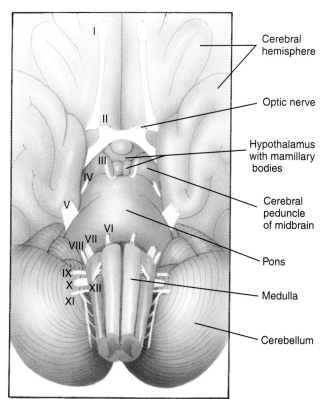

Cerebral hemisphere

Optic nerve

Hypothalamus with mamillary bodies

Cerebral peduncle of midbrain

Pons

Medulla

Cerebellum

FIGURE 1–7

Inferior surface of the brain, showing attachments of cranial nerves.

TABLE 1–2
CRANIAL NERVES

Number	Name	Function
I	Olfactory	Smell
II	Optic	Vision
III	Oculomotor	Moves eyes up, down, medially, raises upper eyelid, constricts pupil
IV	Trochlear	Moves eye medial and down
V	Trigeminal	Facial sensation, chewing, sensation from temporomandibular joint
VI	Abducens	Abducts eye
VII	Facial	Facial expression, closes eyes, tears, salivation, and taste
VIII	Vestibulocochlear	Sensation of head position relative to gravity and head movement; hearing
IX	Glossopharyngeal	Swallowing, salivation, and taste
X	Vagus	Regulates viscera, swallowing, speech, taste
XI	Accessory	Elevates shoulders, turns head
XII	Hypoglossal	Moves tongue

and vestibulocochlear (VIII) nerves. Other cranial nerves are principally motor but also contain some sensory fibers that respond to muscle and tendon movement. These include the oculomotor (III), trochlear (IV), abducens (VI), accessory (XI), and hypoglossal (XII). The remaining cranial nerves are mixed nerves, containing both motor and sensory fibers. Table 1–2 lists the cranial nerves and their functions.

Cerebellum Not part of the brain stem but connected to the posterior brain stem by large bundles of fibers called peduncles is the cerebellum (Fig. 1–8). Fiber tracts joining the midbrain, pons, and medulla with the cerebellum are the superior, middle, and inferior peduncles. The cerebellum consists of two large cerebellar hemispheres and a midline vermis. *Vermis* means "worm," a fitting description for the appearance of the cerebellar midline. Internally, the cerebellar hemispheres are composed of the cerebellar cortex on the surface, underlying white matter, and centrally located deep nuclei. The cerebellum's function is to coordinate movements.

CEREBRUM

Diencephalon The diencephalon consists of four structures (Fig. 1–9):

- Thalamus
- Hypothalamus
- Epithalamus
- Subthalamus

The thalamus is a large, egg-shaped collection of nuclei in the center of the cerebrum. The other three structures are named for their anatomical relationship

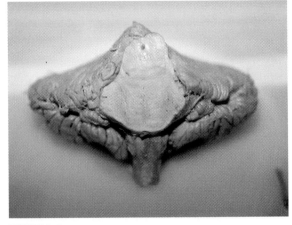

FIGURE 1–8

An anterior view of the cerebellum and the brain stem. The midbrain and pons have been partially dissected to show the fiber tracts.

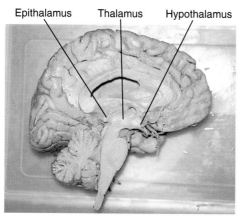

FIGURE 1–9

The parts of the diencephalon that are visible in a midsagittal section are the thalamus, hypothalamus, and epithalamus. The subthalamus is lateral to the plane of section.

to the thalamus: the hypothalamus is inferior to the thalamus, the epithalamus is located posterosuperior to the thalamus, and the subthalamus is inferolateral to the thalamus. The epithalamus consists primarily of the pineal gland.

Thalamic nuclei relay information to the cerebral cortex, process emotional and some memory information, integrate different types of sensation (i.e., touch and visual information), or regulate consciousness, arousal, and attention. The hypothalamus maintains body temperature, metabolic rate, and the chemical composition of tissues and fluids within an optimal functional range. The hypothalamus also regulates eating, reproductive, and defensive behaviors, expression of emotions, growth, and the function of reproductive organs. The pineal gland influences the secretion of other endocrine glands, including the pituitary and adrenal glands. The subthalamus is part of a neural circuit controlling movement.

Cerebral Hemispheres The longitudinal fissure divides the two cerebral hemispheres. The surfaces of the cerebral hemispheres are marked by rounded elevations called gyri (singular: gyrus) and grooves called sulci (singular: sulcus). Each cerebral hemisphere is subdivided into six lobes (Fig. 1–10):

- Frontal
- Parietal
- Temporal
- Occipital
- Limbic
- Insular

The first four lobes are named for the overlying bones of the skull. The limbic lobe is on the medial aspect of the cerebral hemisphere. The insula is a section of the hemisphere buried within the lateral sulcus, revealed by separating the temporal and frontal lobes.

The distinctions among the lobes are clearly marked in only a few cases; the remainder are approximate. Clear distinctions include the following:

- The boundary between the frontal lobe and the parietal lobe, marked by the central sulcus
- The boundary between the parietal lobe and the occipital lobe, clearly marked only on the medial hemisphere by the parieto-occipital sulcus
- The temporal lobe, inferior to the lateral sulcus
- The limbic lobe, on the medial surface of the hemisphere, bounded by the cingulate sulcus and by the margin of the parahippocampal gyrus

The entire surface of the cerebral hemispheres is composed of gray matter, called the cerebral cortex. Deep to the cortex is white matter, composed of axons connecting the cerebral cortex with central nervous system areas. Several collections of these fibers are of particular interest: the commissures and the internal capsule. The commissures are bundles of axons that connect the cortices of the left and right cerebral hemispheres. The corpus callosum is a huge commissure that connects most areas of the cerebral cortex. The much smaller anterior commissure connects the temporal lobe cerebral cortices. The internal capsule consists of axons projecting from the cerebral cortex to subcortical structures and from subcortical structures to the cerebral cortex. The internal capsule is subdivided into anterior and posterior limbs, with a genu between them (Fig. 1–11).

The cerebral cortex processes sensory, motor, and memory information and is the site for reasoning, language, nonverbal communication, intelligence, and personality. The commissures convey information between the left and right cerebral cortices. The internal capsule links cortical and subcortical structures.

Within the white matter of the hemispheres are additional areas of gray matter, the most prominent being the basal ganglia. The basal ganglia nuclei in the cerebral hemispheres are the caudate, putamen,

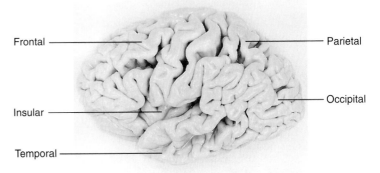

Frontal

Parietal

Insular

Occipital

Temporal

A

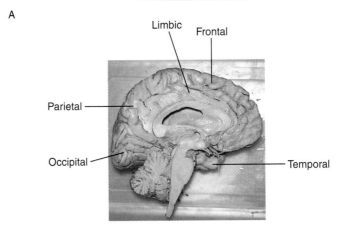

Limbic   Frontal

Parietal

Occipital

Temporal

FIGURE 1–10

Lobes of the cerebral hemispheres.

B

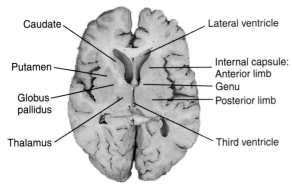

Caudate

Lateral ventricle

Putamen

Internal capsule:
Anterior limb

Genu

Globus pallidus

Posterior limb

Thalamus

Third ventricle

FIGURE 1–11

Horizontal section showing the internal capsule. The internal capsule is the white matter bordered by the head of the caudate, the thalamus, and the lenticular nucleus.

and globus pallidus. The putamen and globus pallidus together are called the lenticular nucleus. The caudate and putamen together are called the corpus striatum. Two additional nuclei, the subthalamic (in the diencephalon) and the substantia nigra (in the midbrain) are also part of the basal ganglia neural circuit. The basal ganglia circuit helps to control movement.

The limbic system is a group of structures in the diencephalon and cerebral hemispheres. The limbic system includes parts of the hypothalamus, thalamus, and cerebral cortex; several deep cerebral nuclei, the most prominent being the amygdala; and the hippocampus, a region of the temporal lobe (Fig. 1–12). The limbic system is involved with emotions and the processing of some types of memory.

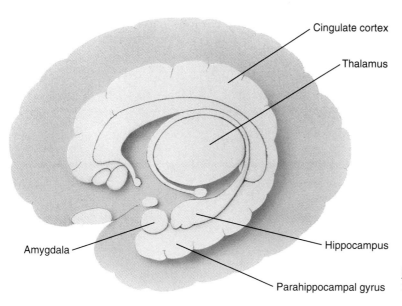

Cingulate cortex

Thalamus

Amygdala

Hippocampus

Parahippocampal gyrus

FIGURE 1–12
Parts of the limbic system.

SUPPORT SYSTEMS

Cerebrospinal Fluid System Cerebrospinal fluid, a modified filtrate of plasma, circulates from cavities inside the brain to the surface of the central nervous system and is reabsorbed into the venous blood system. The cavities inside the brain are the four ventricles: paired lateral ventricles in the cerebral hemispheres, the third ventricle a midline slit in the diencephalon, and the fourth ventricle between the pons and medulla anteriorly and the cerebellum posteriorly (Fig. 1–13). The ventricular system continues through the medulla and spinal cord as the central canal and ends blindly in the caudal spinal cord. Within the ventricles, cerebrospinal fluid is secreted by the choroid plexus. The lateral ventricles are connected to the third ventricle by the interventricular foramina. The third and fourth ventricles are connected by the cerebral aqueduct. Cerebrospinal fluid leaves the fourth ventricle through the lateral foramina and the medial foramen to circulate around the central nervous system.

The meninges, membranous coverings of the brain and spinal cord, are also part of the cerebrospinal fluid system. From outmost to inmost, the meninges are the dura, arachnoid, and pia. Only the first two can be observed in gross specimens; the pia is a very delicate membrane adhered to the surface of the central nervous system. Arachnoid, also a delicate membrane, is named for its resemblance to a spider's web.

The dura, named for its toughness, has two projections that separate parts of the brain: the falx cerebri separates the cerebral hemispheres, and the tentorium cerebelli separates the posterior cerebral hemispheres from the cerebellum (Fig. 1–14). Within these dural projections are spaces, called dural sinuses, that return cerebrospinal fluid and venous blood to the jugular veins. The cerebrospinal fluid system regulates the contents of the extracellular fluid and also provides buoyancy to the central nervous system by suspending the brain and spinal cord within fluid and membranous coverings.

Vascular System Two pairs of arteries supply blood to the brain (Fig. 1–15):

- Two internal carotid arteries
- Two vertebral arteries

The internal carotid arteries provide blood to most of the cerebrum, while the vertebral arteries provide blood to the occipital and inferior temporal lobes and to the brain stem region. To supply the cerebrum, the internal carotid artery divides into the anterior and middle cerebral arteries. The anterior cerebral artery extends anteriorly, then sweeps up and back over the corpus callosum to supply most of the midline cerebral circulation. The middle cerebral artery extends laterally from the internal carotid artery, through the lateral sulcus, to supply most of the lateral surface of the cerebral hemisphere.

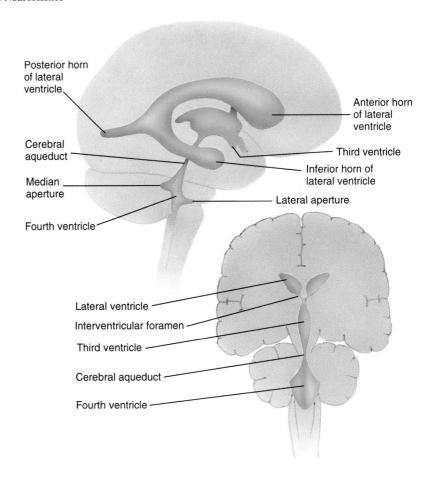

Posterior horn
of lateral
ventricle

Anterior horn
of lateral
ventricle

Cerebral
aqueduct

Third ventricle

Inferior horn of
lateral ventricle

Median
aperture

Lateral aperture

Fourth ventricle

Lateral ventricle

Interventricular foramen

Third ventricle

Cerebral aqueduct

Fourth ventricle

FIGURE 1–13

The four ventricles: two lateral ventricles, the third ventricle, and the fourth ventricle.

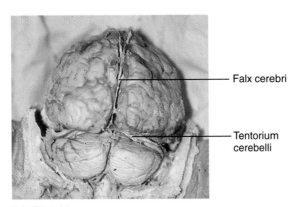

Falx cerebri

Tentorium
cerebelli

FIGURE 1–14

The dura mater covering the posterior brain has been removed to reveal the dural projections: the falx cerebri and the tentorium cerebelli.

The vertebral arteries supply the medulla and upper spinal cord. These arteries then unite at the junction between the pons and medulla to form the basilar artery. The basilar artery provides blood to the pons and most of the cerebellum, then divides to become the posterior cerebral arteries that supply the midbrain and the inferoposterior cerebrum.

The major arteries that supply the cerebrum, the anterior, middle, and posterior cerebral arteries, together with three small arteries, form an anastomosis. The anastomosis, called the circle of Willis, surrounds the hypothalamic area on the inferior surface of the cerebrum. One of the small arteries, the anterior communicating artery, is unpaired and connects the right and left anterior cerebral arteries. Two posterior communicating arteries, one on each side, connect the middle and posterior cerebral arteries.

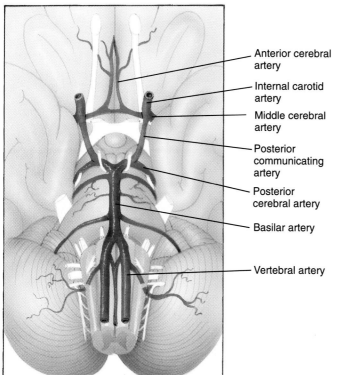

— Anterior cerebral artery

— Internal carotid artery

— Middle cerebral artery

— Posterior communicating artery

— Posterior cerebral artery

— Basilar artery

— Vertebral artery

FIGURE 1–15
Arterial supply of the brain.

In contrast with other parts of the body that have major veins corresponding to the major arteries, venous blood from the cerebrum drains into dural (venous) sinuses. Dural sinuses are canals between layers of dura mater. In turn, the dural sinuses drain into the jugular veins. The vascular system provides oxygen, ionic exchange, and nourishment for the cells of the nervous system.

CLINICAL APPLICATION OF LEARNING NEUROSCIENCE

For therapists, the main purpose in studying the nervous system is to understand the effects of nervous system lesions. A lesion is an area of damage or dysfunction. Signs and symptoms following a lesion of the nervous system depend on the location of the lesion. For example, complete destruction of a specific area of cerebral cortex severely interferes with hand function. The cause of the damage could be blood supply interruption, a tumor, or local inflammation, but damage to that area of the cerebral cortex com-

promises the dexterity of the hand. Depending on their distribution in the nervous system, lesions can be categorized as follows (Daube et al., 1986):

- Focal: limited to a single location
- Multifocal: limited to several, nonsymmetrical locations
- Diffuse: affecting bilaterally symmetrical structures and does not cross the midline as a single lesion

A tumor in the spinal cord is an example of a focal lesion. A tumor that has metastasized to several locations is multifocal. Alzheimer's disease, a memory and cognitive disorder, is diffuse because it affects cerebral structures bilaterally but does not cross the midline as a single lesion.

> Regardless of the cause of nervous system dysfunction, the resulting signs and symptoms depend on the site of the lesion(s).

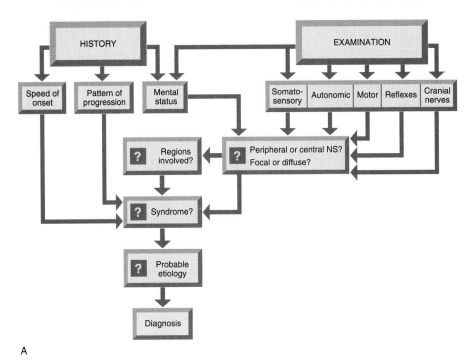

A

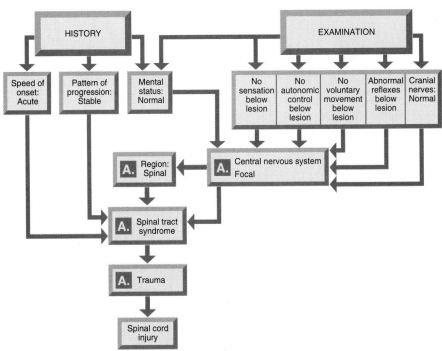

B

FIGURE 1–16

Flow charts illustrating the process of neurological evaluation. *A,* The generalized process. "?" indicates a question that can be answered by analyzing the information that flows into that box. *B,* Application of the neurological evaluation process. The findings of the history and examination are indicated, as are the subsequent steps to reach a diagnosis. "A" indicates an answer to a question posed in *A.* In this case, the diagnosis is spinal cord injury.

Neurological Evaluation

The neurological evaluation has two parts:

- History
- Examination

The purpose of the neurological evaluation is to determine the probable etiology of the neurological problems so appropriate care can be provided. The etiologies that affect the nervous system include the following:

- Trauma
- Vascular disorders
- Inflammation
- Degenerative
- Neoplasms
- Immunologic
- Toxic or metabolic

HISTORY

A history is essentially a structured interview to determine the symptoms that lead the person to seek physical or occupational therapy. Knowing the typical speed of onset and the expected pattern of progression for each category of pathology is critical for recognizing when a specific client's signs and symptoms necessitate referral to a medical practitioner.

The speed of onset and pattern of progression provide important clues to the etiology, or cause, of nervous system dysfunction. Speed of onset is classified as follows:

- Acute, indicating minutes or hours to maximal signs and symptoms
- Subacute, progressing to maximal signs and symptoms over a few days
- Chronic, gradual worsening of signs and symptoms continuing for weeks or years

Acute onset usually indicates a vascular problem, subacute onset frequently indicates an inflammatory process, and chronic onset often suggests either a tumor or degenerative disease. In cases of trauma, the etiology is usually obvious, and in cases of toxic or metabolic disorders, the speed of onset varies according to the specific cause. The pattern of progression can be stable, improving, worsening, or fluctuating.

While discussing the person's history, the therapist can often obtain adequate information about the person's mental status:

- Is the person awake?
- Is the person aware?
- Is the person able to respond appropriately to questions?

EXAMINATION

Specific tests are performed to assess the function of the sensory, autonomic, and motor systems. These tests are described in subsequent chapters. If indicated, additional tests that assess function within specific regions of the nervous system may be performed.

DIAGNOSIS

By synthesizing the information from the history and the examination, the therapist begins to answer the following questions:

- Is the lesion in the peripheral or central nervous system?
- Is the lesion focal, multifocal, or diffuse?
- Does the pattern of signs and symptoms indicate a syndrome?
- What region or regions of the nervous system are involved?
- What is the probable etiology?
- What is the diagnosis?

Figure 1–16 is a flow chart illustrating the integration of information to reach a diagnosis. In many cases the therapist is able to reach a diagnosis. In other cases, the therapist may not be able to answer several of the diagnostic questions, or the diagnosis is beyond the scope of physical therapy practice. In such cases, the person must be referred to the appropriate medical practitioner.

Summary

To understand the nervous system, each level of analysis is essential. As noted by Joaquin Fuster (1994), a brain researcher and psychiatrist, "the problem with the molecular approach to higher neurophysiology is that it proceeds at the wrong (i.e., impractical) level of discourse and analysis (like trying to understand the written message by studying the chemistry of the ink)." To extend Fuster's analogy, if there is a problem with the ink, then studying the ink's chemistry is appropriate. If there is a problem at the molecular level, with the supply of particular ions or molecules required by the nervous system, then the molecular level is the appropriate level of analysis. However, a molecular-level approach to understanding language is not practical; a cognitive-level analysis is appropriate. Some dysfunctions in the nervous system interfere with cellular-level processes, other dysfunctions interfere with the processing of one type of information, and still others interfere with all functions processed in a specific area. To understand each type of dysfunction, the appropriate levels of analysis must be applied.

Scientific investigations at each level of analysis have revealed many details regarding neural function. These details have promoted an improved understanding of function and provided new insight into treatment of neurological disorders. Continued neuroscience research and continued development of new treatment regimes can only bring us closer to a full understanding of nervous system function in health, disease, and recovery.

References

Daube, J. R., Sandok, B. A., Reagan, T. J., and Westmoreland, B F. (1978). Medical neurosciences: An approach to anatomy, pathology, and physiology by systems and levels. Boston: Little, Brown and Co.

Fuster, J. M. (1994). Brain systems have a way of reconciling "opposite" views of neural processing; the motor system is no exception. In: Cordo, P., and Harnad, S. (Eds.). Movement control, p. 139.

2

Physical and Electrical Properties of Cells in the Nervous System

I am a 37-year-old female college professor and physical therapist. As a full-time physical therapist from 1982 to 1988, I worked primarily with adult neurologically impaired individuals in the rehabilitation setting. Since 1988, I have been teaching physical therapy.

I was 28 years old when in 1987 I experienced my first symptoms (looking back now). Numbness in my right arm persisted for about 3 days. A few weeks after the numbness subsided, I experienced a right foot drop. This continued to progress over 24 hours, and I was seen in an emergency room for assessment. The initial tests were a lumbar puncture and myelogram, evoked potentials, and a CT scan, all of which were normal. I continued to have weakness on my right side and mildly slurred speech. These symptoms resolved over a period of about 10 days. I underwent an MRI, which confirmed the diagnosis of multiple sclerosis (MS) secondary to the discovery of a lesion in the cortex. Approximately 6 weeks later, I suffered rapid onset (about 2 hours) of symptoms of left-sided weakness, inability to swallow, unclear speech, and sensory deficits on the left side. Additionally, I experienced Lhermitte's sign* and had (and continue to have) a perfect midline cut (up to but not including the face) in which the right side of my body feels on fire. Every minute of every day.

In the 10 years that I have had MS, I have experienced approximately nine attacks (although none in the past 27 months). Each time the attack was different in nature. I have had two that have been pure sensory involving both lower limbs, two that were purely autonomic in which I vomited for many hours, and one that was a visual field cut only. All the others had elements of sensory, motor, visual, and vestibular problems. I have not experienced any bowel or bladder dysfunction.

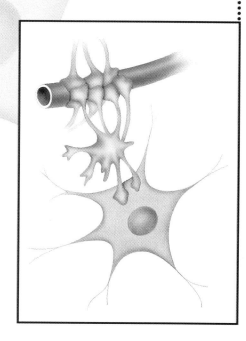

*Lhermitte's sign: abrupt electric-like shocks traveling down the spine upon flexion of the head.

I have had nearly full return of function following every attack, with the only symptoms that remain being the persistent sensory hypersensitivity on my right side (greater in the limbs than in the trunk), mild visual disturbances including hypersensitivity to light and diminished night-driving ability, impaired vibratory sensation, and minor balance deficits. None of the unresolved symptoms have altered my life in a major way. I participate in many activities and have just made some minor accommodative changes. I do not suffer from increased levels of fatigue or have difficulty with heat, unlike many people with MS.

Following each attack, I get almost full return of function, so I consider my condition to be fairly static. I maintain my level of fitness via multiple types of aerobic and anaerobic activities.

I have never had any type of therapy except as a participant in research studies. As a regular participant in research studies in the Portland, Oregon area, I have been involved in a cell-cloning study and a study using the drug Betaseron. I am currently midway through a 2-year study of Avonex, which is an interferon treatment. Prior to the Avonex study, I would typically have one attack per year, including during the 2-year period of the Betaseron study. During that time, I was receiving a placebo and continued having my one attack per year. I have not had an attack in 27 months. Part of that time period includes times when I was receiving Betaseron treatments via subcutaneous injections, and part incorporates the time period since I initiated the Avonex interferon study protocol of once a week intramuscular injections. Because the course of MS is unpredictable and the Avonex study is incomplete, conclusions cannot be drawn regarding the effectiveness of the treatment. I also attribute my continued health to other healthful practices such as diet, exercise, stress management, and purpose in my life. I believe all these factors play a role in maintaining health and preventing or minimizing the disease state.

—Lori Avedisian

Sensation, movement, and mental processes involve a chemical and electrical interaction among cells within the nervous system. Sensory information from peripheral receptors is conveyed to the spinal cord and brain, where it is then analyzed to provide perception of the environment. On the basis of this sensory information, a motor command may be issued for the coordinated movement of the muscles. Our memory of our experiences and movements also results from an array of chemical and electrical interactions within

the brain. This vast network of cells within the nervous system, with its capability to form new interactions and modify its output based on the input to the system, is more complex than any computer. The nervous system has two distinct classes of cells: **glia** (supporting cells) and **neurons** (nerve cells). Although support and nerve cells can be classified into as many as 10,000 different types, there are many common features shared between different cell types.

SUPPORTING CELLS

Glial cells form a critical support system for neurons. The term *glia* is derived from the Greek word for glue because these cells were first recognized as supporting elements that provided firmness and shape to the brain. Unlike neurons, which transmit information, glia do not convey signals. The role of glia is to provide essential support for the conduction of electrical impulses throughout the nervous system. Glia are characterized by size and by function. The large glial cells are called **macroglia**, and the small glial cells are called **microglia**.

Macroglia are generally classified into three groups:

- Oligodendrocytes
- Schwann cells
- Astrocytes

Oligodendrocytes and Schwann cells form a myelin sheath to insulate the electrical conduction region of neurons, the axon. The **myelin** formed by the support cells is composed of fats and proteins and contributes both to the structure of the nervous system and to the conduction of electrical signals. The primary difference between the two types of myelin-producing support cells is their location in the nervous system. **Oligodendrocytes** are found within the central nervous system and form myelin sheaths that envelop several axons from several neurons. Unlike oligodendrocytes, **Schwann cells** are found in the peripheral nervous system. Schwann cell myelin sheaths may envelop only one single neuron's axon, or they may partially surround several axons (Fig. 2–1). **Myelinated** axons are those that are completely enveloped by the myelin sheath. **Unmyelinated** axons include those that are only *partially* enveloped by myelin.

Astrocytes, named for their star-shaped body with radiating projections, play a critical role in nutritive

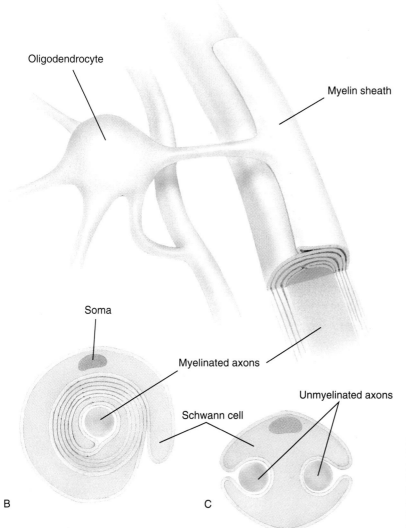

A

Oligodendrocyte

Myelin sheath

Soma

Myelinated axons

Schwann cell

Unmyelinated axons

B

C

FIGURE 2–1

Myelination. *A,* Oligodendrocytes provide myelin sheaths in the central nervous system. *B, C,* Schwann cells provide insulation to peripheral nerves. Myelinated axons are completely enveloped by Schwann cells. Unmyelinated axons are partially surrounded.

and cleanup functions within the central nervous system. Because astrocytes have unique end-feet that connect both to neurons and blood capillaries (Fig. 2–2), they can have a nutritive function and can also function as a component of the blood-brain barrier. Astrocytes do the following:

- Help remove chemical transmitters from the synaptic cleft between neurons

- Take up and buffer excess K^+ released by electrically active neurons
- Contribute to cleanup of neuronal debris
- Help confine the damage of neuronal tissue after injury to the central nervous system

The small glial cells, **microglia,** normally function as phagocytes (cells that ingest and destroy bacteria and cells) that are activated and mobilized after injury, in-

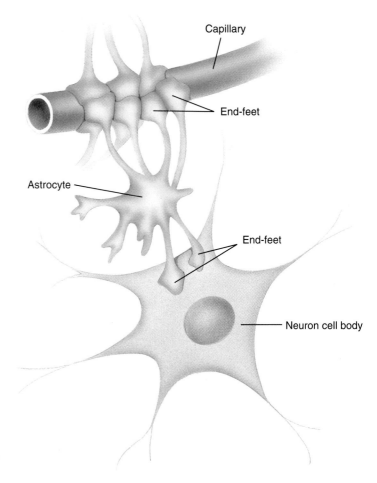

FIGURE 2–2

Astrocyte end-feet. Astrocytes form a connection between neurons and capillaries, providing nutrition.

fection, or disease, acting as the brain's immune system. Normally, microglia clean the neural environment and contribute to the destruction of injured or aging neurons. Microglia have also recently been identified as important in the process of brain development. As cells die during the natural development of the nervous system, they secrete proteins that attract immune system cells into the nervous system. These immune cells transform into macrophages to clean up debris from the dying cells. After cleanup, some of these macrophages leave the nervous system, while others remain and may transform into microglia. Recently, abnormal activation of microglia has been associated with various disease states (Pennisi, 1993). In diseases associated with aging, such as Alzheimer's disease, microglia become activated

and may promote destruction of neurons. Also, the human immunodeficiency virus (HIV), associated with acquired immunodeficiency syndrome (AIDS), can activate microglia and stimulate a cascade of cellular breakdown. Clearly, there exists a delicate balance between the normal, protective roles of microglia and the more recently identified destructive roles.

Oligodendrocytes and Schwann cells contribute to the myelination of neurons throughout the nervous system. Astrocytes and microglia contribute to the nutritive and cleanup functions throughout the central nervous system.

STRUCTURE OF NEURONS

Neurons differ from glia and are unique cells because they can receive input and generate output of information based on the ability to undergo rapid changes in the electrical potential of the cell membrane. A typical neuron has four components (Fig. 2–3):

- Cell body or soma
- Dendrites
- Axon
- Presynaptic terminals

The **cell body (soma)** is the metabolic center of the neuron, containing the nucleus, neurotransmitter-synthesizing mechanisms, and the energy-producing/storing apparatus. **Dendrites** are branchlike extensions from the cell body. They are the input units of the cell, specialized to receive information from other neurons.

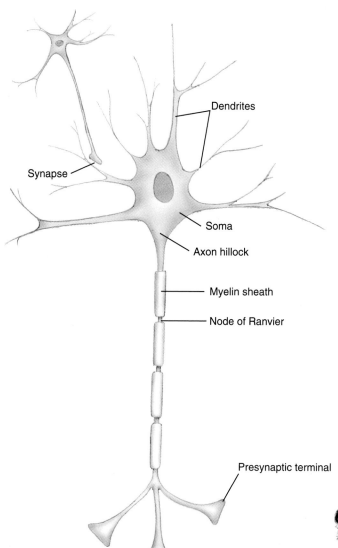

Dendrites
Synapse
Soma
Axon hillock
Myelin sheath
Node of Ranvier
Presynaptic terminal

FIGURE 2–3

Parts of a neuron. Showing are the cell body (soma), the input units (dendrites), and the output unit (axon) with its presynaptic terminals. The axon hillock and nodes of Ranvier contribute to electrical signaling within the neuron. Also shown is a synapse, where a presynaptic terminal of one neuron contacts with a dendrite of another, postsynaptic neuron.

The **axon** is a process that extends from the cell body to target cells. The axon is the output unit of the cell, specialized to transmit information to other neurons. Each neuron has a single axon that arises from a specialized region of the cell, called the **axon hillock.** The axon varies in length, with some axons, such as those transmitting sensory information from the foot to the spinal cord, being up to a meter in length. Axons are called myelinated when completely surrounded by either oligodendrocytes or Schwann cells and unmyelinated when not surrounded. Partially surrounded axons are called unmyelinated, although the term *partially myelinated* would be more accurate.

Presynaptic terminals located at the end of the axon in fingerlike projections are the transmitting elements of the neuron. A presynaptic terminal belongs to the neuron transmitting information, and the postsynaptic terminal belongs to the region of a cell receiving information. A **synapse** is the region of chemical transmission between neurons or between a neuron and a muscle cell. The **synaptic cleft** is the space between the two terminals. Each neuron transmits information about its own activity via the release of chemicals, called **neurotransmitters,** from its presynaptic terminal into the synaptic cleft associated with the postsynaptic terminal of another neuron, muscle cell, or gland. The presynaptic terminal may synapse with the dendrite, cell body, or axon of another neuron. Information can be transferred in only one direction at the synapse: from the presynaptic cell to the postsynaptic cell.

> The basic functions of a neuron are reception, integration, transmission, and transfer of information.

Types of Neurons

The morphology of neurons in vertebrates is classified into two groups of cells:

- Bipolar
- Multipolar

BIPOLAR CELLS

This classification is based on the number of processes that directly arise from the cell body (Fig. 2–4). **Bipolar cells** have two primary processes that extend from the cell body:

- Dendritic root
- Axon

The dendritic root divides into multiple dendritic branches, and the axon projects to form its presynaptic terminals. A typical bipolar cell is the retinal bipolar cell in the eye. A special subclass of bipolar cells is called **pseudounipolar.**

Pseudounipolar cells have one projection from the cell body, which later divides into two axonal roots. Thus, there are no true dendrites in pseudounipolar cells. The most common pseudounipolar cells are sensory neurons that bring information from the body into the spinal cord (see Fig. 2–4). The peripheral axons conduct sensory information from the periphery to the cell body, while the central axons conduct information from the cell body to the spinal cord.

MULTIPOLAR CELLS

The second major category of neurons are the **multipolar cells.** Multipolar cells have a single axon and multiple dendrites arising from many regions of the cell body. These are the most common cells in the vertebrate nervous system, with a variety of different shapes and dendritic organizations throughout the nervous system. Multipolar cells are specialized to receive and accommodate huge amounts of synaptic input to their dendrites. An example of a multipolar cell is the spinal motor neuron, projecting from the spinal cord to innervate skeletal muscle fibers. A typical spinal motor cell receives approximately 8000 contacts on its dendrites and 2000 contacts on the cell body itself. Multipolar cells in the cerebellum, called Purkinje cells, receive as many as 150,000 contacts on their expansive dendritic tree.

Cellular Components of Neurons

Despite their unique functional abilities, neurons contain the structures, called organelles, common to all animal cells. Certain organelles, the mitochondria and smooth endoplasmic reticulum, are distributed throughout neurons. Other organelles, including the nucleus, Golgi apparatus, and rough endoplasmic reticulum, are restricted to the cell body. The functions of the organelles are listed in Table 2–1.

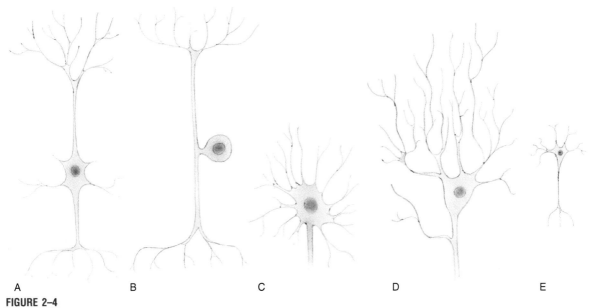

A   B   C   D   E

FIGURE 2–4

Morphology of neurons. *A*, Bipolar cell of the retina. *B*, Pseudounipolar cell of the dorsal root ganglion. *C*, Multipolar cell representative of a spinal motor neuron. *D*, Multipolar cell typical of the cerebellum. *E*, Interneuron distributed throughout the nervous system. Although myelination is not indicated in these diagrams, all of the represented neurons except for the interneuron are normally myelinated.

TABLE 2–1 FUNCTION OF CELLULAR ORGANELLES	
Organelle	**Function**
Nucleus	Control center, contains the neuron's genetic material, directs the metabolic activity of the neuron
Mitochondria	Convert nutrients into an energy source the neuron can use; i.e., synthesize adenosine triphosphate
Endoplasmic reticulum	Rough endoplasmic reticulum (called Nissl substance in neurons): synthesizes and transports proteins Smooth endoplasmic reticulum: synthesizes and transports lipids
Ribosomes	Protein synthesis: free ribosomes (not attached to endoplasmic reticulum) synthesize proteins for the neuron's own use; ribosomes attached to rough endoplasmic reticulum synthesize neurotransmitters
Golgi apparatus	Packages neurotransmitter

Internal structural support for neurons is provided by protein strands called microtubules, microfilaments, and neurofilaments. In addition to providing a skeleton for neurons, microtubules also provide the substrate for axoplasmic transport. All proteins required to nourish the axon and all components of neurotransmitters are supplied to the axon by the soma. These substances are moved along the axon toward the axon terminal by a process called **anterograde transport.** Some substances are recycled by being transported from the axon back to the soma by the process of **retrograde transport** (Fig. 2–5). In general, the movement of substances with the axon is termed **axoplasmic transport.**

Neurons are electrically active cells with structural and functional specializations of dendrites, axons, and synaptic terminals. Cellular organelles within neurons make and transport neurotransmitters for cell-to-cell interaction.

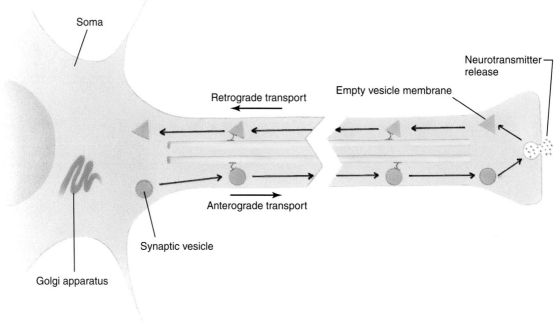

FIGURE 2–5

Axoplasmic transport. Substances required by the axon are delivered from the soma via antero-grade transport. Retrograde transport moves substances from the axon to the soma. The proteins that "walk" the vesicles along the microtubules are shown in red.

DIRECTION OF INFORMATION FLOW IN NEURONS

Information within a neuron can only be transferred in one direction. Neurons are classified into three functional groups, dependent on the neuron's role in the direction of information transfer:

- Afferent neurons
- Efferent neurons
- Interneurons

Afferent means conducting toward; efferent means conducting away from. Interneurons make local connections between many different neurons. When considering the nervous system as a whole, **afferent neurons** bring information into the central nervous system. These neurons convey sensory information. **Efferent neurons** relay commands from the central nervous system to the smooth and skeletal muscles of the body. **Interneurons** are the largest class of neurons and act throughout the nervous system either to process information locally or to convey information short distances from one site in the nervous system to another. For example, interneurons in the spinal cord control the activity of local reflex circuits within the spinal cord.

The terms *afferent* and *efferent* can also refer to the direction of information conveyed by a particular group of neurons within the central nervous system. For example, neurons called thalamocortical convey information from the thalamus to the cerebral cortex. This information is efferent from the thalamus and afferent to the cerebral cortex. Neuronal pathways within the central nervous system are commonly named by combining the name of efferent (i.e., site of origin) and afferent (i.e., site of termination) regions.

Information flow can be from the periphery to the central nervous system, from the central nervous system to the periphery, or within the central nervous system.

TRANSMISSION OF INFORMATION BY NEURONS

The unique ability of neurons to undergo rapid changes in the electrical potential across the cell membrane is the key to neuron function. An electrical potential across a membrane is established by the distribution of ions creating a difference in electrical charge on each side of the cell membrane. Four types of membrane channels allow flow of ions through the membrane:

- Nongated
- Modality-gated
- Ligand-gated
- Voltage-gated

The nongated channels allow diffusion of a small number of ions through the membrane. The other channels are called gated because they open in response to a stimulus and close when the stimulus is removed.

Modality-gated channels, specific to sensory neurons, open in response to mechanical forces (i.e., stretch, touch, pressure), temperature changes, or chemicals. **Ligand-gated channels** open in response to a neurotransmitter binding to the surface of a channel receptor on a postsynaptic cell membrane. When open, the channels allow the flow of electrically charged ions between the extracellular and intracellular environments of the cell, resulting in the generation of local potentials. A third category of gated membrane channels is the **voltage-gated channels.** These channels open in response to changes in electrical potential across the cell membrane. These channels are important in the release of neurotransmitters (see Chap. 3) and the formation of action potentials, as will be discussed later in this chapter.

A rapid change in electrical potential transmits information along the length of an axon and elicits release of chemical transmitters to other neurons or to the electrically excitable membrane of a muscle. Three types of electrical potentials in neurons are essential for transmission of information:

- Resting membrane potential
- Local potentials
- Action potentials

Resting Membrane Potential

The **resting membrane potential** is the difference in electrical potential across the cell membrane of a neuron when the neuron is not transmitting information.

The resting membrane potential is maintained by the cell membrane serving as a capacitor that allows the separation and storage of electrical charge. An unequal distribution of ionic charge across the membrane is essential to the electrical activity of neurons. Two forces acting on each type of ion determine the distribution of each ion. The forces are the concentration gradient and the electrical gradient for that ion.

As a simplified example, consider what would happen if only KCl is inside the cell and the membrane ion channels are permeable only to K^+. Some K^+ diffuses through the membrane from the inside to the outside of the cell, due to the concentration gradient of K^+: the ion diffuses down its concentration gradient from high concentration to low concentration (inside to outside). The diffusion is limited by the electrical gradient. As the K^+ diffuses out of the cell, the increase in positive charge outside the membrane opposes the outward movement, while the negative charge remaining inside the membrane attracts the K^+. In effect, the opposing chemical and electrical forces control the ion movement. Equilibrium of the distribution of a specific ion is reached when there is no net movement of that ion across the membrane. Individual ions continue to diffuse through the membrane, but equal numbers of the ion enter and leave the cell. The membrane **equilibrium potential,** that is, the voltage difference across the membrane due to unequal distribution of an ion, can be determined using the **Nernst equation.** This equation was derived from basic thermodynamic principles and is used in electrophysiological investigations of neuronal cell function.

A typical resting membrane potential is −70 mV, indicating that the inside of the neuron contains more negative charge than the outside (Fig. 2–6). The resting membrane potential is maintained by the following:

- Negatively charged ions (anions) trapped inside the neurons because they are too large to move through the openings in the cell membrane
- Active transport of Na^+ and K^+
- Passive diffusion of Na^+, K^+, and Cl^- ions through nongated ion channels in the cell membrane

The anions are primarily large proteins and cannot diffuse through the membrane, so they remain inside the cell. The concentrations of Na^+ and Cl^- are kept higher on the *outside* compared to the inside of the cell, while the concentrations of K^+ and organic anions are kept higher on the *inside* compared to the outside of the cell. This unequal distribution of ions

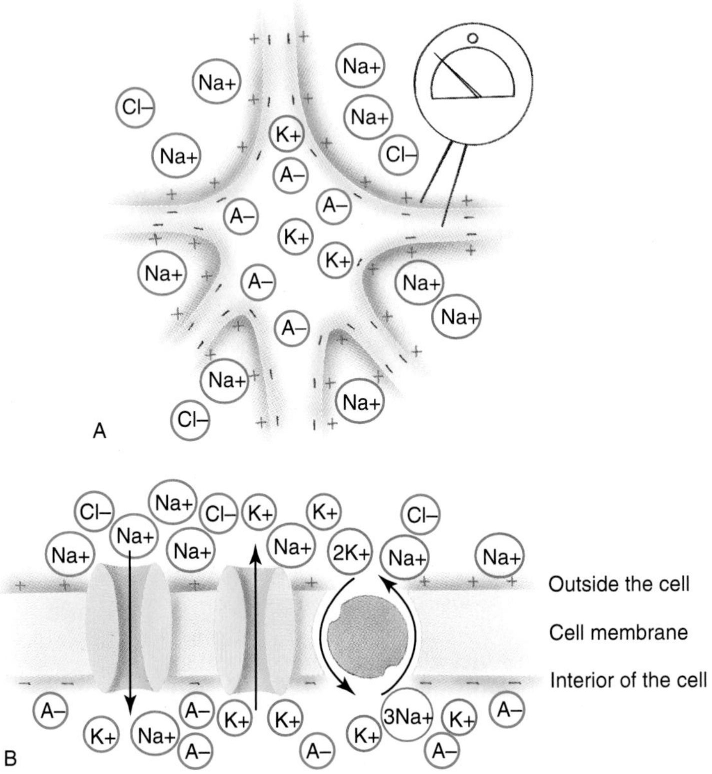

FIGURE 2–6

Resting membrane potential. *A,* Resting membrane potential is measured by comparing the electrical difference between the inside and the outside of the cell membrane. At rest, the inside of the cell membrane is approximately 70 mV more negative than the outside of the cell membrane. *B,* The resting membrane potential is maintained via passive diffusion of ions across the cell membrane and via active transport of Na^+ and K^+ by a Na^+-K^+ pump. The concentrations of Na^+ and Cl^- are kept higher on the outside compared to the inside of the cell, while the concentration of K^+ is kept higher on the inside compared to the outside of the cell. High concentrations of unneutralized negative charges inside the cell also contribute to the negative resting membrane potential.

establishes both a concentration gradient and an electrical gradient that regulates the passive movement of ions through nongated ion channels. Active transport uses an adenosine triphosphate–driven Na^+-K^+ pump that transports Na^+ out of the cell and K^+ into the cell. Three Na^+ ions are transported out for every two K^+ ions transported into the cell. Small amounts of Na^+, K^+, and Cl^- can diffuse through nongated channels in the membrane. At rest, the potential of the membrane is maintained negative on the inside of the cell compared to the outside, since the net transport of Na^+ and diffusion of K^+ out of the neuron leave behind a cloud of unneutralized negative charge inside.

> The unequal distribution of ionic electrical charge across a neuron's cell membrane establishes a membrane potential. The distribution of a specific ion depends on (1) the concentration gradient of the ion and (2) the electrical gradient acting on the ion.

Changes from Resting Membrane Potential

The resting membrane potential is significant because it prepares the membrane for changes in electrical potential. Sudden changes in membrane potential result from the flow of electrically charged ions through gated channels spanning the cell membrane (Fig. 2–7). The cell is **depolarized** when the membrane potential becomes less negative than the resting potential. Depolarization increases the likelihood that the neuron will generate a transmittable electrical signal and is thus considered excitatory. Conversely, when the cell is **hyperpolarized**, the membrane potential becomes more negative than the resting potential. Hyperpolarization decreases the neuron's ability to generate an electrical signal and is thus considered inhibitory.

In addition to the sudden, brief changes in membrane potential lasting only milliseconds, gradual and longer-lasting changes in membrane potential are referred to as **modulation.** Modulation involves small

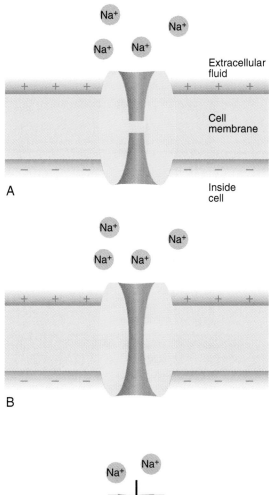

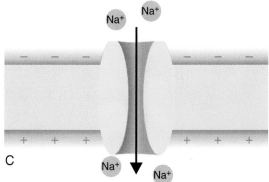

FIGURE 2–7

A sodium ion channel. Ion channels are proteins that span the cell membrane. *A,* When the ion channel is closed, ions cannot pass through the channel. *B,* Application of voltage to the cell membrane causes the channel to change configuration. *C,* This allows ions to pass through the gate. Because the concentration of Na^+ is greater outside the neuron than inside, opening the Na^+ channels produces an influx of Na^+ into the neuron.

changes in the membrane's electrical potential that alter the flow of ions across a cell membrane.

> Alteration in membrane potential occurs when ion channels open to selectively allow the passage of specific ions.

Local Potentials and Action Potentials

The initial change in membrane potential is called a **local potential** because it spreads passively only a short distance along the membrane. **Action potentials** are much larger changes in electrical potential than the local potentials. An action potential involves a brief, large depolarization that can be repeatedly regenerated along the length of an axon. Because it can be regenerated, an action potential *actively* spreads long distances to transmit information down the axon to the presynaptic chemical release sites of the presynaptic terminal. Electrical potentials within each neuron conduct information in a predictable and consistent direction. The conduction originates with local potentials at the receiving sites of the neuron: in sensory neurons, the receiving sites are the sensory receptors; in motor and interneurons, the receiving sites are on the postsynaptic membrane. If the change in local potential results in sufficient depolarization of the cell membrane, then an action potential will be generated and actively spread along the length of the cell axon. The sufficient level of depolarization for generation of an action potential is called the **threshold level**. Only when the electrical potential exceeds the threshold level is an action potential generated.

Figure 2–8 illustrates the sequence of local and action potentials in the transmission of sensory information. The sequence of activity is as follows:

1. Deformation of a peripheral pressure receptor
2. Receptor potential in the sensory ending
3. Action potential in the sensory axon
4. Release of transmitter from the sensory neuron presynaptic terminal
5. Binding of transmitter to ligand-gated receptor on the postsynaptic cell membrane
6. Synaptic potential in the postsynaptic membrane

The specific features of local and action potentials are summarized in Table 2–2 and are discussed in the following sections.

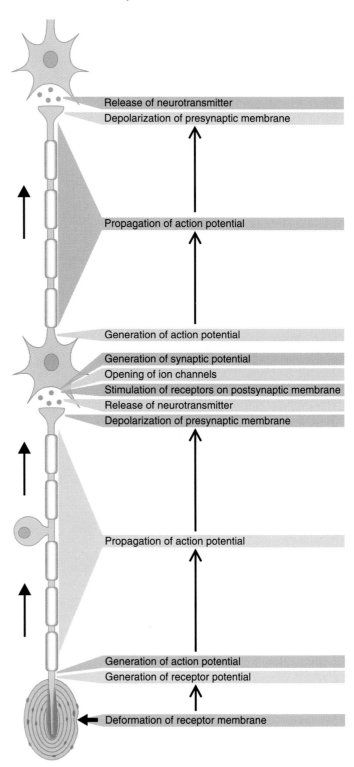

FIGURE 2–8

Sequence of events following stimulation of a sensory receptor. The flow of information via the interaction among receptor potentials, action potentials, and synaptic potentials is shown. A receptor potential is generated by mechanical change (stretch) of the end receptor. An action potential is generated by opening of sodium channels. The action potential propagates along the axon of the sensory neuron from the periphery to the spinal cord. Release of chemical transmitters at the synapse with the second neuron generates a synaptic potential in the second neuron. If sufficient stimuli are received by the second neuron, an action potential is generated in this neuron. The action potential propagates along the axon. When the action potential reaches the axon terminal, chemical transmitter is released from the terminal. The transmitter then binds to receptors on the membrane of the third neuron and opening of membrane channels generates a synaptic potential.

Release of neurotransmitter
Depolarization of presynaptic membrane

Propagation of action potential

Generation of action potential

Generation of synaptic potential
Opening of ion channels
Stimulation of receptors on postsynaptic membrane
Release of neurotransmitter
Depolarization of presynaptic membrane

Propagation of action potential

Generation of action potential
Generation of receptor potential

Deformation of receptor membrane

TABLE 2–2
FEATURES OF LOCAL AND ACTION POTENTIALS

Characteristic	Local Potential	Action Potential
Amplitude	Small, graded	Large, all or none
Effect on membrane	Either depolarizing or hyperpolarizing	Depolarizing
Propagation	Passive	Active and passive
Ion channels responsible for the change in membrane potential	Sensory neuron end-receptor: modality-gated channel Postsynaptic membrane: ligand-gated channel	Voltage-gated channels

LOCAL POTENTIALS

Local potentials are categorized as either **receptor potentials** or **synaptic potentials** depending on whether they are generated at an end-receptor of a sensory neuron or at a postsynaptic membrane. The generation of these local potentials is dependent on the characteristics of gated ion channels within the cell membrane. End-receptors have modality-gated channels, while postsynaptic membranes have ligand-gated channels.

Local receptor potentials are generated when the peripheral end-receptors of a sensory neuron are stretched, compressed, deformed, or exposed to thermal or chemical agents. These changes in the protein structure of the membrane cause modality-gated ion channels to open, encoding the sensory information into a flow of ionic current. For example, stretch of a muscle causes opening of ion channels in the membrane of sensory nerve endings imbedded in the muscle. Opening of the channels results in ion flow and generation of receptor potentials that are graded in both amplitude and duration. The larger or longer-lasting the stimulus, the larger and longer-lasting the resulting receptor potential. Most receptor potentials are depolarizing and therefore excitatory. However, sensory stimulation can also cause a receptor potential that is hyperpolarizing and therefore inhibitory. A receptor potential is purely localized to the receptive surface of the sensory neuron and can only spread passively a very short distance along the axon. Within 1 mm of travel, the receptor potential is only one-third its original amplitude.

Local synaptic potentials are generated in motor neurons and interneurons when they are stimulated by input from other neurons. When a presynaptic neuron releases its neurotransmitter, the chemical travels across the synaptic cleft and interacts with the chemical receptor sites on the membrane of the post-synaptic cell. Binding of the neurotransmitter to receptors on the postsynaptic cell results in opening of ligand-gated ion channels, locally changing the resting membrane potential of the cell. The change in potential constitutes a synaptic potential that can be either depolarizing (excitatory) or hyperpolarizing (inhibitory), depending on the action of the neurotransmitter on the membrane channel. Similar to receptor potentials, the synaptic potentials can only spread passively and are graded in both amplitude and duration. The greater the amount of neurotransmitter and the longer the time over which it is available, the larger and longer-lasting the resulting synaptic potential.

Because local potentials can only spread passively along their receptor or synaptic membrane, the potentials generally travel only 1 to 2 mm. In addition, the amplitude decreases with the distance traveled. The strength of the local potentials can be increased and multiple potentials integrated via the processes of **temporal summation** and **spatial summation** (Fig. 2–9). Temporal summation is the cumulative effect of a series of either receptor potentials or synaptic potentials that occur within milliseconds of each other and are added together. Spatial summation is the process by which either receptor or synaptic potentials generated in different regions of the neuron are added together. Via summation, a sufficient number of potentials occurring within a short period of time cause significant changes in the membrane potential and either promote or inhibit the generation of an action potential.

> Neurons undergo rapid changes in the electrical potential of the membrane to conduct electrical signals. Receptor and synaptic potentials are graded in amplitude and duration and conduct local electrical information in the neuron.

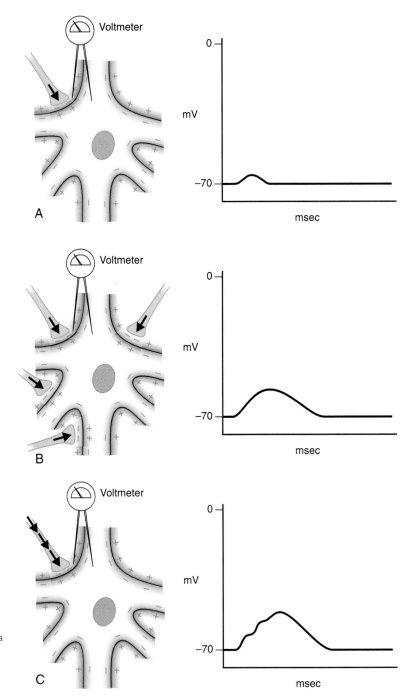

FIGURE 2–9

Integration of local signals. *A,* A single weak input to a cell results in only a slight depolarization of the membrane. *B,* Spatial summation of several different inputs results in a significant depolarization of the membrane. *C,* Temporal summation of several inputs in rapid succession results in significant depolarization of the membrane.

ACTION POTENTIALS

Because the receptor and synaptic potentials only spread passively over short distances, another cellular mechanism, the action potential, is essential for rapid movement of information. The action potential is a large depolarizing signal that is actively propagated along an axon by the repeated generation of the signal. Because they are actively propagated, action potentials transmit information over longer distances than receptor or synaptic potentials. The meaning of the signal is determined not by the signal itself but by the neural pathway along which it is conducted. Unlike the local input signals, which are graded, the action potential is **all-or-none.** This means that every time even minimally sufficient stimuli are provided to generate an action potential, an action potential will be produced. Stronger stimuli will produce action potentials of the same voltage and duration as do the minimally sufficient stimuli. The initiation or firing of an action potential is similar to the striking of a key on a computer keyboard. Regardless of whether the key is struck gently and slowly, or rapidly and hard, the letter will be inscribed when the sufficient amount of pressure is achieved. The shape of the letter is not influenced by how hard the key is pressed.

The generation of action potentials involves a sudden influx of Na^+ through voltage-gated Na^+ channels. Although voltage-gated Na^+ channels are generally absent in the region of the receptor terminal and the synaptic membrane, there is a dense distribution of these Na^+ channels within approximately 1 mm of the input regions. In a sensory nerve, the region closest to the receptor with a high density of Na^+ channels is the trigger zone, and in interneurons and motor neurons, the regions closest to the synapse with a high density of Na^+ channels is the axon hillock. The receptor or synaptic potentials that have passively traveled a short distance toward the trigger zone or axon hillock are both spatially and temporally summated. If the summation of local potentials depolarizes the membrane beyond a voltage threshold level, then an action potential is generated by the opening of many voltage-dependent Na^+ channels. If the summation does not result in depolarization exceeding the threshold, then there will be no action potential.

The stimulus intensity that is just sufficient to produce an action potential is called the **threshold stimulus intensity.** Typically, a 15-mV depolarization,

that is, a change in membrane potential from -70 mV to -55 mV, is sufficient to trigger an action potential. When the voltage across the membrane reaches -55 mV, many voltage-dependent Na^+ channels open. Na^+ flows rapidly into the cell, propelled by the high extracellular Na^+ concentration and attracted by the negative electrical charge inside the membrane. When the K^+ channels open later, K^+ leaves the cell, repelled by the positive electrical charge inside the membrane (created by the influx of Na^+) and by the K^+ concentration gradient. The membrane becomes temporarily more polarized than when at rest. This state is called hyperpolarization. The resting membrane potential is restored by the diffusion of ions and the action of the Na^+-K^+ pump. Figure 2–10 illustrates the change in membrane potential during an action potential. The peak of the spike occurs at about 35 mV, and then the potential quickly drops back toward the resting membrane potential. In summary, an action potential is produced by a sequence of three events (see Fig. 2–10):

1. A rapid depolarization due to opening of the voltage-gated Na^+ channels
2. A decrease in Na^+ conduction due to closing of the channels
3. A rapid repolarization due to opening of voltage-gated K^+ channels

The opening of many Na^+ channels and repolarization of the active region following the spike of an action potential creates a refractory period. The refractory period can be divided into two distinct periods:

- Absolute refractory period
- Relative refractory period

During the **absolute refractory period,** no stimulus, no matter how strong, will elicit another action potential. The absolute refractory period corresponds to the time the firing level is reached until repolarization is approximately one-third complete (Fig. 2–11). During the **relative refractory period,** only a stronger than normal stimulus can elicit another action potential. The relative refractory period corresponds to the time immediately following the absolute refractory period until the membrane potential returns to the resting level (see Fig. 2–11). The refractory period prevents the backward flow of the action potential and promotes forward propagation of the

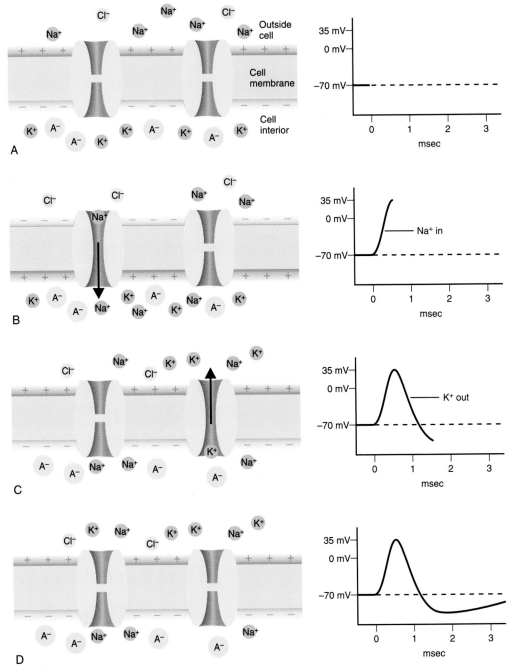

FIGURE 2–10

Action potential. *A,* The resting membrane potential of the cell is approximately −70 mV and membrane channels are closed. *B,* Initiation of the action potential begins with opening of voltage-sensitive Na⁺ channels and a rapid influx of Na⁺, causing the cell membrane to become less negative (i.e., depolarized). *C,* Closing of the Na⁺ channels and opening of K⁺ channels then causes a reversal of membrane potential. Ultimately, a brief hyperpolarization of the membrane results in the potential becoming more negative than the resting potential. *D,* The cell membrane returns to resting potential with the closure of all membrane channels.

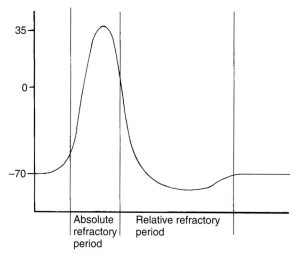

FIGURE 2–11

Refractory periods. During and immediately following the action potential are two refractory periods. The absolute refractory period corresponds to the time the firing level is reached until repolarization (reversal of potential) is one-third complete. The relative refractory period corresponds to the time immediately following the absolute refractory period until the membrane potential returns to the resting level.

action potential. If there were no refractory period, the passive flow of ions associated with an action potential could spread both forward and backward along the length of the axon. Although the flow of K$^+$ out of the cell decreases the electrical potential across the membrane, the resting levels of ion concentration must be restored over time by the Na$^+$-K$^+$ pump actively moving Na$^+$ out of the neuron and K$^+$ into the neuron.

> Action potentials are an all-or-none electrical response to the local depolarization of a membrane. Depolarization of the membrane to the threshold level depends on opening of voltage-gated Na$^+$ channels. Repolarization depends on the opening of K$^+$ channels.

PROPAGATION OF ACTION POTENTIALS

Once an action potential has been generated, the change in electrical potential passively spreads along the axon to the adjacent region of the membrane. When the depolarization of the adjacent, inactive region reaches threshold, another action potential is generated. This process, the passive spread of depolarization to adjacent membrane and generation of new action potentials, is repeated along the entire length of the axon (Fig. 2–12). The propagation of an action potential is dependent on both passive properties of the axon and active opening of ion channels distributed along the length of the axon. Three basic properties of the cell body and axon contribute to the passive conduction of the electrical potential:

- **Axoplasmic resistance** (Ra)—the resistance of the inside cellular fluid
- **Membrane resistance** (Rm)
- **Membrane capacitance** (Cm)—the storage of electrical charge across a membrane

Because rapid propagation of an action potential is functionally very important, two mechanisms have evolved to promote the rapid propagation:

- Increased diameter of the axon
- Myelination

The larger the diameter of the axon, the smaller the Ra and therefore the smaller its resistance to current flow along the length of the axon. For instance, larger axons have lower Ra; therefore, there is a larger current flow with less time required to change the electrical charge of the adjacent membrane. Myelination of the membrane, by glial cells, increases the insulation. This insulation creates a higher Rm to prevent the flow of current across the myelinated axonal membrane and reduces Cm. When Cm is small, fewer charges must be deposited to depolarize the membrane to a threshold level; therefore, current flow for a shorter period of time can result in membrane depolarization.

In addition to promoting a greater distance of passive current spread, myelination also increases the speed of action potential propagation. The heavier the myelin, the greater the potential for propagation and the faster the speed of conduction. Interruptions in the myelin sheath leave small patches of axon unmyelinated. These small patches, called **nodes of Ranvier,** contribute to the speed of action potential conduction. Nodes of Ranvier are distributed approximately every 1 to 2 mm along the axon and contain a high density of Na$^+$ channels and K$^+$ channels. A depolarizing potential spreads rapidly along myelinated regions because of low Cm, then slows when crossing the high-capacitance, unmyelinated region of the node of Ranvier. As a node becomes depolarized, the

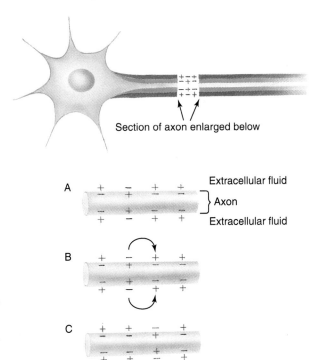

Section of axon enlarged below

A   Extracellular fluid
}   Axon
Extracellular fluid

B

C

FIGURE 2–12

Propagation of action potential. *A,* A depolarizing current passively spreads down the axon, causing the interior of the axon to become more positive than when the membrane is resting. *B,* In the adjacent membrane, when the depolarizing current reaches threshold level, Na^+ channels open, causing rapid depolarization of the membrane. *C,* An action potential is generated, and the depolarizing current continues to propagate down the axon.

opening of voltage-gated Na^+ channels results in the generation of a new action potential and the spread of ionic current along the axon to the next node (Fig. 2–13). Consequently, as the action potential propagates down a myelinated axon, it appears to

quickly jump from node to node. The process by which an action potential jumps from node to node is called **saltatory conduction.** Because the nodal regions also contain K^+ channels, the generation of a refractory period again plays a critical role in the for-

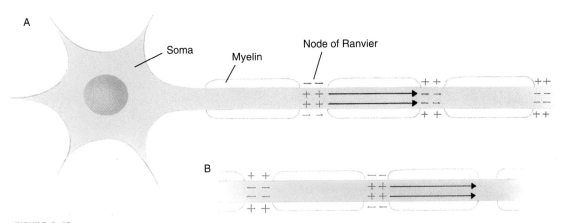

A

Soma   Myelin   Node of Ranvier

B

FIGURE 2–13

Saltatory conduction. The process by which an action potential appears to jump from node to node down an axon is called saltatory conduction. *A,* A depolarizing potential spreads rapidly along the myelinated regions of the axon, then slows when crossing the unmyelinated node of Ranvier. *B,* When an action potential is generated at a node of Ranvier, the depolarizing potential again spreads quickly over myelinated regions, appearing to jump from node to node.

ward propagation of the action potential, by preventing the backward flow of electrical potential.

> Action potentials are propagated down the length of an axon via both passive and active membrane properties.

INTERACTIONS BETWEEN NEURONS

The diversity as well as specificity of function within the nervous system can be attributed to neuronal divergence and neuronal convergence. **Divergence** refers to the process whereby a single neuronal axon may have many branches that terminate on a multitude of cells. **Convergence,** on the other hand, refers to the process by which multiple inputs from a variety of different cells terminate on a single neuron. Via temporal and spatial summation, a sufficient number of convergent inputs occurring within a short period of time cause significant changes in the membrane potential and either promote or inhibit the generation of an action potential.

Through the processes of divergence and convergence, a single stimulus results in a substantial response. For example, a pinprick activates the end-receptors of a sensory neuron that transmits information about tissue damage. The message is conveyed to multiple neurons in the spinal cord, eliciting a motor response to move the body part away from the stimulus. Simultaneously, the information is conveyed to other neurons that relay information to the conscious awareness. Diverse information from a variety of sources may result in either the inhibition or facilitation of a specific response (Fig. 2–14).

> Divergent and convergent synaptic connections contribute to the distribution of information throughout the nervous system.

DISEASES AFFECTING CELLS OF THE NERVOUS SYSTEM

A multitude of diseases affect the nervous system. Diseases causing destruction of neurons are often either limited to the motor neurons of the spinal cord or to specific neuronal populations within the brain. These diseases will be discussed in subsequent chapters. Disease processes that affect the glial cell myelin

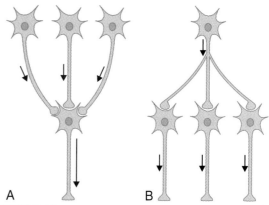

FIGURE 2–14

Convergence and divergence. *A,* Convergent input to interneurons and motor neurons in the spinal cord includes afferent input from the musculoskeletal system and input from the brain. *B,* Divergent output includes the activation of several neurons by single inputs. Only a few of the actual connections are shown.

sheath surrounding the nerve cell and its axons disrupt the conduction of action potentials (Fig. 2–15). Only a few representative diseases are discussed in this section to illustrate the impact of disruption of glial cells.

Myelin is critical to the conduction of all sensory, motor, and integration signals throughout the nervous system. As an action potential travels from a myelinated region to a region where myelin has been damaged, the membrane resistance decreases. This allows leakage of current and disrupts the current flow

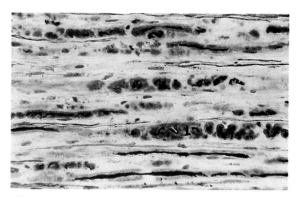

FIGURE 2–15

A nerve biopsy showing peripheral demyelination and axon degeneration that occurs in severe Guillain-Barré syndrome. (Courtesy of Dr. Melvin J. Ball.)

toward the next nodal region, where a high density of Na⁺ channels can generate a new action potential. With demyelination, the conduction of action potentials will be slowed and potentially even blocked, resulting in motor weakness and sensory abnormalities.

Peripheral Nervous System Demyelination

Peripheral neuropathy is any pathological change involving peripheral nerves, whether due to disease or trauma. Peripheral neuropathies often involve destruction of the myelin surrounding the largest, most myelinated sensory and motor fibers, resulting in disrupted proprioception (i.e., awareness of limb position) and weakness. Peripheral demyelination can be caused by metabolic abnormalities, viruses, or toxic chemicals.

Guillain-Barré syndrome involves an acute inflammation and demyelination of peripheral sensory and motor fibers. Guillain-Barré syndrome is thought to be associated with an autoimmune attack, in which a person's own immune system generates antibodies that attack Schwann cells. In severe cases, segmental demyelination is so extreme that the underlying axons also degenerate (Fig. 2–15). Frequently Guillain-Barré syndrome occurs 2 to 3 weeks after a mild infection.

The demyelination in Guillain-Barré syndrome results in decreased sensation and motor paralysis. Cranial nerves are often affected, causing difficulty with chewing, swallowing, speaking, and facial expression. Respiratory paresis and autonomic dysregulation (primarily cardiac rate and blood pressure fluctuations) can occur but are seldom fatal. Pain is prominent in some cases. Most common is deep aching pain or hypersensitivity to touch.

For almost all patients, rapid onset is followed by a plateau, then gradual, complete recovery. People with axonal degeneration tend to have greater residual disorders than people who only have demyelination. Guillain-Barré syndrome can affect people of any age and is rapidly progressive. Most patients experience good recovery over a period of weeks to 2 years; however, approximately 3% of patients with Guillain-Barré syndrome die of cardiac or respiratory failure. (See the box on Guillain-Barré syndrome.)

Medical treatment often includes **plasmapheresis,** the process of filtering the blood plasma to remove

GUILLAIN-BARRÉ SYNDROME

Pathology
Demyelination

Etiology
Probably autoimmune

Speed of onset
Acute

Signs and symptoms
Weakness > sensory signs

Consciousness
Normal

Cognition, language, and memory
Normal

Sensory
Abnormal sensations (tingling, burning); pain

Autonomic
Blood pressure fluctuation, irregular cardiac rhythms

Motor
Paresis or paralysis

Cranial nerves
Motor cranial nerves most affected (eye and facial movements, chewing, swallowing)

Region affected
Peripheral nervous system

Demographics
Affects all ages, no gender preference

Prognosis
Progressively worse for 2–3 weeks, then gradual improvement; 3% mortality rate

circulating antibodies. Physical therapy is commonly directed toward stretching and range of motion during the acute phase of the disease. In the rehabilitation phase, physical therapy is directed toward strengthening and functional mobility. When voluntary movement begins to recover, exercise should be gentle to avoid overwork damage in partially denervated muscles (Zelig et al., 1988). Vigorous exercise may cause damage to muscles if less than one-third of the motor units are intact (Reitsma, 1969). Occupational therapy is directed toward activities of daily living, including self care.

Destruction of Schwann cells impedes conduction of electrical signals along sensory and motor pathways of the peripheral nervous system.

Central Nervous System Demyelination

In contrast to peripheral demyelination, central nervous system demyelination, as in multiple sclerosis (MS), involves damage to the myelin sheaths in the brain and spinal cord. Multiple sclerosis is thought to be an autoimmune disease in which the myelin is attacked by the person's own immune system. However, the actual cause of the immune disorder is currently unknown. **Plaques,** or patches of demyelin-

MULTIPLE SCLEROSIS

Pathology
Demyelination

Etiology
Probably autoimmune

Speed of onset
Can be acute, subacute, or chronic

Time course
Exacerbations and remission

Signs and symptoms

Consciousness
Normal

Cognition, language, and memory
Infrequently affects thinking and/or memory

Sensory
Tingling, numbness, pins and needles

Autonomic
Bladder disorders, sexual impotence in men, genital anesthesia in women

Motor
Weakness, uncoordination, reflex changes

Cranial nerves
Partial blindness in one eye, double vision, dim vision, eye movement disorders

Region affected
Central nervous system

Demographics
Typical age at onset is 20–40 years; affects 3 times as many women as men

Prognosis
Variable course; very rarely fatal; most people with MS live a normal life span

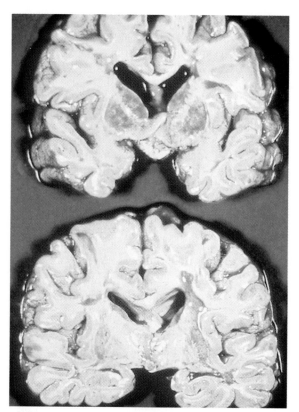

FIGURE 2–16

A coronal section of the cerebrum showing central demyelination. The abnormal areas in the white matter are plaques characteristic of multiple sclerosis. (Courtesy of Dr. Melvin J. Ball.)

ation due to destruction of the oligodendrocytes, form throughout the central nervous system (Fig. 2–16). The demyelination results in delayed and sometimes absent nerve transmission. Signs include weakness, lack of coordination, impaired vision, double vision, and impaired sensation. In addition, there can be disruption of memory and emotional affect. Diagnosis is difficult because MS usually presents with only one sign initially, which may completely resolve. A long period of time without any other signs may follow the initial sign, or the disease may progress intermittently or steadily. In most cases, onset occurs between the ages of 20 and 40 years. Approximately three times as many women as men are affected. Persons with MS and other demyelinating diseases are often told to avoid extreme exertion, immersion in hot tubs, and high temperatures, since it is thought that increases in body temperature alter the activity of the voltage-gated Na^+ channels, further disabling action potential conduction. (See the box on multiple sclerosis.)

> Destruction of oligodendrocytes impedes conduction of electrical signals along pathways of the central nervous system.

CLINICAL NOTES

CASE 1

I.D., a 19-year-old man, suffered from severe flu symptoms requiring him to stay home from work for 2 days. Four days after his return to work, I.D. noted tingling and feelings of numbness in his fingers. By the end of the day, he noted that his hand movements were clumsy. The following day, I.D. returned to work, and at about midday, he was unable to stand and could not use his hands. At the hospital, nerve conduction studies for both motor and sensory pathways were conducted. To test the peripheral sensory nerve pathways, an electrical stimulus is given to the skin on a distal point over a nerve and recorded with surface electrodes at a more proximal point over the same nerve. The time required for transmitting the signal between the two points is the conduction velocity. Peripheral motor conduction studies are similar, except the electrical stimulus is given proximally over the nerve and recorded from the skin over an associated muscle. The studies for I.D. indicated that peripheral sensory and motor conduction times were significantly prolonged bilaterally.

I.D. had suffered peripheral nerve demyelination presumably due to an autoimmune response to some form of viral infection. With the loss of myelin, the nerve conduction was severely impaired. I.D. had sensory loss and muscular weakness that significantly impaired his ability to move. Following medical treatment, he was referred to physical therapy for range of motion, strengthening exercises, and functional mobility training.

Questions

1. The disease was confirmed to involve the peripheral nervous system. Did the loss of myelin involve oligodendrocytes or Schwann cells?

2. How does a loss of myelin along peripheral sensory fibers affect the resistance properties (Ra and Rm) of the axon?

3. Would the generation of local receptor potentials or the propagation of action potentials be impaired?

CASE 2

J.R. is a 27-year-old woman with MS, admitted to the hospital two times in the past year with complaints of bilateral lower-extremity weakness and blurred vision. Upon examination, she exhibited about 30% of normal muscle strength of the left lower extremity and about 50% of normal strength on the right lower extremity. She exhibited mild left foot drop during the swing phase of gait and slight knee hyperextension during the stance phase. At the hospital, visual evoked potentials were performed to assess nerve conduction velocity along the visual tracts. Evoked potentials are extracted from an electroencephalogram (EEG) recorded during repetitive presentation of a flash of light. The time from the stimulus to the appearance of the potential on the EEG indi-

(continued)

cates the central conduction time. For J.R., decreased visual sensory conduction times were determined. J.R. was referred to physical therapy for strengthening exercises and gait training with an ankle-foot orthosis. The physician's orders specify low-repetition exercises and avoidance of physical overexertion.

Questions

1. Delayed conduction times for the evoked potentials suggest a problem with sensory conduction within the central nervous system. What nervous system abnormality can explain the delayed sensory nerve conduction times?

2. What mechanism related to generation of the action potential may be directly impaired by increases of body temperature associated with overexertion?

REVIEW QUESTIONS

1. List two ways in which glial cells differ from nerve cells.

2. To what critical function do both oligodendrocytes and Schwann cells contribute in the nervous system?

3. Do dendritic projections function as input units or output units for a neuron?

4. Name one example of a pseudounipolar cell. Why is it called pseudounipolar?

5. What is the specialized function of multipolar cells?

6. What are the three major ions that contribute to the electrical potential of a cell membrane in its resting state?

7. Define the terms *depolarization* and *hyperpolarization* with respect to the resting membrane potential.

8. If a membrane channel opens when it is bound by a neurotransmitter, what category of membrane channel is it?

9. What does the term *graded* mean with respect to the generation of local receptor and synaptic potentials?

10. List two types of local potential summation that can result in depolarization of a membrane to the threshold level.

11. The generation of an action potential requires the influx of what ion? Is the influx mediated by a voltage-gated channel?

12. Do large-diameter or small-diameter axons promote faster conduction velocity of an action potential?

13. What are the unique features of the nodes of Ranvier that promote generation of an action potential?

14. Are networks composed of interneuronal convergence and divergence found throughout the nervous system or only in the spinal cord?

References

Pennisi, E. (1993). Microglial madness: When the brain's immune system turns from friend to foe. Sci. News 144:378–379.

Reitsma, W. (1969). Skeletal muscle hypertrophy after heavy exercise in rats with surgically reduced muscle function. Am. J. Phys. Med. 48:237.

Zelig, G., Ohry, A., Shemesh, Y., et al. (1988). The rehabilitation of patients with severe Guillain-Barré syndrome. Paraplegia 26(4):250–254.

Suggested Readings

Aidley, D. J. (1989). The physiology of excitable cells (3rd ed.). Cambridge, Mass.: Cambridge University Press.

Armstrong, C. M. (1981). Sodium channels and gating currents. Physiol. Rev. 61:644–683.

Hodgkin, A. L., and Huxley, A. F. (1952). A quantitative description of membrane current and its application to conduction and excitation in the nerve. J. Physiol. (Lond.) 117:500–544.

Kandel, E. R. (1991). Nerve cells and behavior. In: Kandel, E. R., Schwartz, J. H., and Jessell, T. M. (Eds.). Principles of neural science (3rd ed.). New York: Elsevier Science Publishing.

Koester, J. (1991). Passive membrane properties of the neuron. In: Kandel, E. R., Schwartz, J. H., and Jessell, T. M. (Eds.). Principles of neural science (3rd ed.). New York: Elsevier Science Publishing.

Schwartz, J. H. (1991). The cytology of neurons. In: Kandel, E. R., Schwartz, J. H., and Jessell, T. M. (Eds.). Principles of neural science (3rd ed.). New York: Elsevier Science Publishing.

3

Synapses and Synaptic Transmission

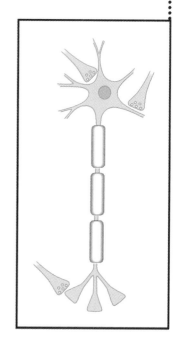

As a child, I was intrigued with the book *Mutiny on the Bounty.* I thought the word *mutiny* sounded like a word I wanted to have in my vocabulary. Little did I know that I would one day use the word in the context of my own body. Today my immune system wages a mutiny of sorts. I have myasthenia gravis (MG).

A year ago I was a 28-year-old male college student engaged in completing the prerequisites for admission to a graduate program in physical therapy. That is when my disease first became apparent. My vision began behaving strangely. Dizziness and disorientation occurred when I tried to scan from one point to another. It was as though one eye couldn't keep up with the other. I visited my ophthalmologist, who suggested everything from a brain tumor to multiple sclerosis. After a battery of tests, including an MRI, all of his theories had been eliminated. Fortunately I was then referred to a neuroophthalmologist who knew what I had before he even examined me. He gave me a Tensilon test, which was positive, and MG was officially diagnosed.

My life has changed significantly over the last year. I am lucky, however, because the disease only affects my eyes at this point. I experience double vision much of the time, and I have difficulty keeping my eyelids open. I have learned that I depended on my eyes in ways I had never before realized. The most noticeable deficit is the absence of binocular vision.

After a quick deterioration at the onset of the disease, my condition stabilized. I take a medication called pyridostigmine bromide (Mestinon), which controls my symptoms to some degree for short periods of time. In addition, I underwent a thymectomy last summer because studies have shown that, for largely unknown reasons, removal of the thymus gland can result in dramatic improvement in patients with MG. These improvements can take up to a year to manifest themselves. I have noticed modest improvements in my condition since the surgery. I have received no physical therapy for my disease because at this point it affects only the oculomotor (eye movement control) portion of my vision.

—David Hughes

INTRODUCTION

In the mature nervous system, neurons make very specific connections at synapses. Synapses are specialized points of contact between a neuron and another neuron, a muscle cell, or a gland. Information is transmitted from one neuron to the postsynaptic cell at the synapse. A **presynaptic terminal** is formed by the axon end-projections of the cell transmitting a signal, and the **postsynaptic terminal** is formed by the membrane region of the receiving cell (Fig. 3–1). The space between the two terminals is referred to as the **synaptic cleft.** The presynaptic terminal contains vesicles (small packets) of chemicals, **neurotransmitters,** used to transmit information across the synaptic cleft. The postsynaptic membrane contains receptors, with specialized molecules designed to bind specific neurotransmitters. When a neurotransmitter is bound

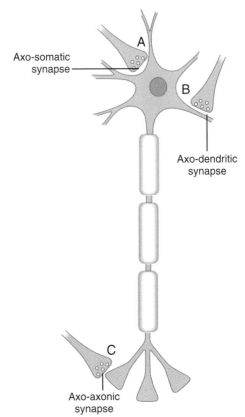

FIGURE 3–2

Types of synapses. *A,* Axosomatic connection between the axon of the presynaptic neuron and the cell body or soma of a postsynaptic neuron. *B,* Axodendritic connection between the axon of the presynaptic neuron and a dendrite of a postsynaptic neuron. *C,* Axoaxonic connection between the axon of the presynaptic neuron and the axon of a postsynaptic neuron.

to the receptor, the receptor changes shape. The change in receptor configuration may

- Open ion channels in the postsynaptic membrane, or
- Initiate changes in activity within the postsynaptic cell

Synaptic contact between neurons can occur on the cell body (axosomatic), the dendrites (axodendritic), and the axon (axoaxonic) of the postsynaptic neuron (Fig. 3–2). A single neuron can have multiple synaptic inputs in each region.

In humans, communication between neurons is accomplished primarily by chemical synaptic transmission. The sequence of events at an active chemical synapse are shown in Figure 3–3 and the list that follows.

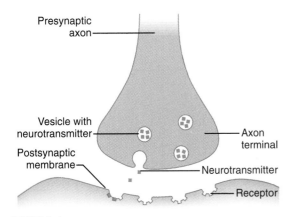

FIGURE 3–1

Synapse. An axon terminal from one neuron communicating via neurotransmitter with any region of membrane on another neuron, muscle cell, or gland forms a synapse.

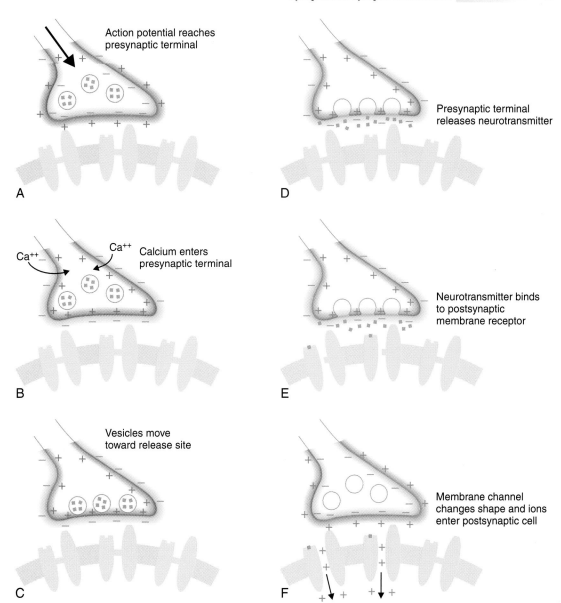

A, Action potential reaches presynaptic terminal

B, Ca^{++} Ca^{++} Calcium enters presynaptic terminal

C, Vesicles move toward release site

D, Presynaptic terminal releases neurotransmitter

E, Neurotransmitter binds to postsynaptic membrane receptor

F, Membrane channel changes shape and ions enter postsynaptic cell

FIGURE 3-3

Series of events at an active chemical synapse. *A,* When the action potential reaches the axon terminal *B,* the change in electrical potential causes the opening of voltage-dependent Ca^{++} channels and the influx of Ca^{++}. *C,* Elevated levels of Ca^{++} then promote the movement of synaptic vesicles to the membrane. *D,* The synaptic vesicles bind with the membrane, then release neurotransmitter into the synaptic cleft. *E,* Neurotransmitter diffuses across the synaptic cleft and activates a membrane receptor. *F,* In this case, the receptor is associated with an ion channel that opens when the receptor site is bound by neurotransmitter.

1. An action potential arrives at the presynaptic terminal.
2. The membrane of the presynaptic terminal depolarizes, causing the opening of voltage-gated Ca^{++} channels.
3. The influx of Ca^{++} into the nerve terminal, combined with the liberation of Ca^{++} from intracellular stores, triggers the movement of synaptic vesicles, containing neurotransmitters, toward a release site in the membrane.
4. Synaptic vesicles bind with the membrane, then rupture, releasing neurotransmitter into the cleft.
5. Neurotransmitter diffuses into the synaptic cleft.
6. Neurotransmitter that contacts receptors on the postsynaptic membrane binds to the receptor.
7. The shape of the membrane receptor changes. This change in receptor shape results in either
   - The opening of an ion channel associated with the membrane receptor, or
   - Activation of intracellular messengers associated with the membrane receptor

The amount of neurotransmitter released from the presynaptic terminal is directly related to the total number of action potentials per unit time reaching the terminal. Although an action potential is all-or-none, strong excitatory stimuli to the presynaptic cell result in a greater number of action potentials reaching the presynaptic terminal per unit time. Also, the longer the duration of the stimulus to the presynaptic cell, the longer the train of action potentials.

An increase in either the strength of a stimulus or the duration of a stimulus to the presynaptic cell results in the release of greater quantities of neurotransmitter.

ELECTRICAL POTENTIALS AT SYNAPSES

The release of neurotransmitters into the synaptic cleft results in the stimulation of membrane receptors on the postsynaptic membrane. The chemical stimulation of these receptors can result in the opening of membrane ion channels. If the synapse is between nerve and muscle, axosomatic, or axodendritic, the flux of ions in the postsynaptic membrane generates a local postsynaptic potential.

Postsynaptic Potentials

Postsynaptic potentials are local changes in ion concentration on the postsynaptic membrane. When neurotransmitter binds to a receptor on the postsynaptic membrane, the effect may be local depolarization or hyperpolarization. A local depolarization is an **excitatory postsynaptic potential** (EPSP). A local hyperpolarization is an **inhibitory postsynaptic potential** (IPSP).

EXCITATORY POSTSYNAPTIC POTENTIAL

An EPSP occurs when a neurotransmitter binds to postsynaptic membrane receptors that open ion channels, allowing a local, instantaneous flow of Na^+ or Ca^{++} into the neuron. The flux of positively charged ions into the cell causes the postsynaptic cell membrane to become depolarized (i.e., less negative). The flux of ions creates an EPSP (Fig. 3–4). Summation of EPSPs can lead to generation of an action potential (see Chap. 2).

Excitatory postsynaptic potentials are common throughout both the central and peripheral nervous system. For example, activation of synapses between nerve and muscle cells at the neuromuscular junction results in EPSPs that lead to excitation of the muscle. At a neuromuscular junction, the action of the neurotransmitter acetylcholine is always excitatory to the muscle cell. Binding of acetylcholine opens membrane channels that allow Na^+ influx into the muscle cell, initiating a series of events leading to mechanical contraction of the muscle cell. Normally, every action potential in a motor neuron (a neuron that innervates muscle) elicits a contraction of the muscle cell because motor neurons release sufficient amounts of transmitter to bind with the many receptors on a muscle cell membrane.

INHIBITORY POSTSYNAPTIC POTENTIAL

An IPSP is a local hyperpolarization of the postsynaptic membrane, decreasing the possibility of an action potential. In contrast to the EPSP, an IPSP involves a local flow of Cl^- and/or K^+ in response to a neurotransmitter binding to postsynaptic membrane receptors (Fig. 3–5). The postsynaptic ion channels open, allowing the flux of Cl^- into the cell or K^+ out of the cell. This causes the local postsynaptic cell membrane

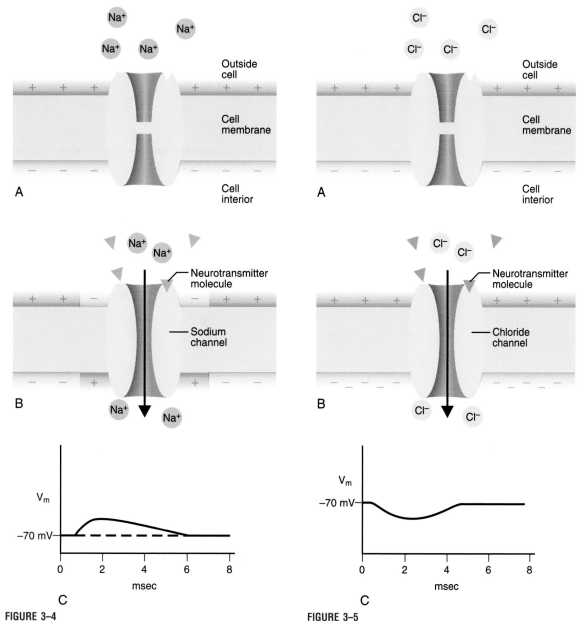

FIGURE 3–4

Excitatory postsynaptic potential. *A*, The resting membrane, with Na⁺ channels closed. *B*, Neurotransmitter released into the synaptic cleft binds with membrane receptors that stimulate the opening of ligand-gated Na⁺ channels. A resulting influx of Na⁺ depolarizes the membrane and thereby causes excitation of the neuron. *C*, The resulting postsynaptic membrane potential is more positive compared to the resting membrane potential.

FIGURE 3–5

Inhibitory postsynaptic potential. *A*, The resting membrane, with Cl⁻ channels closed. *B*, Neurotransmitter released into the synaptic cleft binds with membrane receptors that stimulate the opening of ligand-gated Cl⁻ channels. A resulting influx of Cl⁻ hyperpolarizes the membrane and thereby causes inhibition of the neuron. *C*, The resulting postsynaptic membrane potential is more negative compared to the resting membrane potential.

to become hyperpolarized (i.e., more negative). Hyperpolarization can inhibit generation of an action potential. If excitatory postsynaptic potentials coincide with inhibitory postsynaptic potentials, summation determines whether an action potential will be generated. If the preponderance of input to a neuron is inhibitory, an action potential is not generated in the postsynaptic neuron. Only if sufficient depolarization occurs to reach threshold is an action potential generated in the postsynaptic cell.

> At the postsynaptic membrane, changes in membrane potential can be either excitatory or inhibitory to the postsynaptic neuron.

Presynaptic Facilitation and Inhibition

Activity at a synapse can be influenced by **presynaptic facilitation** and **presynaptic inhibition**. Presynaptic effects occur when the amount of neurotransmitter released by a neuron is influenced by previous activity in an axoaxonic synapse. Neurotransmitter released from the axon terminal of one neuron binding with receptors on the axon terminal of a second neuron alters the membrane potential of the second terminal.

Presynaptic facilitation occurs when a presynaptic axon releases neurotransmitter that slightly depolarizes the axon terminal of a second neuron. This causes a small Ca^{++} influx into the second neuron's postsynaptic terminal. Subsequently, the duration of an action potential in the second neuron is increased when it arrives at the axon terminal. The prolonged action potential allows more Ca^{++} than normal to enter the second postsynaptic terminal. The increased Ca^{++} concentration causes more vesicles of neurotransmitter than usual to move to the cell membrane and rupture. Accordingly, the facilitated neuron releases more transmitter to its target postsynaptic cell (Fig. 3–6A).

In contrast, presynaptic inhibition occurs when an axon releases neurotransmitter that slightly hyperpolarizes the axonal region of another neuron. When an action potential occurs in the second neuron, the duration of the action potential in the axon terminal is decreased due to the local inhibition of the axon terminal membrane. As a result of the decreased duration of the action potential, Ca^{++} influx is reduced.

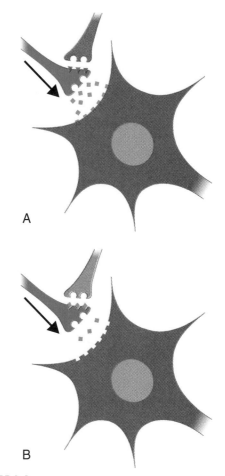

FIGURE 3–6

A, Presynaptic facilitation. *B*, Presynaptic inhibition. In both panels, the presynaptic neuron is green, and the postsynaptic neuron is blue. In *A*, Neurotransmitter from the interneuron (yellow) is bound to receptors on the axon terminal of the presynaptic neuron. This neurotransmitter binding facilitates an action potential reaching the presynaptic terminal, leading to increased Ca^{++} entering the terminal and increased release of neurotransmitter by the green neuron. The result is increased stimulation of the postsynaptic (blue) cell due to an increased release of neurotransmitter. *B* illustrates the opposite effect. The interneuron (gray) releases a neurotransmitter that inhibits the axon terminal of the presynaptic neuron. This inhibitory action depresses the action potential reaching the presynaptic terminal, leading to decreased Ca^{++} entering the terminal and decreased release of neurotransmitter. The result is decreased stimulation of the postsynaptic cell membrane due to decreased release of neurotransmitter into the synaptic cleft between the presynaptic neuron and postsynaptic neuron.

Accordingly, the inhibited neuron releases less transmitter onto its target postsynaptic cell (Fig. 3–6B).

> The release of neurotransmitters from an axon terminal can be either facilitated or inhibited by the chemical action at an axoaxonic synapse.

SYNAPTIC RECEPTORS

The precise action of a neurotransmitter on the postsynaptic cell is dependent on the properties of the receptors with which the transmitter binds. Receptors for neurotransmitters are structural proteins that span the postsynaptic cell membrane. There are two types of synaptic receptors:

- Ligand-gated ion channels
- G-protein mediated receptors

Ligand-Gated Ion Channels

Ligand-gated ion channels consist of proteins that function like a gate. The gate of these channels opens in response to a chemical transmitter binding to the receptor surface (see Figs. 3–4 and 3–5). If neurotransmitter is not bound to the receptor of a ligand-gated channel, the gate is closed, and no ions pass through. However, if neurotransmitter is bound to the receptor, the gate opens, and specific ions are allowed to flow through the gate according to their electrochemical gradient. Electrochemical gradient includes the effects of both the distribution of electrical charge and the concentration gradient of a specific ion on the tendency of an ion to diffuse (see Chap. 2).

For example, the ligand-gated channels that allow Na^+ to diffuse also allow K^+ to diffuse, so these channels are called Na^+-K^+ channels. When a ligand-gated Na^+-K^+ channel is open, Na^+ flows into the neuron and K^+ flows out of the neuron, according to their electrochemical gradients between the inside and outside of the cell. The net result of opening a ligand-gated Na^+-K^+ channel is local depolarization. This occurs because much more Na^+ diffuses into the cell, due to the high concentration of Na^+ outside the cell and the electrical attraction by negative ions inside the cell membrane (Fig 3–7A).

Gated ion channels are open for only a few milliseconds. The brief opening allows a rapid flux of ions

that either facilitate or inhibit the development of synaptic potentials. The receptors on ligand-gated channels are inactivated when the neurotransmitter diffuses away from the synaptic cleft, is broken down by enzymes in the synaptic cleft, or is taken up by the presynaptic axon terminal.

> Rapid and short-lasting opening of membrane channels results from the direct binding of a neurotransmitter to the extracellular membrane receptor site of the channel.

G-Protein Mediated Receptors

G-protein mediated receptors have more complex actions than the ligand-gated channels. Some of the same neurotransmitters that act on ligand-gated ion channel receptors can also act on G-protein mediated receptors. Unlike the rapid, brief activity of gated ion channels, activity of the G-protein mediated receptors is slower and longer lasting and can alter a variety of cellular functions. Activation of G-protein mediated receptors may lead to the opening of membrane ion channels or to a variety of more long-lasting cellular responses. The effects of neurotransmitter binding to ligand-gated channels versus G-protein mediated receptors are contrasted in Figure 3–7B.

The most well-understood pathways use G proteins acting as cytoplasmic shuttles moving between the receptor and target effector proteins on the internal surface of the cell membrane. The G proteins consist of two subunits: the α and βγ chains. When the membrane receptor is not activated, the αβγ complex is bound to the receptor. However, when the receptor is bound by a neurotransmitter, the ensuing sequence is as follows (Chambre, 1987; Casey and Gilman, 1988):

1. The receptor protein changes shape.
2. The G protein becomes activated (via replacement of guanosine diphosphate by guanosine triphosphate).
3. The α chain breaks free to act as a cytoplasmic shuttle.
4. The α chain activates a target protein.
5. The α chain becomes deactivated and reassociates with the βγ chain.

The target protein activated by the α chain may be either a membrane ion channel or an intracellular tar-

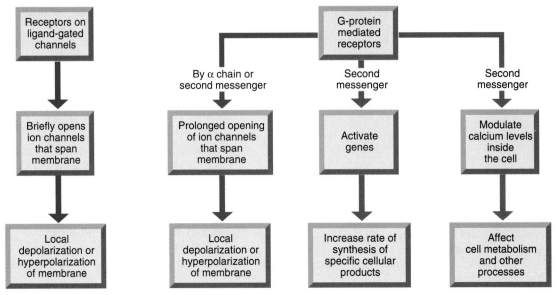

FIGURE 3–7

The effects of neurotransmitter binding to a receptor. *A,* The effects of neurotransmitter binding to ligand-gated channels. *B,* The effects on G-protein mediated receptors.

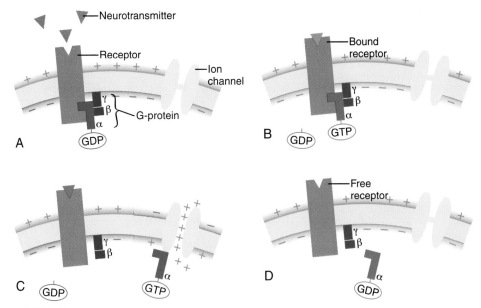

FIGURE 3–8

G-protein gated ion channel. *A,* In the nonstimulated state, the αβγ G-protein complex is associated with a membrane receptor. *B,* Neurotransmitter binds to the membrane receptor, causing a conformational change and activation of the G protein. The α chain detaches from the membrane receptor. *C,* The α–guanosine triphosphate (GTP) complex binds to a membrane-spanning protein channel. A conformational change in the protein channel causes the channel to open, and ions flow into the cell. *D,* The α chain is inactivated and released from the protein channel. The channel closes, and the α chain returns to its host membrane receptor to bind with the βγ chain. GDP, guanosine diphosphate.

get protein. If a G protein activates a membrane ion channel, the change in configuration of the channel protein opens the channel (Fig. 3–8).

> Binding of a neurotransmitter to an extracellular membrane receptor having an associated intracellular G protein results in slowly developing and long-lasting cellular events and/or persistent opening of membrane channels.

SECOND-MESSENGER SYSTEMS

If the G protein activates an intracellular target protein, the target protein generates a diffusible second messenger, which either modulates a membrane channel directly or, more commonly, affects enzyme activity. Via the G-protein pathway, a single neurotransmitter molecule may elicit a cascade of cellular responses through the indirect activation of the second messenger (Fig. 3–9). In second-messenger systems, the neurotransmitter is the first messenger, delivering the message to the receptor but remaining outside the cell. The second messenger, produced inside the cell, conveys the message and activates responses inside the cell. The second messenger may initiate one of the following:

- Opening of membrane ion channels
- Activation of genes, causing increased synthesis of specific cellular products
- Modulation of calcium concentrations inside the cell

Internal stores of Ca^{++} liberated in response to second-messenger systems regulate metabolism and other cellular processes. In this case, Ca^{++} acts as a third messenger. Three second messengers have been identified in neurons:

- Cyclic adenosine monophosphate (cAMP)
- Arachidonic acid
- Inositol triphosphate

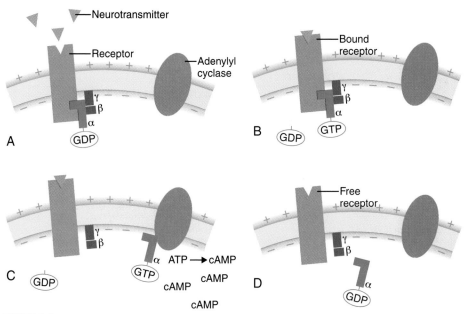

FIGURE 3–9

G-protein second messenger. *A* and *B,* The cellular events are the same as in Figure 3–8*A* and *B.* *C,* The α–guanosine triphosphate (GTP) complex binds to a membrane-spanning protein enzyme called adenylyl cyclase. A conformational change in the protein enzyme causes the enzyme to produce adenosine triphosphate (ATP), which is converted to cyclic adenosine monophosphate (cAMP). The cAMP acts as a second messenger, activating a variety of cellular proteins including kinases and protein channels. *D,* The α chain is inactivated and released from the protein enzyme. The enzyme is inactivated, and the α chain returns to its host membrane receptor to bind with the βγ chain. GDP, guanosine diphosphate.

Cyclic Adenosine Monophosphate One example of a G-protein mediated second-messenger system involves the activation of the enzyme adenylyl cyclase, which converts adenosine triphosphate to cAMP (Figs. 3–9 and 3–10A). Cyclic cAMP plays the role of the second messenger. Increases in cAMP levels modulate membrane receptors or cAMP-dependent cytoplasmic protein enzymes. Membrane channels may be opened, regulating ionic flux into the cell, and a variety of cell regulation and gene expression pathways may be activated by the cytoplasmic proteins. For example, the transmission of pain information in the peripheral nervous system is thought to involve a cAMP-dependent ion channel. Administering morphine may provide relief of pain when morphine binds to a receptor that activates a G protein, which then inhibits adenylyl cyclase. Inhibition of adenylyl cyclase and subsequent inhibition of a cAMP-dependent ion channel may inhibit sensory afferents conveying information about tissue damage from the periphery to the central nervous system and thus relieve pain (Ingram and Williams, 1994).

Arachidonic Acid Another second-messenger system uses G-protein activation of an enzyme called phos-pholipase A_2. This system results in the liberation of arachidonic acid as its second messenger and the initiation of a metabolic cascade that leads to the production of prostaglandins (Fig. 3–10B). Prostaglandins are substances that regulate vasodilation and increase inflammation. Aspirin and other nonsteroidal anti-inflammatory drugs act to reduce pain by inhibiting one of the enzymes in this G protein–initiated cascade.

Inositol Triphosphate In addition to playing a role in the management of pain, G proteins also play a role in many other nervous system functions. The second-messenger pathway of inositol triphosphate (IP_3) acts in the release of Ca^{++} from internal stores so that it can be used in a variety of cellular metabolic processes (Stehno-Bittel et al., 1995) (Fig. 3–10C). When a neurotransmitter binds to a receptor, it activates a G protein, which in turn activates the enzyme phospholipase C. Activation of the enzyme results in the production of the second messenger IP_3, which diffuses into the cytoplasm to stimulate release of Ca^{++} from the endoplasmic reticulum.

Recently, G proteins have been the subject of much scientific investigation, and several new families of G

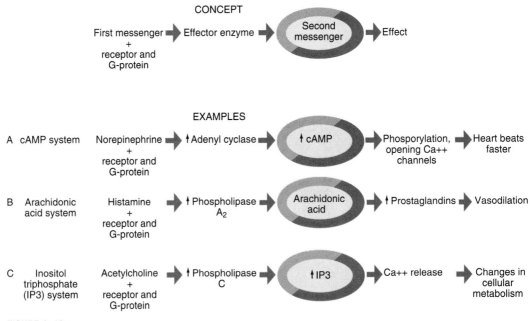

FIGURE 3–10

G-protein mediated second-messenger systems. All three systems shown involve (1) binding of a neurotransmitter to a G-protein associated membrane receptor, (2) activation of an effector enzyme, (3) increased levels of a second messenger, and (4) a cellular and physiological event.

proteins and functions of their multitude of second messengers are being identified throughout the entire nervous system.

NEUROTRANSMITTERS

Most chemicals that transmit information across the synapse are classified as amino acids or their derivatives, amines and peptides. A major exception to the typical chemical composition of transmitters is acetylcholine. Acetylcholine (ACh) is the only neurotransmitter that is not an amino acid or derived from an amino acid; rather it is derived from acetyl coenzyme A and choline. Common neurotransmitters are listed in Table 3–1. The effects of a neurotransmitter depends on the action of the postsynaptic receptors to which the transmitter binds. For example, the neurotransmitter acetylcholine binds with two different classes of receptors. These receptors, identified based on their ability to bind certain drugs, are as follows:

- Muscarinic
- Nicotinic

Muscarinic Receptors

The muscarinic receptors were so named because experimentally, Muscarine, a poison derived from mushrooms, activates only the muscarinic receptors. Muscarinic receptors are found on neurons in the cortex and midbrain and on certain effectors of the autonomic nervous system. The actions of ACh on muscarinic receptors contributes to the regulation of cardiac muscle, smooth muscle, and glandular activity (see Chap. 7). Acetylcholine binding to muscarinic receptors initiates a G-protein mediated response causing either an EPSP or an IPSP.

Nicotinic Receptors

The nicotinic receptors were so named because experimentally, Nicotine, derived from tobacco, activates only the nicotinic receptors. Nicotinic receptors are found on postsynaptic neurons in autonomic ganglia and on the postsynaptic membrane of skeletal muscles. Acetylcholine binding to nicotinic receptors initiates opening of a direct, ligand-gated channel causing a fast EPSP in the postsynaptic membrane.

Neurotransmitter Agonists and Antagonists

The action of neurotransmitters can be mimicked or blocked by drugs. Drugs that affect the nervous system usually bind with neurotransmitter receptors or prevent the release of neurotransmitters. If the drug binds to the receptor and promotes the effects of the naturally occurring neurotransmitter, the drug is called an agonist. If, on the other hand, the drug prevents the release of neurotransmitter or binds to the receptor and impedes the effects of the naturally occurring transmitter, the drug is called an antagonist. For example, because nicotine binds to certain ACh receptors and elicits the same effects as the neurotransmitter would, nicotine is an ACh agonist.

Botulinum toxin A (Botox) is an example of a neurotransmitter antagonist that has recently been used to improve the functional abilities of people with movement abnormalities due to central nervous system disorders. Botulinum toxin A is naturally produced by a family of bacteria and, when ingested, causes widespread paralysis by inhibiting the release of ACh at the neuromuscular junction (Borg-Stein and Stein, 1993). When small doses of botulinum toxin A are therapeutically injected directly into an overactive muscle, the local effect is muscle paralysis (Cromwell and Paquette, 1996). This paralysis lasts for up to 12 weeks (Borg-Stein and Stein, 1993) and results in improved range of motion, resting limb position, and functional movements.

TABLE 3–1	
COMMON NEUROTRANSMITTERS	
Type	**Transmitter**
Cholinergic	Acetylcholine (ACh)
Amino acid	γ-Aminobutyric acid (GABA)
	Glutamate (Glu)
	Glycine (Gly)
Amines	Dopamine (DA)
	Epinephrine
	Histamine
	Norepinephrine (NE)
	Serotonin (5-HT)
Peptides	Endorphins
	Enkephalins
	Substance P

A variety of other drugs acting as either agonists or antagonists are used in the treatment of movement disorders and in psychological and memory disorders. Medications used for the treatment of movement disorders and/or the control of involuntary movements may be either injected locally or taken orally. As our understanding of nervous system transmission and pharmacology improve, new treatments may be developed.

Neurotransmitters are the chemicals released from an axon terminal. Their effect is dependent on the type of receptor they bind with on the postsynaptic cell.

Diseases Affecting the Neuromuscular Junction

Disease processes that disrupt the synaptic connection generally involve the junction between an efferent nerve terminal and a muscle cell, preventing the relay of a chemical signal. Some patients with cancer, such as small cell cancers of the lung, show weakness associated with a neuromuscular synaptic junction disorder. This disorder, called Lambert-Eaton syndrome, involves antibodies to voltage-gated Ca^{++} channels in the presynaptic terminal at the junction between a motor neuron axon and muscle fiber (Rowland, 1991). The blockage of Ca^{++} influx to the terminal results in a decreased release of neurotransmitter and decreased excitation of the muscle.

Another disease, myasthenia gravis, causes severe weakness due to antibodies created against the receptor on the postsynaptic membrane of the muscle. Normal amounts of neurotransmitter are released but have few places to bind. Thus the transmitters are less effective in stimulating the postsynaptic membrane. In myasthenia gravis, increasing weakness occurs with repetitive use of the muscles. For example, more than half of people with myasthenia gravis have difficulty opening their eyelids or with eye movements as the first signs of the disease. Other muscles commonly affected control facial expression, swallowing, proximal limb movements, and respiration. Typically the proximal limb weakness causes difficulty reaching overhead, climbing stairs, and rising from a chair. Drugs that inhibit the breakdown of ACh usually improve function because ACh is available longer to bind with

MYASTHENIA GRAVIS

Pathology
Decreased number of muscle membrane acetylcholine receptors

Etiology
Autoimmune

Speed of onset
Chronic

Signs and symptoms
Usually affect eye movements or eyelids first

Consciousness
Normal

Cognition, language, and memory
Normal

Sensory
Normal

Autonomic
Normal

Motor
Fluctuating weakness; weakness increases with muscle use

Cranial nerves
Cranial nerves are normal; however, skeletal muscles innervated by cranial nerves show fluctuating weakness (because the disorder affects the muscle membrane receptors)

Region affected
Peripheral

Demographics
Can occur at any age, women more often affected than men

Prognosis
Stable or slowly progressive; with medical treatment, >90% survival rate

---

the remaining ACh receptors. The autoimmune assault of ACh receptors can be countered by the following:

- Removal of the thymus gland, an immune organ that functions abnormally in myasthenia gravis, contributing to the damage of ACh receptors
- Immunosuppressive drugs
- Plasmapheresis, the process of removing blood from the body, centrifuging the blood to separate the plasma from the cells, then returning the blood cells and replacing the plasma with a plasma substitute

The previous treatments produce a relatively good prognosis in myasthenia gravis; the survival rate is better than 90%. Onset in women typically occurs between ages 20 and 30 years, while in men the onset most commonly occurs between ages 60 and 70. Occasionally, remissions occur in the course of the disease, but stabilization or progression is the more frequent outcome. (See the box on Myasthenia gravis.)

> Diseases affecting the neuromuscular junction generally impede the transmission of a signal by decreasing the release of neurotransmitter at the synapse or preventing the transmitter from activating the postsynaptic membrane receptor.

TOXINS AFFECTING NEURAL SIGNALING AT SYNAPSES

Several naturally occurring toxins have detrimental effects on neuronal function by interfering with ion channel activation. As previously discussed, botulinum toxin produced by bacteria prevents the release of neurotransmitter at the neuromuscular junction. Tetrodotoxin, a virulent nerve toxin found in Japanese puffer fish, blocks the voltage-dependent Na^+ channels but has no effect on the K^+ channels. This prevents action potentials and causes paralysis. Strychnine blocks the inhibitory synapses on motor neurons, causing convulsions. Tetanus toxin prevents the release of inhibitory neurotransmitters by interneurons, causing tetanic contractions and convulsions. Although these toxins can be detrimental to the nervous system, they are frequently used in scientific laboratories to improve our understanding of synaptic transmission and develop new drug therapies. These toxins are rarely encountered as the causes of disease in the environment.

> A variety of naturally occurring toxins can interfere with nervous system function by blocking synapses or impeding the release of neurotransmitters.

CLINICAL NOTES

CASE 1

M.J., a 54-year-old woman, suffers from small cell cancer of the lung and exhibits generalized, progressive muscle weakness. On medical evaluation, it is determined that M.J.'s weakness is related to a neuromuscular junction disorder consistent with Lambert-Eaton syndrome. In this syndrome, the voltage-gated Ca^{++} channels in the axon terminals at the synapse between the motor nerve and muscle are disrupted. Plasmapheresis, the process of removing blood from the body, centrifuging the blood to separate the plasma from the cells, then returning the blood cells and replacing the plasma with a plasma substitute, effectively reduces M.J.'s weakness. The benefit from plasmapheresis supports the hypothesis that the disease involves circulating antibodies to the Ca^{++} channels in the motor axon terminals, because the circulating antibodies are removed with the plasma.

Questions

1. The neurotransmitter released at the synaptic junction between the motor axon and the muscle is ACh. Why would destruction of the Ca^{++} channels in the axon terminal disrupt the release of ACh from the axon terminal?

2. Would physical therapy be beneficial for increasing M.J.'s strength if the antibodies to the Ca^{++} channel continue to circulate?

(continued)

CASE 2

S.B., a 12-year-old girl, has significant gait abnormalities resulting from cerebral palsy. When ambulating, she walks on her toes and exhibits a scissor gait, with her legs strongly adducted with each step. S.B. has shown no significant improvements in gait with standard physical therapy exercise, gait training, and range of motion exercise. Her physicians now want to inject a small amount of botulinum toxin into the gastrocnemius and adductor magnus muscles of both legs in an effort to reduce involuntary muscle activity and improve gait.

Questions

1. By what mechanism could the injection of botulinum toxin reduce the involuntary muscle activity?

2. At the neuromuscular junction, ACh acts via a ligand-gated receptor. Is the action of ACh on the nicotinic, ligand-gated receptor the same as the action on the muscarinic, G-protein mediated receptor?

REVIEW QUESTIONS

1. What is the difference between postsynaptic inhibition and presynaptic inhibition? Which one results in a decreased release of neurotransmitters?

2. The release of neurotransmitter from synaptic vesicles is dependent on the influx of what ion into the presynaptic terminal?

3. Does direct activation of a membrane ion channel by a neurotransmitter or indirect activation via second-messenger systems result in the faster generation of a synaptic potential?

4. How does binding of a neurotransmitter to the receptor of a ligand-gated ion channel cause the channel to open?

5. How do G proteins contribute to a cascade of cellular events?

6. Is the effect of a neurotransmitter determined by the transmitter itself or by the type of receptor?

References

Borg-Stein, J., and Stein, J. (1993). Pharmacology of botulinum toxin and implications for use in disorders or muscle tone. J. Head Trauma Rehabil. 8:103–106.

Casey, P. J., and Gilman, A. G. (1988). G-protein involvement in receptor-effector coupling. J. Biol. Chem. 263:2577–2580.

Chambre, M. (1987). The G-protein connection: Is it in the membrane or the cytoplasm? Trends Biochem. Sci. 12:213–215.

Cromwell, S. J., and Paquette, V. L. (1996). The effect of botulinum toxin A on the function of a person with poststroke quadriplegia. Phys. Ther. 76(4):395–402.

Ingram, S. L., and Williams, J. T. (1994). Opioid inhibition of Ih via adenylyl cyclase. Neuron 13:179–186.

Rowland, L. P. (1991). Diseases of chemical transmission at the nerve-muscle synapse: Myasthenia gravis. In: Kandel, E. R., Schwartz, J. H., and Jessell, T. M. (Eds.). Principles of neural science (3rd ed.). Elsevier Science Publishing, pp. 235–243.

Stehno-Bittel, L., Luckhoff, A., and Clapman, D. E. (1995). Calcium release from the nucleus by InsP₃ receptor channels. Neuron 14:163–167.

Suggested Readings

Aidley, D. J. (1989). The physiology of excitable cells (3rd ed.). Cambridge, Mass.: Cambridge University Press.

Armstrong, C. M. (1981). Sodium channels and gating currents. Physiol. Rev. 61:644–683.

Kandel, E. R. (1991). Nerve cells and behavior. In: Kandel, E. R.,

Schwartz, J. H., and Jessell, T. M. (Eds.). Principles of neural science (3rd ed.). New York: Elsevier Science Publishing.

Kim, Y. I., and Neher, E. (1988). IgG from patients with Lambert-Eaton syndrome blocks voltage-dependent calcium channels. Science 239:405–408.

4

Neuroplasticity

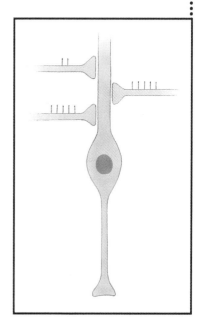

INTRODUCTION

Improved understanding of the mechanisms involved in the electrical potentials of neurons and synaptic transmission have led to a greater understanding of how the brain adapts to changes in input and recovers from injury. The knowledge of the mechanisms of neuroplasticity and various disease processes has subsequently led to the development of new therapeutic treatments for patients with neurological disorders.

Although mature neurons are not capable of division and replication, the central nervous system is capable of an immense degree of neuroplasticity. By definition, neuroplasticity is any change in the nervous system that is not periodic and has a duration of more than a few seconds. These changes include

- Habituation
- Learning and memory
- Recovery from injury

HABITUATION

Habituation is one of the simplest forms of plasticity. In studies of animal posture and locomotion, Charles Sherrington observed that certain reflex behaviors, such as withdrawing a limb from a mildly painful stimulus, ceased to be elicited after several repetitions of the same stimulus. Sherrington proposed that the decreased responsiveness, or habituation, resulted from a functional decrease in the synaptic effectiveness of the stimulated pathways to the motor neuron. Later, it was determined that habituation of the flexor reflex was indeed due to a decrease in synaptic activity between the sensory neurons and interneurons. The decrease is thought to be, in part, due to a decrease in the amount of neurotransmitter released from the presynaptic terminal of the sensory neuron. Generally, after several seconds of rest, the effects of habituation are no longer present, and a reflex can be elicited in response to sensory stimuli. However, with prolonged repetition of stimulation, more permanent, structural changes occur: the number of synaptic connections decrease.

Clinically, the term *habituation* is applied to techniques and exercises used in occupational and physical therapy that are intended to decrease the neural response to a stimulus. For example, some children are extremely reactive to stimulation on their skin. This abnormal sensitivity to tactile stimulation is called tactile defensiveness. The treatment for tactile defensiveness is stimulating the child's skin with gentle stimulation, then gradually increasing the intensity of the stimulation, in an effort to achieve habituation to the tactile stimulation. In people with specific types of vestibular disorders, movements that induce dizziness and nausea are repeatedly performed, again with the purpose of achieving habituation to the movements.

> Short-term changes in neurotransmitter release and postsynaptic receptor sensitivity result in a decreased response to specific, repetitive stimuli.

LEARNING AND MEMORY

Unlike the short-term, reversible effects of habituation, learning and memory involve persistent, long-lasting changes in the strength of synaptic connections. Neuroimaging techniques reveal that during the initial phases of motor learning, large and diffuse regions of the brain show synaptic activity. With repetition of a task, there is a reduction in the number of active regions in the brain. Eventually, when a task has been learned, only small distinct regions of the brain show increased activity during performance of the task.

Long-term memory requires the synthesis of new proteins and the growth of new synaptic connections. With repetition of a specific stimulus, the synthesis and activation of new proteins alter the neuron's excitability and promote the growth of new synaptic connections. A cellular mechanism for the formation of memory, called **long-term potentiation** (LTP), has been identified in the hippocampus. The hippocampus, in the temporal lobe, is essential for processing memories that can easily be verbalized. For example, the hippocampus is important in remembering names but not in remembering motor acts like riding a bike. Experimental stimulation studies demonstrate that a brief high-frequency train of stimuli to the afferent pathways of the hippocampus produces an increase in the amplitude of excitatory postsynaptic potential (EPSP) of an associated hippocampal cell.

The EPSP facilitation can last up to several days in the intact animal. This facilitation, or long-term potentiation, has three requirements (Fig. 4–1):

1. More than one nerve fiber must be activated (cooperativity).

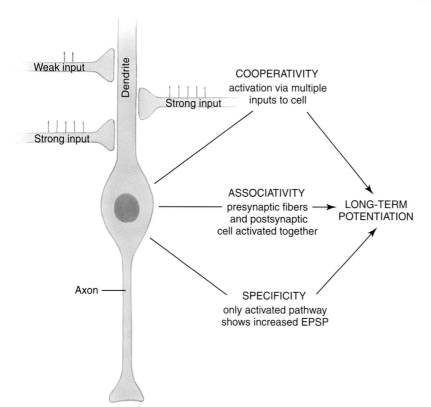

FIGURE 4–1

Long-term potentiation. Increased amplitude of an excitatory postsynaptic potential (EPSP) results from multiple inputs to a neuronal pathway. The properties of cooperativity, associativity, and specificity contribute to the enhancement of the EPSP and the formation of memory.

2. The contributing fibers and the postsynaptic cell must be activated together (associativity).
3. The potentiation must be specific to only the activated pathway (specificity).

The cooperative and associative properties are important, since they imply that multiple inputs may have an additive effect, and a weak input will become potentiated if it is activated in association with a strong input. Experiments suggest that the processing of easily verbalized memory is related to an increase in synaptic activity and metabolic changes that increase the efficiency of cell firing. Once again, Ca^{++} influx plays an important role in synaptic transmission. The initiation of LTP depends on postsynaptic cell depolarization, the influx of Ca^{++}, and Ca^{++} activation of cellular second messengers. The maintenance of LTP requires increases in the calcium-mediated release of neurotransmitter from the presynaptic terminal.

> Long-term changes, including the synthesis of new proteins and growth of new synaptic connections, result in a maintained response and memory of specific, repetitive stimuli.

RECOVERY FROM INJURY

Injuries that damage or sever the axons of neurons cause degenerative changes but may not necessarily result in death of the cell. Some neurons have the ability to regenerate their axon. In contrast to injury to the axon, injuries that destroy the cell body of a neuron invariably lead to death of the cell. When neurons in the adult nervous system die, they are not replaced. However, alterations in synapses, functional reorganization in the central nervous system, and activity-related changes in neurotransmitter release promote recovery from injury.

Axonal Injury

When an axon is severed, the part connected to the cell body is referred to as the proximal segment, and the part isolated from the cell body is called the distal segment. Immediately after injury, protoplasm leaks out of the cut ends, and the segments retract away from each other. Once isolated from the cell body, the distal segment of the axon undergoes a process called **wallerian degeneration** (Fig. 4–2). When the distal segment of an axon degenerates, the myelin sheath pulls away from the segment. The axon swells and breaks into short segments. The axon terminals rapidly degenerate, and their loss is followed by death of the entire distal segment. Glial cells scavenge the area, cleaning up the debris from the degenerating axon. In addition to the axonal degeneration, the associated cell body undergoes degenerative changes called **central chromatolysis,** occasionally leading to cell death. Simultaneously, the postsynaptic cell once in contact with the axon terminal also degenerates and may die.

The regrowth of damaged axons is called **sprouting.** Sprouting takes two forms: collateral and regenerative (Fig. 4–3). Collateral sprouting occurs when a denervated target is reinnervated by branches of intact axons. Regenerative sprouting occurs when an axon and its target have been damaged. The injured axon sends out side sprouts to a new target. Functional regeneration of axons occurs most frequently in the peripheral nervous system, partly because the production of nerve growth factor (NGF) by the Schwann cells contributes to the recovery of peripheral axons. The recovery is slow, with approximately 1 mm of growth per day, or approximately 1 inch of recovery over 1 month. Unlike injuries to the peripheral axons, damage to axons within the central nervous system is irreversible, since central neurons cannot functionally regenerate their axons under normal physiological conditions. The lack of functional regeneration is due to lack of NGF, inhibition of growth by oligodendrocytes, and interference with the cleanup activities of microglia.

Sprouting by peripheral axons can cause problems when an inappropriate target is innervated. For example, after peripheral nerve injury, motor axons may innervate different muscles than previously, resulting in unintended movements when the neurons fire. The unintended movements, called synkinesis, are usually short-lived, as the individual relearns muscle control. Similarly, in the sensory systems, innervation of sensory receptors by axons that previously innervated a different type of sensory receptor can cause confusion of sensory modalities.

> Damaged peripheral axons can recover from injury, and targets deprived of input from the damaged axons can attract new inputs to maintain nervous system function.

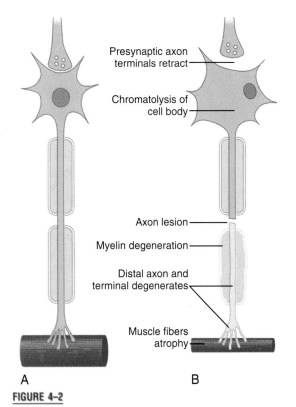

Presynaptic axon
terminals retract

Chromatolysis of
cell body

Axon lesion

Myelin degeneration

Distal axon and
terminal degenerates

Muscle fibers
atrophy

A   B

FIGURE 4–2

Wallerian degeneration. *A,* Normal connections before an axon is severed. *B,* Degeneration following severance of an axon. Degeneration following axonal injury involves several changes: (1) the axon terminal degenerates, (2) myelin breaks down and forms debris, and (3) the cell body undergoes metabolic changes. Subsequently, (4) presynaptic terminals retract from the dying cell body, and (5) postsynaptic cells degenerate.

Synaptic Changes

Following central nervous system injury, synaptic changes include recovery of synaptic effectiveness, denervation hypersensitivity, synaptic hypereffectiveness, and unmasking of silent synapses (Fig. 4–4 on p. 62).

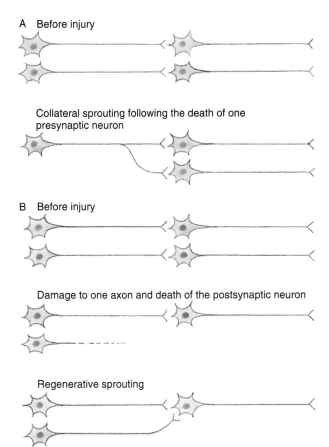

A Before injury

Collateral sprouting following the death of one presynaptic neuron

B Before injury

Damage to one axon and death of the postsynaptic neuron

Regenerative sprouting

FIGURE 4–3

Axonal sprouting. The new growth of axons following injury involves two types of sprouting: *A*, collateral sprouting, in which a denervated neuron attracts side sprouts from nearby undamaged axons; and *B*, regenerative sprouting, in which the injured axon issues side sprouts to form new synapses with undamaged neurons.

Local edema causes some synapses to become inactive owing to compression of the presynaptic neuron's cell body or axon. Once the edema is resolved, **synaptic effectiveness** returns. **Denervation hypersensitivity** occurs when presynaptic axon terminals are destroyed. New receptor sites develop on the postsynaptic membrane in response to transmitter released from other nearby axons (Creese et al., 1977). **Synaptic hypereffectiveness** occurs when only some branches of a presynaptic axon are destroyed. The remaining axon branches receive all of the neurotransmitter that would normally be shared among the terminals, resulting in larger than normal amounts of transmitter being released onto postsynaptic receptors. A final synaptic change is **unmasking (disinhibition) of silent synapses.** In the normal nervous system, many synapses seem to be unused unless injury to other pathways results in increased use of the previously silent synapses (Chollet et al., 1991).

Functional Reorganization

In the adult brain, cortical areas routinely adjust the way they process information and retain the ability to develop new functions. Maps of functional areas of the cerebral cortex are produced by recording the activity of neurons in response to sensory stimulation or during active muscle contractions. Reassignment of neuron function in adults has been shown with magnetic resonance imaging (MRI) of the cortex. For example, cortical mapping in humans shows a large area of somatosensory cortex representing the hand. In people with upper-arm amputations, much of the cortical area that would normally be devoted to the missing hand becomes reorganized for representation of the face (Yang et al., 1994).

Reorganization of representations in the cortex has also been demonstrated by having amputees report where they feel referred sensations. Referred sensa-

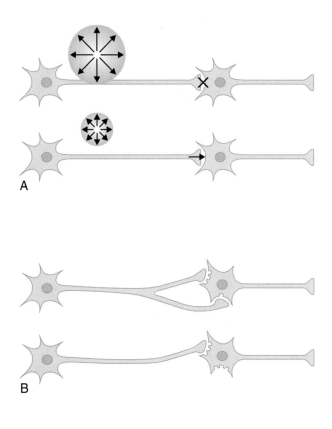

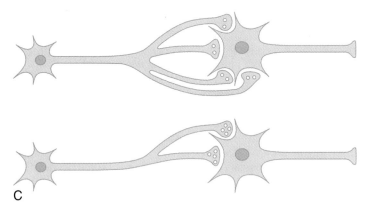

FIGURE 4–4

Synaptic changes following injury. *A,* Recovery of synaptic effectiveness occurs with the reduction of local edema that interfered with action potential conduction. *B,* Denervation hypersensitivity occurs when destruction of presynaptic neurons results in formation of new receptors deprived of adequate supply of neurotransmitter from remaining terminals. *C,* Synaptic hypereffectiveness occurs when neurotransmitter accumulates in the undamaged axon terminals, resulting in excessive release of transmitter and stimulation of the postsynaptic receptors.

tions are sensations felt in one place in response to a stimulus applied in another place. For example, when the chin is touched, a tingling sensation in the missing fifth finger is reported. A precise relationship between the stimulated points and the location of phan-tom sensation is found consistently. Because cortical reorganization has been observed as early as 4 weeks following amputation, the cortical reorganization probably results from unmasking of silent synapses. Amazingly, the correspondence between points stimulated on the chin and the referred sensations to the missing limb shifts when a person imagines pronating the missing forearm, indicating that motor commands have an effect on the cortical representation (Ramachandran, 1993). Merzenich et al. (1990) present evidence that the separations between representations in the cortical map (i.e., the distinction between the index and middle finger) are dynamic and depend on use of the body part. They also report that, in monkeys, changes in cortical organization can be recorded one month after amputation of a finger (Merzenich and Jenkins, 1993; Fig. 4–5).

Functional reorganization after injury to nerves is probably also a factor in some chronic pain syndromes, where pain persists despite the apparent healing of the precipitating injury. This type of plasticity is discussed in Chapter 7.

Thus, although nervous system plasticity research is in its infancy, mechanisms of learning and of recovery from injury are beginning to be explained. Plasticity makes recovery from nervous system injury possible; however, activity is crucial for optimizing recovery.

Cortical areas routinely adjust to changes in sensory input and develop new functions dependent on required motor output.

Activity-Related Changes in Neurotransmitter Release

Activity levels play a role in neuroplasticity because neurotransmitter production and release are regulated by neuronal activity. Repeated stimulation of somatosensory pathways can cause increases of inhibitory neurotransmitters, resulting in decreased sensory cortex response to overstimulation. In contrast, understimulation can have the opposite effect, resulting in the cortex being more responsive to even weak sensory inputs. Reduced activity may also promote axonal growth to restore normal levels of neuronal activity. For example, there is evidence that spinal cord damage in the adult rat promotes increased synaptic connections in the cortex (Ganchrow and Bernstein, 1981). Improved understanding of cellular mechanisms involved in plasticity may lead to improved clinical rehabilitation of peripheral and central nervous system disorders in both children and adults.

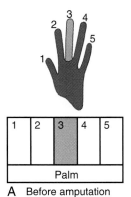

Areas of somatosensory cortex responding to stimulation of digits and palm

A Before amputation

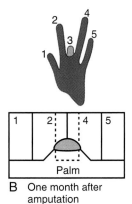

B One month after amputation

Skin surfaces now represented in the cortex where the third digit was formerly represented

C

FIGURE 4–5

Plasticity is demonstrated by changes that occur in the cerebral cortex following amputation of a digit. The cortical territory formerly occupied by representation of the amputated digit is dedicated to the intact adjacent digits and palm. (Modified with permission from Merzenich, M. M., and Jenkins, W. M. [1993]. Reorganization of cortical representations of the hand following alterations of skin inputs. J. of Hand Therapy, pp. 89–104.)

One potentially beneficial treatment of neurochemical disorders uses genetic manipulation to influence neural plasticity. Although research is still in its basic stages, procedures are being developed to genetically modify existing neurons so that they can make and secrete chemicals that are deficient in the brain. Laboratory studies have shown that transfer of a gene for NGF into neurons that secrete the neurotransmitter acetylcholine can protect these neurons from degenerative changes (Mitiguy, 1990). Furthermore, increased levels of NGF and other neurotropic factors may be associated with protective mechanisms that help make specific neurons more resistant to injury and promote survival and plasticity of neurons (Lindval et al., 1994).

METABOLIC EFFECTS OF BRAIN INJURY

When the brain suffers an injury such as a stroke, neurons deprived of oxygen for a prolonged period die and do not regenerate. Unfortunately, the damage is not confined to the oxygen-deprived neurons. Recent evidence indicates that the oxygen-deprived neurons are responsible for the death of adjacent neurons. Neurons deprived of oxygen release large quantities of **glutamate,** an excitatory neurotransmitter, from their axon terminals. Glutamate is one of the principal excitatory neurotransmitters, which at normal concentrations is crucial for central nervous system function. However, excessive concentrations of glutamate can be toxic to neurons. Overexcitation of a neuron, leading to cell death, is called **excitotoxicity.**

The processes involved in excitotoxicity are diagrammed in Figure 4–6. The process begins with the persistent binding of glutamate to the N-methyl-D-aspartate (NMDA)–type glutamate receptor in the cell membrane (Choi, 1988; Stehno-Bittel et al., 1995). Stimulation of this receptor results in the influx of Ca^{++} into the cell and indirectly facilitates the release

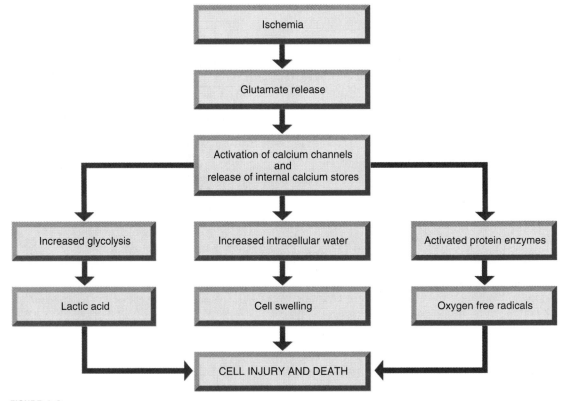

FIGURE 4–6

Schematic process of excitotoxicity. Following an initial ischemic insult, excessive intracellular calcium concentrations result in three pathways of cellular destruction: increased glycolysis, increased intracellular water, and activated protein enzymes.

of internal Ca^{++} stores. Excessive levels of Ca^{++} cannot be removed from the cytoplasm via sequestering into the internal storage depots of the endoplasmic reticulum or external transport via membrane channels. With the increase in Ca^{++} inside the cell, more K^+ diffuses out of the cell, requiring increased glycolysis to provide energy for the Na^+-K^+ pump to actively transport K^+ into the cell. Together, the increased glycolysis and the increased Ca^{++} lead to several destructive consequences for neurons:

- Increased glycolysis liberates excessive amounts of lactic acid, lowering the intracellular pH and resulting in acidosis that can break down the cell membrane.
- High intracellular Ca^{++} levels activate calcium-dependent digestive enzymes called proteases. These activated proteases break down cellular proteins.
- Ca^{++} activates protein enzymes that liberate arachidonic acid, producing substances that cause cell inflammation and produce oxygen free radicals. Oxygen free radicals are charged oxygen particles that are detrimental to mitochondrial functions of the cell.
- In addition to the direct effects of Ca^{++}, an influx of water associated with the ionic influx also causes cell edema.

Ultimately, these cellular events lead to cell death and potential propagation of neural damage if the dying cell releases glutamate and overexcites its surrounding cells.

Excitotoxicity caused by glutamate is thought to be associated with the neuronal damage associated with a cerebrovascular accident (stroke) and traumatic brain injury, as well as with the progression of neural degenerative diseases. Recently, glutamate receptors and some Ca^{++} channels have also been implicated in the neuronal disruption associated with acquired immunodeficiency syndrome (AIDS). Future pharmaceutical treatment of stroke, head injury, and neural diseases may be directed toward blocking the NMDA-type glutamate receptor and thus preventing the cascade of cell death related to excitotoxicity.

In addition to the possible pharmaceutical treatment of blocking NMDA-type glutamate receptors for management of an ischemic insult to the brain, other treatments may be directed specifically toward blocking the effects of Ca^{++} and the effects of free radicals. In animals, when oxygen and blood glucose levels are diminished by an occlusion of blood flow to the brain, levels of an intracellular messenger, inositol triphosphate (IP_3), are increased. This increase stimulates the release of Ca^{++} from intracellular storage sites and promotes a variety of cellular activities. With high cellular activity in the absence of adequate glucose, there is an increase in lactose, free radicals, and other metabolic end products that poison the cell. Also, when IP_3 is broken down by the cell, its by-product, diacylglycerol, breaks down into free fatty acid metabolites that are also poisonous to the cell. Future pharmaceutical treatment of stroke and traumatic brain injury may also be directed toward blocking the IP_3 pathway or administering drugs that act as scavengers for oxygen free radicals, thus preventing the cascade of cell death related to the production of fatty acid metabolites.

> In response to ischemia, cells can die either directly from lack of oxygen or indirectly from the cascade of events resulting from increased stimulation of the glutamate receptors.

EFFECTS OF REHABILITATION ON PLASTICITY

Following brain injury, both the intensity of rehabilitation and the amount of time between the injury and initiation of rehabilitation influence the recovery of neuronal function. Prolonged lack of active movement following cortical injury may lead to subsequent loss of function in adjacent, undamaged regions of the brain. However, a recent study (Nudo et al., 1996) demonstrated that the subsequent damage in adjacent areas of cortex could be prevented by retraining finger movements. A small part of the motor cortex associated with the control of hand movements was damaged in monkeys. When retraining was initiated 5 days following the original injury, loss of function in undamaged adjacent cortical regions was prevented. In some cases, neural reorganization took place, and the hand representation of the cortex extended into regions of the cortex formerly occupied by shoulder and elbow representations (Nudo et al., 1996). Because the functional reorganization coincided with the recovery of fine finger movements, it is proposed that rehabilitation has a direct effect on the integrity and reorganization of adjacent, undamaged regions of motor cortex.

However, too early initiation of vigorous rehabilitation of motor function may be counterproductive. The effects of forced use of a limb, beginning immediately after an experimental lesion of the cortex in adult rats, have been shown to result in a dramatic increase in neuronal injury (Kozlowski et al., 1996). Forced use is the technique of restraining or casting an unimpaired limb and thereby forcing the use of the impaired limb for functional activities. Kozlowski et al. (1996) showed severe, long-lasting behavioral deficits in animals forced to use the impaired limb immediately after a lesion to the sensorimotor cortex. These deficits included poor limb placement, decreased response to sensory stimulation, and defective use of the limb for postural support. Furthermore, the cortex of these animals showed a large increase in the volume of the lesion and an absence of dendritic growth or sprouting. In contrast, in other animals with the same lesion, early casting of the impaired forelimb caused no neural changes and only a slight decrease in later use of the impaired limb. These results demonstrate that immediate, forced use of an impaired limb may expand brain injury. Preliminary data indicate that excitotoxicity, caused by use-dependent increases in cortical activity, is a possible explanation for the increase in lesion size (Kozlowski, 1996).

In contrast to the negative effects of forced use in the period immediately following an injury to the sensorimotor cortex, positive effects are associated with forced use of a limb in the chronic stages of recovery. Movement improvements have been demonstrated in people with strokes who have had chronic dysfunction of their upper extremity for more than 1 year prior to undergoing a period of forced use (Ostendorf and Wolf, 1981; Wolf et al., 1989; Taub et al., 1993). Forced use was implemented by restraining the unaffected upper extremity of people with chronic dysfunction resulting from strokes and then practicing functional movements with the affected upper extremity. The subsequent recovery of function in the affected limb suggests that the movement deficits may have been in part related to learned nonuse. During the initial recovery period, people may have learned that the affected arm was not functional. Therefore, they compensated by using the unaffected arm. The inability to successfully move during the acute phase may be associated with the edema following brain injury and with a loss of function and electrical activity in other brain areas neuronally connected to the damaged area. Recovery from cortical injury occurs slowly, over many months, and increased motor abilities should become increasingly possible. Unfortunately, by the time the initial effects of cortical injury have subsided, people have "learned" that the affected arm is not functional. If people are motivated to use the limb after the stage of acute recovery, then unmasking of intact neuronal abilities may lead to functional improvements. Forced use following chronic disruption of movement will be discussed further in Chapter 10.

NEUROTRANSPLANTATION AND STEREOTAXIC SURGICAL APPROACHES

Using information about neuronal pathways and neuronal plasticity, recent scientific and surgical advances have contributed to new experimental approaches in treatment of neurological disorders. These approaches include deep-brain stimulation, neuronal transplantation or implantation, and stereotaxic surgery. For the purpose of this discussion, techniques used in the treatment of two signs of a basal ganglia disorder involving deficient dopamine, Parkinson's disease, are presented. The signs are tremors and akinesia. Tremors are involuntary, rhythmical shaking movements of the limbs. By strict definition, akinesia is the absence of movement. However, in clinical use, the term *akinesia* is used to describe paucity of movement. Despite the focus on Parkinson's disease, similar treatments are being tested for other movement disorders and for chronic intractable pain.

Deep-brain stimulation for the treatment of tremor in Parkinson's disease involves the high-frequency stimulation of neurons in the thalamus (Benabid et al., 1991). The basic concept is to inhibit the firing of an overactive set of neurons by using maintained electrical stimulation. Neuronal transplantation for treatment of Parkinson's disease involves the placement of fetal-donor dopamine-producing cells into the basal ganglia. In contrast, implantation involves the relocation of an individual's own adrenal cells into the basal ganglia, where they will synthesize dopamine. These approaches are based on the hypothesis that if the transplanted or implanted cells thrive in the brain environment, they will provide a new internal source of dopamine.

In some specialized treatment centers, stereotaxic surgery is currently performed for the treatment of severe tremor (Ohye et al., 1982) and akinesia (Iacono et al., 1994) associated with Parkinson's disease. Stereotaxic surgery involves the precise destruction of a small region of cells in the thalamus for the treatment of tremor and in the globus pallidus of the basal ganglia for the treatment of akinesia. The surgeries are thalamotomy and pallidotomy, respectively. Because these cell populations are thought to become overactive in the disease process, destruction of the overac-

tive cells may result in functional improvement. Although no long-term follow-up studies are available, the initial assessment of these stereotaxic approaches and fetal tissue implants are promising and seem to offer therapeutic benefits.

> Neural activity can be altered using a variety of neurosurgical approaches aimed at restoring more normal transmitter levels and neuron activity levels.

CLINICAL NOTES

CASE 1

B.G., a 37-year-old woman, suffered a compound fracture of her right distal radius and ulna following a fall while ice-skating. Internal fixation of the fracture was required, and B.G. was restricted to very limited use of her dominant right arm and hand. Postoperatively, B.G. reported decreased sensations in the fourth and fifth fingers of her right hand. Due to the severity of the fracture, some of the ulnar nerve fibers had been damaged. Six weeks after injury, B.G. was referred to therapy for range of motion of the right wrist and hand and low-resistance exercise. Grip strength in the right hand was two-thirds the strength of the left hand. During therapy, B.G. reported "burning sensations" and "pins and needles" in the digits of her right hand.

Questions

1. Is it possible for the damaged and/or severed ulnar nerve axons to recover after injury?

2. Should the therapist anticipate the abnormal sensory sensations to diminish over the course of a few months?

CASE 2

K.S., a 52-year-old man, experienced some right-sided weakness and then collapsed while working in the fields at his farm. Several hours passed before K.S. was found. He was transported to the local hospital, where it was determined that he had suffered a stroke. The stroke resulted from sudden blockage of an artery, preventing blood flow to a region of the brain. K.S. was referred to occupational and physical therapy and presented with a right facial droop, inability to move his right arm and leg, and decreased sensation on the right side of the body. K.S. required maximal assistance for all mobility.

Questions

1. Was the brain damage associated with the stroke most likely confined only to the cells that were deprived of oxygen due to decreased blood flow?

2. If excitotoxicity was in part responsible for the severity of the stroke, which principal excitatory neurotransmitter would be involved?

REVIEW QUESTIONS

1. What are three requirements for long-term potentiation (LTP)?

2. Define wallerian degeneration.

3. Can cortical motor and sensory maps change even in the adult mammal?

4. Define the term *excitotoxicity*.

5. Name one end product of glycolysis that contributes to cell death.

6. Identify two mechanisms by which excessive levels of intracellular calcium promote cell death.

7. Can some of the brain damage associated with strokes, traumatic injury, and degenerative diseases potentially be reduced with the administration of pharmaceutical agents?

References

Benabid, A. L., Pollack, P., Gervason, C., et al. (1991). Long-term suppression of tremor by chronic stimulation of the ventral intermediate thalamic nucleus. Lancet 337:403–406.

Choi, D. W. (1988). Glutamate neurotoxicity and diseases of the central nervous system. Neuron 1:623–634.

Chollet, F., DiPiero, V., Wise, R. J., et al. (1991). The functional anatomy of motor recovery after stroke in humans: A study with positron emission tomography. Ann. Neurol. 29(1):63–71.

Creese, I., Burt, D. R., and Snyder, S. H. (1977). Dopamine receptor binding enhancement accompanies lesion-induced behavioral supersensitivity. Science 197(4303):596–598.

Ganchrow, D., and Bernstein, J. H. (1981). Bouton renewal patterns in rat hindlimb cortex after thoracic dorsal funicular lesions. J. Neurosci. Res. 6:525–537.

Iacono, R. P., Lonser, R. R., Mandybur, G., Morenski, J. D., Yamada, S., and Shima, F. (1994). Stereotaxic pallidotomy results for Parkinson's exceed those of fetal graft. Amer. Surg. 60: 777–782.

Kozlowski, D. A., James, D. C., and Schallert, T. (1996). Use-dependent exaggeration of neuronal injury after unilateral sensorimotor cortex lesions. J. Neurosci. 16(15):4776–4786.

Lindvall, O., Kokaia, Z., Bengzon, J., Elmer, E., and Kokaia, M. (1994). Neurotrophins and brain insults. TINS 17(11): 490–497.

Merzenich, M. M., and Jenkins, W. M. (1993). Reorganization of cortical representations of the hand following alterations of skin inputs. J. of Hand Therapy, pp. 89–104.

Merzenich, M. M., Recanzone, G. H., Jenkins, W. M., et al. (1990). Adaptive mechanisms in cortical networks underlying cortical contributions to learning and nondeclarative memory. Cold Spring Harb Symp Quant Biol LV:873–887.

Mitiguy, J. (1990, winter). Brain trauma—why damage continues after injury: Today's promise, tomorrow's practice? Headlines, pp. 3–9.

Nudo, R. J., Wise, B. M., SiFuentes, F., et al. (1996). Neural substrates for the effects of rehabilitative training on motor recovery after ischemic infarct. Science 272:1791–1794.

Ohye, C., Hirai, T., Miyazaki, M., Shibazaki, T., and Nakajima, H. (1982). VIM thalamotomy for the treatment of various kinds of tremor. Appl. Neurophysiol. 45:275–280.

Ostendorf, C. G., and Wolf, S. L. (1981). Effect of forced use of the upper extremity of a hemiplegic patient on changes in function. Phys. Ther. 61:1022–1028.

Ramachandran, V. S. (1993). Behavioral and magnetoencephalographic correlates of plasticity in the adult human brain. Proc. Natl. Acad. Sci. 90:10413–10420.

Stehno-Bittel, L., Luckhoff, A., and Clapman, D. E. (1995). Calcium release from the nucleus by InsP$_3$ receptor channels. Neuron 14:163–167.

Taub, E., Miller, N. E., Novack, T. A., et al. (1993). Technique to improve chronic motor deficit after stroke. Arch. Phys. Med. Rehabil. 74:347–354.

Wolf, S. L., Lecraw, D. E., Barton, L. A., and Jann, B. B. (1989). Forced use of hemiplegic upper extremities to reverse the effect of learned nonuse among chronic stroke and head injured patients. Exp. Neurol. 104:125–132.

Yang, T. T., Gallen, C. C., Ramachandran, V. S., et al. (1994). Noninvasive detection of cerebral plasticity in adult human somatosensory cortex. Neuroreport 5:701–704.

Suggested Readings

Kaas, J. H. (1995). The reorganization of sensory and motor maps in adult mammals. In: Gazzaniga, M. S. (Ed.). The cognitive neurosciences. Cambridge, Mass.: MIT Press, pp. 51–72.

Kandel, E. R. (1991). Cellular mechanisms of learning and the biological basis of individuality. In: Kandel, E. R., Schwartz, J. H., and Jessell, T. M. (Eds.). Principles of neural science (3rd ed.). New York: Elsevier Science Publishing, pp. 1009–1031.

Stehno-Bittel, L. (1995). Calcium signalling in normal and abnormal brain function. Neurol. Rep. 19(2):12–17.

Sun, G. Y. (1992). Cerebral ischemia and poly phosphoinositide metabolism. In: Bazan, N.G., and Ginsburg, M. D. (Eds.). Neurochemical correlates of cerebral ischemia. New York: Plenum Press.

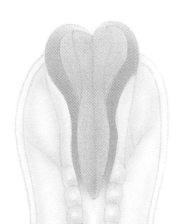

5

Development of the Nervous System

I am a 22-year-old student. Next year I will complete my master's degree in physical therapy, and I plan to specialize in pediatrics. Helping children with neurological deficits is very important to me, as I was diagnosed with cerebral palsy at 2½ years of age. At that time, a friend asked my parents if they would let me be seen by a pediatric specialist because the friend noticed that I was still crawling while all the children I was playing with were walking. I had no other signs of delayed development, verbally, cognitively, or socially, but motorically I was far behind my peers. Unlike the pediatricians that I had seen previously, who said that I would outgrow my motoric delay, this specialist confirmed what my parents had suspected. A diagnosis of mild spastic diplegic cerebral palsy* was made, and my parents searched for things they could do to encourage my development.

I have yet to understand why my doctors did not tell my parents about physical therapy. Fortunately, I started school 3 years later, and my physical education teacher took an interest. To the best of his abilities, he used his skills as an educator and read extensively over the next 6 years to provide opportunities for me to develop motor skills. My first formal therapy session came in eighth grade, when I was referred by the school to an occupational therapist for an evaluation and to develop a physical education program that I could do independently. That visit sparked my interest in rehabilitation, shaping my choice of career.

As I mentioned, my cerebral palsy is mild. My cognitive skills are not affected, and my upper limb coordination is near normal. One physician's record states that there was some involvement of my left upper limb, but I do not notice any problems except when my reflexes are tested. I am inclined to think that any decrease in

*Bilateral excessive muscle tone with weakness, usually affecting the lower limbs.

upper limb coordination is due to lack of challenges at a younger age, but I cannot confirm this suspicion. The most significant physical impact cerebral palsy has had on my life is on my gait pattern and recreational activities. As a child, motor dysfunction was more a daily problem than it is now because I could not keep up with my friends. I still struggle at times. Most recently, I struggled with learning to perform dependent-patient transfers in physical therapy school. Personally, I think that the greatest impact cerebral palsy has had on my life is a psychological one. There are still some things I would like to learn to do, but failing with motor activities as a child has influenced what I am willing to try now. On the other hand, that is why I am becoming a physical therapist: I want children and adults to know that physical limitations do not have to prevent them from enjoying life as much as anyone else.

—Heidi Boring

INTRODUCTION

From a single fertilized cell, an entire human being can develop. How is the exquisitely complex nervous system generated during development? Genetic and environmental influences act on cells throughout the developmental process, stimulating cell growth, migration, differentiation, and even cell death and axonal retraction to create the mature nervous system. Some of these processes are completed in utero, while others continue during the first several years after birth. Understanding the beginnings of the nervous system is vital for comprehending developmental disorders and helpful in understanding the anatomy of the adult nervous system.

DEVELOPMENTAL STAGES IN UTERO

Humans in utero undergo three developmental stages:

- Preembryonic
- Embryonic
- Fetal

Preembryonic Stage

The preembryonic stage lasts from conception to 2 weeks. Fertilization of the ovum usually occurs in the uterine tube. The fertilized ovum, a single cell, begins cell division as it moves down the uterine tube and

into the cavity of the uterus (Fig. 5–1). By repeated cell division, a solid sphere of cells is formed. Next, a cavity opens in the sphere of cells. At this stage of development, the sphere is called a blastocyst. The outer layer of the blastocyst will become the fetal contribution to the placenta, and the inner cell mass will become the embryo. The blastocyst implants into the endometrium of the uterus. During implantation, the inner cell mass develops into the embryonic disk, consisting of two cell layers: ectoderm and endoderm. Soon, a third cell layer, mesoderm, forms between the other two layers.

Embryonic Stage

During the embryonic stage, from the second to the end of the eighth week, the organs are formed (Fig. 5–2). The ectoderm develops into sensory organs, epidermis, and the nervous system. Mesoderm develops into dermis, muscles, skeleton, and the excretory

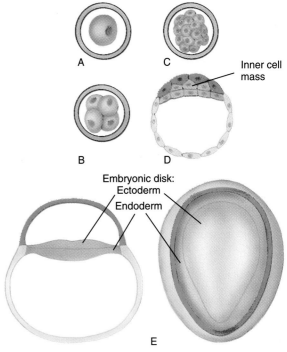

FIGURE 5–1

A, Fertilized ovum, a single cell. *B*, Four-cell stage. *C*, Solid sphere of cells. *D*, Hollow sphere of cells. The inner cell mass will become the embryonic disk. *E*, The two-layered embryonic disk, shown in cross section (left) and from above (right). The upper layer of the disk is the ectoderm, and the lower layer is the endoderm.

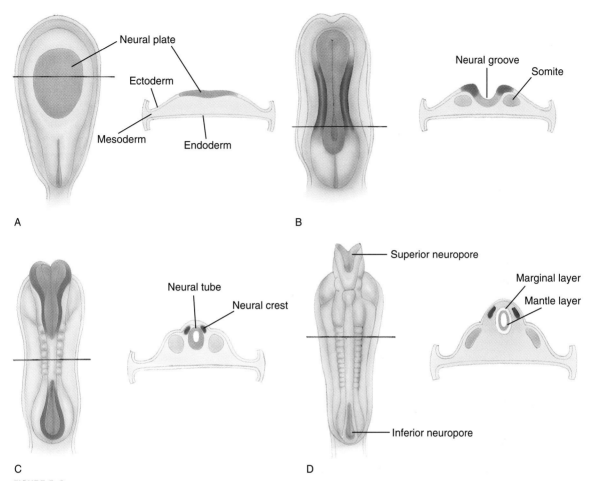

FIGURE 5–2

A, Cross sections through the embryo are shown on the right. On the left, the view is from above the embryo (day 16). Compare with figure 5–1*E*. *B,* The midline section of the neural plate moves toward the interior of the embryo, creating the neural groove (day 18). *C,* The folds of the neural plate meet, forming the neural tube. The neural crest separates from the tube and from the remaining ectoderm (day 21). *D,* The open ends of the neural tube are neuropores. The neural tube differentiates into an inner mantle layer and an outer marginal layer.

and circulatory systems. Endoderm differentiates to become the gut, liver, pancreas, and respiratory system.

> The nervous system develops from ectoderm, the outer cell layer of the embryo.

Fetal Stage

The fetal stage lasts from the end of the eighth week until birth. The nervous system develops more fully and myelination (insulation of axons by fatty tissue) begins.

FORMATION OF THE NERVOUS SYSTEM

Formation of the nervous system occurs during the embryonic stage and has two phases. First, tissue that will become the nervous system coalesces to form a tube running along the back of the embryo. When the ends of the tube close, the second phase, brain formation, commences.

Neural Tube Formation (Days 18 to 26)

The nervous system begins as a longitudinal thickening of the ectoderm, called the **neural plate** (see Fig. 5–2*A*). The plate forms on the surface of the embryo, extending from the head to the tail region, in contact with amniotic fluid. The edges of the plate fold to create the **neural groove,** and the folds grow toward each other (see Fig. 5–2*B*). When the folds touch (day 21), the neural tube is formed (see Fig. 5–2*C*). The neural tube closes first in the future cervical region. Next, the groove rapidly zips closed rostrally and caudally, leaving open ends called neuropores (see Fig. 5–2*D*). Cells adjacent to the neural tube separate from the tube and the remaining ectoderm to form the **neural crest.** When the crest has developed, the neural tube moves inside the embryo as the overlying ectoderm (destined to become skin) closes over the tube and neural crest. The superior neuropore closes by day 27, and the inferior neuropore closes about 3 days later.

By day 26, the tube differentiates into two concentric rings (see Fig. 5–2*D*). The **mantle layer** (inner wall) contains cell bodies and will become gray matter. The **marginal layer** (outer wall) contains processes of cells, whose bodies are located in the mantle layer. The marginal layer develops into white matter, consisting of axons and glial cells.

> The brain and spinal cord develop entirely from the neural tube.

Relationship of Neural Tube to Other Developing Structures

As the neural tube closes, the adjacent mesoderm divides into spherical cell clusters called **somites** (see Fig. 5–2*B*). Developing somites cause bulges to appear on the surface of the embryo (Fig. 5–3). The somites first appear in the future occipital region, and new somites are added caudally. The anteromedial part of a somite, the **sclerotome,** becomes the vertebrae and the skull. The posteromedial part of the somite, the **myotome,** becomes skeletal muscle. The lateral part of the somite, the **dermatome,** becomes dermis (Fig. 5–4).

FIGURE 5–3

Photographs of embryos early in the fourth week. In *A*, the embryo is essentially straight, whereas the embryo in *B* is slightly curved. In *A*, the neural groove is deep and is open throughout its entire extent. In *B*, the neural tube has formed opposite the somites but is widely open at the rostral and caudal neuropores. The neural tube is the primordium of the central nervous system (brain and spinal cord). (Courtesy of Professor Hideo Nishimura.)

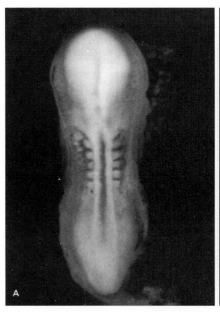
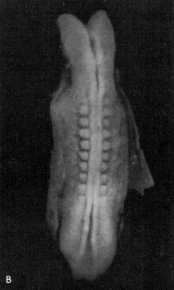

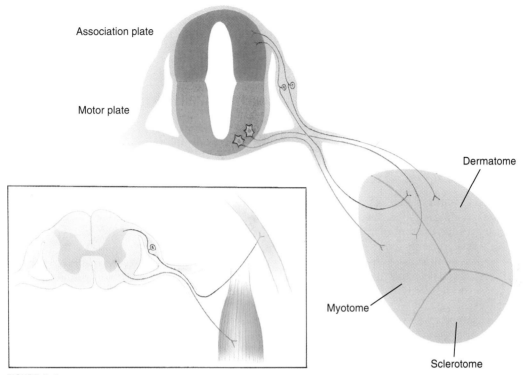

Association plate

Motor plate

Dermatome

Myotome

Sclerotome

FIGURE 5–4

The neurons connecting the neural tube with the somite are shown. The mantle layer of the neural tube has differentiated into a motor plate (ventral) and an association plate (dorsal). The inset illustrates the same structures in maturity. The following changes have occurred: part of the neural plate → spinal cord, motor plate → ventral horn, association plate → dorsal horn, myotome → skeletal muscle, and dermatome → dermis.

As the cells of the mantle layer proliferate in the neural tube, grooves form on each side of the tube, separating the tube into ventral and dorsal sections (see Fig. 5–4). The ventral section is the **motor plate** (also called basal plate). Axons from cell bodies located in the motor plate grow out from the tube to innervate the myotome region of the somite. As development continues, this association leads to the formation of a **myotome**: a group of muscles derived from one somite and innervated by a single spinal nerve. Thus myotome has two meanings: (1) an embryologic section of the somite and (2) after the embryonic stage, a group of muscles innervated by a segmental spinal nerve. Neurons whose cell bodies are in the basal plate become motor neurons, which innervate skeletal muscle, and interneurons. In the mature spinal cord, the gray matter derived from the basal plate is called the ventral horn.

The dorsal section of the neural tube is the **association plate** (also called alar plate). In the spinal cord, these neurons proliferate and form interneurons and projection neurons. In the mature spinal cord, the gray matter derived from the association plate is called the dorsal horn (see Fig. 5–4).

Neurons in the dorsal region of the neural tube process sensory information. Neurons with cell bodies in the ventral region innervate skeletal muscle.

The **neural crest** separates into two columns, one on each side of the neural tube. The columns break up into segments that correspond to the dermal areas of the somites. Neural crest cells form peripheral sensory neurons, myelin cells, autonomic neurons, and endocrine organs (adrenal medulla and pancreatic islets). The cells that become peripheral sensory neurons grow two processes; one connects to the spinal cord, and the other innervates the region of the somite that will become dermis. Like the term *myotome,* **dermatome** has two meanings: (1) the area of the somite that will become dermis and (2) after the embryonic stage, the dermis innervated by a single spinal nerve. The peripheral sensory neurons, also known as primary sensory neurons, convey information from sensory receptors to the association plate. The cell bodies of the peripheral sensory neurons are outside the spinal cord, in the dorsal root ganglion.

Until the third fetal month, spinal cord segments are adjacent to corresponding vertebrae, and the roots of spinal nerves project laterally from the cord. As the fetus matures, the spinal column grows faster than the cord. As a result, the adult spinal cord ends at the L1–L2 vertebral level. Caudal to the thoracic levels, roots of the spinal nerves travel inferiorly to reach the intervertebral foramina (Fig. 5–5). Within the adult spinal cord, the neural tube cavity persists as the central canal.

> The peripheral nervous system, with the exception of motor neuron axons, develops from the neural crest.

Brain Formation (Begins Day 28)

When the superior neuropore closes, the future brain region of the neural tube expands to form three enlargements (Fig. 5–6): the **hindbrain** (rhombencephalon), **midbrain** (mesencephalon), and **forebrain** (prosencephalon). Soon two additional enlargements appear, providing the brain with five distinct regions. The enlargements, like their precursor neural tube, are hollow. In the mature nervous system, the fluid-filled cavities are called ventricles.

The hindbrain divides into two sections; the lower section becomes the **myelencephalon,** and the upper section becomes the **metencephalon.** These later differentiate to become the medulla, pons, and cerebellum. In the upper hindbrain, the central canal expands to form the fourth ventricle. The pons and upper

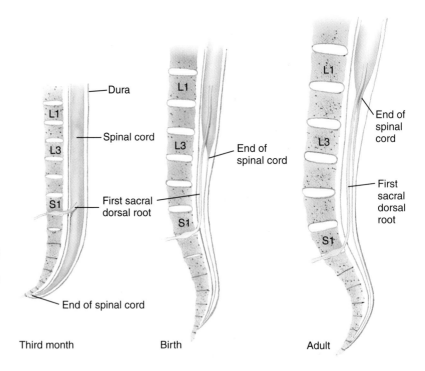

FIGURE 5–5

After the third month in utero, the rate of growth of the vertebral column exceeds that of the spinal cord. The passage of the nerve roots through specific vertebral foramina is established early in development, so the lower nerve roots elongate within the vertebral canal to reach their passage. For simplicity, only the first sacral nerve root is illustrated.

Dura

Spinal cord

First sacral dorsal root

End of spinal cord

L1

L3

S1

Third month

L1

L3

S1

End of spinal cord

Birth

L1

L3

S1

End of spinal cord

First sacral dorsal root

Adult

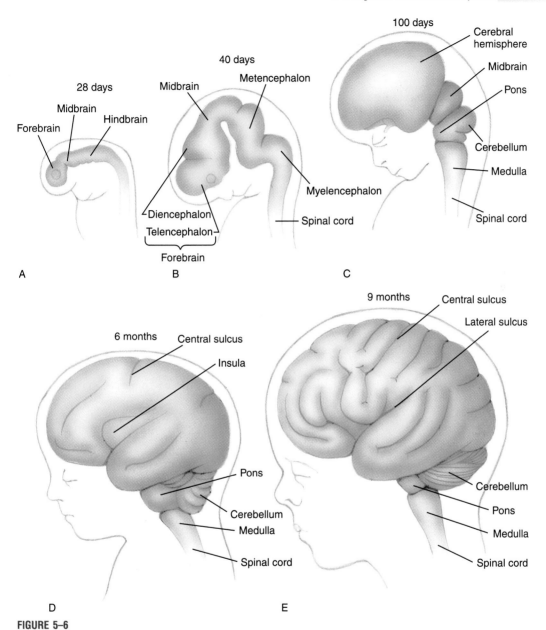

FIGURE 5–6

Brain formation. *A,* Three-enlargement stage. *B,* Five-enlargement stage. *C,* Telencephalon has grown so extensively that the diencephalon is completely covered in a lateral view. *D,* The insula is being covered by continued growth of adjacent areas of the cerebral hemisphere. *E,* Folding of the surface of the cerebral and cerebellar hemispheres continues.

medulla are anterior to the fourth ventricle, and the cerebellum is posterior. In the cerebellum, the mantle layer gives rise to both deep nuclei and the cortex. To become the cortex, the mantle layer cell bodies migrate through the white matter to the outside.

The **midbrain** enlargement retains its name, midbrain, throughout development. The central canal becomes the cerebral aqueduct in the midbrain, connecting the third and fourth ventricles.

The posterior region of the forebrain stays in the midline to become the **diencephalon.** The major structures are the **thalamus** and the **hypothalamus.** The midline cavity forms the third ventricle.

The anterior part of the forebrain becomes the **telencephalon.** The central cavity enlarges to form the lateral ventricles (Fig. 5–7). The telencephalon becomes the **cerebral hemispheres;** the hemispheres expand so extensively that they envelop the dien-

cephalon. The cerebral hemispheres consist of deep nuclei, including the basal ganglia (groups of cell bodies); white matter (containing axons); and the cortex (layers of cell bodies on the surface of the hemispheres). As the hemispheres expand ventrolaterally to form the temporal lobe, they attain **C** shape. As a result of this growth pattern, certain internal structures including the caudate nucleus (part of the basal ganglia) and lateral ventricles also become **C**-shaped (Fig. 5–8).

Continued Development During Fetal Stage

Lateral areas of the hemispheres do not grow as much as the other areas, with the result that a section of cortex becomes covered by other regions. The covered region is the **insula,** and the edges of the folds that cover the insula meet to form the lateral sulcus. In the mature brain, if the lateral sulcus is pulled open, the insula is revealed. The surfaces of the cerebral and cerebellar hemispheres begin to fold, creating sulci, grooves into the surface, and gyri, elevations of the surface.

Table 5–1 summarizes normal brain development.

CELLULAR-LEVEL DEVELOPMENT

The progressive developmental processes of cell proliferation, migration, and growth, extension of axons to target cells, formation of synapses, and myelination of axons are balanced by the regressive processes that extensively remodel the nervous system during development.

Epithelial cells that line the neural tube divide to produce neurons and glia. The neurons migrate to their final location by one of two mechanisms: (1) sending a slender process to the brain surface and then hoisting themselves along the process (Brittis and Silver, 1994) or (2) climbing along radial glia (long cells that stretch from the center of the brain to the surface). The neurons differentiate appropriately after migrating to their final location. The function of each neuron—visual, auditory, motor, and so on—is not genetically determined. Instead, function depends on the area of the brain where the neuron migrates. Daughter cells of a specific mother cell may assume totally different functions depending on the location of migration (Walsh and Cepko, 1992).

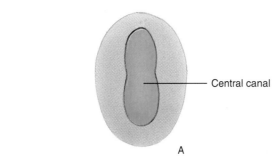

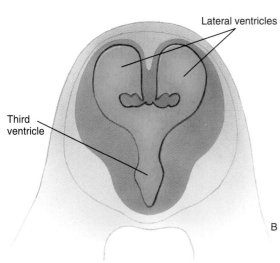

FIGURE 5–7

Formation of ventricles. *A,* Central canal in neural tube. *B,* Coronal section of developing telencephalon.

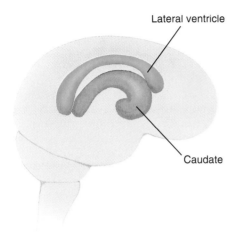

Lateral ventricle

Caudate

A

TABLE 5-1		
SUMMARY OF NORMAL BRAIN DEVELOPMENT		
Hindbrain →	Metencephalon →	Pons, upper medulla, cerebellum, fourth ventricle
	Myelencephalon →	Lower medulla
Midbrain →	Midbrain →	Midbrain, cerebral aqueduct
Forebrain →	Diencephalon →	Thalamus, hypothalamus, third ventricle
	Telencephalon →	Cerebral hemispheres, including basal ganglia, cerebral cortex, lateral ventricles

How do neurons in one region of the nervous system find the correct target cells in another region? For example, how do neurons in the cortex direct their axons down through the brain to contact specific neurons in the spinal cord? Processes grow outward from the neuron cell body. The forward end of the process expands to form a **growth cone** that samples the environment, contacting other cells and chemical cues. The growth cone recoils from some chemicals it encounters and advances into other regions where the chemical attractors are specifically compatible to the growth cone characteristics.

When the growth cone contacts its target cell, synaptic vesicles soon form, and microtubules that formerly ended at the apex of the growth cone project to the presynaptic membrane. With repeated release of neurotransmitter, the adjacent postsynaptic membrane develops a concentration of receptor sites. In early development, many neurons develop that do not survive. **Neuronal death** claims as many as half of the neurons formed during the development of some brain regions. The neurons that die are probably ones that failed to establish the optimal connections with their target cells or were too inactive to maintain their connection. Thus development is partially dependent on activity. Some neurons that survive retract their axons from certain target cells while

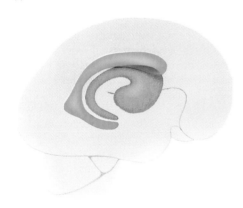

B

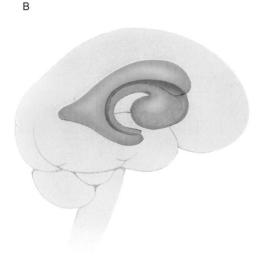

C

FIGURE 5-8

The growth pattern of the cerebral hemispheres results in **C** shape of some of the internal structures. The changing shapes of the caudate nucleus and lateral ventricle are shown.

leaving other connections intact. For example, in the mature nervous system, a muscle fiber is innervated by only one axon. During development, several axons may innervate a single muscle cell. This polyneuronal innervation is eliminated by the 25th week in developing humans (Hesselmans et al., 1993). These two regressive processes, neuronal death and **axon retraction,** sculpt the developing nervous system.

Neuronal connections also sculpt the developing musculature. Experiments that change motor neuron connections to a muscle fiber demonstrate that muscle fiber type (fast or slow twitch) is dependent on innervation. **Fast twitch muscle** is converted to slow twitch if innervated by a slow motor neuron, and **slow twitch muscle** can be converted to fast twitch if innervated by a fast motor neuron (Buller et al., 1960).

Before neurons with long axons become fully functional, their axons must be insulated by a **myelin sheath,** composed of lipid and protein. The process of acquiring a myelin sheath is **myelination.** The process begins in the fourth fetal month; most sheaths are completed by the end of the third year of life. The process occurs at different rates in each system. For instance, the motor roots of the spinal cord are myelinated at about 1 month of age, but tracts sending information from the cortex to activate motor neurons are not completely myelinated, and therefore not fully functional, until a child is approximately 2 years old. Thus, if neurons that project from cerebral cortex to motor neurons were damaged perinatally, motor deficits might not be observed until the child was older. For example, if some of the cortical neurons controlling lower limb movements were damaged at birth, the deficit might not be recognized until the child is more than a year old and has difficulty standing and walking. This is an example of **growing into deficit:** nervous system damage that occurred earlier is not evident until the systems damaged would normally have become functional.

DEVELOPMENTAL DISORDERS: IN UTERO AND PERINATAL DAMAGE OF THE NERVOUS SYSTEM

The central nervous system is most susceptible to major malformations between day 14 and week 20, as the fundamental structures of the central nervous system are forming. After this period, growth and remodeling continue; however, insults cause functional disturbances and/or minor malformations.

Neural Tube Defects

Anencephaly, formation of a rudimentary brain stem without cerebral and cerebellar hemispheres, occurs when the cranial end of the tube remains open, and the forebrain does not develop. The skull does not form over the incomplete brain, leaving the malformed brain stem and meninges exposed. Anencephaly can be detected by maternal blood tests, amniotic fluid tests, and ultrasound imaging. The causes include chromosomal abnormalities, maternal nutritional deficiencies, and maternal hyperthermia. Most fetuses with this condition die before birth, and almost none survive more than a week after birth.

A developmental malformation of the hindbrain is the **Arnold-Chiari deformity:** the inferior cerebellum and medulla are elongated and protrude into the vertebral canal (Fig. 5–9). Both the medulla and pons are small and deformed. The deformity is associated with enlargement of the cranium due to blockage of the flow of cerebrospinal fluid (hydrocephalus; see Chap. 18); sensory and motor disorders may result. Problems with tongue and facial weakness, deafness, weakness of lateral eye movements, and coordination of move-

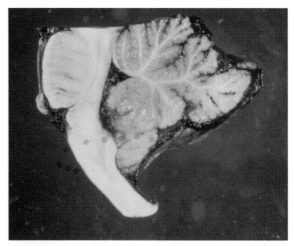

FIGURE 5–9

The Arnold-Chiari malformation consists of malformation of the pons, medulla, and inferior cerebellum. The medulla and inferior cerebellum protrude into the foramen magnum.

ments are common due to the malformation of lower cranial nerves and of the cerebellum. If the deficits are stable, no medical treatment is indicated. If the deficits are progressing, surgical removal of the bone immediately surrounding the malformation may be indicated. (See the box on Arnold-Chiari malformation.)

If the inferior neuropore does not close, the resulting neural tube defect is called **spina bifida** (Fig. 5–10). Developing vertebrae do not close around an incomplete neural tube, resulting in a bony defect at the distal end of the tube. Maternal nutritional deficits

are associated with higher incidence of the disorder. The severity of the defect varies; if neural tissue does not protrude through the bony defect (spina bifida occulta), usually spinal cord function is normal. Meningocele is a more severe malformation, with protrusion of the meninges through the bony defect. Spinal cord function may be impaired. In meningomyelocele, neural tissue with the meninges protrudes outside the body. Meningomyelocele always results in abnormal growth of the spinal cord and some degree of lower-extremity dysfunction; often, bowel and bladder control is impaired. No consensus exists on proper medical management of meningomyelocele. Myeloschisis is the most severe defect, consisting of a malformed spinal cord open to the surface of the body, which occurs when the neural folds fail to close. (See the box on spina bifida, p. 81.)

Forebrain Malformation

The prosencephalon normally divides into two cerebral hemispheres. Rarely, this division does not occur, resulting in a single cerebral hemisphere, often associated with facial abnormalities: a single eye (or no eye), a deformed nose, and cleft lip and palate. The defect is called holoprosencephaly. Genetic factors have been implicated in the disorder. The disorder can be identified in utero by genetic tests and by ultrasound.

Exposure to Alcohol or Cocaine in Utero

What are the consequences of maternal substance abuse? Fetal alcohol syndrome (consisting of impairment of the central nervous system, growth deficiencies before and/or after birth, and facial anomalies) and the milder syndrome of alcohol-related birth defects are examples of substance abuse interfering with development during gestation. Both syndromes are due to maternal alcohol intake, with mental retardation, an abnormally small head, eyes set abnormally far apart, and aberrant neurobehavioral development being common deficits. Malformation of the cerebellum, cerebral nuclei, corpus callosum, neuroglia, and neural tube lead to cognitive, movement, and behavioral problems. See Osborn et al. (1993) for a review of the etiology and effects of fetal alcohol syndrome.

ARNOLD-CHIARI MALFORMATION

Pathology
Developmental abnormality

Etiology
Unknown

Speed of onset
Unknown

Signs and symptoms

Consciousness
Normal

Cognition, language, and memory
Normal

Sensory
May have loss of pain and temperature sensation on shoulders and lateral upper limbs if upper central spinal cord is abnormal (see Chap. 12)

Autonomic
Vomiting secondary to hydrocephalus

Motor
Uncoordinated movements

Cranial nerves
Vertigo (sensation of spinning); deafness; tongue, facial muscle, and lateral eye movement weakness

Region affected
Upper spinal cord, brain stem, and cerebellum

Demographics
Only affects developing nervous system

Prognosis
Defect is stable; symptoms are stable or progressive

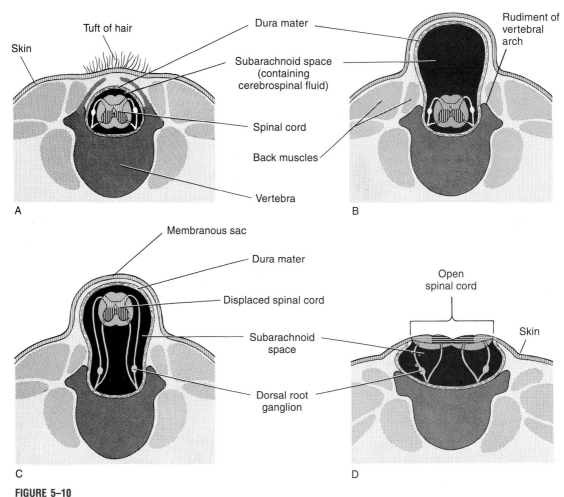

FIGURE 5–10

Various types of spina bifida and the commonly associated malformations of the nervous system. *A,* Spina bifida occulta. About 10% of people have this vertebral defect in L5, S1, or both. It usually causes no back problems. *B,* Spina bifida with meningocele. *C,* Spina bifida with meningomyelocele. *D,* Spina bifida with myeloschisis. The types illustrated in *B* to *D* are often referred to collectively as spina bifida cystica because of the cystlike sac that is associated with them. (Courtesy of Professor Hideo Nishimura.)

The effects of in utero exposure to cocaine depend on the stage of development. Disturbance of neuronal proliferation is the most frequent consequence of cocaine exposure during neural development, but interference with other neurodevelopmental processes also occurs. Why does cocaine have more severe effects than many other abused drugs? Because cocaine easily crosses the placenta, and because cocaine remains in fetal circulation for prolonged times due to immaturity of the enzymes for its metabolism and the incomplete development of the renal excretion system. In a recent meta-analysis of the research on the effects of cocaine use during pregnancy, the difficulties with reaching firm conclusions are due to the confounding effects of many other variables that tend to correlate with cocaine use. The other variables include low socioeconomic status, use of other illegal drugs, smoking, lack of prenatal care, poor maternal nutrition, and infection with sexually transmitted diseases that transfer to the fetus (Lutiger et al., 1991).

SPINA BIFIDA

Pathology
Developmental abnormality

Etiology
Some cases due to maternal nutritional deficits

Speed of onset
Unknown

Signs and symptoms
Occulta: *no signs or symptoms*
Meningocele: *signs and symptoms vary depending on location and severity of the malformation*

Consciousness
Normal

Cognition, language, and memory
Normal

Sensory
Meningomyelocele: *lack of sensation in the lower limbs*

Autonomic
Meningomyelocele: *lack of bladder and bowel control*

Motor
Meningomyelocele: *paralysis of the lower limbs*

Cranial nerves
Normal

Region affected
Inferior spinal cord

Demographics
Only affects developing nervous system

Prognosis
Defect is stable

Abnormal Location of Cells

What happens when the process of cell migration goes awry? Cells fail to reach their normal destination. In the cerebral cortex, this results in abnormal gyri, due to abnormal numbers of cells in the cortex, and heterotopia, the displacement of gray matter, commonly into the deep cerebral white matter. Seizures are often associated with heterotopia.

Mental Retardation

At autopsy, approximately half of the individuals with severe mental retardation were shown to have major defects in the morphology of dendrites and dendritic spines (Huttenlocher, 1991). Dendritic spines are projections from the dendrites, common in cerebral and cerebellar cortex projection neurons, that are the preferential sites of synapses.

Cerebral Palsy

Cerebral palsy is a movement and postural disorder caused by permanent, nonprogressive damage of a developing brain. The etiology of cerebral palsy is diminished blood or oxygen supply to the developing brain. Cerebral palsy in approximately 12% to 24% of full-term infants is secondary to brain damage occurring between the 28th week of gestation and 1 week after birth (Volpe, 1992). In premature infants, the brain damage usually occurs postnatally. Cerebral palsy is classified according to the type of motor dysfunction. The most common are as follows:

- Spastic
- Athetoid
- Ataxic
- Mixed

In spastic cerebral palsy, the damaged neurons are around the ventricles. Muscle shortening in spastic cerebral palsy often results in toe walking and a scissor gait. In scissor gait, one leg swings in front of the other instead of straight forward, producing a crisscross motion of the legs during walking. In athetoid cerebral palsy, the neuronal damage is in the basal ganglia. Athetoid cerebral palsy is characterized by slow, writhing movements of the extremities and/or trunk. In ataxic cerebral palsy, the damage is in the cerebellum. Ataxic cerebral palsy consists of incoordination, weakness, and shaking during voluntary movement. If more than one type of abnormal movement coexist in a person, the disorder is classified as mixed type. Cognitive, somatosensory, visual, auditory, and speech deficits are frequently associated with cerebral palsy. Although the nervous system damage is not progressive, new problems appear as the child reaches each age for normal developmental milestones: that is, when the child reaches the age when most children walk, the inability of the child with cerebral palsy to walk independently at the usual age will become apparent. (See the box on cerebral palsy.)

CEREBRAL PALSY

Pathology
Developmental abnormality

Etiology
Insufficient blood or oxygen supply to developing brain

Speed of onset
Unknown

Signs and symptoms

Consciousness
Normal

Cognition, language, and memory
Frequently associated with mental retardation, although some people with cerebral palsy have above normal intelligence and memory

Sensory
May be normal or impaired

Autonomic
Usually normal

Motor
Spastic type: muscle shortening, increased muscle resistance to passive movements
Athetoid type: slow, writhing movements
Ataxic type: incoordination, weakness, shaking during voluntary movements

Cranial nerves
Not directly affected; however, due to abnormal neural input, the output of motor cranial nerves is impaired

Region affected
Brain; some abnormalities in spinal cord

Demographics
Only developing nervous system affected

Prognosis
The abnormality is stable, but functional disabilities may only become obvious as the person grows

Even at birth, the infant brain is far from its adult form. Thus, damage during development has different consequences than injury of a fully developed brain. These differences are reviewed in an article by Leonard (1994). He contrasts the functional changes, neural substrates, and types of movement disorders in individuals with prenatal or perinatal brain damage (cerebral palsy) and adult-onset brain damage.

Summary of Developmental Disorders

Major deformities of the nervous system occur before week 20 because the gross structure is developing during this time. After 20 weeks of normal development, damage to the immature nervous system causes minor malformations and/or disorders of function. Table 5–2 summarizes the processes of development and the consequences of damage during the peak time of each process.

NERVOUS SYSTEM CHANGES DURING INFANCY

Many experiments have investigated the consequences of sensory deprivation on the infant nervous system. These experiments indicate that **critical periods** during development are crucial for normal outcomes. Critical periods are the time when neuronal projections compete for synaptic sites; thus, the nervous system optimizes neural connections during the critical period.

One example of changing the functional properties of the nervous system was demonstrated in infant monkeys. Monkeys raised with one eyelid sutured shut from birth to 6 months were permanently unable to use vision from that eye, even after the sutures were removed. Recordings indicated that the retinal cells responded normally to light and the information was relayed correctly to the visual cortex, but the visual cortex did not respond to the information (Hubel and Wiesel, 1977). Occluding vision in one eye in an adult monkey for an equivalent period of time had relatively little effect on vision once visual input was restored. Thus the critical period for tuning the visual cortex is during the first 6 months of development in monkeys.

Changes analogous to the functional disuse in the monkeys may explain the results of Spitz and Wolf's 1946 pioneering study of infants raised in an orphanage versus infants cared for by their mothers in prison. The infants in the orphanage had less human contact and also had cribs covered by sheets, preventing them from watching activity around them. In the prison center, the infants were held and attended to by their mothers, and the infants were able to see outside of their cribs. On developmental tests at 4 months, the infants in the orphanage performed better than those in the prison center. However, by the second year, the prisoners' babies all walked and talked, performing near age norms. In contrast, only

TABLE 5–2

SUMMARY OF DEVELOPMENTAL PROCESSES AND THE CONSEQUENCES OF INTERFERENCE WITH SPECIFIC DEVELOPMENTAL PROCESSES

Developmental Processes	Peak Time of Occurrence	Disorders Secondary to Interference with Developmental Process
Neural tube formation	Weeks 3–4	Anencephaly, Arnold-Chiari malformation, spina bifida occulta, meningocele, meningomyelocele
Formation of brain enlargements	Months 2–3	Holoprosencephaly
Cellular proliferation	Months 3–4	Fetal alcohol syndrome, cocaine-affected nervous system
Neuronal migration	Months 3–5	Heterotopia, seizures
Organization (differentiation, growth of axons and dendrites, synapse formation, selective neuron death, retraction of axons)	Month 5 to early adulthood	Mental retardation, trisomy 21, cerebral palsy*
Myelination	Birth to 3 years after birth	Unknown

*In cases of cerebral palsy where a cause can be identified, bleeding or hypoxia occurred in the central nervous system during the perinatal period; however, in many other cases, no cause can be readily identified, and thus the time of onset is unknown.

2 of the 26 orphans could walk and speak, and these 2 performed far below their age norms, suggesting a lack of necessary experience during critical periods in human development of speech and walking.

Interruption of development during a critical period may explain some of the differences in outcome between perinatal and adult brain injury (Leonard, 1994). In children with cerebral palsy, damage to fibers descending from the cerebrum to the spinal cord during fetal development or at birth may eliminate some competition for synaptic sites during a critical period, resulting in persistence of inappropriate connections. These inappropriate connections, in addition to the deficiency of descending control, result in abnormal movement. The adult with brain damage loses descending control, but because development is complete, dysfunction is not compounded by inappropriate connections.

> Critical periods are times when axons are competing for synaptic sites. Normal function of neural systems is dependent on appropriate experience during the critical period.

CLINICAL NOTES

CASE 1

A 2-year-old boy has no reaction to any stimulation below the level of the umbilicus. He does not voluntarily move his lower limbs, his lower limb muscles are atrophied, and he has no voluntary control of his bladder or bowels. His mother reports that he had surgery on his back 2 days after birth. Above the level of the umbilicus, sensation and movement are within normal limits.

Questions

1. The nervous system deficits affect which of the systems: sensory, autonomic, or motor?

2. The lesion is in what region of the nervous system: the peripheral, spinal, brain stem, or cerebral region?

REVIEW QUESTIONS

1. When do the organs form during development?

2. List the steps in formation of the neural tube.

3. What is a myotome?

4. What part of the nervous system does the neural crest become?

5. Describe the changes in the neural tube that lead to formation of the brain.

6. List the progressive processes of cellular-level development.

7. Describe the regressive processes of cellular-level development.

8. Explain the concept of "growing into deficit."

9. List and define the five defects of neural tube formation.

10. About half of the cases of severe mental retardation are associated with what developmental defect?

11. What is cerebral palsy? List the major types of cerebral palsy.

12. What are critical periods?

References

Brittis, P. A., and Silver, J. (1994). Exogenous glycosaminoglycans induce complete inversion of retinal ganglion cell bodies and their axons within the retinal neuroepithelium. Proc. Nat. Acad. Sci. U. S. A. 91:7539–7542.

Buller, A. J., Eccles, J. C., and Eccles, R. M. (1960). Interactions between motoneurons and muscles in respect of the characteristic speeds of the responses. J. Physiol. 150:417–439.

Hesselmans, L. F., Jennekens, F. J., Van den Oord, C. J., et al. (1993). Development of innervation of skeletal muscle fibers in man: Relation to acetylcholine receptors. Anat. Rec. 236(3):553–562.

Hubel, D. H., and Wiesel, T. N. (1977). Ferrier lecture: Functional architecture of macaque monkey visual cortex. Proc. R. Soc. Lond. B. Biol. Sci. 198:1–59.

Huttenlocher, P. R. (1991). Dendritic and synaptic pathology in mental retardation. Pediatr. Neurol. 7:79–85.

Leonard, C. T. (1994). Motor behavior and neural changes following perinatal and adult-onset brain damage: Implications for therapeutic interventions. Phys. Ther. 74:753–767.

Lutiger, B., Graham, K., Einarson, T. R., et al. (1991). Relationship between gestational cocaine use and pregnancy outcome: A meta-analysis. Teratology 44:405–414.

Osborn, J. A., Harris, S. R., and Weinberg, J. (1993). Fetal alcohol syndrome: Review of the literature with implications for physical therapists. Phys. Ther. 73:599–607.

Spitz, R. A., and Wolf, K. M. (1946). Anaclitic depression: An inquiry into the genesis of psychiatric conditions in early childhood, II. Psychoanal. Study Child 2:313–342.

Volpe, J. J. (1992). Value of MR in definition of the neuropathology of cerebral palsy in vivo. A.J.N.R. Am. J. Neuroradiol. 13:79–83.

Walsh, C., and Cepko, C. L. (1992). Widespread dispersion of neuronal clones across functional regions of the cerebral cortex. Science 255(5043):434–440.

Suggested Readings

Gilbert, S. G. (1989). Pictorial human embryology. Seattle: University of Washington Press.

Hamilton, W. J., Boyd, J. D., and Mossman, H. W. (1962). Human embryology (prenatal development of form and function). Baltimore: Williams & Wilkins.

Moore, K. L. (1989). Before we are born: Basic embryology and birth defects. Philadelphia: W. B. Saunders.

Netter, F. H. (1983). Embryology. Nervous system, part I: Anatomy and physiology. West Caldwell, N.J.: CIBA Pharmaceutical Company.

6

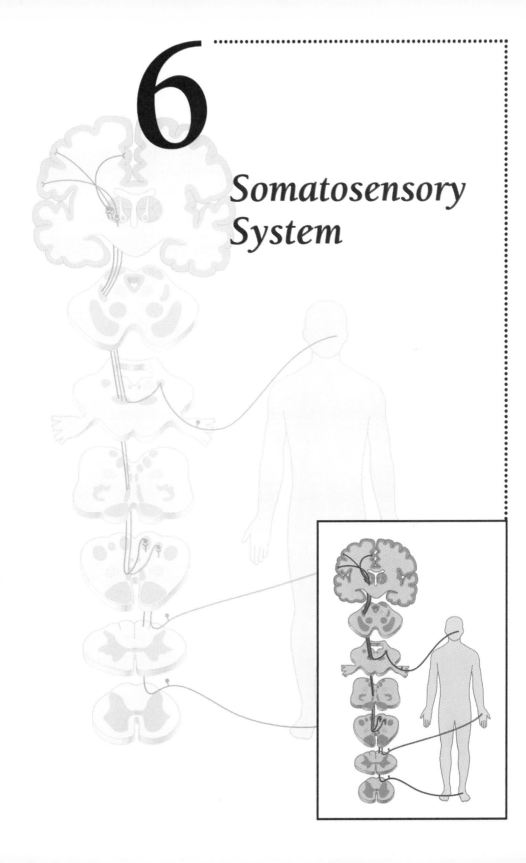

Somatosensory
System

INTRODUCTION

Sensation allows us to investigate the world, move accurately, and avoid or minimize injuries. This chapter discusses somatosensation, the sensory information from the skin and musculoskeletal systems. The special senses of smell, vision, hearing, and equilibrium and the sensations from the viscera are discussed in subsequent chapters.

Sensory information from the skin is called superficial or cutaneous. Superficial sensory information includes touch, pain, and temperature. Touch sensation includes superficial pressure and vibration. In contrast, sensory information from the musculoskeletal system includes proprioception and pain. Proprioception provides information regarding stretch of muscles, tension on tendons, position of joints, and deep vibration. Proprioception includes both static joint position sense and kinesthetic sense, sensory information about movement.

All pathways that convey somatosensory information share similar anatomical arrangements. Receptors in the periphery encode the mechanical, chemical, or thermal stimulation received into receptor potentials (see Chap. 2). If the receptor potentials exceed the threshold of the trigger zone, an action potential is generated in a peripheral axon. The action potential is conducted along a peripheral axon, to a soma in a dorsal root ganglion, then along the proximal axon into the spinal cord. Within the spinal cord, the information ascends via axons in the white matter to various regions of the brain. The information is transmitted through a series of neurons and synapses.

> Information in the somatosensory system proceeds from the receptor through a series of neurons to the brain.

The diameter of the axons, the degree of axonal myelination, and the number of synapses in the pathway determine how quickly the information is processed. Much somatosensory information is not consciously perceived but is processed at the spinal level in local neural circuits or by the cerebellum to adjust movements and posture. The distinction between sensory information (nerve impulses generated from the original stimuli) and sensation (awareness of stimuli from the senses) should be noted throughout this

chapter. Perception, the interpretation of sensation into meaningful forms, occurs in the cerebral cortex and is discussed in Chapters 16 and 17.

PERIPHERAL SOMATOSENSORY NEURONS

Sensory Receptors

Sensory receptors are located at the distal ends of peripheral nerves. Each type of receptor is specialized, responding only to a specific type of stimulus, the adequate stimulus, under normal conditions. Based on the characteristics of the adequate stimulus, somatosensory receptors are classified as follows:

- Mechanoreceptors, responding to mechanical deformation of the receptor by touch, pressure, stretch, or vibration
- Chemoreceptors, responding to substances released by cells, including damaged cells following injury or infection
- Thermoreceptors, responding to heating or cooling

A subset of each type of somatosensory receptors is classified as nociceptors. Nociceptors are preferentially sensitive to stimuli that damage or threaten to damage tissue. Stimulation of nociceptors results in a sensation of pain. For example, when pressure mechanoreceptors are stimulated by stubbing a toe, the sensation experienced is pain rather than pressure. The receptors that encode the pain message are nociceptors, not the lower-threshold pressure receptors that convey information experienced as nonpainful pressure. Information from each of these types of receptors may reach awareness, but much of the information is used to make automatic adjustments and is selectively prevented from reaching consciousness by descending and local inhibitory connections.

Receptors that respond as long as a stimulus is maintained are called **tonic receptors.** For example, some stretch receptors in muscles, the tonic stretch receptors, fire the entire time a muscle is stretched. Receptors that adapt to a constant stimulus and stop responding are called **phasic receptors.** Muscles also contain phasic stretch receptors, which respond only briefly to a quick stretch.

Somatosensory Peripheral Neurons

Cell bodies of most peripheral sensory neurons are located outside the spinal cord in the dorsal root ganglia or outside the brain in cranial nerve ganglia. Peripheral sensory neurons have two axons:

- Distal axons conduct messages from receptor to the cell body.
- Proximal axons project from the cell body into the spinal cord or brain stem.

Some proximal axons that enter the spinal cord extend as far as the medulla before synapsing.

Peripheral axons, also called afferents, are classified according to axon diameter. The most commonly used system for classifying peripheral sensory axons designates the axons in order of declining diameter: Ia, Ib, II or Aβ, Aδ, and C (Fig. 6–1). The diameter of an axon is functionally important: larger-diameter axons transmit information faster than smaller-diameter axons. The faster conduction occurs because axoplasmic resistance is lower in large-diameter axons and because the large-diameter axons are myelinated, reducing membrane capacitance (see Chap. 2).

Cutaneous Innervation

The area of skin innervated by a single afferent neuron is called the receptive field for that neuron (Fig. 6–2). Receptive fields tend to be smaller distally and

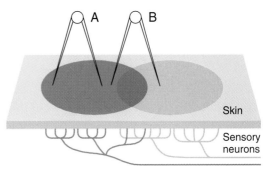

FIGURE 6–2

Receptive fields. The areas of skin innervated by each neuron are indicated on the surface of the skin. *A*, The caliper points touching the skin would be perceived as one point because both points are within the receptive field of one neuron. *B*, The caliper points would be perceived as two points because the points are contacting the receptive field of two neurons.

larger proximally. Distal regions of the body also have a greater density of receptors than the proximal areas. The combination of smaller receptive fields and greater density of receptors distally enables us to distinguish between two closely applied stimuli on a fingertip, while the same stimuli cannot be distinguished on the trunk.

SKIN SENSATIONS

Sensations from skin include the following:

- Touch
- Pain
- Temperature

Touch information is categorized as fine touch or coarse touch. **Fine touch** includes a variety of receptors (Fig. 6–3) and subsensations. The superficial fine touch receptors have small receptive fields, allowing resolution of closely spaced stimuli. The superficial fine touch receptors are Meissner's corpuscles, sensitive to light touch and vibration, and Merkel's disks, sensitive to pressure. Hair follicle receptors, sensitive to displacement of a hair, also have small receptive fields. The subcutaneous fine touch receptors have large receptive fields, providing less localization and discrimination of stimuli. Subcutaneous fine touch receptors are pacinian corpuscles, responsive to touch and vibration, and Ruffini's corpuscles, sensitive to

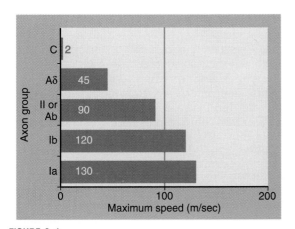

FIGURE 6–1

Conduction velocity of sensory axons.

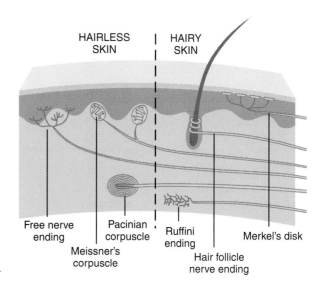

HAIRLESS SKIN | HAIRY SKIN

Free nerve ending
Meissner's corpuscle
Pacinian corpuscle
Ruffini ending
Hair follicle nerve ending
Merkel's disk

FIGURE 6–3
Cutaneous receptors.

stretch of the skin. All of the fine touch receptors transmit information on Aβ afferents.

Coarse touch is mediated by free endings throughout the skin (Fig. 6–3). These free nerve endings provide information perceived as crudely localized touch or pressure and the sensations of tickle and itch. **Nociceptors** are free nerve endings located in the skin, responsive to stimuli that damage or threaten tissue. Nociceptors provide information perceived as pain. **Thermal receptors,** also free nerve endings, respond to either warmth or cold within the temperature range that does not damage tissue. The information from all of the free nerve endings is conveyed by Aδ and C afferents. Although the various tactile receptors respond to different types of stimuli, natural stimuli typically activate several types of tactile receptors (Vallbo, 1995).

As noted in Chapter 5, the area of skin innervated by axons from cell bodies in a single dorsal root is a dermatome. In the brachial and lumbosacral plexus, sensory axons innervating specific parts of the limbs are separated from other axons arising in the same dorsal root and regrouped to form peripheral nerves. Thus peripheral nerves, such as the median nerve, have a different pattern of innervation than the dermatomes. Dermatomes and the cutaneous distribution of peripheral nerves are illustrated in Figure 6–4.

Although cutaneous receptors are not proprioceptors, the information from cutaneous receptors contributes to our sense of joint position and movement. The contribution of cutaneous receptors is primarily kinesthetic, responding to stretching of or increasing pressure on the skin. However, Ruffini's corpuscles discharge in response to static joint angles.

> Cutaneous receptors respond to touch, pressure, vibration, stretch, noxious stimuli, and temperature.

Musculoskeletal Innervation

MUSCLE SPINDLE

The sensory organ in muscle is the **muscle spindle,** consisting of muscle fibers, sensory endings, and motor endings (Fig. 6–5). The sensory endings of the spindle respond to stretch, that is, changes in muscle length and the velocity of length change. Quick and tonic stretch of the spindle is registered by type Ia afferents. Tonic muscle stretch is monitored by type II afferents. Small efferent fibers to the ends of muscle spindle fibers adjust spindle fiber stretch so the spindle is responsive through the physiologic range of muscle lengths.

Intrafusal and Extrafusal Fibers Muscle spindles are embedded in skeletal muscle. Because the spindle is fusiform (tapered at the ends), the specialized muscle fibers inside the spindle are designated **intrafusal fibers;** the ordinary skeletal muscle fibers outside the spindle are **extrafusal.** The ends of the intrafusal fibers connect to extrafusal fibers, so stretching the

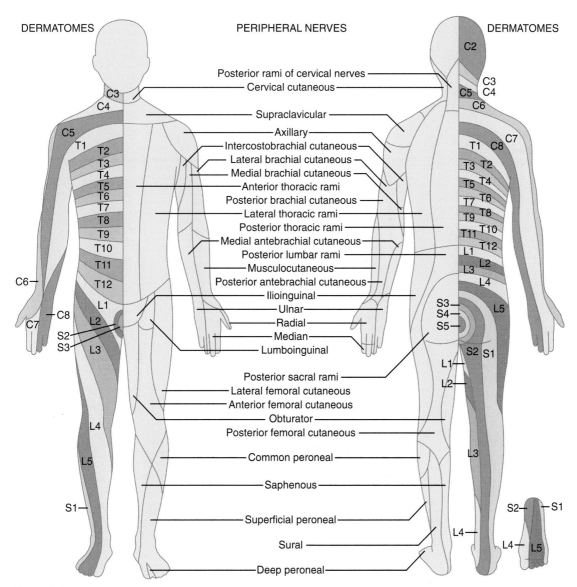

FIGURE 6–4

Dermatomes and cutaneous distribution of peripheral nerves.

muscle stretches the intrafusal fibers. To serve the dual purposes of providing information about the length and rate of change in length of the muscle, the spindle has two types of muscle fibers, two types of sensory afferents, and two types of efferents.

Intrafusal fibers are contractile only at their ends; the central region cannot contract. The arrangement of nuclei in the central region characterizes the two types of intrafusal fibers:

- **Nuclear bag fibers** have a clump of nuclei in the central region.
- **Nuclear chain fibers** have nuclei arranged single file.

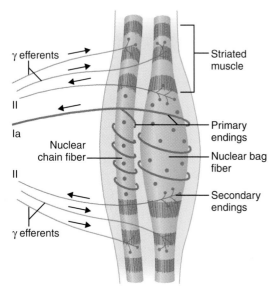

FIGURE 6–5

Simplified illustration of a muscle spindle. The intrafusal muscle fibers are nuclear chain and nuclear bag fibers. Stretch of the central region of intrafusal fibers is sensed by primary and secondary endings. The sensory information is conveyed to the central nervous system by group Ia and group II afferents. Efferent control of intrafusal fibers is via gamma motoneurons.

For spindles to monitor muscle length and rate of change in length, two different sensory endings are required:

- **Primary endings** of Ia neurons wrap around the central region of each intrafusal fiber.
- **Secondary endings** of group II afferents end mainly on nuclear chain fibers adjacent to the primary endings.

Because of their appearance, primary endings are also known as annulospiral endings, and secondary endings are called flower-spray endings. The discharge of primary endings is both phasic and tonic. The phasic discharge is maximal during quick stretch and fades quickly, as when a tendon is tapped with a reflex hammer. The tonic discharge is sustained during constant stretch; the rate of firing is proportional to the stretch of spindle fibers. Secondary endings respond only tonically.

If a muscle is passively stretched, the muscle spindles respond to the stretch (Fig. 6–6A). If the ends of intrafusal fibers were not contractile, the sensory endings would only register change when the muscle was

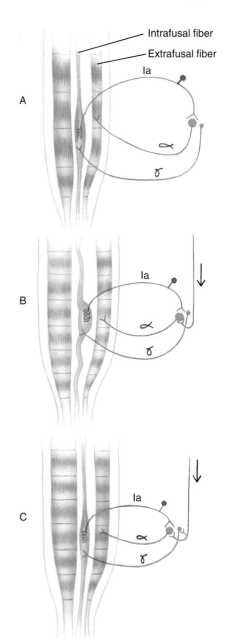

FIGURE 6–6

A, During passive stretch, spindles are elongated as the muscle is stretched. This activates the spindle sensory receptors. B, If the gamma motoneurons did not fire when the alpha motoneurons to the extrafusal muscles fire, the intrafusal central region would be relaxed and the afferent neurons inactive. C, Normally, during active muscle contraction, alpha and gamma motoneurons are simultaneously active. The firing of gamma motoneurons causes the ends of intrafusal fibers to contract, thus maintaining the stretch on the intrafusal central region and preserving the ability of the sensory endings to indicate stretch.

fully elongated; if the muscle were contracted even slightly, the spindle would be slack, rendering the sensory endings insensitive to stretch (Fig. 6–6*B*). To maintain sensitivity of the spindle throughout the normal range of muscle lengths, **gamma motoneurons** fire, causing the ends of the intrafusal fibers to contract. Contracting the ends of the intrafusal fibers stretches the central region, thus maintaining sensory activity from the spindle (Fig. 6–6*C*). Gamma efferent control is dual, with gamma **dynamic axons** ending on nuclear bag fibers to adjust the sensitivity of primary afferents and gamma **static axons** innervating both types of intrafusal fibers to tune the sensitivity of both primary and secondary afferents.

> Muscle length is signaled by type Ia and II afferents, reflecting stretch of the central region of both types of intrafusal fibers. Spindle sensitivity to changes in length is adjusted by gamma static efferents.

> Velocity of change in muscle length is signaled only by type Ia afferents, with information mainly from nuclear bag fibers whose sensitivity is adjusted by gamma dynamic efferents.

GOLGI TENDON ORGANS

Tension in tendons is relayed from Golgi tendon organs, encapsulated nerve endings woven among the collagen strands of the tendon near the musculotendinous junction (Fig. 6–7, *left*). Golgi tendon organs are sensitive to very slight changes (< 1 g) in the tension on a tendon and respond to tension exerted both by active contraction and by passive stretch of muscle. Information is transmitted from Golgi tendon organs into the spinal cord by Ib afferents.

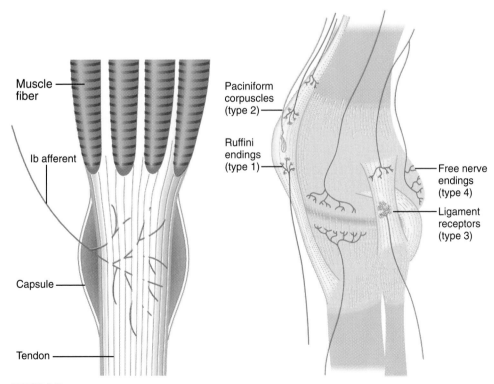

FIGURE 6–7

Left, Golgi tendon organ. *Right,* Joint receptors. A numerical classification of joint receptors, developed by Wyke (1967), is indicated in parentheses after the name of each receptor.

JOINT RECEPTORS

Joint receptors respond to mechanical deformation of the capsule and ligaments (Fig. 6–7, *right*). Ruffini's endings in the joint capsule signal the extremes of joint range and respond more to passive than active movement. Paciniform corpuscles respond to dynamic movement but not when joint position is constant. Ligament receptors are similar to Golgi tendon organs and signal tension. Free nerve endings are most often stimulated by inflammation. The afferents associated with the joint receptors are as follows:

- Ligament receptors—Ib
- Ruffini's and paciniform endings—II
- Free nerve endings—Aδ and C

Fully normal proprioception requires muscle spindles, joint receptors, and cutaneous mechanoreceptors. This redundancy probably reflects the importance of proprioception to the control of movement. People with total hip joint replacements retain good proprioception in midrange (Barrack et al., 1983). This indicates that joint receptors are not essential for proprioception.

> Muscle spindles respond to quick and to prolonged stretch of the muscle. Tendon organs signal the force generated by muscle contraction or by passive stretch of the tendon. Joint receptors respond to mechanical deformation of joint capsules and ligaments.

Summary of the Function of Different-Diameter Axons

Large-diameter afferents transmit information from specialized receptors in muscles, tendons, and joints. Medium-sized afferents transmit information from joint capsules, muscle spindles, and cutaneous touch, stretch, and pressure receptors. The smallest-diameter afferents convey crude touch, nociceptive, and temperature information from both the musculoskeletal system and the skin. Table 6–1 summarizes the axon types, associated receptors, and adequate stimuli for the somatosensory system.

PATHWAYS TO THE BRAIN

Three types of pathways bring sensory information to the brain (Table 6–2):

- Conscious relay pathways
- Divergent pathways
- Unconscious relay pathways

An important distinction among the types of pathways is the fidelity of information conveyed. Pathways that transmit signals with high fidelity provide accurate details regarding the location of the stimulation. For example, high-fidelity signals from the fingertips allow people to recognize two points separated by as little as 1.6 mm as being distinct points and to identify precisely where on the fingertip the stimulation occurred. The ability to identify the location of stimulation is achieved by the anatomical arrangement of axons in the pathways. In high-fidelity pathways, a somatotopic arrangement of information is created. Somatotopic arrangement means that axons from one part of the body are close to axons carrying signals from adjacent parts of the body and are segregated from axons carrying information from distant parts of the body. For example, axons carrying information from the thumb are near axons carrying information from the index finger and relatively distant from axons carrying information from the toes.

In describing pathways in the nervous system, only the neurons with long axons that connect distant regions of the nervous system are counted. These neurons with long axons are called projection neurons. The convention for numbering or naming only the projection neurons omits the small, integrative interneurons interposed between the projection neurons. Thus a three-neuron pathway means three projection neurons, but a number of interneurons may also be linked in the pathway.

The first type of pathways, **conscious relay pathways,** bring information about location and type of stimulation to the cerebral cortex. The information in conscious relay pathways is transmitted with high fidelity, thus providing accurate details regarding the stimulus and its location. Because the information in these pathways allows us to make fine distinctions about stimuli, the term *discriminative* is used to describe the sensations conveyed in conscious relay pathways. Discriminative touch and proprioceptive information ascends ipsilaterally in the posterior

TABLE 6–1

AXON TYPES, ASSOCIATED RECEPTORS, AND ADEQUATE STIMULI FOR THE SOMATOSENSORY SYSTEM

Axon Size	Proprioception			Cutaneous and Subcutaneous Touch and Pressure			Pain and Temperature		
	Axon Type	*Receptors*	*Stimulus*	*Axon Type*	*Receptors*	*Stimulus*	*Axon Type*	*Receptors*	*Stimulus*
Large myelinated	Ia	Muscle spindles	Muscle stretch	—			—		
	Ib	Golgi tendon organs	Tendon tension	—			—		
		Ligament receptors	Ligament tension						
Medium myelinated	II	Muscle spindles	Muscle stretch	Aβ	Meissner's	Touch, vibration	—		
					Pacinian	Touch vibration			
		Paciniform & Ruffini's type receptors in joint capsules	Joint movement		Ruffini's	Skin stretch			
					Merkel's	Pressure			
					Hair follicle	Pressure			
Small myelinated							Aδ	Free nerve ending	Pain, temperature, coarse touch*
Unmyelinated							C	Free nerve ending	Pain, temperature, itch, tickle*

*The axon size correlates with speed of conduction; thus the large, myelinated fibers conduct fastest, and the unmyelinated fibers have the slowest conduction speeds.

TABLE 6–2

SOMATOSENSORY PATHWAYS

Type	Information Conveyed	Anatomical Name	Termination
Conscious relay	Discriminative touch and conscious proprioception	Dorsal column/medial lemniscus	Primary sensory area cerebral cortex
	Discriminative pain and temperature	Neospinothalamic	Primary sensory area cerebral cortex
Divergent	Slow, aching pain	Spinomesencephalic	Midbrain
		Spinoreticular	Reticular formation
		Paleospinothalamic	Many areas of cerebral cortex
Unconscious relay	Movement-related information	Spinocerebellar	Cerebellum

spinal cord. Discriminative pain and temperature information crosses the midline soon after entering the cord and then ascends contralaterally.

The second type of pathways, **divergent pathways,** transmit information to many locations in the brain stem and cerebrum and use pathways with varying numbers of neurons. The sensory information is used at both conscious and unconscious levels. Aching pain is a form of sensation that is transmitted via divergent pathways in the central nervous system.

The third type of pathways, **unconscious relay pathways,** bring unconscious proprioceptive and other movement-related information to the cerebellum. This information plays an essential role in automatic adjustments of our movements and posture.

> Conscious relay pathways convey high-fidelity, somatotopically arranged information to the cerebral cortex. Divergent pathways convey information that is not somatotopically organized to many areas of the brain. Unconscious relay pathways convey movement-related information to the cerebellum.

Conscious Relay Pathways to Cerebral Cortex

All four types of somatosensation reach conscious awareness:

- Touch
- Proprioception
- Pain
- Temperature

The pathways involve three projection neurons. The pathways to consciousness travel upward in the spinal cord via two routes:

- Dorsal columns
- Anterolateral tracts

The routes in the spinal cord are composed of white matter, because myelin promotes rapid conduction along the axons. The dorsal columns carry sensory information about discriminative touch and conscious proprioception; discriminative pain and temperature information travels in the anterolateral tracts. To be aware of sensory information, the information must reach the thalamus, where crude awareness is possible (Newman, 1995). For discriminative perception,

stimuli localized with fine resolution, information must be processed by the cerebral cortex.

If peripheral afferent information is absent, awareness of body parts can be lost. Oliver Sacks, a neurologist, recounts his strange experience of believing that he had lost his leg following severe damage to several nerves in a climbing accident. The complete loss of sensation from his leg led to a lack of awareness of the limb. Although he was not paralyzed, he was unable to voluntarily take a step until his physical therapist moved his leg passively, giving him the concept of how to move the injured leg (Sacks, 1984).

DISCRIMINATIVE TOUCH AND CONSCIOUS PROPRIOCEPTION PATHWAY

Discriminative touch includes localization of touch and vibration and the ability to discriminate between two closely spaced points touching the skin. **Conscious proprioception** is the awareness of the movements and relative position of body parts. Cerebral cortical integration of dorsal column information allows identification of an object by touch and pressure information. **Stereognosis** is the ability to use touch and proprioceptive information to identify an object; for example, a key in the hand can be identified without vision. The information conveyed in this pathway is important for controlling fine movements and for making movements smooth.

The pathway for discriminative touch and conscious proprioception uses a three-neuron relay (Fig. 6–8):

- The primary, or first-order, neuron conveys information from the receptors to the medulla.
- The secondary, or second-order, neuron conveys information from the medulla to the thalamus.
- The tertiary, or third-order, neuron conveys information from the thalamus to the cerebral cortex.

Dorsal Column/Medial Lemniscus System Stimulation of receptors at the distal end of the primary neuron is conveyed to the cell body in the dorsal root ganglion. The primary neuron's proximal axon enters the spinal cord via the dorsal root, then ascends in the ipsilateral dorsal column. Axons from the lower limb occupy the more medial section of the dorsal column, called the **fasciculus gracilis.** Axons from the upper limb occupy the lateral section of the dorsal column, called the **fasciculus cuneatus.** This pattern

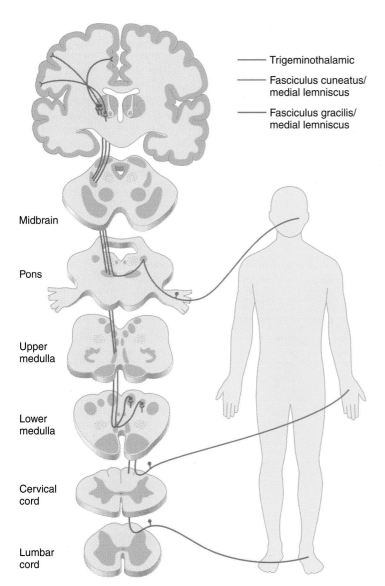

Trigeminothalamic

Fasciculus cuneatus/
medial lemniscus

Fasciculus gracilis/
medial lemniscus

Midbrain

Pons

Upper
medulla

Lower
medulla

Cervical
cord

Lumbar
cord

	First-order neuron cell body	Second-order neuron cell body	Third-order neuron cell body
From the body	Dorsal horn	Nucleus cuneatus or gracilis	VPL nucleus of thalamus
From the face (touch pathway only)	Trigeminal ganglion	Main sensory nucleus of trigeminal nerve	VPM nucleus of thalamus

FIGURE 6–8

Discriminative touch and conscious proprioceptive
information pathways.

occurs because nerve fibers entering the dorsal column from higher segments are added laterally to the fibers already in the dorsal column from lower segments.

The axons that ascend in the fasciculus gracilis synapse with second-order neurons in the **nucleus gracilis** of the medulla. The axons in the fasciculus cuneatus synapse with second-order neurons in the **nucleus cuneatus** of the medulla. Thus, in a tall person a primary neuron could be 6 feet long, extending from a toe to the medulla.

Throughout the spinal cord, primary neurons of the discriminative touch/conscious proprioception pathway have many collaterals entering the gray matter. Some collaterals contribute to motor control, some influence activity in neurons in other sensory systems, and others influence autonomic regulation.

Cell bodies of the second-order neurons are located in the nucleus gracilis or cuneatus. Axons from the second-order neurons cross the midline as the internal arcuate fibers, then ascend to the thalamus as the **medial lemniscus.** The second-order neurons end in an area of the thalamus named for its location, the **ventral posterolateral (VPL)** nucleus.

Third-order neurons connect the thalamus to the sensory cortex. The axons form part of the thalamocortical radiations, fibers connecting the thalamus to the cerebral cortex. Thalamocortical axons travel through the internal capsule.

Discriminative Touch Information from the Face

Sensory innervation for the face is supplied by the three divisions of the **trigeminal nerve** (see Fig. 6–8) (see Chap. 13 for more details on the trigeminal nerve). Neurons in the trigeminal nerve are the first-order neurons for discriminative touch information from the face. Their cell bodies are in the trigeminal ganglion, and the proximal axons end in the **trigeminal main sensory nucleus.** Second-order neuron cell bodies are located in the trigeminal main sensory nucleus, and the axons cross the midline in the pons and end in the **ventral posteromedial (VPM)** nucleus of the thalamus. Third-order axons continue to the cerebral cortex. Details regarding sensation from the face are presented in Chapter 13. The effects of lesions in the discriminative touch/conscious proprioception pathways are illustrated in Figure 6–9.

Somatotopic Arrangement of Information

Although the axons in the dorsal column are arranged segmentally as they enter the dorsal columns, the axons are rearranged into a somatotopic organization as they ascend. The somatotopic arrangement is maintained throughout the second- and third-order neurons, so that the area of cerebral cortex devoted to discriminative somatosensation, the **primary sensory (primary somatosensory) cortex,** receives somatotopically organized information. The primary sensory cortex is located in the gyrus posterior to the central sulcus, that is, the postcentral gyrus.

The size of the area of primary sensory cortex devoted to a specific part of the body is represented by the **homunculus** surrounding the cortex in Figure 6–10. The homunculus is a map, developed by recording the responses of awake individuals during surgery. Small areas of the cerebral cortex are electrically stimulated, and the people report what they feel. When the sensory cortex is stimulated, people report feeling sensations that seem to originate from the surface of the body. For example, stimulation of the medial postcentral gyrus elicits sensations that seem to originate in the contralateral lower limb. Another method of testing is to stimulate areas on the body and record from the cerebral cortex. For example, touching a fingertip activates neurons in the superolateral postcentral gyrus. The homunculus illustrates the proportions and arrangement of cortical areas that contain representations of the surface of the body. The fingers and lips of the homunculus are much larger than their proportion of the body would indicate. The large cortical representation corresponds to the relatively high density of receptors in these regions and the associated degree of fine motor control.

Somatosensory Areas of the Cerebral Cortex

The primary sensory cortex discriminates the size, texture, or shape of objects. Another area of the cerebral cortex, the **somatosensory association area,** analyzes the information from the primary sensory area and the thalamus and provides stereognosis and memory of the tactile and spatial environment.

Sensory information essential for identifying objects by palpation, distinguishing between closely spaced stimuli, and controlling fine movement and smoothness of movement travels in the dorsal columns, then in the medial lemniscus, to the primary sensory cortex. Tactile information from the face travels in the trigeminal nerve, then to the thalamus, to the sensory cortex.

LESION IN:

EFFECT ON DISCRIMINATIVE
TOUCH AND CONSCIOUS
PROPRIOCEPTIVE INFORMATION:

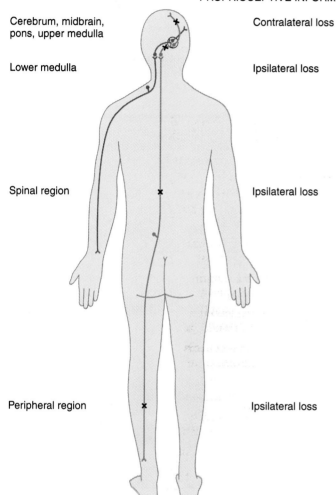

Cerebrum, midbrain,
pons, upper medulla

Contralateral loss

Lower medulla

Ipsilateral loss

Spinal region

Ipsilateral loss

Peripheral region

Ipsilateral loss

FIGURE 6–9

The effect of lesion location on transmission of
discriminative touch and conscious proprioceptive information.

DISCRIMINATIVE PAIN AND TEMPERATURE, COARSE TOUCH

Anterolateral Columns The anterolateral white matter in the spinal cord contains axons that transmit discriminative information about pain and temperature and also about coarse touch. Coarse touch is not considered further here because less information is conveyed than by the dorsal column/medial lemniscus system, and thus coarse touch cannot be tested independently when discriminative touch is intact.

Several parallel tracts ascend in the anterolateral spinal cord. One of the tracts, the **neospinothalamic tract,** is part of a three-neuron conscious relay pathway, called the neospinothalamic pathway. The term *neospinothalamic* distinguishes the more recent phylogenetic appearance of the fast conduction pathway from an older, slower conducting divergent pathway, the paleospinothalamic pathway.

Temperature Sensation Warmth and cold are detected by specialized free nerve endings of small

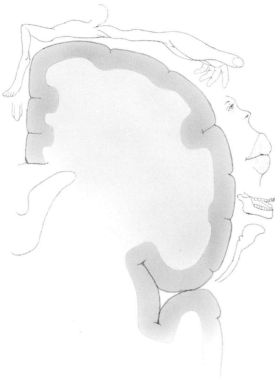

FIGURE 6–10

Primary sensory cortex. The areas of cortex responding to somatosensory stimulation are indicated by the homunculus.

myelinated and unmyelinated neurons. Aδ fibers carry impulses produced by cooling, and C fibers carry information regarding heat. In the neospinothalamic pathway, the first-order neuron synapses with the second-order neuron in the dorsal horn. The second-order axons cross the midline one to five spinal segments above or below their level of entrance into the spinal cord. The axons of second-order neurons ascend contralaterally to project to the VPL nucleus of the thalamus. The axons of third-order neurons project from the thalamus to the sensory cortex.

Pain Pain is an extremely complex phenomenon. Persistent pain affects emotional, autonomic, and social functioning. Pain is composed of both protective sensation and the emotional response to this sensation. Nociceptive information travels in several different pathways.

A common patient report of back pain secondary to lifting a heavy object consists of an initial immediate sharp sensation that indicates the location of the injury; this is called fast or **neospinothalamic pain.** Fast pain is often followed by a dull, throbbing ache, which is not well localized. The later pain is called slow or **paleospinothalamic pain.** Both types of pain occur in acute pain. Impulses conveying both fast and slow pain travel together in the anterolateral section of the spinal cord, and then their paths become separate in the brain (Fig. 6–11). Fast pain uses a conscious relay pathway and therefore is discussed in this section; slow pain is discussed in a subsequent section on divergent pathways.

Fast, Localized Pain Fast pain uses a three-neuron system (see Fig. 6–11A):

- The first-order neuron brings information into the spinal cord.
- The axon of the second-order neuron crosses the midline and projects from the spinal cord to the thalamus.
- The third-order neuron projects from the thalamus to the cerebral cortex.

The primary neuron in the fast pain pathway is a small myelinated Aδ fiber. Aδ fibers transmit information from free nerve endings in the periphery to the spinal cord. The endings respond to noxious mechanical stimulation (high-threshold mechanoreceptor afferents) or to mechanical or thermal stimulation (mechanothermal afferents). The peripheral axon brings an impulse to the cell body in the dorsal root ganglion. The central axon enters the cord, then travels several levels in the **dorsolateral tract** (Lissauer's marginal zone) before entering and terminating in lamina I, II, and/or V of the dorsal horn. The neurotransmitter released is believed to be glutamate.

The cell body of the second-order neuron is in lamina I, II, or V of the dorsal horn. The axon of the second-order neuron crosses the midline in the anterior white commissure, then ascends to the thalamus via the **neospinothalamic tract.** Second-order neurons are **nociceptive specific,** which only receive information from Aδ nociceptive fibers. Most neospinothalamic tract neurons end in the VPL nucleus of the thalamus.

The third-order neurons arise in the VPL nucleus and project to the primary sensory cortex. A lesion in the VPL nucleus interrupts the pathway to the cortex,

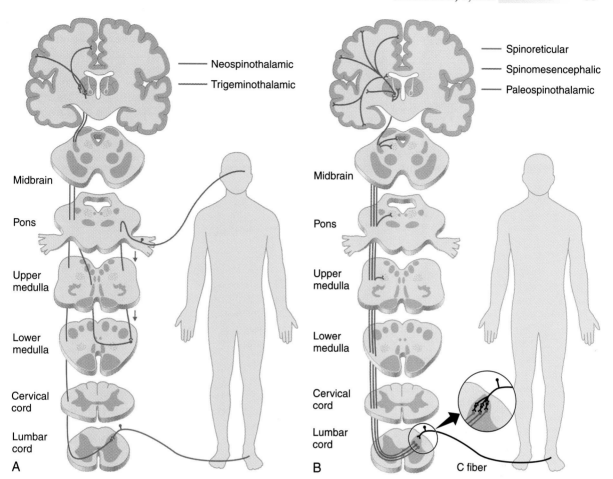

	First-order neuron cell body	Second-order neuron cell body	Third-order neuron cell body
From the body	Dorsal root ganglion	Dorsal horn of spinal cord	VPL nucleus of thalamus
From the face	Trigeminal ganglion	Spinal nucleus of trigeminal nerve	VPM nucleus of thalamus

FIGURE 6–11

A, Sharp, localized pain information travels in a three-neuron pathway. *B,* Slowly conducted pain information from the body travels in the spinoreticular, spinomesencephalic, and paleospinothalamic tracts; synapses occur in the reticular formation, superior colliculus, periaqueductal gray, and midline and intralaminar nuclei of the thalamus. Efferents from the thalamic nuclei spread project to widespread areas of the cerebral cortex and to the striatum.

causing inability to localize painful stimuli despite feeling the affective aspects of pain.

> Information that enables people to localize noxious sensations and to consciously distinguish between warmth and cold travels to the cerebral cortex via neospinothalamic pathways.

Comparison of the Dorsal Column/Medial Lemniscus and Neospinothalamic Systems The neospinothalamic and the dorsal column systems are anatomically similar, consisting of three neuron relay pathways. Unlike the dorsal columns, containing axons of primary neurons and ascending ipsilaterally, the ascending axons in the anterolateral columns are second-order neurons, and most ascend contralaterally. In both the dorsal column paths and the neospinothalamic tract, the second-order axon crosses the midline. However, in the dorsal column path the crossing occurs in the medulla, while in the neospinothalamic pathway the crossing occurs in the spinal cord before the axon ascends. The second neuron in both dorsal column and neospinothalamic paths ends in the VPL nucleus of the thalamus. In both pathways, third-order neurons project from the thalamus to the primary sensory cortex, where the information can be localized.

Comparison of Dorsal Column and Anterolateral Column Information In contrast to the discriminative touch and conscious proprioceptive information traveling in the dorsal columns, the anterolateral white matter contains axons transmitting information about pain, temperature, and coarse touch. However, functions of the dorsal and anterolateral columns are not rigidly segregated; information about nondiscriminative (coarse) touch travels in the anterolateral system, and some pain and temperature information ascends in the dorsal columns (Willis and Coggeshall, 1991).

Fast Pain Information from the Face Afferent information interpreted as fast pain from the face travels in the **trigeminal nerve.** Fibers in this pathway travel down into the medulla and upper cervical cord before synapsing. Second-order fibers cross the midline and ascend to the VPM nucleus of the thalamus. Third-order neurons project to the cerebral cortex. Figure 6–12 summarizes the effect of various lesions on transmission of fast pain and discriminative temperature information.

Fast Versus Slow Pain When fast pain information reaches the primary sensory cortex, a person is consciously aware of sharp pain in a specific location. If tissue damage has occurred, the fast pain is followed by slow, aching pain. The onset of slow pain is later than fast pain because the impulses travel on smaller, unmyelinated axons. The difference in conduction velocities results in a C fiber's requiring about 2 seconds to transmit information to the spinal cord, while Aδ fibers require as little as 0.03 second.

Divergent Pathways

SLOW, ACHING PAIN

Unlike fast pain, the affective dimension of pain depends on a divergent network of neurons. Slow pain information uses several pathways with variable numbers of projection neurons, not a three-neuron pathway like fast pain. The information from the slow pain system is not somatotopically organized, so slow pain cannot be precisely localized.

First Neuron The first neuron is a small, unmyelinated C fiber. The receptors are free nerve endings, sensitive to noxious heat, chemical, or mechanical stimulation (polymodal afferents). High-threshold C fiber endings become **sensitized** with repeated stimulation. Thus, after injury, these neurons can be fired with less stimulation than is usually required. Tissue damage also results in release of chemicals—histamine, prostaglandins, and others—that sensitize pain receptors. For example, a gentle touch on sunburned skin can be painful.

Information from free nerve endings in the periphery travels in peripheral axons to the cell body in the dorsal root ganglion. The central axon enters the cord, branches in the dorsolateral tract, and then synapses with interneurons in lamina I, II, and/or V of the dorsal horn. Lamina I is also called the marginal layer, and lamina II is also called the substantia gelatinosa. The neurotransmitter is **substance P.** Axons from the interneurons synapse with cell bodies of ascending projection neurons in laminae V to VIII.

Ascending Projection Neurons The ascending projection neurons are **wide dynamic-range** neurons, which receive information from cutaneous, musculoskeletal, and visceral receptors. Transmission of pain information in wide dynamic-range neurons can be modified by the activity of larger diameter afferents

LESION IN:

EFFECT ON FAST PAIN
AND TEMPERATURE
SENSATION:

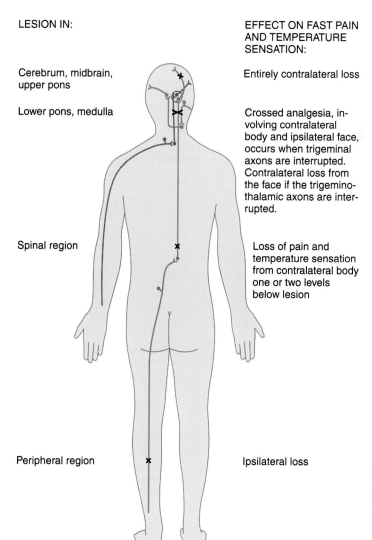

Cerebrum, midbrain,
upper pons

Entirely contralateral loss

Lower pons, medulla

Crossed analgesia, in-
volving contralateral
body and ipsilateral face,
occurs when trigeminal
axons are interrupted.
Contralateral loss from
the face if the trigeminothalamic axons are inter-
rupted.

Spinal region

Loss of pain and
temperature sensation
from contralateral body
one or two levels
below lesion

Peripheral region

Ipsilateral loss

FIGURE 6–12

Effects of lesion location on the transmission of fast
pain and discriminative temperature information.
Crossed analgesia occurs with lesions in the lower
pons and medulla because axons conveying pain information from the face (compare with Fig. 6–11A)
descend ipsilaterally near the spinothalamic tract carrying pain information from the contralateral body.

and by descending central influences. This will be
discussed in Chapter 7. The ascending axons reach
the midbrain, reticular formation, and thalamus via
three tracts in the anterolateral spinal cord (see Fig.
6–11B):

• Spinomesencephalic
• Spinoreticular
• Paleospinothalamic

These three tracts are parallel ascending tracts. Of
these tracts, only information in the paleospinothala-

mic tract is believed to be perceived as pain. Information in the other tracts serves arousal, motivational,
and reflexive functions and/or activates descending
projections that control the flow of sensory information.

The **spinomesencephalic tract** carries pain sensory information to the superior colliculus and to an
area surrounding the cerebral aqueduct, the periaqueductal gray. The spinomesencephalic tract is involved
in turning the eyes and head toward the source of
noxious input and in activating descending tracts that

control pain. The periaqueductal gray is part of the descending pain control system (discussed in Chap. 7). The **spinoreticular** ascending neurons synapse in the brain stem reticular formation. Neurons from the reticular formation then project to the midline and intralaminar nuclei of the thalamus.

The **paleospinothalamic tract** neurons carrying slow pain information project to the **midline** and **intralaminar nuclei** of the thalamus. The neurons located in these thalamic nuclei have large receptive fields, sometimes from the entire body. These thalamic nuclei project to areas of the cerebral cortex involved with emotions, sensory integration, personality, and movement, as well as to the basal ganglia. Activity in the spinoreticular and paleospinothalamic tracts results in arousal, withdrawal, autonomic, and affective responses to pain.

If someone breaks a bone in the hand, the divergent pain pathways provide information that contributes to automatically directing the eyes and head toward the injury, automatically moving the hand away from the cause of injury, becoming pale, and feeling faint, nauseous, and emotionally distressed. The information provided by the divergent pathways is not well localized so that the entire hand seems to hurt.

> The slow pain pathways provide information that produces automatic and emotional responses to noxious stimuli.

Slow pain information from the face proceeds in the **trigeminoreticulothalamic pathway.** The first neurons are C fibers in the trigeminal nerve, which synapse in the reticular formation with ascending projection neurons. These neurons project to the intralaminar nuclei. The projections from the intralaminar nuclei are similar to the paleospinothalamic pathway, with projections to many areas of the cerebral cortex.

Although an intact sensory cortex is required for the localization of pain, crude awareness of slow pain can be achieved in many cortical areas and possibly in the thalamus and basal ganglia.

TEMPERATURE INFORMATION

Temperature information is also transmitted in phylogenetically older pathways to the reticular formation, to the nonspecific nuclei of the thalamus, to subcortical nuclei, and to the hypothalamus. This temperature information that does not reach conscious awareness contributes to arousal, provides gross localization, and contributes to autonomic regulation.

Unconscious Relay Tracts to the Cerebellum

Information from proprioceptors and information about activity in spinal interneurons are transmitted to the cerebellum via the **spinocerebellar tracts.** Information relayed by these tracts is critical for adjusting movements. For example, one of the complications of diabetes is dysfunction of proprioceptive neurons. If, as often occurs in diabetes, proprioceptive information from the ankle is decreased, sway during stance increases. Inadequate proprioceptive input can also cause ataxia (uncoordinated movement) because the loss of sensory feedback disrupts movement control (see Chap. 10 regarding types of ataxia).

Two of the spinocerebellar pathways deliver information from receptors in muscles, tendons, and joints from peripheral neurons to the cerebellum. These two neuron pathways relay high-fidelity, somatotopically arranged information to the cerebellar cortex. In contrast, two other spinocerebellar tracts are specialized to provide feedback to the cerebellum about the activity in spinal interneurons and in the descending motor tracts. These one-neuron internal feedback tracts do not directly convey information from any peripheral receptors.

HIGH-FIDELITY PATHWAYS

There are two pathways that relay high-fidelity, somatotopically arranged information to the cerebellar cortex (Fig. 6–13):

- Posterior spinocerebellar pathway
- Cuneocerebellar pathway

Posterior Spinocerebellar Pathway The **posterior** (dorsal) **spinocerebellar pathway** transmits information from the legs and the lower half of the body. The proximal axon of the first-order neuron travels in the dorsal column to the thoracic or upper lumbar spinal cord, then synapses in the area of the dorsal gray matter called the **nucleus dorsalis** (Clarke's nucleus). The nucleus dorsalis extends vertically from spinal

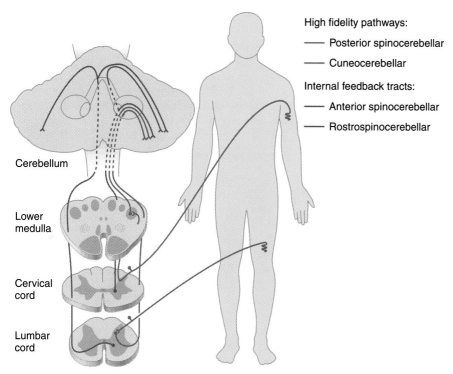

High fidelity pathways:

—— Posterior spinocerebellar

—— Cuneocerebellar

Internal feedback tracts:

—— Anterior spinocerebellar

—— Rostrospinocerebellar

Cerebellum

Lower
medulla

Cervical
cord

Lumbar
cord

FIGURE 6–13

Tracts transmitting unconscious proprioceptive information. The high-fidelity pathways are the posterior spinocerebellar from the lower body and the cuneocerebellar from the upper body. Internal feedback tracts are the anterior spinocerebellar from the lower spinal cord and the rostrospinocerebellar from the cervical cord.

segments T1 to L2. Second-order axons form the posterior spinocerebellar tract. The tract remains ipsilateral and projects to the cerebellar cortex via the inferior cerebellar peduncle.

Cuneocerebellar Pathway The **cuneocerebellar pathway** begins with primary afferents from the arm and upper half of the body; the central axons travel via the posterior columns to the lower medulla. The synapse between the first- and second-order neurons occurs in the **lateral cuneate nucleus,** a nucleus in the medulla analogous to the nucleus dorsalis in the spinal cord. The second-order neurons form the cuneocerebellar tract, enter the ipsilateral inferior cerebellar peduncle, and end in the cerebellar cortex. The target neurons for both the posterior spinocerebellar and cuneocerebellar tracts are arranged somatotopically in the cerebellar cortex.

INTERNAL FEEDBACK TRACTS

The two internal feedback tracts monitor the activity of spinal interneurons and of descending motor signals from the cerebral cortex and brain stem (see Fig. 6–13):

- Anterior spinocerebellar tract
- Rostrospinocerebellar tract

Anterior Spinocerebellar Tract The **anterior spinocerebellar tract** transmits information from the thoracolumbar spinal cord. The tract begins with cell bodies in the lateral and ventral horns, in the area of the spinal cord containing the most interneurons. The axons cross to the opposite side and ascend in the contralateral anterior spinocerebellar tract to the midbrain. Leaving the midbrain, the axons enter the cerebellum via the superior cerebellar peduncle. Most fibers recross the midline before entering the cerebellum, so each side of the cerebellum gets information from both sides of the lower body. The bilateral projection may reflect the normally automatic coordination of lower limb activities, as opposed to the typically more voluntary control of the upper limbs.

Rostrospinocerebellar Tract The **rostrospinocerebellar tract** transmits information from the cervical spinal cord to the ipsilateral cerebellum and enters the cerebellum via both the inferior and superior cerebellar peduncles.

The anterior and rostrospinocerebellar tracts apprise the cerebellum of the descending commands delivered to the neurons that control muscle activity via interneurons located between descending motor tracts and motor neurons that innervate muscles. Information about the activity of spinal reflex circuits is also conveyed by the internal feedback tracts.

FUNCTION OF SPINOCEREBELLAR TRACTS

Information that travels in the spinocerebellar tracts is not consciously perceived. Damage to the spinocere-

bellar tracts cannot be differentiated from lesions of the cerebellum by clinical tests, although imaging can distinguish between these lesions. The information in the spinocerebellar tracts is used for unconscious adjustments to movements and posture. Because the internal feedback tracts convey descending motor information to the cerebellum prior to the information reaching the motor neurons, and the high-fidelity pathways convey information from muscle spindles, tendon organs, and cutaneous mechanoreceptors, the cerebellum obtains information about movement commands and about the movements or postural adjustments that followed the commands. Thus the cerebellum can compare the intended motor output with the actual movement output. The cerebellum uses this information to make corrections to the neural commands via its connections with other brain areas (see Chap. 9).

> Information in the spinocerebellar tracts is from proprioceptors, spinal interneurons, and descending motor pathways. This information, which does not reach conscious awareness, contributes to automatic movements and postural adjustments.

Summary

Somatosensory pathways provide information about the external world, information used in movement control and to prevent or minimize injury. Conscious information about external objects can be provided by all four types of discriminative sensation: touch, proprioception, pain, and temperature. Discriminative sensations require analysis of sensory signals by the somatosensory area of the cerebral cortex. The dorsal column/medial lemniscus and neospinothalamic pathways deliver the high-fidelity, somatotopically arranged information to the cerebral

cortex. This conscious information contributes to our understanding of the physical world and to the control of fine movements. Unconscious information that contributes to the control of posture and movement is delivered to the cerebellum by the spinocerebellar tracts. Unconscious nociceptive information provides information about stimuli that threaten to damage or have damaged tissue. Paleospinothalamic, spinoreticular, and spinomesencephalic tracts deliver information to the thalamus, reticular formation, and midbrain that elicits automatic responses to nociceptive stimuli.

REVIEW QUESTIONS

1. What are the three types of somatosensory receptors?

2. What are nociceptors?

3. To what do the primary and secondary sensory endings in muscle spindles respond?

4. How is the sensitivity of sensory endings in a muscle spindle maintained when the muscle is shortened?

5. What type of information is transmitted by large-diameter Ia and Ib axons?

6. What classes of axons convey nociceptive and temperature information?

7. What are the three types of pathways that convey information to the brain?

8. High-fidelity, somatotopically arranged somatosensory information is conveyed to what area of the cerebral cortex?

9. Neural signals that are interpreted as dull, aching pain travel in what pathway?

10. All of the unconscious relay pathways end in what part of the brain?

11. Where do synapses occur between neurons conveying discriminative touch information from the left lower limb?

12. Where do synapses occur between neurons conveying discriminative pain information from the left lower limb?

13. Name the tracts that relay unconscious proprioceptive information to the cerebellum. Name the tracts that provide unconscious information about activity in spinal interneurons and descending motor commands.

References

Barrack, R. L., Skinner, H. B., Cook, S. D., et al. (1983). Effect of articular disease and total knee arthroplasty on knee joint-position sense. J. Neurophysiol. 50(3):684–687.

Newman, J. (1995). Thalamic contributions to attention and consciousness. Conscious Cogn. 4(2):172–193.

Sacks, O. (1984). A leg to stand on. New York: Harper & Row.

Vallbo, A. B. (1995). Single-afferent neurons and somatic sensation in humans. In: Gazzaniga, M. S. (Ed.). The cognitive neurosciences. Cambridge, Mass.: M.I.T. Press, pp. 237–252.

Willis, W. D., and Coggeshall, R. E. (1991). Sensory mechanisms of the spinal cord (2nd ed.). New York: Plenum Press.

Wyke, B. (1967). The neurology of joints. Ann. R. Coll. Surg. 41(1):25–50.

Suggested Readings

Gandevia, S. C., and Burke, D. (1994). Does the nervous system depend on kinesthetic information to control natural limb movements? In: Cordo, P., and Harnad, S. (Eds.). Movement control. Cambridge, England: Cambridge University Press, pp. 12–30.

Kandel, E. R., and Jessell, T. M. (1991). Touch. In: Kandell, E. R., Schwartz, J. H., and Jessell, T. M. (Eds.). Principles of neural science (3rd ed.). New York: Elsevier Science Publishing, pp. 367–383.

Martin, J. H., and Jessell, T. M. (1991). Modality coding in the somatic sensory system; and Anatomy of the somatic sensory system. In: Kandell, E. R., Schwartz, J. H., and Jessell, T. M. (Eds.). Principles of neural science (3rd ed.). New York: Elsevier Science Publishing, pp. 341–365.

7

Somatosensation:
Clinical
Applications

I am 69 years old, retired from working for the county, and the mother of three children. Nine years ago I awoke with sciatica, a severe pain extending from the left buttock, down the back of my leg, and into my big toe. I could not bend over to put on shoes or socks. A myelogram, an x-ray study of the spinal region in which dye is injected into the spinal region, showed a herniated intervertebral disk. I developed an excruciating headache secondary to the myelogram, and the scheduled surgery to remove part of the disk was cancelled. After two months of bed rest, I recovered.

One year later I again developed sciatic pain in my left leg that rapidly intensified. I couldn't walk at all because of the pain. I had to crawl. The pain was unbelievable. This time, magnetic resonance imaging revealed that two intervertebral disks had herniated. One month later, surgery was performed and when I awoke the sciatic pain was completely gone. Two years later, I was vacuuming and abruptly developed agonizing pain in my left leg. Surgery again repaired the disk. Since then I have had several deep cortisone shots that effectively relieved the pain.

Throughout this time, I didn't have any lack of sensation, weakness, or other problems. My ability to move was curtailed during the periods when I had sciatica. I could only move in ways that didn't hurt, so I couldn't drive or use stairs. The only time I wasn't in pain was when I was lying down, perfectly still. The pain completely dominated my life.

In physical therapy following the second surgery, I learned two exercises that I do daily. The first exercise is back extension. I lie on the floor on my stomach, my palms on the floor under my shoulders, then slowly push with my arms to raise my head and upper trunk off the floor. I hold this position for 20 seconds, then lie flat again. The other exercise is done lying on my side. If I am lying on my left side, I clasp my hands in front of me, then slowly raise both arms in an arc toward the

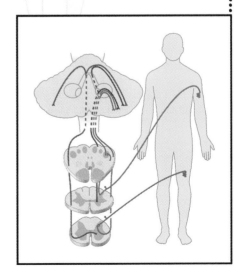

ceiling and then to the floor on my right. When I began this exercise I could only move through half of the arc with my arms, but now I can reach across to the opposite side. This rotation of my spine works very well. I am much more limber now than when I began these exercises. I am free of pain now, except for a dull ache upon awakening that is relieved by the exercises. Also, I am careful not to lift more than ten pounds and I've learned to take breaks when I'm gardening.

—Pauline Schweizer

INTRODUCTION

Somatosensation contributes to smooth, accurate movements, to the prevention or minimization of injury, and to our understanding of the external world. The first two of these topics are considered in this chapter. Perception, or the ability to interpret somatosensation as meaningful information, is discussed in Chapters 16 and 17.

Contribution of Somatosensory Information to Movement

The role of sensation in movement is complex. At the turn of the century, Sherrington performed an experiment on a monkey to determine the effect of loss of information conveyed by the dorsal roots. He cut the dorsal roots entering the spinal cord from one arm, severing the sensory axons. Sherrington found that even after recovery from the surgery, the monkey avoided using that limb. This experimental outcome reinforced the assumption that sensation is essential for movement. This is consistent with observations in humans. People who lack sensation in one upper limb tend to avoid using the limb, substituting with the unimpaired limb whenever possible.

    However, Taub et al. (1973) tested the effects of bilateral deafferentation. His group deafferented both forearms in newborn monkeys and sewed their eyelids closed to eliminate vision. These monkeys were able to ambulate, climb, and pick up some objects, although their performance was very clumsy and motor development was retarded. The monkeys were unable to use their fingers individually, for example, to pick up small objects.

    The functional problems of a person with a severe peripheral sensory loss were reported by Rothwell

et al. (1982). Motor power was nearly normal, the subject could move individual fingers separately, and without vision, he could move his thumb accurately at different speeds and with different levels of force. Yet he could not write, hold a cup, or button his shirt. These difficulties were due to lack of somatosensation, depriving him of normal, automatic corrections to movement. When he tried to hold a pen to write, his grip did not automatically adjust because unconscious information about the appropriate changes in pressure was not available.

> Sensation is not required for effective gross movements, but sensation is required for effective fine movement and makes gross movements more efficient.

Somatosensory Information Protects Against Injury

The deafferented monkeys studied by Taub et al. (1973) self-inflicted serious wounds by biting their own limbs. Therefore the deafferented monkeys were fitted with wire-mesh visors to prevent self-injury. Similarly, people with somatosensory deficits are prone to pressure-induced skin lesions, burns, and joint damage because of their lack of awareness of excessive pressure, temperature, or stretch.

TESTING SOMATOSENSATION

Clinically, the sensory examination covers conscious relay pathways:

- Discriminative touch
- Conscious proprioception
- Fast pain
- Discriminative temperature

These pathways are tested because the findings give information that can be used to localize a lesion.

    The purpose of the sensory examination is to establish if there is sensory impairment and, if so, its location, type of sensation affected, and severity of the deficit. The following guidelines serve to improve the reliability of sensory testing.

1. Administer the tests in a quiet, distraction-free setting.
2. The subject should be seated or lying supported by a firm, stable surface to avoid challenging balance during the testing.

3. Explain the purpose of the testing.
4. Demonstrate each test prior to administering the test. During the demonstration, the subject should be allowed to see the stimulus.
5. During testing, block the subject's vision by having the subject close the eyes or wear a blindfold or by placing a barrier between the part being tested and the subject's eyes.
6. Apply stimuli near the center of the dermatomes being tested. Record results after each test. The time interval between stimuli should be irregular to avoid prediction of stimulation by the subject. Comparing the subject's responses on the left and right sides is often informative, especially if one side of the body or face is neurologically intact.

Quick Screening

Quick screening for sensory impairment consists of testing proprioception and vibration in the fingers and toes and testing fast pain sensation in the limbs, trunk, and face with pinprick (Adams and Victor, 1993). This quick screening evaluates the function of some large-diameter and some small-diameter axons. Because of the possibility of spreading bloodborne diseases, care should be taken during pinprick testing to prevent puncturing the skin, and each pin should be discarded after use on a single person. A paper clip or plastic toothpick may serve as an alternative to the pin and is less likely to puncture the skin. If loss or impairment of sensation is found, additional testing is performed to determine the precise pattern of sensory loss.

Indications for more thorough testing include the following:

- Any complaints of sensory abnormality or loss
- Nonpainful skin lesions
- Localized weakness or atrophy

Complete Sensory Evaluation

A complete sensory evaluation includes measuring **sensitivity**, and **thresholds** for stimulation of each conscious sensation (except that proprioceptive thresholds are not measured). For example, a measure of conscious touch sensitivity is the ability to distinguish between two closely applied points on the skin; threshold is the lowest intensity of a stimulus that can be perceived, as when the person barely perceives being

touched. Figure 7–1 shows normal values for touch sensitivity (two-point discrimination) on various parts of the body. Table 7–1 explains the testing procedures. A sensory assessment form is shown in Figure 7–2.

Interpreting Test Results

The results from these testing procedures can be used to map a person's pattern of sensory loss. The resulting map can be compared with standardized maps of **peripheral nerve distribution** and of dermatome

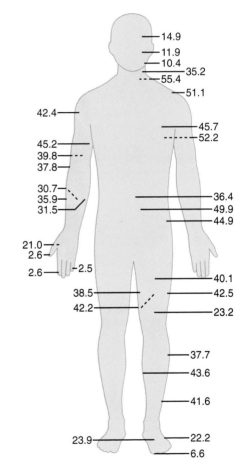

FIGURE 7–1

Normal two-point discrimination values for various locations on the body. (Values from Physical Therapy. Nolan M. F. Limits of two-point discrimination ability in the lower limbs of young adult men and women. 1983;63:1424. Two-point discrimination assessment in the upper limbs in young adult men and women. 1982; 62:965. Quantitative measure of cutaneous sensation: Two-point discrimination values for the face and trunk. 1985;65[2]:181–185. Reprinted with permission of the APTA.)

TABLE 7-1

TESTING THE SOMATOSENSORY SYSTEM

System	Perception	Instructions to Subject	Test Procedure
Discriminative touch	Location of touch	Before testing touch, ask subject to say "yes" when the touch is felt and then to point to or report where the touch was felt.	Touch subject lightly with a wisp of cotton.
	Tactile thresholds*	Ask subject to say "yes" if a stimulus is felt.	Touch a Frey's hair (nylon filaments available in sets of 10; bending pressure ranges from 0.02–40.0 g) to the subject's skin; press so that the filament bends.
	Two-point discrimination	Ask subject to report whether one point or two is felt.	Using calipers, apply light, equal pressure to two points. Begin with the points of the calipers further apart than the mean value for the body part being tested. With each trial, move the points closer together until the subject cannot distinguish two points as separate. Measure the distance between the points on a ruler. To prevent anticipation, randomly stimulate with a single point. Normal sensitivity ranges are indicated in Fig. 7–1.
	Bilateral simultaneous touch	Ask subject to say "yes" if the touch is felt. If subject says "yes," ask where the touch was felt.	Lightly touch both sides of the body simultaneously.
	Directional cutaneous kinesthesia	Ask subject to say "yes" if the movement is felt. If subject says "yes," ask the direction of movement.	Draw a line (with a dull point) on the subject's skin; on the fingertip, normals can tell direction when line >1 cm; on other areas of the body, normals are accurate when line >6 cm (Lindblom and Tegner, 1989); people are more sensitive to moving touch stimuli than to stationary touch.
	Graphesthesia	Ask subject to identify the letter you draw in the palm of the hand.	Using a key or similar object, draw a letter in the palm of subject's hand.

Conscious proprioception	Joint movement	Ask subject to report whether joint is bending or straightening during the movement.	Firmly hold the sides of the joint (usually big toe or a finger), and passively flex or extend the joint.
	Joint position	Tell subject you are going to move a joint. After the movement has stopped, either ask the subject to match the final joint position with the opposite limb or to report the position of the joint.	Passively flex or extend the joint (usually elbow or ankle). Maintain a static position before asking subject to respond.
	Vibration	1. Ask subject to report when vibration stops, or 2. Ask subject to report if tuning fork is vibrating or not.	1. Touch a vibrating tuning fork to a bony prominence, or 2. Randomly apply a vibrating or nonvibrating tuning fork to a bony prominence.
	Vibration perception threshold*	Ask subject to report whether vibration is felt.	Use a biothesiometer to administer controlled vibration.[†]
Discriminative touch and conscious proprioception	Stereognosis	Ask subject to identify an object you place in the hand. Tell subject that manipulation of the object is allowed.	Place an object (key, paper clip) in subject's hand.
	Fast pain	Ask subject to report "sharp" or "dull." If subject reports feeling the stimulus, ask where the stimulus was felt.	Gently poke subject with a pin or touch with blunt end of a pin (or use a toothpick). To map an area of decreased or lost sensation, drag a pin lightly along the skin to determine regions of normal and abnormal sensitivity.
	Pain threshold*	Ask subject to report when pain is first experienced.	Press with a blunt probe with a strain gauge attached.
Discriminative temperature	Heat or cold	Ask subject to report temperature as either hot or cold.	Touch subject with test tubes filled with warm (40°C) and cool (10°C) water. Maintain contact with subject's skin for about 3 seconds before asking for a response.
	Thermal threshold*	Ask subject to report whether the second of a pair of stimuli is warmer or colder than the first.	Use a thermal testing device.[†]

*Tests used in cases where greater accuracy and reproducibility are required (e.g., documenting progressive sensory loss in a diabetic patient) or in research. More information about these tests is available in Munsat (1989).

[†]See LeQuesne et al. (1989).

Diagnosis _____

Limb _____

Sketch the distribution of signs and symptoms.

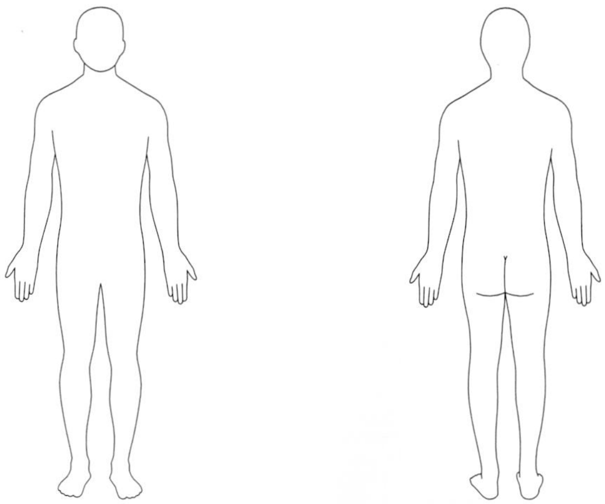

SYMPTOMS	Right	Left
Numbness		
Abnormal sensations (paresthesia or dysesthesia)		
SIGNS		
Light touch		
Two-point discrimination		
Static proprioception		
Kinesthesia		
Vibration		
Pinprick		
Warmth		
Cold		

For symptoms, record the person's report. For signs, record as WNL (within normal limits), I (impaired), or A (absent).

FIGURE 7–2

Sensory assessment form.

distributions to determine if the person's pattern of sensory loss is consistent with a peripheral nerve or a spinal region pattern (see Fig. 6–4). Because every individual is unique and adjacent dermatomes overlap one another, the maps presented represent common but not definitive nerve distributions. The overlap of adjacent dermatomes also ensures that if only one sensory root is severed, there is not complete loss of sensation in any area.

For predicting hand function from sensory tests, only two-point discrimination scores correlate well with hand function. Pressure sensitivity scores using monofilaments do not correlate with hand function (Dellon and Kallman, 1983).

People on ventilators or with communication disorders may present unusual challenges to sensory testing. The therapist may be able to establish a communication system using eye blinks (one for yes, two for no) or finger movements with cooperative people.

ELECTRODIAGNOSTIC STUDIES

Recording the electrical activity from nerves reveals the location of pathology and is often diagnostic. Two methods of examining sensory nerve function are as follows:

• Nerve conduction velocity (NCV) testing
• Somatosensory evoked potentials (SEPs)

Nerve conduction velocity studies only evaluate the function of peripheral nerves. Somatosensory evoked potentials test both peripheral and central nerve transmission. In both NCV and SEP testing, an electrical stimulation is applied to the peripheral nerve so that all axons are depolarized simultaneously. In electrodiagnostic studies, measurements of the latencies, amplitudes, and conduction velocities obtained can be compared with unaffected nerves in the same patient or with published normal values. Some physical therapists specialize in performing NCVs, but therapists typically do not perform SEPs.

Sensory Nerve Conduction Velocity Studies

To test NCV, surface recording electrodes are placed along the course of a peripheral nerve, and then the nerve is electrically stimulated. Nerve conduction velocity tests only quantify the function of the fastest-conducting axons. Because large-diameter axons conduct fastest normally, in intact nerves NCV testing measures only the performance of the large-diameter fibers. Because the velocity of nerve conduction depends on an intact myelin sheath, NCV is slowed throughout a nerve that has been demyelinated. If myelin has only been damaged by a focal injury, conduction is slowed only at the injured segment.

The function of the sensory fibers in the median nerve can be tested by electrically stimulating the skin of the middle finger and recording the electrical activity evoked in the median nerve at the wrist, elbow, and axilla (see Fig. 7–2). To obtain the NCV, the distance between electrodes is divided by the amount of time from the stimulus to the first depolarization at the recording electrode. The amplitude of the depolarization is also measured. Amplitude serves as an indicator of the number of axons conducting. Often the results from two recording sites are compared; for example, the amplitude and latency recorded at the wrist are compared with measurements at the elbow.

To determine if a NCV study is normal, three numerical values are compared:

• Distal latency
• Amplitude of the evoked potential
• Conduction velocity

Distal latency is the time required for the depolarization evoked by the stimulus to reach the distal recording site. The results of an NCV test in a normal nerve and in an abnormally functioning nerve are illustrated in Figure 7–3. Sensory nerve conduction may also be studied by stimulating proximal to the recording site. In this method, the recording electrode is picking up impulses that were propagated in the direction opposite to the normal physiological direction of sensory nerve impulse propagation.

Somatosensory Evoked Potentials

Somatosensory evoked potentials evaluate the function of the pathway from the periphery to the upper spinal cord or to the cerebral cortex. The skin over a peripheral nerve is electrically stimulated, and the resulting electrical activity is recorded from the skin over the upper cervical spinal cord or from the scalp over the primary somatosensory cortex. Again the velocity is determined by dividing the distance between

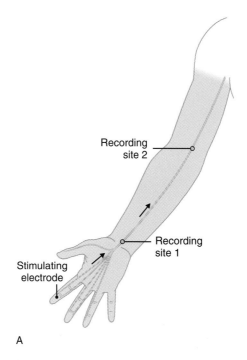

A

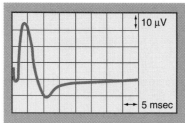

B

	Latency (m/s)	Amplitude (µV)	Distance (cm)	Velocity (m/s)
Normal	2.44	34	12	49.1
Abnormal	6.5	18	13	20

C

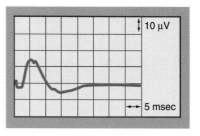

D

FIGURE 7–3

Sensory nerve conduction velocity test: median nerve. *A,* Stimulation of index finger, recording from skin over the median nerve at the wrist and elbow. *B,* At the wrist, recording from a normal nerve. *C,* Numerical results of the NCV tests. *D,* Recording at the wrist from a demyelinated nerve.

the stimulating and recording electrodes by the time required for the action potential to be transmitted. Somatosensory evoked potentials are used to verify subtle signs and locate lesions of the dorsal roots, posterior columns, and brain stem. For example, SEPs may be used in people with multiple sclerosis to determine the location of a lesion.

SENSORY SYNDROMES

Peripheral Nerve Lesions

The general term for dysfunction or pathology of one or more peripheral nerves is **neuropathy.** Peripheral nerves are subject to trauma and disease. Complete severance of a peripheral nerve results in lack of sensation in the distribution of the nerve, pain may occur, and the sensory changes are accompanied by motor and reflex loss. Compression of a nerve affects large myelinated fibers preferentially, with initial relative sparing of the smaller pain, thermal, and autonomic fibers. For example, when one stands up after prolonged sitting with the legs crossed, occasionally one finds that part of a limb has "fallen asleep." The sensory loss proceeds in the following order:

1. Conscious proprioception and discriminative touch
2. Cold
3. Fast pain
4. Heat
5. Slow pain

When the compression is relieved, abnormal sensations called **paresthesias** occur as the blood supply increases. Paresthesias include burning, pricking, and tingling sensations. After compression is removed, sensations return in the reverse order that they were lost. Thus aching pain occurs first, then a sensation of warmth, then sharp, stinging sensations, then cold, and finally a return of discriminative touch and conscious proprioception.

Demyelination of axons in a peripheral nerve often affects proprioception and vibratory sense most severely because the large axons are the most heavily myelinated, resulting in diminished or loss proprioception. Neuropathy is discussed further in Chapter 11.

> Neuropathy is dysfunction or pathology of one or more peripheral nerves.

Spinal Region

Common causes of dysfunction of the spinal region include the following:

- Trauma to the spinal cord, completely or partially severing the cord
- Diseases that compromise the function of specific areas within the spinal cord (These diseases are discussed in Chap. 12.)
- A virus infecting the dorsal root ganglion

COMPLETE TRANSECTION OF THE SPINAL CORD

Complete transection of the cord prevents all sensation one or two levels below the level of the lesion from ascending to higher levels in the cord. Clinically, the observed complete loss of sensation begins in dermatomes one or two levels below the level of the lesion because of the overlap of nerve endings in adjacent dermatomes. Voluntary motor control below the lesion is also lost.

HEMISECTION OF THE SPINAL CORD

A hemisection, that is, damage to the right or left half of the cord, interrupts pain and temperature sensation from the contralateral body because the axons transmitting pain and temperature information cross to the opposite side of the cord soon after entering the cord. As a result of collateral branching of nociceptive axons in the dorsolateral tract (Lissauer's marginal zone), the complete loss of pain sensation occurs two to three dermatomes below the level of the lesion. Because discriminative touch and conscious proprioception information ascends on the same side of the cord as it entered, these sensations are lost ipsilateral to the lesion. Paralysis also occurs ipsilaterally. The pattern of loss is called Brown-Séquard's syndrome.

In posterior columns lesions, conscious proprioception, two-point discrimination, and vibration sense are lost below the level of the lesion. Immediately after the lesion, movements are uncoordinated, that is, ataxic. If the lesion is above C6, the person may be unable to recognize objects by palpation because of loss of ascending sensory information from the hand.

INFECTION

An infection of a dorsal root ganglion or cranial nerve ganglion with varicella-zoster virus causes **varicella zoster,** also called shingles or herpes zoster. The varicella-zoster virus causes chickenpox. After a chickenpox infection, the sensory ganglia hold latent components of the varicella-zoster virus. Occasionally, some of the virus reverts to infectiousness. If the level of circulating antibodies is inadequate, the virus begins to multiply and is transported antidromically down sensory peripheral axons. The virus irritates and inflames the nerve, causing pain. The virus is released into the skin around the sensory nerve endings, causing painful eruptions on the skin. The infec-

tion is usually limited to one dermatome or trigeminal nerve branch (Fig. 7–4). If treated effectively very early, the duration and severity of varicella zoster can usually be limited. However, in severe or inadequately treated cases, **postherpetic neuralgia** develops. Postherpetic neuralgia is severe pain that persists more than 1 month after the zoster infection. The etiology of postherpetic neuralgia is currently not understood. (See the box on varicella zoster.)

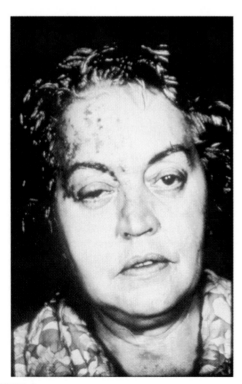

FIGURE 7–4
Varicella zoster (shingles) eruptions on the skin. One branch of the trigeminal nerve is affected.

VARICELLA ZOSTER

Pathology
Infection of sensory root cell bodies, causing inflammation of sensory neurons

Etiology
Varicella-zoster virus

Speed of onset
Acute or subacute

Signs and symptoms

Consciousness
Normal

Communication and memory
Normal

Sensory
Itching, burning, or tingling may precede eruption of vesicles; pain is often severe

Autonomic
Normal

Motor
Normal

Region affected
Usually limited to one dermatome (often thoracic) or one branch of trigeminal nerve

Demographics
Both genders equally affected; incidence increases with aging

Prognosis
Pain usually lasts 1–4 weeks but may persist longer and may progress to postherpetic neuralgia; ultimately, the pain resolves

Brain Stem Region Lesions

Because the axons carrying sensory information from the body and face cross the midline at various levels, lesions in the brain stem usually cause a mix of ipsilateral and contralateral signs. Only in the upper midbrain, after all discriminative sensation tracts have crossed the midline, will sensory loss be entirely contralateral. Throughout the brain stem, a lesion of trigeminal nerve proximal axons or of the trigeminal nerve nuclei cause an ipsilateral loss of sensation from the face.

A lesion in the posterolateral medulla or lower pons can cause a mixed sensory loss, consisting of ipsilateral loss of pain and temperature sensation from the face combined with contralateral loss of pain and temperature information from the body (Fig. 7–5). This occurs because the trigeminal nerve pain information is uncrossed in the medulla and lower pons,

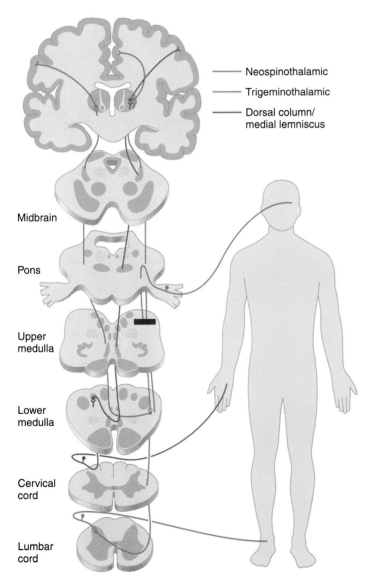

Neospinothalamic

Trigeminothalamic

Dorsal column/
medial lemniscus

Midbrain

Pons

Upper
medulla

Lower
medulla

Cervical
cord

Lumbar
cord

FIGURE 7–5

Mixed sensory loss due to a lesion in the posterolateral medulla. Pain and temperature information are lost from the ipsilateral face and the contralateral body. Discriminative touch and conscious proprioception are not affected by the lesion because the medial lemniscus is medial to the site of the lesion.

while ascending pain information from the body crosses in the spinal cord. Discriminative touch and proprioceptive information from the body is not affected because the tracts conveying this information travel in the medial medulla and pons. Discriminative touch and proprioceptive information from the face is not affected because these tracts and nuclei are superior to the medulla.

A lesion in the medial medulla or lower pons may cause impairment of pain sensation from the contralateral face owing to interruption of some second-order axons conveying information from the trigeminal nerve, combined with the loss of discriminative touch and conscious proprioceptive information from the contralateral body. The contralateral loss occurs because the medial lemniscus axons have crossed the midline in the lower medulla.

A lesion in the posterolateral upper pons or midbrain, after the trigeminothalamic tracts (except proprioceptive) and all of the tracts from the body have crossed the midline, causes contralateral sensory loss from the face (except proprioceptive) and entirely contralateral loss from the body because all tracts have crossed the midline below the lesion.

> Lesions in the brain stem often cause mixed sensory impairments.

Thalamic Lesions

Lesions in the ventral posterolateral (VPL) or ventral posteromedial (VPM) nucleus of the thalamus result in decreased or lost sensation from the contralateral body or face. Rarely, people who have strokes that affect the VPL or VPM nucleus have severe pain in the contralateral body or face. The etiology of such thalamic pain is not currently understood.

Somatosensory Cortex

The sensory effects of a cortical lesion are contralateral and include decrease or loss of discriminative sensations:

- Conscious proprioception
- Two-point discrimination
- Stereognosis
- Localization of touch and fast pain stimuli

Cortical processing is essential for discriminative sensation, although crude awareness of sensation is possible at the thalamic level.

In cases of **sensory extinction** (also called sensory inattention), the loss of sensation is only evident when symmetrical body parts are tested bilaterally. For example, if both hands are touched or pricked simultaneously, the person may only be aware of stimulation on the same side of the body as the cortical lesion. If stimuli are not simultaneous, people with sensory extinction are aware of stimulation on either side of the body. Sensory extinction is a form of unilateral neglect because the person neglects stimuli on one side of the body if the other side of the body is stimulated simultaneously. Unilateral neglect is discussed in Chapter 17.

CLINICAL PERSPECTIVE ON PAIN

How Is Pain Controlled?

What is a typical response to hitting one's thumb with a hammer? A common sequence is to withdraw the thumb, yell (via limbic connections), and then apply pressure to the injured thumb. The first scientific explanation of how pressure and other external stimuli inhibit pain transmission was the **gate theory of pain,** proposed by Melzack and Wall in 1965. They hypothesized that information from first-order low-threshold mechanical afferents and from first-order nociceptive afferents converges onto the same second-order neurons. They proposed that the preponderance of activity in the primary afferents determines the pattern of signals the second-order neuron transmits. Thus, if the low-threshold mechanical afferents are more active than the nociceptive afferents, the mechanoreceptive information is transmitted and the nociceptive information inhibited. According to their theory, transmission of pain information is blocked in the dorsal horn, closing the gate to pain. Lamina II, the substantia gelatinosa, was suggested as the site of interference with pain message transmission.

Although later investigations demonstrated that some details of the original gate theory proposal are incorrect, the gate theory is important because it inspired inquiry into the mechanics and control of pain. One result of these investigations was the clinical application of transcutaneous electrical nerve stimulation (TENS). TENS uses electrical current applied to the skin to interfere with the transmission of pain information.

COUNTERIRRITANT THEORY

A theory that has incorporated findings from research stimulated by the gate theory is the **counterirritant theory.** According to the counterirritant theory, inhibition of nociceptive signals by stimulation of non-nociceptive receptors occurs in the dorsal horn of the spinal cord (Fig. 7–6). For example, pressure stimulates mechanoreceptive afferents. Theoretically, proximal branches of the mechanoreceptive afferents activate interneurons that release the neurotransmitter **enkephalin.** Enkephalin binds with receptor sites on both the primary afferents and interneurons of the pain system. Enkephalin binding depresses the release of substance P and hyperpolarizes the interneurons, thus inhibiting the transmission of nociceptive signals.

ANALGESIC SYSTEMS

Analgesia is an absence of pain in response to stimulation that would normally be painful (Merskey and Bogduk, 1994). The endogenous, or naturally occurring, substances that activate analgesic mechanisms are called **endorphins.** Endorphins include enkephalins, dynorphin, and β-endorphin. Opiates, analgesic drugs that block nociceptive signals without affecting other sensations, bind to the same receptor sites as endorphins. Because opiates bind to the receptor sites, the receptors are sometimes called opiate receptors.

The transmission of nociceptive information can also be inhibited by activity of supraspinal levels of the nervous system (Fig. 7–7). Brain stem areas that provide intrinsic analgesia form a neuronal descending system, arising in the following:

- **Raphe nuclei** in the medulla
- **Periaqueductal gray** (PAG) in the midbrain
- **Locus ceruleus** in the pons

When the raphe nuclei are electrically stimulated, axons projecting to the spinal cord release the neurotransmitter serotonin in the dorsal horn, inhibiting the wide dynamic range tract neurons (both by direct action and via interneurons), thus interfering with

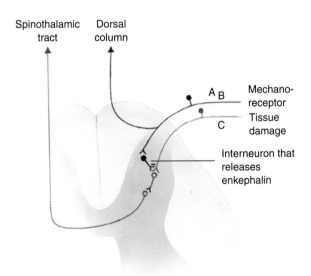

FIGURE 7–6

Circuits in the dorsal horn that may produce inhibition of nociceptive signals. Collaterals of mechanoreceptive afferents stimulate interneurons that release enkephalins. Enkephalin binding inhibits the transmission of pain messages by primary afferents and interneurons in the nociceptive pathway.

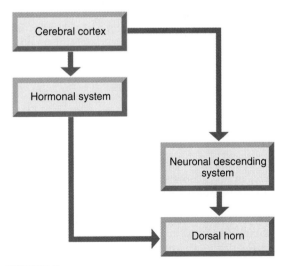

FIGURE 7–7

Flow chart of supraspinal analgesic systems. Cerebral cortical output activates the hormonal and neuronal descending systems that inhibit the transmission of nociceptive information in the dorsal horn.

transmission of nociceptive messages. Stimulation of PAG produces analgesia via activation of the raphe nuclei. The third descending system, the ceruleospinal (from locus ceruleus), inhibits spinothalamic activity in the dorsal horn but is non–opiate mediated; instead, binding of the transmitter norepinephrine on the primary afferent neuron directly suppresses the release of substance P (Lipp, 1991).

Narcotics, drugs derived from opium or opium-like compounds, bind to opiate receptor sites in the PAG, raphe nuclei, and dorsal horn of the spinal cord. By activating the receptor sites, narcotics induce analgesia and stupor (a state of reduced consciousness). If the descending tracts from the raphe nuclei are severed, administration of morphine or other opiates results only in slight analgesia because the lesion of the raphespinal tracts blocks descending inhibition. The slight analgesia that occurs is the result of morphine binding to opiate receptors in the dorsal horn.

The pain-inhibiting centers do not lie dormant waiting for an electrode or a drug to stimulate them. How are they normally activated? People injured in accidents, disasters, or athletic contests sometimes don't feel pain until after the emergency or game is over. Stress during an emergency or competition may trigger the pain inhibition systems. **Stress-induced analgesia** requires activation of the raphe nuclei descending tracts plus release of hormonal endorphins from the pituitary gland (β-endorphins) and adrenal medulla (similar to enkephalins). The hormonal endorphins bind to opiate receptors in the brain and spinal cord. β-Endorphins are the most potent endorphins, and their effects last for hours. Stress-induced analgesia may be triggered by cortical input to the descending pain inhibition systems.

TRANSMISSION OF NOCICEPTIVE INFORMATION

The transmission of nociceptive information can be altered at several locations in the nervous system. The phenomenon of **pain inhibition** is summarized with a five-level model (Fig. 7–8).

- **Level I** occurs in the **periphery**. Non-narcotic analgesics (e.g., aspirin) decrease the synthesis of prostaglandins, preventing prostaglandins from sensitizing pain receptors.
- **Level II** occurs in the **dorsal horn**, via local inhibitory neurons releasing enkephalin or dynor-

phin. This is the level of counterirritant effects; examples are superficial heat and high-frequency, low-intensity TENS. Activity in collateral branches of non-nociceptive afferents decreases or prevents the transmission of pain information to the second-order neuron in the spinal cord.
- **Level III** is the fast-acting **neuronal descending system,** involving the PAG, raphe nuclei, and locus ceruleus.
- **Level IV** is the **hormonal system**, involving the periventricular gray matter (PVG) in the hypothalamus, the pituitary gland, and the adrenal medulla. Direct electrical stimulation of the PVG results in analgesia with a 10-minute latency, and the effect lasts for hours after the stimulation has stopped. Low-frequency TENS may act on this level because its pattern of action has a similar latency and lasting effect.
- **Level V** is the **cortical level.** Here, expectations, excitement, distraction, and placebos all play a role.

Transmission of nociceptive information can be inhibited by binding of endorphins or of analgesic drugs to receptor sites in the dorsal horn, PAG, PVG, and raphe nuclei. Norepinephrine binding to primary afferents in the dorsal horn also inhibits transmission of nociceptive information. In the periphery, signals from nociceptors can be inhibited by non-narcotic analgesics.

Pain transmission can also be intensified at several levels. Edema and endogenous chemicals can sensitize free nerve endings in the periphery. For example, following a minor burn injury, stimuli that would normally be innocuous can cause exquisite pain. Fear and anxiety can also exacerbate pain.

SPECIFIC TYPES OF PAIN

Pain from Injured Muscles and Joints

Both Aδ and C fibers are found in skeletal muscle and joints, so signals interpreted as both fast and slow pain can occur with musculoskeletal injuries. In muscle, the C fibers are highly reactive when muscle tissue is ischemic. In an inflamed joint, nociceptors that

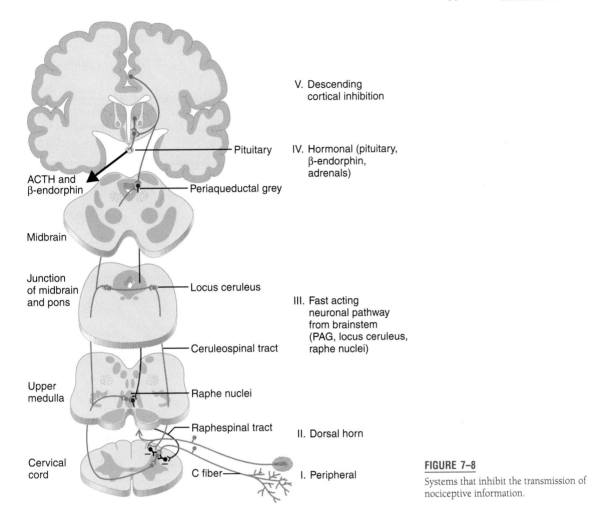

V. Descending
   cortical inhibition

Pituitary

IV. Hormonal (pituitary,
    β-endorphin,
    adrenals)

ACTH and
β-endorphin

Periaqueductal grey

Midbrain

Junction
of midbrain
and pons

Locus ceruleus

III. Fast acting
     neuronal pathway
     from brainstem
     (PAG, locus ceruleus,
     raphe nuclei)

Ceruleospinal tract

Upper
medulla

Raphe nuclei

Raphespinal tract

II. Dorsal horn

Cervical
cord

C fiber

I. Peripheral

FIGURE 7–8

Systems that inhibit the transmission of
nociceptive information.

would normally be silent are sensitized. The sensitized neurons fire in response to normally innocuous stimuli, even with slight movements, and may even fire spontaneously. For example, after an ankle sprain, partial weight bearing may be painful, and the ankle may ache even at rest.

Unlike superficial pain, which encourages withdrawal (movement to escape the source of pain), deep pain usually occurs after tissue has been damaged. The function of deep pain may be to encourage rest of the damaged tissue. After a lower limb injury, the pain on weight bearing often produces a modified gait. The modified gait is called antalgic and is characterized by a shortened stance phase on the affected side.

Referred Pain

Referred pain is perceived as coming from a site distinct from the actual site of origin. Usually pain is referred from visceral tissues to skin. For example, during a heart attack, the brain may misinterpret the nociceptive information as arising from the skin or the medial left arm. Similarly, gallbladder pain is often referred to the right subscapular region.

Referred pain is explained by convergence of nociceptive information from different sources. Referred pain occurs when

- Branches of nociceptive fibers from an internal organ and branches from nociceptive fibers from the skin converge on the same second-order neuron in the spinal cord or in the thalamus.
- A single dorsal root neuron has two peripheral axons, one that innervates skin and one that innervates viscera. Stimulation of the visceral branch of

dual receptive neurons may be the source of misinterpretation (Taylor et al., 1984).

Common patterns of referred pain are illustrated in Figure 7–9.

> Identifying referred pain is important in preventing misdiagnoses and malpractice, so that people with disorders not amenable to physical therapy can be referred to the appropriate practitioner.

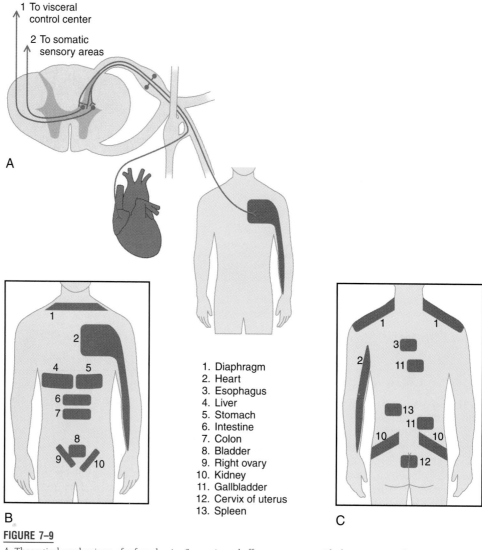

1. Diaphragm
2. Heart
3. Esophagus
4. Liver
5. Stomach
6. Intestine
7. Colon
8. Bladder
9. Right ovary
10. Kidney
11. Gallbladder
12. Cervix of uterus
13. Spleen

FIGURE 7–9

A, Theoretical mechanisms of referred pain. Some visceral afferents synapse with the same second-order neurons as somatosensory afferents. Other neurons have dual axons, innervating both viscera and somatosensory structures. *B* and *C,* Common patterns of referred pain.

Chronic Pain

Therapists must distinguish between acute pain and chronic pain and between pain and disability so appropriate treatment can be administered (see DeRosa and Porterfield [1992] regarding appropriate treatment). Pain is an unpleasant subjective experience, while disability is the lack of ability to perform normal tasks. Characteristics of acute and chronic pain are compared in Table 7–2.

Chronic pain can be classified according to etiology as follows:

- Nociceptive
- Neuropathic
- Chronic pain syndrome

NOCICEPTIVE CHRONIC PAIN

Nociceptive chronic pain is due to stimulation of nociceptive receptors. Examples are chronic pain resulting from tissue damage by cancer or by a vertebral tumor pressing on pain receptors in the meninges surrounding the spinal cord.

NEUROPATHIC CHRONIC PAIN

Neuropathic pain can be due to abnormal activity in different locations in the nervous system:

- The periphery (e.g., nerve compression in carpal tunnel syndrome)
- The central nervous system (deafferentation or phantom limb pain)
- The sympathetic nervous system (sympathetically maintained pain)

Studies of rats demonstrate that a tendency to develop neuropathic pain may be genetic (Devor and Raber, 1990).

Neuropathic Chronic Pain in the Periphery Injury or disease of peripheral nerves often results in sensory abnormalities. A complete nerve section results in lack of sensation from that nerve's receptive field, but sometimes **paresthesia** (abnormal sensation) and pain also occur in the denervated region. Partial damage to a nerve can result in **allodynia** (pain in response to normally nonpainful stimuli) and sensations like electric shock.

These unusual sensations are the result of aberrant activity in the peripheral nervous system, evoking abnormal responses in the central nervous system. Peripheral abnormalities causing neuropathic pain include the development of ectopic foci and ephaptic transmission in an injured nerve. **Ectopic foci** are sites that are abnormally sensitive to mechanical stimulation. The foci can occur at the nerve stump, in areas of myelin damage, or in the dorsal root ganglion cells. Ectopic foci occur because the proteins normally synthesized by the cell body and transported to the distal peripheral axon are obstructed by the lesion, resulting in abnormal excitability of the sites.

TABLE 7–2		
CHARACTERISTICS OF ACUTE AND CHRONIC PAIN		
	Acute Pain	**Chronic Pain**
Causes	Threat of or actual tissue damage	Continuing tissue damage Environmental factors (operant conditioning) Sensitization of nociceptive pathway neurons Dysfunction of endogenous pain control systems
Client report	Clear description of location, pattern, quality, frequency, and duration	Vague description
Function	Warning of tissue damage, enforce rest of healing tissue	If tissue damage is not continuing, no biological benefit; may have social or psychological benefit
Consequences	Excessive autonomic activity Excessive neuroendocrine activation If not adequately treated, can be as harmful as disease (Liebeskind, 1991) and may progress to chronic pain	Severe financial, emotional, physical, and/or social stresses on the person and family Physiological consequences of inactivity

These sites can become so sensitive to mechanical stimulation that tapping on an injured nerve can elicit pain or tingling (**Tinel's sign**). The sensitivity of these sites to circulating catecholamines may explain sympathetically maintained pain (see Chap. 8). The damage to myelin may also result in **ephaptic transmission** (cross-excitation) among neurons. As a result of the lack of insulation between neurons, excitation of one neuron may induce activity in another neuron.

Neuropathic Chronic Pain due to Abnormal Central Nervous System Activity When peripheral sensory information is completely absent, as occurs in people with deafferentation or amputation, neurons in the central nervous system that formerly received information from the body part may become abnormally active. Avulsion of dorsal roots from the spinal cord, as sometimes occurs in motorcycle accidents, causes people to feel burning pain in the area of sensory loss.

Almost all people with amputations report sensations that seem to originate from the missing limb, called phantom limb sensation. Much less frequently, people with amputations report that their phantom sensation is painful. This condition is called phantom limb pain. Phantom limb pain must be differentiated from stump pain because some causes of stump pain can be successfully treated. Stump pain is caused peripherally by neuropathy, neuroma, a poorly fitting prosthesis, or nerve compression.

Neuropathic Chronic Pain due to Abnormal Activity in the Sympathetic Nervous System Sympathetically maintained pain begins after an injury or disease. Sometimes it arises weeks or months after an injury has healed. As pain increases in intensity, trophic changes occur in the skin, fingernails, toenails, and bone. See Chapter 8 for a discussion of the mechanism.

> Neuropathic pain results from changes in neuronal activity secondary to injury.

SURGICAL TREATMENT OF CHRONIC PAIN

Theoretically, cutting selected dorsal roots (**dorsal rhizotomy**) or the spinothalamic tracts should eliminate pain sensation in people with pain that is resistant to other treatments. In practice, however, dorsal rhizotomy often fails to alleviate pain. The persistence of pain following spinothalamic tractotomy may be due to central nervous system changes in response to the original maintained pain or to pain-mediating fibers traveling in the dorsal columns (see Chap. 6).

CHRONIC PAIN SYNDROME

In many orthopedic physical therapy clinics, the majority of clients are evaluated and treated for low back pain (DeRosa and Porterfield, 1992). DeRosa and Porterfield (1992) define chronic low back pain syndrome as a lack of a direct relationship between mechanical stresses on the low back and the pain response. The transition from acute low back pain after injury to chronic low back pain has been characterized by Waddell et al. (1993) as a change in pain etiology from tissue damage to a physiological impairment consisting of the following:

- Muscle guarding
- Abnormal movement
- Disuse syndrome

Figure 7–10 emphasizes the elements contributing to acute and chronic low back pain. In discussing chronic low back pain, Waddell (1987) cautions that "physical treatment directed to a supposed but unidentified and possibly nonexistent nociceptive source is not only understandably unsuccessful but failed treatment may both reinforce and aggravate pain, distress, disability, and illness behavior." For example, when a person complains of low back pain and magnetic resonance imaging (MRI) shows a bulging intervertebral disk, treatment may be directed toward the disk. However, Jensen and Brant-Zawadzki (1994) found that 64% of people *without* low back pain had abnormal findings on MRI of the lower spine, leading to the conclusion that disk bulges or protrusions may be coincidental rather than causative of low back pain. The fact that active exercise programs have consistently been shown to be the most effective treatment for chronic low back pain syndrome (DeRosa and Porterfield, 1992) supports Waddell's concept of a physiological impairment etiology. (See the box on chronic low back pain syndrome on p. 126.)

Chronic pain that does not fit the nociceptive/neuropathic model is exceedingly difficult to investigate because good animal models have not been achieved and obvious ethical issues are involved in trying to induce chronic pain in humans. Hypothesized mecha-

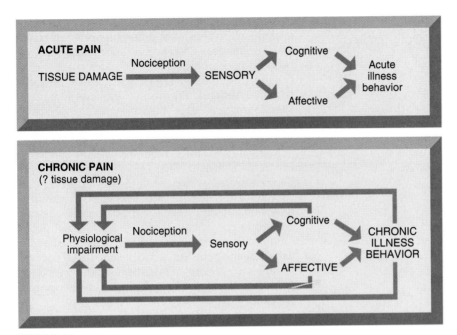

FIGURE 7–10

The differences between factors in acute and chronic low back pain. (Redrawn with permission from Waddell et al. [1993]. A fear–avoidance beliefs questionnaire [FABQ] and the role of fear–avoidance beliefs in chronic low back pain and disability. Pain. 52[2]:157–168.)

nisms are abnormally low levels of endorphins or the presence of reverberating circuits, in which neurons restimulate themselves. An additional possible central source of chronic pain syndrome is suggested by Yamashiro et al. (1991). Yamashiro's group found sites of epileptic activity in the thalamic nuclei during surgery on people with chronic pain.

When pain persists, depression, sleep disturbance, preoccupation with pain, decreased activity, and fatigue are all common.

Fibromyalgia Another example of a chronic pain syndrome is fibromyalgia. The person with **fibromyalgia** has tenderness of muscles and adjacent soft tissues, stiffness of muscles, and aching pain. Mental or physical stress, trauma, and sleep disorders may contribute to the disorder. The painful area shows a regional rather than dermatomal or peripheral nerve distribution. Multiple reproducible tender points are found on palpation. Education and exercise have been demonstrated to benefit people with fibromyalgia (Burckhardt et al., 1994).

Myofascial Pain Although **myofascial pain** is also reported to involve tender points, myofascial pain is a controversial diagnosis without good empirical support (Wolfe et al., 1992; Nice et al., 1992). According to Travell and Simons (1983), pressure on sensitive points (called trigger points) reproduces the person's pattern of referred pain, and diagnosis is confirmed by inactivating the trigger point, either by stretch or by injecting local anesthetic into the trigger points.

> Chronic pain syndromes are physiological impairments consisting of muscle guarding, abnormal movement, and disuse syndrome.

PSYCHOLOGICAL EXPLANATIONS OF CHRONIC PAIN SYNDROME

Sometimes chronic pain syndrome is attributed to psychological problems. However, research evidence for psychogenic etiology is lacking. In fact, the major-

CHRONIC LOW BACK PAIN SYNDROME

Pathology
Unknown

Etiology
Deconditioning

Speed of onset
Chronic

Signs and symptoms

   Consciousness
   Normal

   Communication and memory
   Normal

   Sensory
   Aching pain

   Autonomic
   Normal

   Motor
   Muscle guarding, disuse, abnormal movement patterns

Region affected
May be central nervous system

Demographics
Unknown

Prognosis
Variable

ity of studies on personality characteristics have failed to find any psychological differences between people with chronic pain who have an organic diagnosis and people with chronic pain who have a psychological diagnosis (see Gamsa [1994] for a critical review of relevant research). An organic diagnosis indicates a specific anatomical or physiological change in the body, in contrast to a psychological diagnosis, where no organic basis for the problem has been identified.

Naliboff et al. (1982) performed a study comparing people with organic versus psychological diagnoses. They compared personality attributes of people with chronic pain to people with chronic nonpainful illnesses, including diabetes and hypertension. Abnormal scores on psychological testing correlated significantly with the person's self-reported limitations in daily activities; the authors concluded that the constraints on activities caused the aberrant psychological scores rather than psychological factors causing the pain.

Although a small subgroup of people with chronic pain may have psychological problems as a primary factor in their pain, this is currently highly speculative. A psychological etiology cannot be generalized to all people with chronic pain in whom a nociceptive or neuropathic cause cannot be identified.

> Available evidence does not support a psychological etiology of chronic pain.

CLINICAL NOTES

CASE 1

A 25-year-old man sustained an incomplete spinal cord injury in an industrial accident. His left leg is paralyzed. With his eyes closed, the following sensory deficits are noted:

- Left lower extremity: cannot report direction of passive joint movement of the hip, knee, ankle, great toe

- Left side, below L2 level: cannot distinguish between two closely spaced points applied to the skin nor detect vibration

- Right lower extremity below L4 level: cannot distinguish between test tubes filled with warm or cold water nor distinguish between sharp and dull stimuli

- Reduced sensation in left L2 dermatome

(continued)

Intact sensation:

- All sensations intact above L2 level

- Left lower extremity: can distinguish between test tubes filled with warm or cold water and can distinguish between sharp and dull stimuli

- Right lower extremity: can distinguish between two closely spaced points applied to the skin, can accurately report direction of passive movements of the joints, and can detect vibration

Question

Explain the pattern of sensory loss seen in this person.

CASE 2

A 42-year-old chef is unable to work because of weakness and altered sensation in her dominant right hand. She reports difficulty lifting heavy skillets, which began 2 months ago. Tingling in the thumb, index and middle fingers, and half of the ring finger began 6 weeks ago. Currently, she is unable to stir anything for more than 5 minutes using her right hand. The thenar muscles are visibly atrophied. The results of motor and sensory testing of the right upper limb are as follows:

- Pinch grip on the right is 30 g (versus 120 g with left hand).

- Diminished sensation in the lateral three and a half digits.

- Strength, reflexes, and sensation are within normal limits throughout the remainder of the right upper extremity.

Question

What is the location and probable etiology of the lesion?

CASE 3

A 73-year-old man was found unconscious on the floor at home 1 week ago. He is now lucid and cooperative. Sensation and movement are within normal limits on the right side of the body and face. He is unresponsive to touch, pinprick, pressure, and passive joint motion on the left side. The lower half of the left side of his face appears to droop, and he cannot actively move the left limbs. When assisted to sitting, he collapses as soon as support is removed.

Question

What is the location and probable etiology of the lesion?

CASE 4

A 24-year-old woman was stopped at a traffic signal when her car was struck from behind. She noted some stiffness of her neck the following day; within a week, her upper back was also stiff and becoming painful. X-rays were normal. During the next 2 weeks, she was free of pain. Now, 5 weeks after the accident, she is seeking treatment because the neck aching has returned, turning her head while driving is painful, and she has dull headaches. She denies having reinjured her neck but reports stressful deadlines at work. She has no motor or sensory loss. Active neck range of motion is as follows:

- 20° forward flexion, 35° extension (30% and 70% of normal values)

(continued)

- Lateral flexion: 15° to left, 24° to right (38% and 60% of normal)
- Rotation: 20° to left, 33° to right (36% and 60% of normal)

Question

What is the probable etiology of the problem?

REVIEW QUESTIONS

1. How is a quick screening for somatosensory impairments performed?

2. What signs and/or symptoms indicate that a thorough somatosensory evaluation should be performed?

3. What precautions are essential to ensure valid results of somatosensory testing?

4. How is a sensory nerve conduction velocity study performed?

5. Which types of somatosensory information are usually most impaired by demyelination? Why?

6. What sensory and motor losses occur with a left hemisection of the spinal cord? Name the resulting syndrome.

7. What is varicella zoster?

8. Describe the sensory loss associated with a lesion in the left posterolateral lower pons.

9. What is sensory extinction?

10. How does the counterirritant theory explain the inhibition of pain messages by the application of pressure to an injured finger?

11. List the origins of the three supraspinal analgesic systems.

12. How do narcotics induce their effects?

13. Name the levels of the pain inhibition model.

14. Define *referred pain*.

15. What is the difference between nociceptive chronic pain and neuropathic chronic pain?

16. List four examples of neuropathic chronic pain.

17. Define *paresthesia* and *dysesthesia*.

18. What are ectopic foci?

19. What is phantom limb pain?

20. Often, people with chronic low back pain have no identifiable tissue damage. Waddell et al. (1993) propose that physiological changes following acute low back injury may cause the chronic pain. What are the changes Waddell et al. cite as responsible for the continued pain? What evidence supports their contentions?

21. Do most chronic pain syndromes have a psychological etiology?

References

Adams, R. D., and Victor, M. (1993). Other somatic sensation. In: Principles of neurology (5th ed.). New York: McGraw-Hill, pp. 130–147.

Burckhardt, C. S., Mannerkorpi, K., Hedenberg, L., and Bjelle, A. (1994). A randomized, controlled clinical trial of education and physical training for women with fibromyalgia. J. Rheumatol. 21:714–720.

Dellon, A. L., and Kallman C. H. (1983). Evaluation of functional sensation in the hand. J. Hand Surg. [Am.] 8(6):865–870.

DeRosa, C. P., and Porterfield, J. A. (1992). A physical therapy model for the treatment of low back pain. Phys. Ther. 72: 261–272.

Devor, M., and Raber, P. (1990). Heritability of symptoms in an experimental model of neuropathic pain. Pain 42:51–67.

Gamsa, A. (1994). The role of psychological factors in chronic pain: Part I. A half century of study; Part II. A critical appraisal. Pain 57:5–29.

Jensen, M. C., and Brant-Zawadzki, M. N. (1994). Magnetic resonance imaging of the lumbar spine in people without back pain. N. Engl. J. Med. 331(2):69–73.

LeQuesne, P. M., Fowler, C. J., and Parkhouse, N. (1989). Quantitative tests to determine peripheral nerve fiber integrity. In: Munsat, T. L. Quantification of neurologic deficit. Boston: Butterworth Publishers.

Liebeskind, J. C. (1991). Pain can kill. Pain 44:3–4.

Lindblom, U., and Tegner, R. (1989). Quantification of sensibility in mononeuropathy, polyneuropathy, and central lesions. In: Munsat, T. L. Quantification of neurologic deficit. Boston: Butterworth Publishers.

Lipp, J. (1991). Possible mechanisms of morphine analgesia. Clin. Neuropharmacol. 14(2):131–147.

Melzack, R., and Wall, P. D. (1965). Pain mechanisms: A new theory. Science 150:971–979.

Merskey, H., and Bogduk, N. (1994). Classification of chronic pain, 2nd ed. IASP Task Force on Taxonomy, IASP Press, Seattle, WA. pp. 209–214.

Munsat, T. L. (1989). Quantification of neurologic deficit. Boston: Butterworth Publishers.

Naliboff, B. D., Cohan, M. J., and Yellen, A. N. (1982). Does the MMPI differentiate chronic illness from chronic pain? Pain 13:333–341.

Nice, D. A., Riddle, D. L., Lamb, R. L., et al. (1992). Intertester reliability of judgments of the presence of trigger points in patients with low back pain. Arch. Phys. Med. Rehabil. 73:893–898.

Nolan, M. F. (1996). Introduction to the neurologic examination. Philadelphia: F. A. Davis.

Rothwell, J. C., Traub, M. M., et al. (1982). Manual motor performance in a deafferented man. Brain 105:515–542.

Taub, E., Perrella, P., and Barro, G. (1973). Behavioral development after forelimb deafferentation on day of birth in monkeys with and without blinding. Science 181(103):959–960.

Taylor, D. C. M., Pierau, F.-K., and Mizutani, M. (1984). Possible bases for referred pain. In: The neurobiology of pain. Manchester, England: Manchester University Press.

Travell, J. G., and Simons, D. G. (1983). Myofascial pain and dysfunction: The trigger point manual. Baltimore: Williams & Wilkins.

Waddell. (1987). A new clinical model for the treatment of low-back pain. Spine 12:632–644.

Waddell, G., Newton, M., Henderson, I., et al. (1993). A fear-avoidance beliefs questionnaire (FABQ) and the role of fear-avoidance beliefs in chronic low back pain and disability. Pain 52(2):157–168.

Wolfe, F., Simons, D. G., et al. (1992). The fibromyalgia and myofascial pain syndromes: A study of tender points and trigger points in persons with fibromyalgia, myofascial pain syndrome, and no disease. J. Rheumatol. 19(6):944–951.

Yamashiro, K., Iwayama, K., et al. (1991). Neurones with epileptiform discharge in the central nervous system and chronic pain: Experimental and clinical investigations. Acta Neurochir. Suppl. (Wien) 52:130–132.

Suggested Readings

Basbaum, A. L. (1984). Endogenous pain control systems: Brainstem spinal pathways and endorphin circuitry. Annu. Rev. Neurosci. 7:309–338.

Bonica, J. J. (1990). The management of pain. Philadelphia: Lea & Febiger, pp. 18–126.

Fields, H. L. (1987). Pain. New York: McGraw-Hill.

8

Autonomic Nervous System

The autonomic nervous system is critical for the survival of the individual and the species because it regulates homeostasis and reproduction. Homeostasis is the maintenance of an optimal internal environment, including body temperature and chemical composition of tissues and fluids. The autonomic nervous system regulates the activity of internal organs and vasculature. Thus the autonomic system regulates circulation, respiration, digestion, metabolism, secretions, body temperature, and reproduction. The aspects of the autonomic nervous system considered in this chapter include sensory receptors, afferent pathways, central regulation, and efferent pathways to the effectors (Fig. 8–1). The autonomic efferent pathways are the sympathetic and parasympathetic divisions of the nervous system.

> The autonomic system regulates the viscera and vasculature.

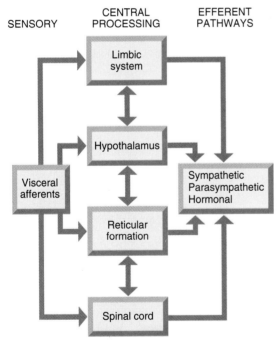

FIGURE 8–1

Flow of information in the autonomic nervous system.

SENSORY RECEPTORS

Receptors of the autonomic system include mechanoreceptors, chemoreceptors, nociceptors, and thermoreceptors. The **mechanoreceptors** respond to pressure and to stretch. Pressure receptors are found in the atria, ventricles, carotid arteries, and lungs. Stretch receptors respond to distention of the veins, bladder, or intestines.

Chemoreceptors sensitive to chemical concentrations in the blood are located in the carotid and aortic bodies (respond to oxygen), medulla (respond to hydrogen ions and carbon dioxide), and hypothalamus (respond to blood glucose levels and to concentration of electrolytes). Chemoreceptors in the stomach, taste buds, and olfactory bulbs also respond to chemical concentrations.

Nociceptors, found throughout the viscera and in the walls of arteries, are typically most responsive to stretch and ischemia. Visceral nociceptors are also sensitive to irritating chemicals. **Thermoreceptors** in the hypothalamus respond to very small changes in the temperature of circulating blood, and cutaneous thermoreceptors respond to external temperature changes.

AFFERENT PATHWAYS

The information from visceral receptors enters the central nervous system by two routes: into the spinal cord via the dorsal roots and into the brain stem via cranial nerves (Fig. 8–2). Cranial nerves conveying autonomic afferent information include the facial (VII), glossopharyngeal (IX), and vagus nerves (X). All three of these cranial nerves transmit taste information, and the glossopharyngeal and vagus transmit information from the viscera.

CENTRAL REGULATION OF VISCERAL FUNCTION

Most visceral information entering the brain stem via cranial nerves converges in the **solitary nucleus,** the main visceral sensory nucleus (Fig. 8–3). In turn, information from the solitary nucleus is relayed to vis-

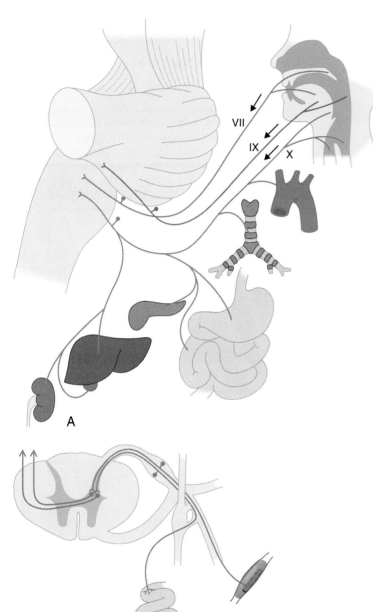

A

B

FIGURE 8–2

Afferent autonomic pathways into the brain stem and spinal cord. *A,* Visceral afferent information from the tongue and soft palate enters the brain stem through cranial nerves VII and IX. Information from the larynx and thoracic and abdominal viscera reaches the brain stem via cranial nerve X. *B,* Stretch of blood vessels in the periphery is registered by free nerve endings in the vessel walls. This information is conveyed via fibers in peripheral nerves into the spinal cord. Information from stretch receptors in the gastrointestinal tract passes through an autonomic ganglion, without synapsing, before entering the spinal cord.

ceral control areas in the pons and medulla and to modulatory areas in the hypothalamus, thalamus, and limbic system. Modulatory areas regulate the activity of areas that directly control a particular function. For example, the limbic system does not directly control respiratory rate but instead influences the activity of respiratory control areas in the pons and medulla.

Visceral afferents entering the spinal cord synapse with visceral efferents (autonomic reflexes; see Chap. 12) and with neurons that ascend to regions of the brain stem, hypothalamus, and thalamus. Visceral no-

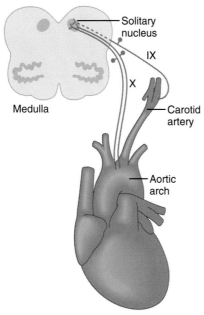

FIGURE 8–3

Visceral information converges in the solitary nucleus of the brain stem. An example is the convergence of blood pressure and blood chemical composition information, monitored by pressure and chemoreceptors in the carotid artery and the aortic arch. The information is transmitted to the solitary nucleus in the medulla.

ciceptive afferents have additional connections with the following:

- Somatosensory nociceptive afferents, contributing to referred pain (see Chap. 7)
- Somatic efferents, to produce muscle guarding (protective contraction of skeletal muscles)

Figure 8–4 illustrates the activity in these pathways during acute appendicitis.

> Afferent autonomic information is processed in the solitary nucleus, spinal cord, and areas of the brain stem, hypothalamus, and thalamus.

Control of Autonomic Functions by the Medulla and Pons

Areas within the medulla regulate heart rate, respiration, vasoconstriction, and vasodilation via signals to autonomic efferent neurons in the spinal cord and by signals conveyed in the vagus nerve. Areas in the pons are also involved in regulating respiration.

Role of the Hypothalamus, Thalamus, and Limbic System in Autonomic Regulation

The hypothalamus, thalamus, and limbic system modulate brain stem autonomic control. Visceral information reaching the hypothalamus, the master controller of homeostasis, is used to maintain equilibrium in the interior of the body. The hypothalamus influences cardiorespiratory, metabolic, water reabsorption, and digestive activity by acting on the pituitary gland, control centers in the brain stem, and spinal cord.

Visceral information reaching the thalamus is projected mainly to the limbic system, a collection of cerebral areas involved in emotions, moods, and motivation. Activation of limbic areas can produce autonomic responses; examples include increased heart rate due to anxiety, blushing with embarrassment, and crying.

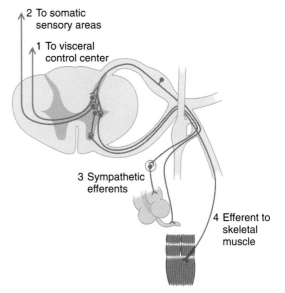

FIGURE 8–4

Pathways of afferent autonomic information in the spinal cord. The early stage of acute appendicis is shown: the nociceptive fibers enter the T10 signal segment. Connections with

1. Autonomic tract fibers convey the information to areas in the brain stem, hypothalamus, and limbic system
2. Somatic afferents result in pain sensation referred to the umbilical region
3. Sympathetic efferents inhibit peristalsis in the intestine
4. Somatic efferents elicit contraction of abdominal muscles

Vital functions are controlled by areas in the medulla and pons. The hypothalamus, thalamus, and limbic system modulate the brain stem control.

Integration of Information

Autonomic regulation is often achieved by integrating information from peripheral afferents with information from receptors within the central nervous system. For example, if peripheral chemoreceptors in the carotid body signal a drop in oxygen content in the blood, the information is conveyed to the solitary nucleus (in the medulla) by the glossopharyngeal nerve. Then signals are sent to autonomic control areas to increase the depth and rate of respiration. If a specific group of neurons in the medulla, directly sensitive to the concentration of carbon dioxide and hydrogen ions (pH) in the blood, sense deviations from the optimum physiological range, respiration is adjusted.

EFFERENT PATHWAYS

Autonomic efferent neurons are classified as sympathetic and parasympathetic. In general, the connections from the central nervous system to autonomic effectors use a two-neuron pathway, with the two neurons synapsing in a peripheral ganglion. The neuron extending from the central nervous system to the ganglion is called **preganglionic**; the neuron connecting the ganglion with the effector organ is called **postganglionic**.

Differences Between the Somatic Motor System and Autonomic Efferent System

All of the central nervous system output is delivered by somatic or autonomic efferent neurons. Somatic efferents innervate only skeletal muscle, and their activation is frequently voluntary. Autonomic efferents supply all other parts of the body that are innervated. The autonomic system is different from the somatic nervous system in three major ways:

1. Unlike the somatic nervous system, regulation of autonomic functions is typically unconscious and can be exerted by hormones.

2. Unlike skeletal muscle, many internal organs can function independently of nervous system input. Examples include independent activity of the heart and the gastrointestinal tract. The heart can continue to beat without neural connections. The gastrointestinal tract is unique in having an intrinsic nervous system, the enteric nervous system, so capable of operating independently of the central nervous system that the system has been called the visceral brain. This system of ganglia and sensory and motor neurons is located entirely within the walls of the digestive system. Because its function is purely digestive, further discussion of the enteric nervous system is beyond the scope of this text.

3. Somatic efferent pathways use one neuron; autonomic efferent pathways usually use two neurons, with a synapse outside the central nervous system.

Neurotransmitters Used by the Autonomic Efferent System

Autonomic neurons secrete the neurotransmitter acetylcholine, norepinephrine, or epinephrine. Neurons that secrete acetylcholine are called **cholinergic**. Neurons secreting norepinephrine or epinephrine are called **adrenergic**.

CHOLINERGIC NEURONS AND RECEPTORS

Autonomic neurons that secrete acetylcholine include the following (Fig. 8–5):

- All of the preganglionic neurons in the autonomic nervous system
- Postganglionic neurons of the parasympathetic system
- Sympathetic postganglionic neurons that innervate sweat glands and some sympathetic postganglionic neurons that innervate vessels in skeletal muscle

The effect of a neurotransmitter depends on the type of receptors activated by the transmitter. This is of particular importance in the autonomic nervous system, where differences in types of receptors are the key to distinct physiological effects of different drugs. Based on their ability to bind certain drugs, two groups of **cholinergic receptors** have been

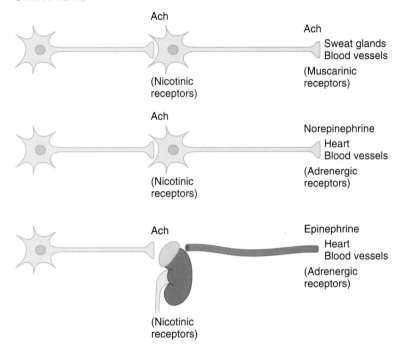

SYMPATHETIC

PARASYMPATHETIC

FIGURE 8–5

Neurotransmitters secreted by autonomic neurons. The transmitter is identified above the synapse, and the type of receptor is listed below the synapse. Ach, acetylcholine.

identified: nicotinic and muscarinic. Nicotine, derived from tobacco, activates only the **nicotinic receptors,** found on postsynaptic neurons in autonomic ganglia. Acetylcholine binding to nicotinic receptors causes a fast excitatory postsynaptic potential (EPSP) in the postsynaptic membrane. Muscarine, a poison derived from mushrooms, activates only the **muscarinic receptors** in the membranes of effectors. Acetylcholine binding to muscarinic receptors initiates a G-protein mediated response, which can be either an EPSP or inhibitory postsynaptic potential (IPSP).

ADRENERGIC NEURONS AND RECEPTORS

The transmitter released by most sympathetic postganglionic neurons is norepinephrine. The adrenal medulla, a part of the sympathetic system, is specialized to release epinephrine and norepinephrine directly into the blood. Receptors that bind norepinephrine or epinephrine are called adrenergic receptors. There are two groups of **adrenergic receptors,** designated α and β; each of these has subtypes, indicated by subscripts: α_1, α_2, β_1, and β_2.

Cholinergic neurons secrete acetylcholine. Cholinergic receptors are nicotinic or muscarinic. Adrenergic neurons secrete norepinephrine. Adrenergic receptors are α or β.

SYMPATHETIC NERVOUS SYSTEM

Sympathetic Efferent Neurons

Cell bodies of the sympathetic preganglionic neurons are in the lateral horn of the spinal cord gray matter (Fig. 8–6). Because the cell bodies are located from the T1 to L2 levels, the sympathetic nervous system is often called the **thoracolumbar outflow.** Sympathetic efferent neurons innervate the adrenal medulla, vasculature, sweat glands, erectors of hair cells, and the viscera.

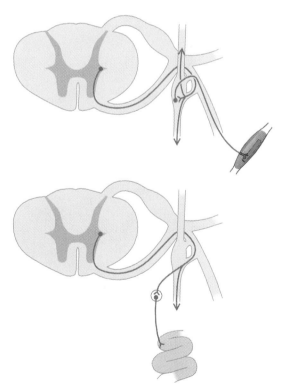

FIGURE 8–6
Sympathetic outflow innervating arterioles in skeletal muscle and in the walls of viscera.

SYMPATHETIC EFFERENTS TO THE ADRENAL MEDULLA

Direct connections are provided from the spinal cord to the adrenal medulla (Fig. 8–7A). The adrenal medulla can be considered a specialized sympathetic ganglion that secretes epinephrine and norepinephrine into the bloodstream.

SYMPATHETIC EFFERENTS TO THE PERIPHERY AND THORACIC VISCERA

Sympathetic efferents to the limbs, face, body wall, heart, and lungs synapse in ganglia alongside the vertebral column, called paravertebral ganglia (Fig. 8–7B). The paravertebral ganglia are interconnected, forming sympathetic trunks. Preganglionic sympathetic axons leave the spinal cord through the ventral root, join the spinal nerve, then travel in a very short connecting branch to the paravertebral ganglia. The connecting branch, called the white ramus communicans, is composed of sympathetic axons transferring from the spinal nerve to the paravertebral ganglion. The preganglionic axons either synapse in the paravertebral ganglion or travel up or down the sympathetic chain before synapsing in a ganglion.

The cell body of the postganglionic neuron is in the paravertebral ganglion. The postganglionic axon enters a peripheral nerve via a connecting branch, the gray ramus communicans, then travels in either the ventral or dorsal ramus to the periphery.

Given that the head, except for the face, and most of the upper limbs are innervated by cervical spinal cord segments, and preganglionic sympathetic fibers arise only from thoracolumbar segments, how do sympathetic signals reach the head and upper limbs? Cervical paravertebral ganglia are supplied by preganglionic fibers that ascend from the upper thoracic cord. The cervical ganglia are named superior, middle, and cervicothoracic (Fig. 8–7C). The cervicothoracic ganglion, often called the stellate ganglion because of its star shape, is formed by the fusion of the inferior cervical and first thoracic ganglion. Postganglionic fibers from the superior and stellate ganglia innervate arteries of the face, dilate the pupil of the eye, and assist in elevating the upper eyelid. Other fibers from the cervicothoracic ganglion descend with fibers from the middle cervical ganglion to supply the heart and lungs.

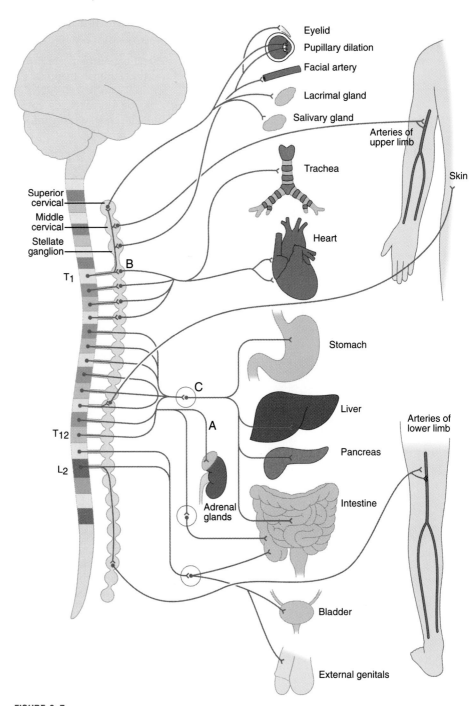

FIGURE 8–7

Efferents from the spinal cord to sympathetic effector organs: *A,* Direct, one-neuron connections to adrenal medulla. *B,* Two-neuron pathways to the periphery and thoracic viscera, with synapses in paravertebral ganglia. *C,* Two-neuron pathways to the abdominal and pelvic organs, with synapses in outlying ganglia.

The lower lumbar and sacral paravertebral ganglia are supplied by preganglionic fibers that descend from the upper lumbar cord. Postganglionic neurons from the lower lumbar and parasacral paravertebral ganglia innervate effectors in the lower limbs.

SYMPATHETIC EFFERENTS TO ABDOMINAL AND PELVIC ORGANS

The preganglionic sympathetic axons to abdominal and pelvic organs pass through the sympathetic ganglia without synapsing, then synapse in outlying ganglia near the organs (see Fig. 8–7C). The preganglionic neurons travel in splanchnic nerves, which are peripheral nerves that innervate the viscera. Sympathetic signals to the gastrointestinal tract slow or stop peristalsis, reduce glandular secretions, and constrict sphincters within the digestive system.

> Sympathetic efferents to the periphery and thoracic viscera synapse in the paravertebral sympathetic ganglia. Sympathetic efferents to abdominal and pelvic organs synapse in outlying ganglia, not in the paravertebral sympathetic ganglia.

Functions of the Sympathetic System

The primary role of the sympathetic nervous system is to maintain optimal blood supply in the organs. Normally, moderate activity of the sympathetic system stimulates smooth muscle in the walls of blood vessels, maintaining some contraction of the vessel walls. Generally, increasing sympathetic activity further constricts the vessels, and decreasing sympathetic activity allows vasodilation. For example, when a person rises from supine to standing, blood pressure needs to be increased to prevent fainting. Firing of certain sympathetic efferents stimulates vasoconstriction in skeletal muscles, thus maintaining blood flow to the brain. However, when vigorous activity is demanded from skeletal muscles, firing of a different set of sympathetic efferents to blood vessels in skeletal muscles produces vasodilation. These opposing ef-

fects of sympathetic activation are possible because the different subtypes of adrenergic receptors elicit different effects.

The role of the sympathetic system is often illustrated by describing the physiological responses to fear. When a person feels threatened, the sympathetic nervous system prepares for vigorous muscle activity, that is, for fight or flight. This is achieved by increasing blood flow to active muscles, increasing blood glucose levels, dilating bronchioles and coronary arteries, increasing blood pressure, and increasing heart rate. Simultaneously, sympathetic firing reduces activity in the digestive system.

REGULATION OF BODY TEMPERATURE

Sympathetic activity regulates body temperature by effects on metabolism and on effectors in the skin. Epinephrine released by the adrenal medulla increases the metabolic rate throughout the body. In the skin, sympathetic signals control the diameter of the blood vessels, secretion of the sweat glands, and erection of hairs. Blood flow in the skin is controlled by α-adrenergic receptors in the smooth muscles of arterioles. Norepinephrine binding to α-adrenergic receptors in skin arterioles also stimulates precapillary sphincters to contract, forcing blood to bypass the capillaries and decreasing the radiation of heat from the skin. When the precapillary sphincters relax, blood enters the capillaries, and heat radiates from the skin. Sweating, activated when acetylcholine binds with muscarinic receptors on sweat glands, helps to dissipate heat. In humans, erection of hair cells probably contributes little to the retention of body heat.

REGULATION OF BLOOD FLOW IN SKELETAL MUSCLE

Control of blood flow in skeletal muscle is more complex than in the skin. Skeletal muscle veins and venules are called **capacitance vessels** because blood pools in these vessels when their walls are relaxed. If pooling of blood in the lower limbs and abdomen is not prevented when a person assumes an upright position, the resulting drop in blood pressure can deprive the brain of adequate blood supply, causing **syncope** (fainting). Normally the pooling of blood is prevented by vasoconstriction of the capacitance vessels, prior to the change in position. This is accomplished by the release of norepinephrine to bind with α-adrenergic receptors in the walls of skeletal muscle venules and veins, causing vasoconstriction.

Arteriole walls in skeletal muscle contain α- and β₂-adrenergic and muscarinic cholinergic receptors. The action of norepinephrine on α-adrenergic receptors causes vasoconstriction of skeletal muscle arterioles. Binding of epinephrine to β₂-adrenergic receptors, or binding of acetylcholine to muscarinic receptors, vasodilates skeletal muscle arterioles during exercise or fight-or-flight situations. Local blood chemistry also affects the diameter of arterioles.

> Sympathetic activity may either vasodilate or vasoconstrict arterioles supplying skeletal muscle. Sympathetic activity vasoconstricts arterioles in the skin.

SYMPATHETIC CONTROL IN THE HEAD

Sympathetic effects on blood flow, sweating, and erection of hair cells of the head are identical to sympathetic actions in the remainder of the body. In addition, sympathetic signals dilate the pupil of the eye and assist in elevating the upper eyelid. The levator palpebrae superioris muscle consists of both smooth and skeletal muscle fibers; only the smooth muscle fibers are innervated by the sympathetic system. The skeletal muscle fibers are innervated by the facial cranial nerve. Sympathetic fibers also innervate salivary glands; their activation causes secretion of thick saliva, which causes a sensation of dryness in the mouth.

REGULATION OF THE VISCERA

Sympathetic effects on the thoracic viscera include increasing heart rate and contractility when β₁-adrenergic receptors are activated in cardiac muscle and dilation of the bronchial tree when β₂-adrenergic receptors are activated in the respiratory tract. Distribution of adrenergic receptor types is illustrated in Figure 8–8.

Drugs that bind with a receptor but do not activate the receptor are called blockers. Drugs that activate receptors are called agonists. α-Blockers are used to decrease high blood pressure by blocking the action

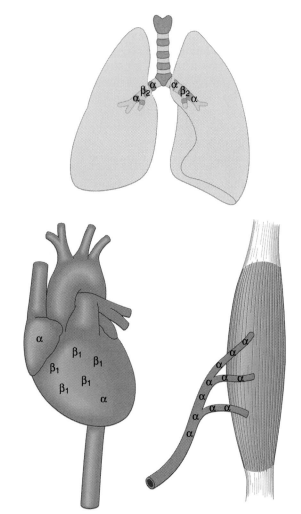

FIGURE 8–8

Distribution of adrenergic receptors. α-Adrenergic receptors are most abundant in arterioles of peripheral smooth muscle but are also found in the heart and bronchial smooth muscle. β₁-Adrenergic receptors are found primarily in the heart. β₂-Adrenergic receptors are most numerous in bronchial smooth muscle but are also found in peripheral smooth muscle.

of norepinephrine on receptors in blood vessels, producing vasodilation. The differences between receptor subtypes (i.e., β₁ and β₂) allow the design of drugs that bind with one subtype of receptor and not another. β₁-Blockers decrease heart rate and contractility without affecting the airways (Peel and Mossberg, 1995). β₂-Agonists prevent constriction of the airways without affecting heart function and thus are useful for treating asthma (Cahalin and Sadowsky, 1995).

In the gastrointestinal tract, sympathetic signals contract sphincters and decrease blood flow, peristalsis, and secretions. Sympathetic stimulation also inhibits contraction of the bladder and bowel walls and contracts internal sphincters.

METABOLISM

When the adrenal medulla releases epinephrine into the bloodstream, the most significant effect is stimulation of metabolism in cells throughout the entire body. Epinephrine release usually coincides with a generalized release of norepinephrine from sympathetic postganglionic neurons because the sympathetic system is often activated as a whole. In addition to its effect on metabolism, epinephrine reinforces the effects of norepinephrine on most target organs.

> The sympathetic nervous system optimizes blood flow to the organs, regulates body temperature and metabolic rate, and regulates the activity of viscera.

PARASYMPATHETIC NERVOUS SYSTEM

The parasympathetic nervous system uses a two-neuron pathway from the spinal cord to the effectors. Because the preganglionic cell bodies are found in nuclei of the brain stem and the sacral spinal cord, this system is often called the **craniosacral outflow** (Fig. 8–9). The ganglia of the parasympathetic nervous system are separate, unlike the interconnected ganglia of the sympathetic trunk. Parasympathetic ganglia are located near or in the target organs.

Parasympathetic information from the brain stem travels in cranial nerves to outlying ganglia. Postganglionic neurons are distributed to the eye, salivary glands, and viscera.

Parasympathetic fibers are distributed in cranial nerves III, VII, IX, and X. Seventy-five percent of the parasympathetic fibers in cranial nerves travel in cranial nerve X, the vagus nerve.

Parasympathetic fibers arising in the sacral spinal cord have cell bodies in the lateral horn of sacral levels S2 to S4. Their axons travel in pelvic splanchnic nerves, distributed to the lower colon, bladder, and external genitalia. In contrast to the sympathetic system, the parasympathetic system does not innervate the limbs or body wall.

Functions of Parasympathetic Nervous System

The principal function of the parasympathetic system is energy conservation and storage. Efferent fibers in the vagus nerve innervate the heart and the smooth muscle of the lungs and digestive system. Vagus nerve activity to the heart can produce either bradycardia (slowing of the heart rate) or decreased cardiac contraction force. Vagus stimulation in the respiratory system causes bronchoconstriction and increases mucus secretion. In the digestive system, vagus activity increases peristalsis, glycogen synthesis in the liver, and glandular secretions.

Fibers in cranial nerves VII and IX, the facial and glossopharyngeal nerves, innervate salivary glands. Other fibers in cranial nerve VII innervate the lacrimal gland, providing tears to moisten the cornea and for crying. Fibers in cranial nerve III, the oculomotor nerve, constrict the pupil and increase the convexity of the lens of the eye for focusing on close objects.

The sacral parasympathetic efferents regulate emptying of the bowels and bladder and the erection of the penis or clitoris. Specific autonomic reflexes are discussed in the context of various regions of the nervous system. For example, reflexive control of the pupil is discussed in Chapter 13, and bladder and bowel reflexes are covered in Chapter 12.

> Parasympathetic activity decreases cardiac activity, facilitates digestion, increases secretions in the lungs, eyes, and mouth, controls convexity of the lens in the eye, constricts the pupil, controls voiding of the bowels and bladder, and controls the erection of sexual organs.

Comparison of Sympathetic and Parasympathetic Functions

In actions on the thoracic and abdominal viscera, the bladder and bowels, and the pupil of the eye, the effects of sympathetic and parasympathetic activity are synergistic: their opposing actions are balanced to provide optimal organ function. For example,

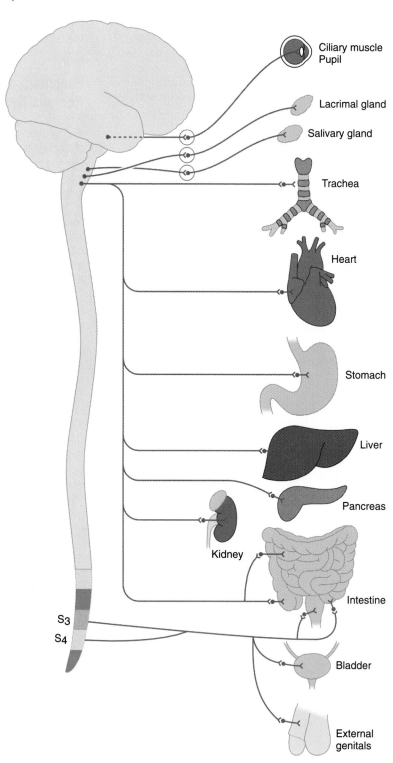

FIGURE 8–9

Parasympathetic outflow through cranial
nerves III, VII, IX, and X and S2 to S4.

immediately before a person begins to exercise, sympathetic signals increase heart rate and contractility, while parasympathetic signals that would slow heart rate decrease. Figure 8–10 illustrates the autonomic areas and pathways that regulate heart rate.

The autonomic efferent systems also have separate, unopposed effects: the sympathetic roles in regulating effectors in the limbs, face, and body wall and assisting elevation of the upper eyelid are not countered by parasympathetic innervation to these effectors. The role of the parasympathetic system in increasing the convexity of the lens of the eye is also unopposed. Tables 8–1 and 8–2 (on p. 144) list the actions of the autonomic efferent systems. Table 8–3 (on p. 145) summarizes the distribution of neurotransmitters and receptors in the autonomic efferent systems.

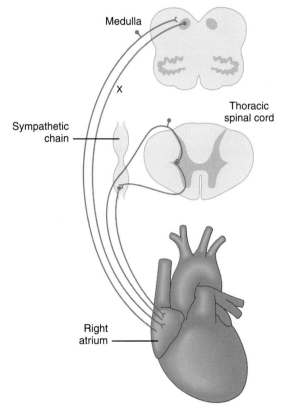

FIGURE 8–10

Autonomic regulation of heart rate. Sensory information enters both the medulla and spinal cord. Regulation is achieved by parasympathetic fibers in the vagus nerve and by sympathetic fibers from the thoracic spinal cord.

CLINICAL CORRELATIONS

Peripheral Region

If a peripheral nerve is severed, interruption of sympathetic efferents causes loss of vascular control, temperature regulation, and sweating in the region supplied by the peripheral nerve. These losses may lead to trophic changes in the skin.

If a lesion affects the stellate ganglion, sympathetic activity in the head and neck is decreased. This leads to drooping of the upper eyelid, constriction of the pupil, and vasodilation, with absence of sweating on the ipsilateral face and neck. This constellation of signs is called **Horner's syndrome** and occurs with lesions of the cervical sympathetic chain or its central pathways.

Spinal Region

A complete spinal cord lesion interrupts all communication between the cord below the lesion and the brain, disrupting ascending and descending autonomic signals at the level of the lesion. The severity of autonomic dysfunction depends on how much of the cord is isolated from the brain. Lower-level lesions allow the brain to influence more of the cord; higher-level lesions isolate more of the cord. Complete lesions above the lumbar level obstruct voluntary control of bladder, bowel, and genital function. Complete lesions above the midthoracic level isolate much of the cord from control by the brain, jeopardizing homeostasis by interfering with blood pressure regulation and the ability to adjust core body temperature. The autonomic consequences of spinal cord injury are discussed more fully in Chapter 12.

Brain Stem Region

Lesions in the brain stem region may interfere with descending control of heart rate, blood pressure, and respiration. Brain stem lesions may also affect cranial nerve nuclei, interfering with constriction of the pupil, production of tears, salivation, or regulation of thoracic and abdominal viscera.

TABLE 8–1
EFFECT OF SYMPATHETIC ACTIVITY ON BLOOD VESSELS

Organ	Transmitter	Receptor	Effect on Vessel Wall	Purpose
Skin	Adrenergic	α	Vasoconstriction of arterioles	↓ radiation of heat from skin
Skeletal muscle	Adrenergic	α	Vasoconstriction of venules and veins	↑ peripheral vascular resistance, ↑ blood pressure
	Adrenergic	β_2	Vasodilation of arterioles	More blood available to muscle
	Acetylcholine	Muscarinic	Vasodilation of arterioles	More blood available to muscle
Heart	Adrenergic	β	Dilation	More blood available to heart

Cerebral Region

Damage to certain nuclei in the hypothalamus disrupts homeostasis, with consequent metabolic and behavioral dysfunctions. Obesity, anorexia, hyperthermia, hypothermia, and emotional displays dissociated from feelings can occur. Activity in other limbic (emotional) areas can also interfere with homeostasis. For example, the response to perceived threat includes sympathetic activity that increases blood flow to skeletal muscles, accelerates cardiac rate, strengthens cardiac contraction, and decreases blood flow to skin, kidneys, and digestive tract.

Sympathetically Maintained Pain

Sympathetically maintained pain is a syndrome of pain, vascular changes, and atrophy. An aberrant response of the sympathetic nervous system to trauma

TABLE 8–2
COMPARISON OF SYMPATHETIC AND PARASYMPATHETIC EFFECTS ON ORGAN FUNCTION

Organ	Function	Sympathetic Effect	Parasympathetic Effect
Eye	Diameter of pupil	↑	↓
	Curvature of lens		↑
Heart	Contraction rate	↑	↓
	Force of contraction	↑	
Blood vessels	See Table 8–1		
Lungs	Diameter of bronchioles	↑	↓
	Diameter of blood vessels	↑	
	Secretions		↑
Sweat glands	Production of sweat	↑	
Salivary glands	Thick secretion	↑	
	Thin, profuse secretion		↑
Lacrimal glands	Production of tears		↑
Adrenal medulla	Secretion of epinephrine	↑	
Gastrointestinal tract	Peristalsis	↓	↑
	Secretions	↓	↑
Liver	Glucose release	↑	
	Glycogen synthesis		↑
Pancreas	Secretions	↓	↑
Bowel and bladder	Emptying	↓	↑
External genitalia	Erection of penis or clitoris		↑

TABLE 8–3
TRANSMITTERS AND RECEPTORS IN THE AUTONOMIC NERVOUS SYSTEM

Transmitter	Site of Transmitter Release	Receptor Type
Acetylcholine	Synapse between preganglionic and postganglionic neurons (both sympathetic and parasympathetic)	Nicotinic
	Parasympathetic postganglionic to smooth muscles and glands	Muscarinic
	Sympathetic postganglionic to sweat glands and some arterioles in skeletal muscle (dilates arterioles)	Muscarinic
Norepinephrine	Sympathetic postganglionic to constrict blood vessels in skeletal muscles, skin, and viscera, dilate pupil	α
	Sympathetic postganglionic to vasodilate arterioles in skeletal muscle, dilate bronchioles, decrease gastrointestinal activity, accelerate heart rate	β
Epinephrine	Adrenal medulla: release transmitter into bloodstream	α and β

Norepinephrine has a greater effect on α-adrenergic receptors than on β; epinephrine is equally effective in activating both α and β.

---

produces the syndrome. The trauma may be quite minor; for example, an ankle sprain that heals within a few days. The primary complaint is severe, spontaneous pain, aggravated by psychological as well as physical stimuli. Early signs of sympathetically maintained pain include red skin, excessive sweating, edema, and skin atrophy. If the condition progresses to its late stage, muscle atrophy, osteoporosis, and arthritic changes occur. Motor signs that may be associated include paresis, spasms, and difficulty initiating movement (Schwartzman and Kerrigan, 1990).

Treatment includes sympathetic blockade by injection of an anesthetic into the sympathetic ganglion affecting the limb. Thus, for lower limb sympathetically maintained pain, the lumbar sympathetic chain is blocked. For the upper limb, the stellate ganglion is blocked. Physical therapy consists of active and active assisted exercise and tactile stimulation to tolerance. Passive manipulation may aggravate sympathetically maintained pain.

The terms *causalgia, Sudeck's atrophy,* and *reflex sympathetic dystrophy* are often used synonymously with *sympathetically maintained pain,* despite the attempts of some authors to distinguish among these terms. Although authorities agree that disuse of the limb is a primary precipitating factor, the precise mechanism of sympathetically maintained pain is controversial. Classical theories postulate excessive activity of sympathetic efferents, but recent findings of low serum levels of norepinephrine in the affected limb indicate that sympathetic efferents are not overactive. A recent theory of sympathetically maintained pain is that the peripheral receptors become abnormally responsive to circulating adrenergic transmitters (Harden et al., 1994). (See the box on sympathetically maintained pain.)

Syncope

Syncope (fainting) is a brief loss of consciousness due to inadequate blood flow to the brain. If the cause of the syncope is powerful emotions, the attack is called vasodepressor syncope or neurogenic shock. Strong emotion can initiate sudden, active vasodilation of intramuscular arterioles, causing a precipitous fall in blood pressure. The blood flow to the head is temporarily reduced, leading to loss of consciousness and paleness of the face. Blood flow to the head is restored when the person is horizontal.

In some cases, particularly when syncope occurs in response to painful stimuli or on standing after prolonged bed rest, vagal stimulation to the heart follows the intramuscular vasodilation. The vagal activity slows the heart, further decreasing blood pressure, and elicits nausea, salivation, and increased perspiration. When vagal signs occur with vasodepressor syncope, called a vasovagal attack, excessive activity is occurring in both the sympathetic and parasympathetic systems.

SYMPATHETICALLY MAINTAINED PAIN

Pathology
Unknown

Etiology
Usually secondary to trauma; may be abnormal receptor sensitivity to circulating adrenergic transmitters

Speed of onset
Chronic

Signs and symptoms

Consciousness
Normal

Communication and memory
Normal

Sensory
Severe, spontaneous pain

Autonomic
Abnormal sweating, vasodilation in skin, atrophy of muscles, joints, and skin

Motor
May have paresis, spasms, difficulty initiating movement

Region affected
Peripheral

Demographics
4 times more frequent in women

Prognosis
Early intervention has best outcome; intensive physical therapy often required; some cases are intractable

---

Although vasodepressor syncope is the most common type of syncope, many different causes can produce syncope. As mentioned earlier in this chapter, assuming an upright posture can cause pooling of blood in the lower body, resulting in syncope. Other causes include insufficient cardiac output, hypoxia, and hypoglycemia.

Tests of Autonomic Function

The ability of the sympathetic nervous system to regulate blood pressure can be evaluated by measuring the person's blood pressure in supine, having the person stand, and measuring the blood pressure 2 minutes later. Abnormal responses include a drop of more than 30 mm Hg in systolic blood pressure or more than 15 mm Hg in diastolic blood pressure.

Sympathetic regulation of the skin can be tested by the sweat test or hand vasomotor test. For the sweat test, sweat is absorbed by small pieces of filter paper placed on the skin, and then the filter paper is weighed to determine the amount of sweat. For the vasomotor test, skin temperature is measured before and after the hands are immersed in cold water. This test is used to assess the amount of vasoconstriction.

CASE 1

C.M., a 42-year-old man, sustained mild soft tissue damage to his right wrist in a fall 4 months ago. Since the injury, he has completely stopped using his right upper limb. C.M. was initially evaluated in physical therapy 1 week ago. At that time he reported that he wears a sling, even while sleeping, and protects his entire right arm as much as possible. He also reported a constant burning sensation and that any stimulation, even putting on clothes or a breeze across the skin, evokes sharp pain. The therapist observed the following abnormalities in his fourth and fifth fingers: excessive sweating; shiny, red skin; and opaque, long fingernails. C.M. stated that he cannot tolerate cutting the nails because of pain. When asked to move his right arm, C.M. refused. He did allow passive flexion of the shoulder to 40°. C.M. refused to allow the therapist to touch his distal arm and refused to actively move his hand or wrist.

The following day, his doctor injected the right stellate ganglion with lidocaine, a drug that reversibly depresses neuronal function. After the injection, C.M.'s upper eyelid drooped, his pupil was constricted, and sweating on the ipsilateral face and neck was absent. These side effects of the block indicate that the injection was successful in blocking sympathetic activity in the stellate ganglion. The intended result of the block, disrupting the pain in the right upper limb, ensued, and C.M. reluctantly moved his wrist and hand actively in therapy the same day. He requires vigorous daily physical therapy to move his hand and continues to protect his arm when not in therapy.

Questions

1. What is the most likely diagnosis? Why?

2. What is the name of the syndrome that affected C.M.'s pupil, eyelid, and sweating on one side of his face and neck?

3. How was the diagnosis confirmed?

CASE 2

R.D. is a 23-year-old professional basketball player. While waiting to play in a championship game, he collapsed on the sidelines. His pulse could not be palpated, his blood pressure was 60/45 mm Hg, breathing was almost imperceptible, his pupils were dilated, and his face was pale. He was unconscious for about 15 seconds; then color began to return to his face, and his breathing and pulse quickly returned to normal. R.D. regained his awareness of the environment on regaining consciousness. He reported feeling fine, although he felt weak; no headache or confusion followed the attack.

Question

What is the most likely diagnosis?

CASE 3

B.H., a 47-year-old man, had a myocardial infarction 3 weeks ago. He has been referred to physical therapy for cardiac rehabilitation. He is taking propranolol, a β-blocker.

Questions

1. What effect does blocking β-adrenergic receptors have on cardiovascular function?

2. Given that aerobic exercise prescriptions are based on a percentage of predicted age-related maximal heart rate, how will the β-blocker effects impact your exercise prescription?

REVIEW QUESTIONS

1. As a cardiac rehabilitation client begins treadmill exercise, which of his visceral and vascular sensory receptors would register changes? To what stimuli would the receptors be responding?

2. What is the function of visceral afferents?

3. What areas of the brain directly control autonomic function? What areas modulate activity in the autonomic control centers?

4. What are the differences between the autonomic and somatic efferent systems?

5. What are the sympathetic trunks? How do sympathetic fibers that leave the paravertebral ganglia reach effectors in the periphery, for example, blood vessels in skeletal muscle and in the skin?

6. What are splanchnic nerves?

7. What is the primary function of the sympathetic nervous system? How can sympathetic activation elicit vasodilation in skeletal muscles under certain conditions and vasoconstriction in the same vessels when conditions change?

8. What are capacitance vessels, and what is their significance? How is the blood flow in skeletal muscle arterioles controlled?

9. What would happen if the sympathetic fibers to the levator palpebrae superioris muscle and the pupil of the eye did not function?

10. What functions does the sympathetic nervous system regulate?

11. What functions are controlled by the parasympathetic system?

References

American College of Sports Medicine (1990). Position statement on the recommended quantity and quality of exercise for developing and maintaining cardiorespiratory and muscular fitness in healthy adults. Med. Sci. Sports Exerc. 22:265–274.

Cahalin, L. P., and Sadowsky, H. S. (1995). Pulmonary medications. Phys. Ther. 75(5):397–414.

Harden, R. N., Duc, T. A., Williams, T. R., et al. (1994). Norepinephrine and epinephrine levels in affected versus unaffected limbs in sympathetically maintained pain. Clin. J. Pain 10(4): 324–330.

Peel, C., and Mossberg, K. A. (1995). Effects of cardiovascular medication on exercise responses. Phys. Ther. 75(5):387–396.

Pollock, M. L., and Wilmore, J. H. (1990). Exercise in health and disease (2nd ed.). Philadelphia: W. B. Saunders.

Schwartzman, R. J., and Kerrigan, J. (1990). The movement disorder of reflex sympathetic dystrophy. Neurology 40:57–61.

9

Motor System

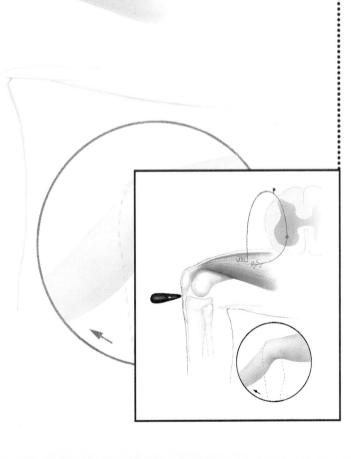

INTRODUCTION

Reading, talking, walking, preparing dinner, playing an instrument, every action we perform, requires the motor system. Movement is orchestrated by the coordinated action of the peripheral, spinal, brain stem/cerebellar, and cerebral regions, shaped by a specific context, and directed by the intentions of the performer. Consider how movement strategies change when ice covers the sidewalk: Cadence, step length, and posture adjust to the new requirements. A young child may choose to sit and scoot rather than risk falling. These alternatives are selected based on sensory information; as we saw in the example of Rothwell's patient (severe peripheral neuropathy [see Chap. 7]), normal motor performance and sensation are inextricably linked. The sensory information required varies with the task and is often used to prepare for movement (feed-forward) in addition to providing information during and after movement (feedback).

My first experience teaching wheelchair-to-car transfers to a person with quadriplegia (C7 level) underscores the complexity of movement. Despite our best efforts, he didn't move from the wheelchair. He asked if he could try it his way, so I guarded as he completed an unorthodox but successful transfer. He placed his forehead on the dashboard, threw his forearm onto the roof using momentum and gravity, and then, contracting neck and forearm flexors, lifted himself into the car. Biomechanics and environmental constraints prevented a conventional car transfer, but he used biomechanics and environmental affordances* to solve the movement problem. As in most normal movements, he initiated and controlled the action; the movement was not in response to any external stimulus.

How does the nervous system direct a simple motor act, like picking up a pen? The sequence of neural activity begins with a decision made in the anterior part of the frontal lobe (Fig. 9–1). Next, the motor planning areas are activated, then the control circuits. The control circuits, consisting of the cerebellum and basal ganglia, regulate the activity in descending motor pathways. The descending motor pathways deliver signals to spinal interneurons and lower motoneurons. Lower motoneurons transmit

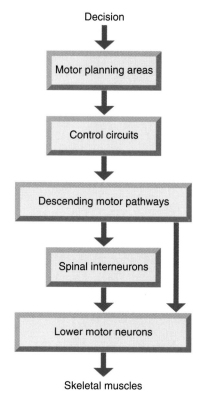

Decision

Motor planning areas

Control circuits

Descending motor pathways

Spinal interneurons

Lower motor neurons

Skeletal muscles

FIGURE 9–1

The neural structures required to produce normal movements. Although sensory information influences each of the neural structures involved in generating movements, the sensory connections have been omitted for simplicity.

signals directly to the skeletal muscles, resulting in contraction of the appropriate muscle fibers to move the upper limb and fingers.

Normal control of movement requires activation of the following:

- Lower motoneurons
- Spinal region connections
- Descending pathways
- Control circuits
- Motor planning areas

Motoneurons are nerve cells that control the activity of skeletal muscles. **Lower motoneurons** directly innervate skeletal muscle fibers. In the **spinal** region, interactions among neurons determine the information conveyed by lower motoneurons to muscles. **Descending pathways** deliver movement information

---

Environmental affordance is a term coined by Gibson (1977) to describe the possibilities offered to a person by the environment.

from the brain to lower motoneurons in the spinal cord or brain stem. The neurons whose axons travel in descending pathways are upper motoneurons. Descending pathways are classified as **postural/gross movement pathways**, controlling automatic skeletal muscle activity, and **fine movement pathways**, controlling skilled, voluntary movements.

The **control circuits** are the basal ganglia and cerebellum. The control circuits adjust activity in the descending pathways, resulting in excitation or inhibition of the lower motoneurons. Thus control circuits partially determine muscle contraction. In all regions of the central nervous system, sensory information adjusts motor activity. Therefore, the contribution of sensation to movement will be covered with each motor section.

> Lower motoneurons have their cell bodies in the spinal cord or brain stem and synapse with skeletal muscle fibers. Connections in the spinal cord or brain stem determine the activity of lower motoneurons. In contrast, upper motoneurons arise in the cerebral cortex or brain stem, and their axons travel in descending pathways to synapse with lower motoneurons and/or interneurons in the brain stem or spinal cord. Control circuits adjust the activity of the descending pathways.

LOWER MOTONEURONS

There are two types of lower motoneurons: **alpha motoneurons** and **gamma motoneurons.** Both alpha and gamma motoneurons have cell bodies in the ventral horn of the spinal cord. Their axons leave the spinal cord via the ventral root, travel through the spinal nerve, then through the peripheral nerve to reach skeletal muscle. Axons of alpha motoneurons project to extrafusal skeletal muscle, whereas the axons of gamma motoneurons project to intrafusal fibers in the muscle spindle (see Chap. 6). Alpha motoneurons have large cell bodies and large, myelinated axons, while gamma motoneurons have medium-sized myelinated axons (Table 9–1). We will consider the alpha motoneurons first.

The axons of alpha motoneurons branch into numerous terminals as they approach muscle. Each terminal ends near a single muscle fiber at a **neuromuscular junction**, a synapse between nerve and muscle. When the alpha motoneuron axon releases its neurotransmitter, acetylcholine, the transmitter binds to

TABLE 9–1		
CHARACTERISTICS OF MOTONEURONS		
Axon Size and Myelination	**Axon Type**	**Innervates**
Large myelinated	Aα	Extrafusal muscle fibers
Medium myelinated	Aγ	Intrafusal muscle fibers

nicotinic receptors on the muscle membrane, causing local depolarization of the muscle cell membrane. As depolarization spreads along the muscle cell membrane, a series of reactions causes muscle fibers to contract.

Motor Units

An alpha motoneuron and the muscle fibers it innervates are called a **motor unit.** Whenever an alpha motoneuron is active, neurotransmitter is released at all of its neuromuscular junctions, and all muscle fibers innervated by that neuron contract (Fig. 9–2). Motor units are classified as slow twitch or fast twitch, depending on the speed of muscle contraction in response to a single electrical shock. The neuron innervating the muscle determines the twitch characteristics of the muscle fibers. Smaller-diameter, slower-

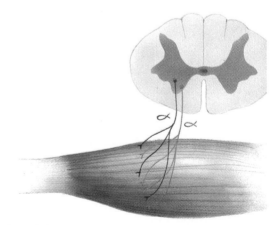

FIGURE 9–2

A motor unit consists of an alpha motoneuron and the muscle fibers it innervates. Two motor units are illustrated to show that muscle fibers innervated by a single neuron are distributed throughout the muscle.

conducting alpha motoneurons innervate slow twitch muscle fibers; larger-diameter, faster-conducting alpha motoneurons innervate fast twitch muscle fibers.

Slow twitch muscle fibers tend to comprise the majority of muscle fibers in postural and slowly contracting muscles. For example, the soleus muscle has primarily slow twitch fibers and is tonically active in standing and phasically active in walking. The gastrocnemius muscle has more **fast twitch muscle fibers** than the soleus, and phasic contraction results in fast, powerful movements, as in sprinting. In most movements, slow twitch muscle fibers are activated first because the small cell bodies of the slow-conducting alpha motoneurons depolarize before the cell bodies of the larger alpha motoneurons. The slow twitch muscle fibers typically continue to contribute during faster actions as the fast twitch units are recruited.

The order of recruitment from smaller to larger alpha motoneurons is called Henneman's size principle. A few exceptions to the rule of slow twitch muscles being recruited first have been noted: rapid alternating movements may only recruit fast twitch units, since slow twitch activation would interfere with the movement; and fast twitch muscles are recruited first in some instances of synergistic rather than prime mover activation, as when the abductor pollicis is used to assist thumb flexion.

Motor units also vary in the number of muscle fibers innervated by a single neuron. The gastrocnemius muscle has more than 1000 muscle fibers innervated by each motoneuron. In contrast, intrinsic muscles of the hand and the extraocular muscles have as few as 10 muscle fibers per motoneuron because precise control is required.

> A motor unit is a single alpha motoneuron and the muscle fibers the alpha motoneuron innervates.

The activity of a motor unit depends on the convergence of information from peripheral sensors, spinal connections, and descending pathways onto the cell body and dendrites of the alpha motoneuron.

Peripheral Sensory Input to Motoneurons

Tension on the muscle tendon is converted to neural impulses by the Golgi tendon organ. The information is conveyed within the spinal cord via collaterals and interneurons to the lower motoneurons. Sensory information from muscle spindles signal muscle length and velocity of changes in muscle length. Afferent information from muscle spindles is used to correct small errors in movement reflexively (spinal region), to make larger corrections via connections in the brain stem, and to provide proprioceptive information to the cerebral cortex and cerebellum. The sensitivity of the muscle spindle is adjusted by gamma motoneurons.

Alpha-Gamma Coactivation

During most movements, the alpha and gamma motoneuron systems function simultaneously, in a pattern of **alpha-gamma coactivation.** This coactivation maintains the stretch on the central region of the muscle spindle intrafusal fibers when the muscle actively contracts. When excitatory signals converging on lower motoneurons (from both peripheral afferent input and descending pathways) are sufficient to cause alpha motoneurons to fire, the gamma motoneurons to spindle fibers in the same muscle also fire. Alpha-gamma coactivation occurs because most sources of input to alpha motoneurons have collaterals that project to gamma motoneurons and because gamma motoneurons, with their smaller cell bodies, require less excitation to reach threshold than do alpha motoneurons.

All normal movements require sensory information, spinal circuitry, and descending commands. Next we will consider spinal reflexes (simple movement responses to peripheral inputs) and spinal region coordination.

SPINAL REGION

In the spinal region, somatosensory information is integrated with descending motor commands to generate motor output. Networks of interneurons act flexibly to produce relatively simple movements, for example, reflexes or the reciprocal activation of flexor and extensor muscles.

Reflexes

Early movement theorists believed that reflexes were the basis of movement. However, reflexes are not the sole basis of movement because by definition reflexes are involuntary responses to external stimuli. When a person decides to reach for a book, what is the stimulus? Most movement is automatic or voluntary and anticipatory, not reflexive. However, reflex examination provides information about the peripheral and spinal circuits and the level of background excitation in the spinal cord. Spinal region reflexes require sensory receptors, primary afferents, connections between primary afferents and lower motoneurons, and effectors (muscles or glands). Spinal region reflexes can operate without supraspinal input; however, normally signals from the brain influence spinal reflexes by adjusting the background level of neural activity in the spinal cord. In this section, the proprioceptive reflexes involving information from muscle spindles and from Golgi tendon organs will be followed by a brief discussion of cutaneous reflexes.

PROPRIOCEPTIVE REFLEXES

Muscle Spindles A brisk tap with a reflex hammer on the quadriceps tendon elicits a reflexive contraction of the quadriceps muscle because quick stretch activates neural connections between muscle spindles and alpha motoneurons to the same muscle. Tapping the tendon delivers a quick stretch to the muscle and the spindles embedded parallel to the muscle fibers. The primary endings of the spindles are stimulated by the quick stretch, then Ia afferents transmit action potentials to the spinal cord and release neurotransmitter at synapses with the alpha motoneurons. The alpha motoneurons depolarize, action potentials are propagated to the neuromuscular junctions, acetylcholine is released and binds with muscle receptors, and the muscle fibers contract. Only one synapse links the afferent and efferent neurons; thus the phasic response to stretch is a monosynaptic reflex (Fig. 9–3). Muscle contraction in response to quick stretch is the **phasic stretch reflex;** the terms *myotatic reflex, muscle stretch reflex,* and *deep tendon reflex* are all synonymous.

Although monosynaptic reflexes involve only one connection between afferent and efferent neurons, normal expression of the reflex requires additional

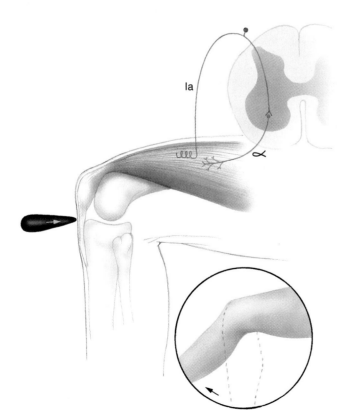

FIGURE 9–3

Phasic stretch reflex. Quick stretch of a muscle, elicited by striking the muscle's tendon, stimulates the Ia afferents from the muscle spindle. Activity of Ia afferents causes monosynaptic excitation of alpha motoneurons to the stretched muscle, resulting in abrupt contraction of the muscle fibers.

spinal connections. The additional spinal connections, mediating **reciprocal inhibition,** ensure that antagonists are inhibited during agonist contractions. For example, a biceps stretch reflex causes shortening of the biceps and abruptly stretches the triceps. Without a mechanism to prevent a triceps stretch reflex, the biceps contraction would be opposed by its antagonist. To avert an antagonist stretch reflex, activity in collateral branches from the Ia afferent stimulates interneurons to inhibit the alpha efferent to the antagonist (Fig. 9–4).

During gait, the amount of reciprocal inhibition is modulated by upper motoneurons. Reciprocal inhibition is also used extensively during voluntary motion to prevent antagonist opposition to the movement.

In contrast to the phasic stretch reflex, another stretch reflex is elicited by maintained stretch of the central region of spindle muscle fibers. This tonic stretch reflex is multisynaptic. The maintained stretch fires the primary and secondary spindle endings, both type Ia and II afferents conduct excitation into the spinal cord, and interneurons link the afferent fiber terminals with alpha motoneurons. Thus, when slow or sustained stretch is applied to the central spindle, the tonic stretch reflex facilitates contraction in the muscle.

Golgi Tendon Organ Reflex Another proprioceptive reflex is the **tendon organ reflex.** Tendon tension is registered by Golgi tendon organs. The information is conveyed into the spinal cord by Ib afferents, stimulating interneurons that inhibit the alpha motoneurons to the same muscle, resulting in **autogenic inhibition** (Fig. 9–5). Synergistic muscles may also be inhibited. The major role of tendon organ information is to adjust muscle activity, in concert with information from muscle spindles and descending control. In rare situations of extreme muscle contraction, inhibition by the tendon reflex may prevent muscles from exerting enough force to tear the tendon. The protective mechanism can fail if force buildup is very rapid. For example, people who play basketball occasionally rupture their Achilles tendon during a jump because the sudden forceful muscle contraction occurs before reflexive inhibition from the Golgi tendon organ can occur.

CUTANEOUS REFLEXES

Cutaneous stimulation can also result in reflexive movements. If someone steps on a tack, the **withdrawal reflex** automatically lifts the foot by flexing the lower limb, even before the person is consciously aware of pain (Fig. 9–6). The withdrawal reflex and related reactions will be examined in Chapter 12.

> Reflexes can be elicited by stimulation of musculoskeletal or cutaneous receptors. Stimulation of the muscle spindle receptors can result in phasic and/or tonic stretch reflexes. Activation of the Golgi tendon organ can inhibit activity of the corresponding muscle. Noxious cutaneous information can result in a withdrawal reflex.

RELATIONSHIP BETWEEN REFLEXIVE AND VOLUNTARY MOVEMENT

Classically, reflexes were considered to be responses to particular types of sensory information exciting only specific, isolated pathways within the spinal cord and resulting in stereotyped output. Voluntary movement was considered entirely separate from reflexes. Further research has not supported these divisions. Most sensory stimuli act in an ensemble fashion, the interneurons stimulated can vary, and appropriate levels of the central nervous system interact to produce context-dependent movement. For example, the

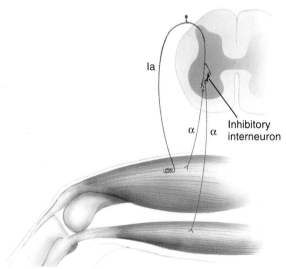

FIGURE 9–4

Reciprocal inhibition. Often, when a muscle is activated, opposition to the movement by antagonist muscles is prevented via inhibitory interneurons.

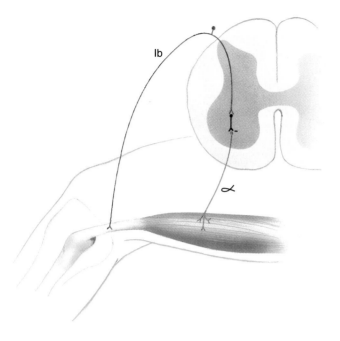

Ib

α

FIGURE 9–5

Autogenic inhibition. Stretch of a tendon activates Ib afferents that synapse with inhibitory interneurons to decrease muscle activity.

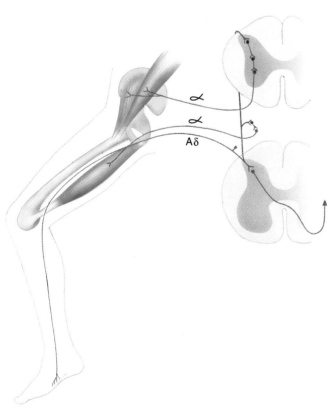

α

α

Aδ

FIGURE 9–6

Withdrawal reflex. Usually in response to a painful cutaneous stimulus, muscles are activated to move the body part away from the stimulation. This action requires polysynaptic connections at multiple levels of the cord because the active muscles are innervated by various spinal segments.

movement response to tendon tap can be modified by increasing a person's arousal, or alertness, level. Arousal changes the level of descending input to the spinal circuitry. Furthermore, muscle spindle output is modified by sensitivity adjustments and by the recent movements and contractions the muscle has undergone (Gandevia, 1993). As a result, muscle spindle output is not linearly related to changes in muscle length or rate of change in length. Spindle information is integrated with other proprioceptive inputs to adjust muscle output.

H-REFLEX

To quantify whether alpha motoneurons are facilitated or inhibited, **H-reflex** testing may be used. H-reflex testing substitutes cutaneous electrical stimulation of a peripheral nerve for the tendon percussion of the myotatic reflex. For example, a stimulating electrode is placed over the tibial nerve in the popliteal fossa, and a recording electrode is placed on the inferomedial gastrocnemius (Fig. 9–7). A weak current, adequate to stimulate only the largest axons (motor axons and group Ia and Ib afferents, which have the lowest electrical threshold) is administered to the skin over the tibial nerve. Action potentials travel both toward the muscle (via motor axons) and toward the spinal cord (group Ia and Ib afferents), resulting in two compound action potentials that can be recorded from the muscle. The shorter latency M wave is produced by the impulses traveling along the motor fibers, causing an almost immediate muscle contraction. The H-reflex is produced by activating the afferent fibers in the tibial nerve. The action potential in each large afferent fiber travels into the spinal cord, resulting in transmission across synapses to facilitate an alpha motoneuron. If the central excitatory state of the alpha motoneurons is near threshold, activation of the alpha motoneurons will cause depolarization of the calf muscle membranes. The H-reflex is slightly faster (by 10 milliseconds) than the tendon tap reflex because activation of spindle receptors is not required.

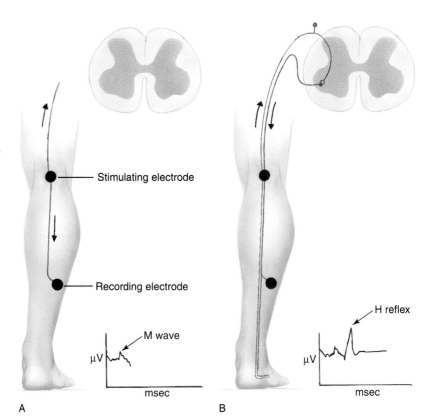

FIGURE 9–7

H-reflex electrode placement for quantifying the level of excitation in tibial nerve alpha motoneurons. *A,* Electrical stimulation applied to the skin in the popliteal fossa over the tibial nerve evokes action potentials in both sensory and motor axons. The action potentials are propagated proximally and distally from the site of stimulation. When the action potentials in alpha motoneurons reach the terminals, acetylcholine is released and binds with receptors on the muscle membrane, and the muscle membrane depolarizes. The depolarization is recorded as the M wave. *B,* Action potentials evoked in the Ia and Ib fibers are propagated into the spinal cord. Via synaptic connections, alpha motoneurons are stimulated. Then action potentials are propagated toward the muscle, and acetylcholine is released at the neuromuscular junction and binds with receptors on the muscle membrane. When the muscle membrane depolarizes, the H-reflex is recorded.

Spinal Region Coordination

In addition to their reflexive effects, connections within the spinal cord contribute to coordination of movement. Interneurons in the spinal cord contribute to the organization of movement by linking motoneurons into functional groups. An example is the functional organization among interneurons excited by group II afferents. Group II afferents deliver information from tonic receptors in muscle spindles, certain joint receptors, and cutaneous and subcutaneous touch and pressure receptors (see Chap. 6). The interneurons excited by group II afferents project to motoneurons controlling muscles acting at other joints, thus providing a spinal region basis for muscle synergies. Muscle synergies are coordinated muscular actions, such as finger and elbow flexion combined with supination of the forearm when bringing a bit of food to the mouth. Motor control researchers typically use the term *synergy* to describe the activity of muscles that are often activated together by a normal nervous system. Clinicians often restrict the use of the term *synergy* to indicate pathological synergies, as when a person with an upper motoneuron lesion (due to head injury or stroke) cannot flex the shoulder without simultaneous, obligatory flexion of the elbow.

Complex patterns of muscle activation have been demonstrated at the spinal region in mammals, and similar mechanisms are believed to exist in the human spinal cord. For example, cats with complete lower thoracic spinal cord lesions are able to walk on a treadmill after recovery from surgery. As the treadmill moves the feet, hind limb muscles contract in near-normal synergy patterns, generating reciprocal hind limb movements (Pearson, 1976) (Fig. 9–8). Thus, even in the absence of descending commands, spinal connections act as **central pattern generators** to produce walking. In this example, afferent input activates the central pattern generators. Central pat-

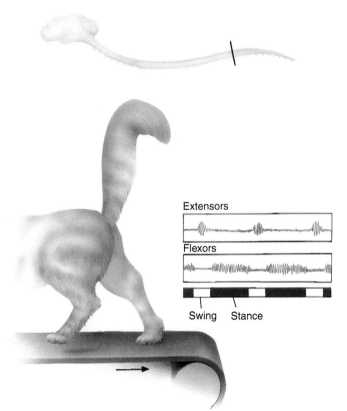

Extensors

Flexors

Swing   Stance

FIGURE 9–8

A cat with complete severance of the thoracic spinal cord is able to perform a near-normal walking pattern with its hind legs, despite lack of connection between the lumbar spinal cord and upper cord and brain. (Modified by permission from Pearson, K. [1976]. The control of walking. Sci. Am. 235[6]:72–86. Copyright © 1976 by Scientific American, Inc. All rights reserved.)

tern generators are flexible networks of interneurons that produce purposeful movement.

Can central pattern generators be activated without afferent information? Grillner and Zangger (1975) studied walking in cats with a lesion separating the brain stem from the cerebrum (the intact spinal cord was still connected to the brain stem) and the limbs deafferented. The cats were supported by a sling, and walking was induced by electrical stimulation of brain stem locomotor areas. The pattern of muscle activation was more variable but otherwise similar to that of intact cats, indicating that the pattern of muscle activation is not dependent on sensory input.

Then what is the role of sensory input in motor function? If a limb of a cat with a lesion completely interrupting the connections between the spinal cord and the brain is prevented from moving during walking, the muscle activity in the limb persists. That is, if limb movement is interrupted during swing phase, the muscle activity appropriate to swing phase will continue until the limb is allowed to move. So afferent information appears to be important in changing from swing phase of ambulation to the stance phase and back. More intriguing is that results of identical cutaneous stimulation during limb movements of cats with complete spinal cord lesions produces different responses depending on the phase of the step cycle. For example, touching the dorsum of the cat's paw during swing phase causes the paw to be raised to clear the obstacle. Touching the same point during stance causes the limb to extend (Forssberg et al., 1975) (Fig. 9–9). This **response reversal** ensures that the movement goals are achieved. If response reversal did not occur, touching the dorsum of the foot during swing phase would result in extension of the limb, which would prevent the limb from moving forward, or the same stimulation would result in raising the paw during stance, destabilizing the cat. Thus, response reversals adapt the ongoing activity of the central pattern generators to the environment.

In humans, the evidence for central pattern generators is based on patterns of lower limb movements in infants. Month-old infants held upright with their feet touching a surface often make leg movements that resemble adult walking. In most 6-week-old infants, this stepping reflex can no longer be elicited. McGraw (1945) attributed the disappearance of the stepping reflex to cortical inhibition. This explanation was generally accepted until Thelen et al. (1984) compared the leg movements of month-old infants supported upright under three conditions: without weights, with

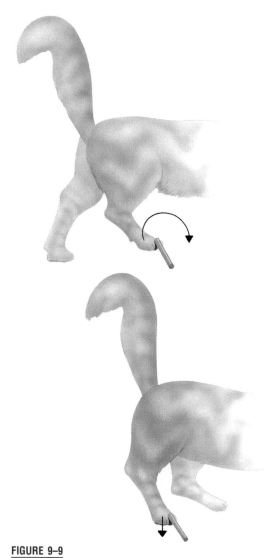

FIGURE 9–9

In a cat with a complete severance of the thoracic spinal cord stimulation of the dorsum of a cat's paw elicits different movement patterns depending on the phase of the gait cycle.

weights on the legs, and with the legs in warm water. In the weighted condition, the extra weight was adjusted to approximate the weight gain that would be expected in the infant between the fourth and sixth weeks. The infants stepped least in the weighted condition and most in the water condition. Thus, normal loss of the stepping reflex is apparently due to weight gain rather than nervous system changes. Thelen's research is consistent with the idea that central pattern generators produce stepping movements in newborns and then biomechanical changes temporarily conceal the activity of the central pattern generators until infants gain strength. Additional evidence for human central pattern generators is the infant stepping pattern elicited in anencephalic children (Forssberg, 1985). Anencephalic children are born without cerebral and cerebellar hemispheres and have only a rudimentary brain stem. However, if an anencephalic child is supported in an upright position with the feet in contact with a firm surface, the child actively reciprocally flexes and extends the legs in a stepping pattern.

If human spinal cords contain central pattern generators, why are people with complete spinal cord transections above the lumbar level unable to walk? Human walking may require more descending control because the postural adjustments necessary for upright walking are more complex than for quadrupedal ambulation.

> Central pattern generators in the spinal cord probably contribute to normal walking in humans but do not provide adequate control to support walking without descending input.

Motoneuron Pools in the Spinal Cord

Within the ventral horn of the spinal cord, cell bodies of motoneurons are arranged in motoneuron pools. Axons from a pool project to a single muscle. The medially located pools innervate axial and proximal muscles; laterally located pools innervate distal muscles. In addition to the medial/lateral organization, the pools are arranged with anteriorly located pools innervating extensors and more posterior pools (still within the ventral horn) innervating flexors (Fig. 9–10). The neurons of the descending pathways are grouped according to their termination in the medial or lateral cord.

DESCENDING MOTOR PATHWAYS (UPPER MOTONEURONS)

Upper motoneurons project from supraspinal centers to lower motoneurons (alpha and gamma) and to interneurons in the brain stem and spinal cord. Upper

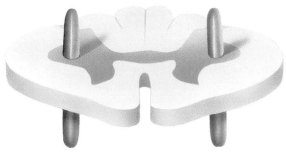

FIGURE 9–10

The cell bodies of motoneurons are arranged in groups corresponding to each muscle innervated. The pools may extend several spinal cord segments.

motoneurons projecting to the spinal cord can be classified according to where they synapse: medially, laterally, or throughout the ventral horn. The group ending medially, the **medial activation system,** control lower motoneurons that innervate postural and girdle muscles. The group ending laterally, the **lateral activation system,** control lower motoneurons that innervate distally located muscles used for fine movement. The group ending throughout the ventral horn, the **nonspecific activating pathways,** contribute to background levels of excitation in the cord and facilitate local reflex arcs.

Postural and Gross Movement Pathways: Medial Activation System

Four tracts from the brain stem and one from the cerebral cortex deliver commands to the medial motoneuron pools. The brain stem tracts arise in the tectum (posterior midbrain), medial reticular formation, and vestibular nuclei (brain stem nuclei concerned with equilibrium); hence the tracts are tectospinal, medial reticulospinal, and medial and lateral vestibulospinal. The tract from the cortex is the medial corticospinal*; in this case, medial refers to the location of the tract in the white matter of the spinal cord.

---

*Evolving terminology: Corticospinal neurons are also known as corticomotoneuronal. Historically, *corticospinal* was an appropriate term because the sensory regulation function of some corticospinal neurons was undiscovered; now, *corticospinal* is a somewhat ambiguous term but remains the most commonly used term to describe upper motoneurons that arise in the cerebral cortex and terminate in the spinal cord.

The activity of the medial activating pathways (Fig. 9–11) can be demonstrated by the reactions generated when a loud noise occurs behind someone. The eyes and face turn toward the sound, and postural adjustments support the movements, even before the person is consciously aware of the stimulus. These coordinated, involuntary reactions occur via circuitry in the brain stem.

The superior colliculus processes visual, auditory, and somatic information; in the example above, the stimulus is auditory. Neural activity in the superior colliculus stimulates neurons that project to the spinal cord in the **tectospinal tract,** activating lower motoneurons in the cervical spinal cord to reflexively turn the head toward the sound.

Quickly turning the head destabilizes a person unless other automatic responses compensate for the change in weight distribution. Loss of balance is prevented by muscles whose motoneurons are activated by neurons in the pontine reticular formation and vestibular nuclei.

The **medial reticulospinal tract** begins in the pontine reticular formation. Stimulation of this tract facilitates ipsilateral lower motoneurons innervating pos-

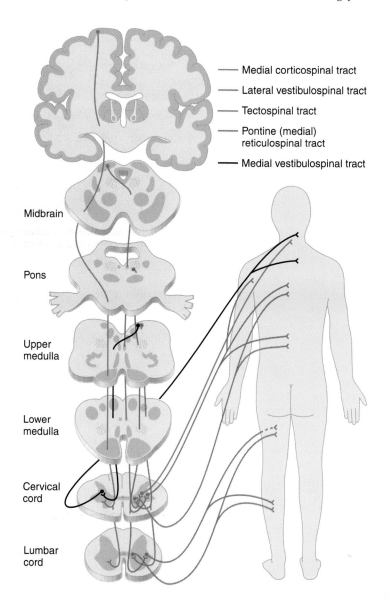

Medial corticospinal tract
Lateral vestibulospinal tract
Tectospinal tract
Pontine (medial) reticulospinal tract
Medial vestibulospinal tract

Midbrain

Pons

Upper medulla

Lower medulla

Cervical cord

Lumbar cord

FIGURE 9–11

Medial activation pathways adjust the activity in axial and girdle muscles.

tural muscles and limb extensors. Medial vestibular nuclei receive information about head movement and position from the vestibular apparatus. Axons projecting from these nuclei to the spinal cord, the **medial vestibulospinal tracts,** project bilaterally to cervical and thoracic levels and affect activity in lower motoneurons controlling neck and upper back muscles. The lateral vestibular nucleus responds to gravity information from the vestibular system. The pathways from the lateral vestibular nucleus, **lateral vestibulospinal tracts,** project ipsilaterally and facilitate lower motoneurons to extensors while inhibiting lower motoneurons to flexors. The activity of these descending tracts is not limited to responding to destabilization; instead, these tracts operate constantly to maintain equilibrium in upright positions. In addition to their activation by sensory input, the reticulospinal and tectospinal neurons are influenced by the cerebral cortex, forming corticoreticulospinal and corticotectospinal pathways.

The direct pathway from the cerebral cortex to the spinal cord, the **medial corticospinal tract,** descends from the cortex through the internal capsule and the anterior brain stem. The medial corticospinal tract ends in the thoracic cord and thus assists in control of neck, shoulder, and trunk muscles.

The medial activating pathways are primarily involved in control of posture and proximal movements. The tracts include the medial corticospinal, medial reticulospinal, medial and lateral vestibulospinal, and tectospinal.

Most supraspinal control of posture and proximal movement is from brain stem centers. Cortical projections probably prepare the postural system for intended movements. In contrast to the postural and gross movement control of the medial activation system, a different group of descending pathways controls distal limb movements.

Fine Movement Pathways: Lateral Activation System

Three pathways controlling distal limb movements descend in the lateral spinal cord and synapse with laterally located motoneuron pools in the ventral horn. These descending pathways are the lateral corticospinal, rubrospinal, and lateral reticulospinal (Fig. 9–12). In monkeys, the only long-term deficit from severing the lateral corticospinal and rubrospinal

tracts is the inability to use the fingers individually to pick small objects out of deep cavities in a board; balance, walking, running, and climbing abilities remain near normal (Lawrence and Kuypers, 1968).

The **lateral corticospinal tract*** arises in motor planning areas and primary motor cortex. The axons project downward, passing through the internal capsule, the basis pedunculi of the midbrain, the anterior pons, the pyramids of the medulla, and lateral spinal cord to synapse with lower motoneurons controlling fine, distal movements. The corticospinal tracts in the lower medulla form the pyramids, where, at the junction of the medulla and spinal cord, the lateral corticospinal axons cross to the contralateral side (the medial corticospinal axons remain ipsilateral) (see Fig. 9–12). The unique contribution of lateral corticospinal neurons is **fractionation** of movement. Fractionation is the ability to activate individual muscles independently of other muscles. Fractionation is essential for normal movement of the hands, enabling us to tie knots, press individual piano or typewriter keys, and pick up small objects.

Some corticospinal axons, the **corticobulbar fibers,** project to cranial nerve nuclei in the brain stem and thus do not reach the spinal cord. Corticobulbar activity controls lower motoneurons innervating the muscles of the face, tongue, pharynx, and larynx (Fig. 9–13). Lower motoneurons to muscles of the lower face are controlled by contralateral corticobulbar fibers. Lower motoneurons to muscles of the upper face are bilaterally controlled.

Corticospinal fibers arise in the primary motor, premotor, and supplementary motor cortex. The corticospinal cell bodies in the primary motor cortex are arranged somatotopically, in an inverted homunculus similar to the cortical sensory representation. Many neurons in the primary motor cortex are specialized for movement execution. Two regions anterior to the primary motor cortex are involved in preparing for movement: the **lateral premotor area** is on the lateral surface of the hemisphere, and the **supplementary**

---

*Clinical terminology: Historically, the corticospinal pathway was considered to be the most important pathway, with other descending pathways playing minor supporting roles. Because the lateral corticospinal path forms the medullary pyramids, this pathway was called the pyramidal system. The remaining motor pathways were called extrapyramidal. The basal ganglia were mistakenly believed to exclusively control the extrapyramidal tracts, and thus in clinical terminology, *extrapyramidal* became synonymous with *basal ganglia.* This terminology remains common in clinical use. The division of motor control into pyramidal/extrapyramidal is a false dichotomy because the basal ganglia are a major influence on cortical motor areas, and thus contribute to the control of the pyramidal tract, and because the cerebral cortex and cerebellum have great influence on the descending tracts.

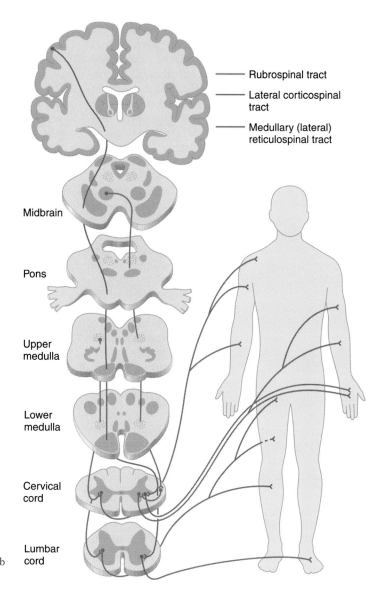

Rubrospinal tract

Lateral corticospinal tract

Medullary (lateral) reticulospinal tract

Midbrain

Pons

Upper medulla

Lower medulla

Cervical cord

Lumbar cord

FIGURE 9–12

Lateral activation pathways adjust the activity in limb muscles.

motor area is on the superior and medial surface (Fig. 9–14). The lateral premotor area is named for its position anterior to the primary motor cortex. Stimulation of the lateral premotor area produces muscle activity that spans several joints. Unlike the lateral premotor cortex, many supplementary motor cortex cells are active prior to bimanual movements (Tanji et al., 1988) and sequential movements (Mushiake et al., 1990).

The **rubrospinal tract** originates in the red nucleus of the midbrain, crosses to the opposite side, then descends through the pons, medulla, and lateral spinal cord to synapse with lower motoneurons primarily innervating upper limb flexor muscles (see Fig. 9–12). The **lateral (medullary) reticulospinal tract** descends bilaterally to facilitate flexor muscles motoneurons and to inhibit extensor motoneurons. However, during movement, the effect of lateral reticu-

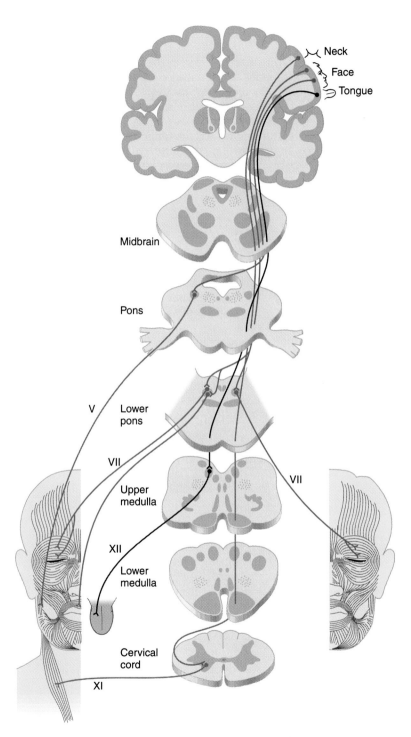

Neck
Face
Tongue

Midbrain

Pons

V   Lower pons

VII

Upper medulla

VII

XII

Lower medulla

Cervical cord

XI

FIGURE 9–13

Corticobulbar pathways. Axons from the cerebral cortex transmit information to cranial nerve cell bodies; the cranial nerves project to muscles controlling movements of the head and neck. All eight cranial nerves that innervate skeletal muscle are influenced by descending input from the cortex. For simplicity, only four of the eight cranial nerves that innervate skeletal muscle are illustrated.

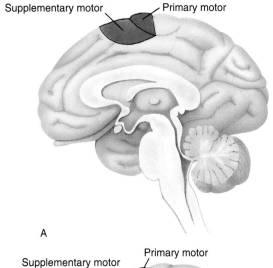

A

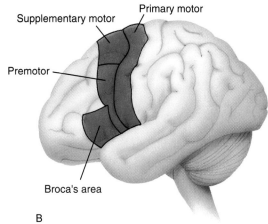

B

FIGURE 9–14

Location of the primary motor, premotor, and supplementary motor cortices. Broca's area plans the movements of speech.

lospinal tract activity can be reversed. That is, during some movements, the lateral reticulospinal tract inhibits flexor muscle motoneurons and facilitates extensor muscle motoneurons. The rubrospinal and lateral reticulospinal pathways both receive input from the cerebral cortex.

The lateral descending pathways control limb movements. The pathways include the lateral corticospinal, rubrospinal, and lateral reticulospinal. The lateral corticospinal tract is unique in providing fractionation of distal movements.

Nonspecific Activating Pathways

Tracts descending from the locus ceruleus and from the raphe nuclei enhance the activity of interneurons and motoneurons in the spinal cord. The locus ceruleus and raphe nuclei are located in the brain stem, and the tracts arising from them are the **ceruleospinal** and the **raphespinal tracts.** The motor effects are due to release of neuromodulators (see Chap. 3). These effects are general, not related to specific movements, and may contribute to changes in motor performance with varying levels of motivation.

CONTROL CIRCUITS

The basal ganglia and cerebellum, via separate neuronal circuits, adjust activity in the descending pathways. Because the basal ganglia and cerebellum control certain aspects of movement, yet do not have any direct connections with lower motoneurons, they function as control circuits (Westmoreland et al., 1994). Both the basal ganglia and cerebellum influence motor areas of the cortex via the thalamus. In addition, both the basal ganglia and cerebellum have inputs to certain brain stem nuclei controlling descending pathways.

Basal Ganglia

The basal ganglia include the following nuclei found in the cerebrum and midbrain: the caudate, putamen, globus pallidus, subthalamic nuclei, and substantia nigra (Fig. 9–15).* Based on anatomical proximity, some of the nuclei have joint names: the globus pallidus and putamen together form the lentiform nucleus; the caudate and putamen together are the striatum. The lentiform nucleus is shaped like a broad, solid cone, with the putamen lateral (wide end of cone) and the globus pallidus medial (point of cone).

---

*Historically, the term *basal ganglia* referred generically to deep gray matter in the cerebrum. Thus the amygdala and the thalamus were considered part of the basal ganglia, and the subthalamic nucleus and substantia nigra were excluded. Because subsequent research has established that the functions of the amygdala and most of the thalamus are not closely related to the functions of the striatum and globus pallidus, most authors now exclude the amygdala and thalamus. Similarly, because essential functional relationships have been confirmed among the substantia nigra, subthalamic nucleus, globus pallidus, and striatum, these structures are currently considered to constitute the basal ganglia.

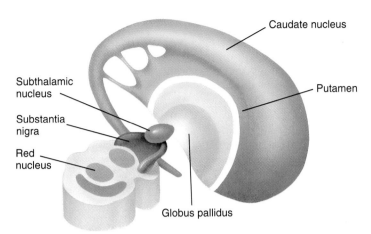

Caudate nucleus

Subthalamic
nucleus

Putamen

Substantia
nigra

Red
nucleus

Globus pallidus

FIGURE 9–15

The basal ganglia include the caudate, putamen, globus pallidus, substantia nigra, and subthalamic nuclei. (Modified with permission from Hendelman, W. J. [1994]. Students atlas of neuroanatomy. Philadelphia: W. B. Saunders, p. 39.)

The globus pallidus has an internus (medial) section and an externus (lateral) section. The caudate is joined with the putamen anteriorly, and during growth of the brain, the caudate assumes a **C** shape adjacent to the lateral ventricle (see Chap. 5). Because the function of the caudate is primarily cognitive, not motor, the caudate will be covered in Chapter 16.

The substantia nigra is a nucleus in the midbrain named for the color of its cells. Some cells contain melanin, making the nucleus appear black. The substantia nigra has two parts: compacta and reticularis. The substantia nigra compacta provides essential dopamine to the striatum. The substantia nigra reticularis and the globus pallidus internus are the output nuclei of the basal ganglia system. The output from these nuclei inhibits the motor thalamus and the pedunculopontine nucleus of the midbrain (Table 9–2). Stimulation of the pedunculopontine nucleus elicits rhythmical behaviors such as locomotor patterns. In addition to the motor circuit, the basal ganglia also have circuits that influence eye movements and emotions. These circuits are beyond the scope of this text.

Although functioning basal ganglia are vital for normal movement, they have no direct connections with lower motoneurons. Their influence is exerted through either of the following:

• Motor planning areas of the cerebral cortex
• Pedunculopontine nucleus of the midbrain

The influence on the supplementary motor area of the cerebral cortex is indirect, via the thalamus. A major basal ganglia circuit connects motor areas of the cortex to the putamen, the putamen to the output nuclei (via a direct and an indirect route), and the output nuclei to the motor thalamus, and the motor thalamus excites the motor areas of the cerebral cortex (Fig. 9–16). The motor control exerted by the basal ganglia on the cortex is transmitted via the corticofugal tracts. Corticofugal tracts are composed of upper motoneurons whose cell bodies are in the cerebral cortex: corticospinal, corticopontine, and corticobulbar tracts. A second important basal ganglia motor circuit consists of connections from the globus pallidus internus to the pedunculopontine nucleus, thence to the reticulospinal and to the vestibulospinal tracts.

TABLE 9–2	
BASAL GANGLIA MOTOR CIRCUIT	
Role	**Structure**
Receive input from premotor and sensorimotor cortex	Putamen
Process information within the basal ganglia circuit	Globus pallidus externus Subthalamic nucleus Substantia nigra compacta
Send output to the motor areas of the cerebral cortex (via the motor thalamus) and pedunculopontine nucleus	Globus pallidus internus Substantia nigra reticularis

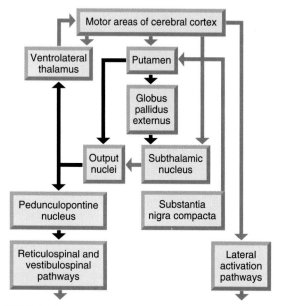

FIGURE 9–16

Basal ganglia functional connections during normal movement. Red arrows indicate that the connection is excitatory. Black arrows indicate that the connection is inhibitory. The direct route within the basal ganglia is from the putamen to the output nuclei. The indirect route is from the putamen to the globus pallidus externus, then to the subthalamic nucleus, and finally to the output nuclei. The output nuclei are the globus pallidus internus and the substantia nigra reticularis.

NEUTROTRANSMITTERS IN THE BASAL GANGLIA CIRCUIT

Of the many neurotransmitters and neuromodulators active in the basal ganglia circuitry, the actions of only a few are well understood. The effects are complex because often one structure will inhibit a second structure whose output is also inhibitory to a third structure. The result of this sequence is to disinhibit, or increase the activity in, the third structure because inhibition of the second structure reduces its inhibitory output.

The cortical motor areas produce excitation of the putamen by delivering the transmitter glutamate. In the direct pathway, the putamen inhibits the output nuclei by γ-aminobutyric acid (GABA) and substance P. Because the output from the output nuclei is inhibitory to the motor thalamus and pedunculopontine nucleus (via GABA), the net effect of increasing input to the putamen is to increase the excitatory output from the motor thalamus to the cortical motor

areas. Therefore the activity in the corticofugal pathways will increase.

The indirect route proceeds from the putamen's inhibiting the globus pallidus externus (via GABA and enkephalins) to inhibition of the subthalamic nucleus (GABA), excitation of the substantia nigra reticularis (glutamate), inhibition of the motor thalamus (GABA), ending with less excitation of the motor areas of the cerebral cortex. Therefore, activity in the indirect route decreases the activity in corticofugal pathways.

Dopamine from the substantia nigra compacta enhances activity in the motor cortex by binding with two types of receptors in the basal ganglia circuit, D_1 and D_2. Binding of dopamine with the D_1 receptor facilitates activity in the direct pathway, while binding of dopamine with the D_2 receptor inhibits activity in the indirect pathway.

FUNCTION OF THE BASAL GANGLIA

The basal ganglia are involved with comparing proprioceptive information and movement commands, sequencing movements, and regulating muscle tone and muscle force. Hallet (1993) proposes that the function of the basal ganglia is to select and inhibit specific motor synergies; the direct path from the putamen to the output nuclei selects synergies, and the indirect pathway (via globus pallidus externus and subthalamic nucleus) inhibits synergies. The role of the basal ganglia in motor learning will be discussed in Chapter 16.

Cerebellum

Massive amounts of sensory information enter the cerebellum, and cerebellar output is vital for normal movement. However, severe damage to the cerebellum does not interfere with sensory perception or with muscle strength. Instead, coordination of movement and postural control are degraded, and muscle tone is decreased. Muscle tone is the resistance of muscle to passive stretch.

ANATOMY OF THE CEREBELLUM

The outer layer of the cerebellum is gray matter, consisting of three cortical layers (Fig. 9–17). The outer and inner layers contain interneurons (granule, Golgi, stellate, and basket cells), and the middle layer contains Purkinje cell bodies. Purkinje cells are the out-

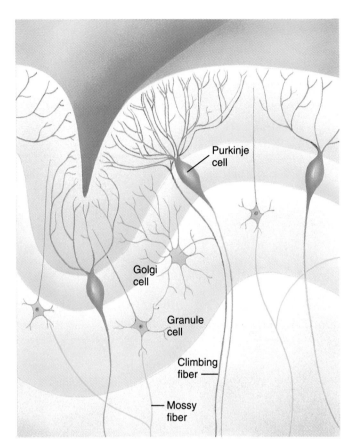

FIGURE 9–17

Three layers of the cerebellar cortex. In the middle layer are cell bodies of Purkinje cells, the output neurons of the cerebellar cortex. Climbing and mossy fibers are the input fibers to the cerebellar cortex.

put cells from the cerebellar cortex, and their projections inhibit the deep cerebellar nuclei and the vestibular nuclei. Two types of afferents enter the cerebellum: mossy fibers, from the spinal cord, reticular formation, vestibular system, and pontine nuclei; and climbing fibers, from the inferior olivary nucleus in the medulla.

Three lobes form the cerebellum (Fig. 9–18):

- **Anterior**
- **Posterior**
- **Flocculonodular**

The anterior lobe is superior and is separated from the larger posterior lobe by the primary fissure. Tucked underneath the posterior lobe, touching the brain stem, is the small flocculonodular lobe.

Vertically, the cerebellum can be divided into sections (see Fig. 9–18 *E*):

- Midline **vermis**
- **Paravermal hemisphere**
- **Lateral hemisphere**

Each of the vertical sections are associated with a specific class of movements, as we will see later. Each vertical section projects either to specific deep cerebellar nuclei or to vestibular nuclei. The deep cerebellar nuclei, from medial to lateral, are the fastigial, globose, emboliform, and dentate nuclei. Fibers connecting the cerebellum with the brain stem form three cerebellar peduncles on each side of the brain stem. The superior cerebellar peduncle attaches to the midbrain and contains most of the cerebellar efferent fibers. Fibers from the cerebral cortex synapse in the pons, and then the information travels via axons in the middle peduncle into the cerebellum. The inferior peduncle brings afferent information from the brain stem and spinal cord into the cerebellum and sends

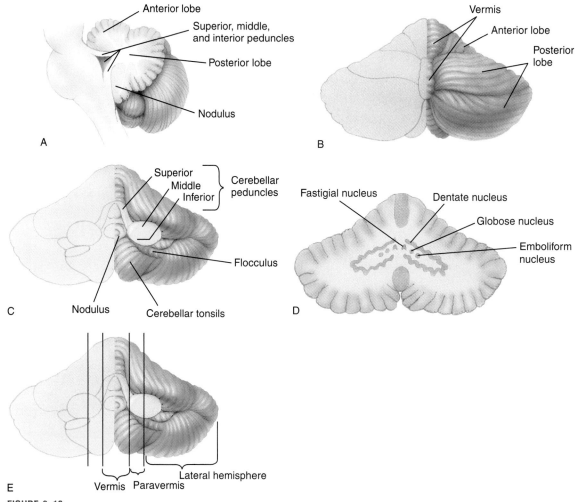

FIGURE 9–18

Anatomy of the cerebellum. *A*, Midsagittal section, showing cerebellar peduncles and lobes of the cerebellum. *B*, Posterior view of cerebellum. *C*, Anterior view of cerebellum with the brain stem removed. *D*, Coronal section of cerebellum, revealing the deep cerebellar nuclei. *E*, Vertical divisions of the cerebellum.

efferents from the cerebellum to the vestibular nuclei and reticular nuclei in the brain stem.

FUNCTIONS OF THE CEREBELLUM

The cerebellum has two major roles: comparing actual motor output to the intended movement and then adjusting the movement as necessary. Information regarding intended movements is delivered to the cerebellum by collaterals of corticospinal and corticobulbar fibers that synapse in the pons with neurons projecting to the cerebellum. The cerebellum also receives information regarding activity in spinal interneurons via the internal feedback pathways (anterior spinocerebellar and rostrospinocerebellar pathways [see Chap. 6]). Information about the actual movement is provided by information from muscle spindles, Golgi tendon organs, and cutaneous mechanoreceptors (posterior spinocerebellar and cuneocerebellar pathways [see Chap. 6]). The cerebellum integrates information from these sources and produces appropriate adjustments in the activity of descending activation pathways.

Human movements can be categorized into three broad classes:

- Equilibrium
- Gross movements of the limbs
- Fine, distal, voluntary movements

The cerebellum has specialized regions for controlling each of these classes of movement. Equilibrium is regulated by the vestibulocerebellum, named for its reciprocal links with the vestibular system. Gross limb movements are regulated by the spinocerebellum, named for its extensive connections with the spinal cord. Distal limb voluntary movements are regulated by the cerebrocerebellum, named for its connections with the cerebral cortex (Fig. 9–19).*

When a person reaches for a book off a high shelf, the vestibulocerebellum provides anticipatory contraction of lower limb and back muscles to prevent loss of balance. Otherwise, as the upper limb moves, altering the position of the body's center of mass, the person would fall forward. The upper limb reaching movement is coordinated by the spinocerebellum. Without the spinocerebellar contribution, the reach would be jerky and inaccurate. The cerebrocerebellum coordinates the finger movements to grasp the book.

Vestibulocerebellum is the functional name for flocculonodular lobe because this area receives information directly from vestibular receptors and connects reciprocally with the vestibular nuclei. The vestibulocerebellum also receives information from visual areas of the brain. Via connections with vestibular nuclei, the vestibulocerebellum influences eye movements and postural muscles.

Spinocerebellum is the functional name for the vermis and paravermal region because of extensive connections with the spinal cord. Somatosensory information, internal feedback from spinal interneu-

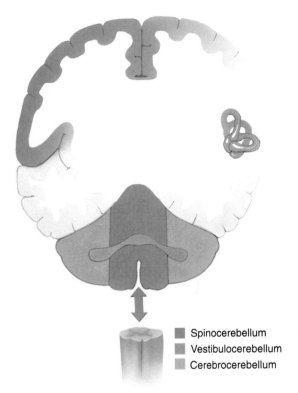

FIGURE 9–19

The three functional divisions of the cerebellum and their connections: vestibulocerebellum, spinocerebellum, and cerebrocerebellum.

Legend:
- Spinocerebellum
- Vestibulocerebellum
- Cerebrocerebellum

rons, and sensorimotor cortex information converge in the spinocerebellum. This information is used to control ongoing movement via the brain stem descending pathways.

The vermis projects to the fastigial nucleus; the nucleus in turn controls medial activation pathways by direct action on brain stem nuclei and on the cerebral cortex via the motor thalamus. The vermis also has connections with cranial nerve nuclei that control speech. The paravermal area projects to the globose and emboliform nuclei; these nuclei influence the lateral activation pathways by action on brain stem nuclei and by projecting to the cerebral cortex via the motor thalamus.

The lateral cerebellar hemispheres connect indirectly with areas of the cerebral cortex controlling distal limb muscles; thus this section of the cerebellum is the **cerebrocerebellum.** Input to the cerebrocerebellum is from cerebral cortex (premotor, sensorimotor,

---

*In addition to the anatomical and functional classification systems, some authorities divide the cerebellum into sections based on phylogenetic considerations. According to this classification system, parts of the cerebellum are designated by when each part appeared in the evolution of species. Thus the part of the cerebellum considered to be phylogenetically the oldest is called the archicerebellum, the old part is called paleocerebellum, and the newest part is called the neocerebellum. Each of these designations is roughly equivalent to the functional classifications: archicerebellum is approximately the same as vestibulocerebellum, paleocerebellum is approximately the same as spinocerebellum, and neocerebellum is approximately the same as cerebrocerebellum.

TABLE 9–3

NEURAL CONNECTIONS OF THE FUNCTIONAL DIVISIONS OF THE CEREBELLUM

Functional Division (Anatomical Location in Parentheses)	Receives Input from	Sends Output to	Output Reaches Lower Motoneurons via
Vestibulocerebellum (flocculonodular lobe)	Vestibular apparatus Vestibular nuclei	Vestibular nuclei	Vestibulospinal tracts and tracts that coordinate eye and head movements (see Chap. 14)
Spinocerebellum Vermal section	Spinal cord (from trunk) Vestibular nuclei Auditory and vestibular information (via brain stem nuclei)	Vestibular nuclei Reticular nuclei (via fastigial nucleus)	Vestibulospinal tracts Reticulospinal tracts
Paravermal section	Spinal cord (from limbs)	Red nucleus (via globose and emboliform nuclei)	Rubrospinal tract
Cerebrocerebellum (lateral cerebellar peduncles)	Cerebral cortex (via pontine nuclei)	Motor and premotor cortices (via dentate nucleus and motor thalamus)	Corticospinal, cortico-bulbar, and rubrospinal tracts

and other cortical areas [see Chap. 16]) fibers that synapse with neurons in the pons. Axons of the pontine neurons project to the cerebrocerebellum. Efferents from the lateral cerebellar hemispheres project to the dentate nucleus. Prior to voluntary movements, alterations in dentate neural activity precede changes in activity in motor areas of the cerebral cortex. Thus the dentate is involved in motor planning. Efferents from the dentate nucleus project to the motor thalamus, efferents from which in turn project to the cerebral cortex. Functions of the cerebrocerebellum and dentate include the following:

• Coordination of voluntary movements via influence on corticofugal pathways

• Planning of movements
• Ability to judge time intervals and produce accurate rhythms (Ivry and Keele, 1989; Lundy-Ekman et al., 1991)

The connections of the functional divisions of the cerebellum are listed in Table 9–3 and illustrated in Figure 9–20.

Summary

For normal movement, the motor planning areas, control circuits, and descending pathways must act in concert with sensory information to provide instructions to lower motoneurons. Only lower motoneurons deliver the signals from the central nervous system to the skeletal muscles that generate movement. Figures 9–21 and 9–22 summarize the complex neural activity required to generate movement.

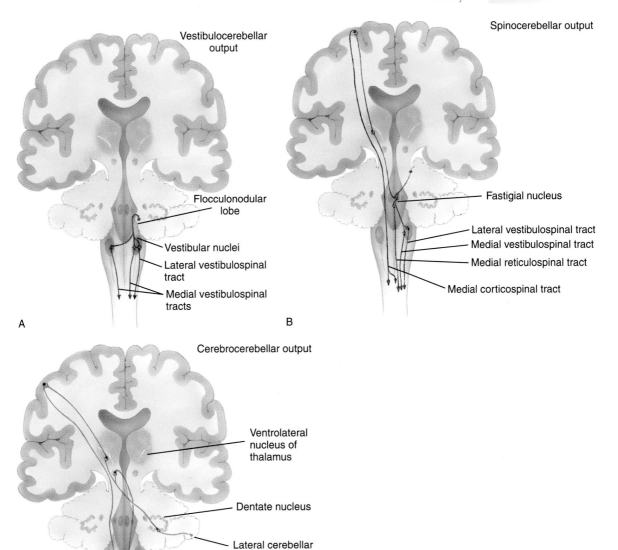

Vestibulocerebellar
output

Flocculonodular
lobe

Vestibular nuclei

Lateral vestibulospinal
tract

Medial vestibulospinal
tracts

A

Spinocerebellar output

Fastigial nucleus

Lateral vestibulospinal tract
Medial vestibulospinal tract
Medial reticulospinal tract

Medial corticospinal tract

B

Cerebrocerebellar output

Ventrolateral
nucleus of
thalamus

Dentate nucleus

Lateral cerebellar
hemisphere

Rubrospinal tract

Lateral
corticospinal tract

C

FIGURE 9–20

The efferents from each functional division of the cerebellum.
A, Vestibulocerebellar output. *B,* Spinocerebellar output.
C, Cerebrocerebellar output.

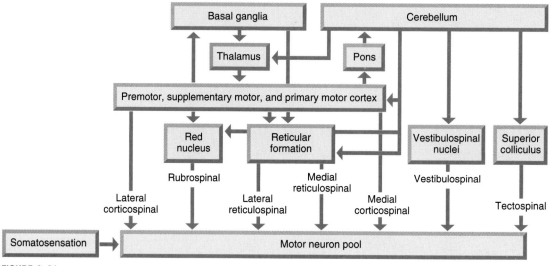

FIGURE 9–21

Flow chart summarizing the relationships among components of the motor system.

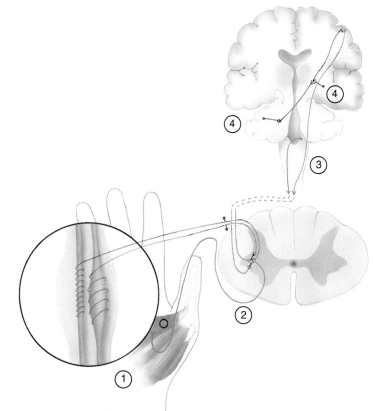

FIGURE 9–22

Simplified schematic diagram of the major influences on muscle activity. *1,* Peripheral region. The alpha motoneurons innervate contractile fibers in skeletal muscle. Proprioception from muscle sensors (muscle spindle shown) affects motor output. *2,* The spinal region integrates information from other spinal cord segments (not shown), from local circuits within the cord, and from the brain. *3,* The descending tracts provide information from the brain. The tracts illustrated are the lateral corticospinal and lateral reticulospinal. *4,* The control circuits (cerebellum, basal ganglia) adjust the level of activity in the descending tracts.

REVIEW QUESTIONS

1. What is the difference between a lower motoneuron and an upper motoneuron?

2. What is the general function of control circuits?

3. Why are slow twitch muscle fibers usually activated before fast twitch muscle fibers?

4. What factors determine the activity of a motor unit?

5. What is alpha-gamma coactivation?

6. What are the differences between the phasic and tonic stretch reflexes?

7. What roles does the Golgi tendon organ play?

8. How does changing a person's arousal level alter the response to a quadriceps tendon tap?

9. How is an H-reflex produced? What information does an H-reflex give?

10. What is the function of reciprocal inhibition?

11. How does the use of the term *synergy* differ between clinicians and motor control researchers?

12. What is a central pattern generator?

13. What does response reversal demonstrate?

14. List each medial activating pathway and its function.

15. List each lateral activating pathway and its function.

16. List each nonspecific activating pathway and its function.

17. What are the two routes for information from the basal ganglia output nuclei to the lower motoneurons?

18. What is the end result of dopamine binding in the basal ganglia direct and indirect pathways?

19. What are the functions of the basal ganglia?

20. What is the major function of the cerebellum?

21. What are the major sources of input to the cerebellum?

22. Which part of the cerebellum coordinates individual finger movements? Gross limb movements? Postural adjustments?

References

Forssberg, H. (1985). Ontogeny of human locomotor control: 1. Infant stepping, supported locomotion and transition to independent locomotion. Exp. Brain Res. 57:480–493.

Forssberg, H., Grillner, S., Rossignol, S., et al. (1975). Phase dependent reflex reversal during walking in chronic spinal cats. Brain Res. 85:103–107.

Gandevia, S. C. (1993). Assessment of corticofugal output: Strength testing and transcranial stimulation of the motor cortex. Science and practice in clinical neurology. Cambridge, England: Cambridge University Press, pp. 76–88.

Gibson, J. J. (1977). The theory of affordances: Perceiving, acting and knowing. Hillsdale, N.J.: Lawrence Erlbaum.

Grillner, S., and Zangger, P. (1975). How detailed is the central pattern generation for locomotion? Brain Res. 88:367–371.

Hallett, M. (1993). Physiology of basal ganglia disorders: An overview. Can. J. Neurol. Sci. 20:177–183.

Ivry, R., and Keele, S. (1989). Timing functions of the cerebellum. J. Cognitive Neurosci. 1(2):136–152.

Lawrence, D., and Kuypers, H. (1968). The functional organization of the motor system in the monkey: I. The effects of bilateral pyramidal lesions. II. The effects of lesions of the descending brainstem pathways. Brain 91(1):1–36.

Lundy-Ekman, L., Ivry, R., Keele, S., et al. (1991). Timing and force control deficits in clumsy children. Cognitive Neurosci. 3:367–376.

McGraw, M. B. (1945). The neuromuscular maturation of the human infant. New York: Hafner Press.

Mushiake, J., Inase, M., and Tanji, J. (1990). Selective coding of motor sequence in the supplementary motor area of the monkey cerebral cortex. Exp. Brain Res. 82:208–210.

Pearson, K. (1976). The control of walking. Sci. Am. 235(6): 72–86.

Tanji, J., Okano, K., and Sato, K. C. (1988). Neuronal activity in cortical motor areas related to ipsilateral, contralateral and bilateral digit movements of the monkey. J. Neurophysiol. 60: 325–343.

Thelen, E., Fisher, D. M., and Ridley-Johnson, R. (1984). The relationship between physical growth and a newborn reflex. Infant Behav. Dev. 7:479–493.

Westmoreland, B. F., Benarroch, E. E., Daube, J. J., et al. (1994). Medical neurosciences. Boston: Little, Brown and Co.

Suggested Readings

Cordo, P., and Harnad, S. (1994). Movement control. Cambridge, England: Cambridge University Press.

Rothwell, J. (1994). Control of human voluntary movement (2nd ed.). London: Chapman and Hall.

Umphred, D. A. (1995). Neurological rehabilitation (3rd ed.). St. Louis: Mosby.

10

Clinical Disorders of the Motor System

I am a 39-year-old woman. Before my stroke, I was very athletic. I ran every day. Two years ago I was at work, filling orders at a shoe warehouse, when I developed an excruciating headache. Prior to this, I never had headaches. My sister, who worked with me, asked if I needed an ambulance. I didn't think I needed an ambulance for a headache, so she drove me to a local emergency medical clinic. In the car on the way to the clinic, I had seizures. An aneurysm (a dilation of part of a wall of an artery, where the arterial wall is abnormally thin) had burst in my brain, causing bleeding into my brain. I underwent surgery to repair the damaged artery. People told me the doctor was amazed that I survived, and he repaired a second aneurysm during the surgery so that it would not rupture later. I had another surgery to insert a shunt about 2 weeks later because fluids were not draining normally from my brain.

I don't remember anything about the 2 months following the surgery. The stroke never affected my sensation or language abilities. The first thing I remember is that I couldn't recall how to chew or swallow. I couldn't plan the movements. I had to slowly figure out by trial and error how to eat by myself. Now I can do many things independently, except transfers and walking. Keeping my balance is difficult and fatiguing. When I am sitting, I use my arms for balance. My legs are very weak; I can move them a little when I'm lying down, but I cannot move them when I'm standing. I have had physical therapy since my hospitalization, focusing on balance, transfers, standing, and assisted walking. I take Dilantin to prevent seizures. I also take the antidepressants amitriptyline hydrochloride and nortriptyline hydrochloride. The amitriptyline hydrochloride also acts as an aid for sleeping, which I need because I am not active enough to get tired.

> The stroke has profoundly altered my life. Before the stroke, I was vigorous, healthy, and independent. Now I live in a convalescent home, I use a wheelchair, and I need help to get into and out of the wheelchair. I think about when I could walk before the stroke, and I am planning on walking again.
>
> —Janet Abernathy

Disorders of the motor system include abnormalities of the following:

- Muscle strength
- Muscle bulk
- Muscle contraction
- Muscle tone
- Reflexes
- Movement efficiency and speed
- Postural control

MUSCLE STRENGTH AND MUSCLE BULK

Although the terms *paresis* and *paralysis* are often used synonymously, technically **paralysis** refers to complete loss of voluntary contraction and **paresis** refers to a partial loss. Decreased muscle strength is commonly described by its distribution: hemiplegia is weakness affecting one side of the body, paraplegia affects the body below the arms, and tetraplegia affects all four limbs. The loss of muscle bulk is called atrophy. Muscle atrophy resulting from lack of use is called disuse atrophy; muscle atrophy resulting from damage to the nervous system is called neurogenic atrophy.

INVOLUNTARY MUSCLE CONTRACTIONS

Spontaneous involuntary muscle contractions include the following:

- Muscle spasms
- Cramps
- Fasciculations
- Fibrillations
- Abnormal movements generated by dysfunctional basal ganglia

Muscle spasms are sudden, involuntary contractions of muscle. Cramps are particularly severe and painful muscle spasms. **Fasciculations** are quick twitches of muscle fibers of a single motor unit that are visible on the surface of the skin. **Fibrillations** are brief contractions of single muscle fibers not visible on the surface of the skin. Abnormal movements due to basal ganglia disorders will be considered within that context.

Of these involuntary contractions, the first three occasionally occur in a healthy neuromuscular system, or they may be signs of pathology. For example, benign muscle spasms and cramps are common following prolonged exercise, particularly if excessive sweating has caused sodium depletion. Benign fasciculations are visible in the eyelid twitches that sometimes accompany anxiety. Pathological cramps, spasms, and fasciculations will be discussed in the context of specific lesions. Only fibrillations are always pathological and can result from either upper or lower motoneuron disorders.

MUSCLE TONE

Muscle tone is the amount of tension in resting muscle. Clinically, muscle tone is usually assessed by passive range of motion. In a person with an undamaged neuromuscular system who is able to completely relax the muscles, resistance to passive muscle stretch is minimal. Intrinsic viscoelastic characteristics of the muscle provide the tension of normal resting muscle tone; stretch reflexes do not contribute. Even in relaxed standing, humans rely primarily on the loading of the skeleton and viscoelastic properties of muscles; muscles are only slightly active or become active intermittently when sway exceeds tolerable limits.

When a healthy person is anxious, the muscle tone increases, and stretch reflexes become more responsive. During passive range of motion in an anxious, neurologically intact subject, two responses can be noted. The first response provides strong, brief resistance to movement, generated by spinal circuits that are activated by muscle spindle and Golgi tendon organ afferents (Burke and Gandevia, 1993). The second response is tonic, provides less resistance, and is adjusted by descending motor commands. Thus, intrinsic viscoelastic properties, descending motor commands, and muscle proprioceptors (spindles and

Golgi tendon organs) function together to regulate muscle tone.

Damage anywhere in the motor system often interferes with the ability to regulate muscle tone. **Hypotonia** (also called flaccidity), or abnormally low resistance to passive stretch, occurs in the following:

- Cerebellar disorders (due to decreased descending facilitation to interneurons and motoneurons)
- Lower motoneuron lesions (due to interruption of afferent and/or efferent axons)
- Temporarily following an acute upper motoneuron lesion

The other extreme, **hypertonia** or abnormally strong resistance to passive stretch, occurs in the following:

- Chronic upper motoneuron lesions
- Some basal ganglia disorders

There are two types of hypertonia:

- Spastic
- Rigid

In **spastic hypertonia**, usually called **spasticity**, the amount of resistance to passive movement depends on the velocity of movement. Therefore spasticity is velocity-dependent hypertonia. In **rigidity**, resistance to passive movement remains constant regardless of the speed of force application. Thus rigidity is velocity-independent hypertonia.

When descending motor commands are interrupted by an acute upper motoneuron lesion, the lower motoneurons affected become temporarily inactive. This condition is called spinal or cerebral shock, depending on the location of the lesion. During nervous system shock, stretch reflexes cannot be elicited, and the muscles are hypotonic; that is, the muscles have abnormally low tone because facilitation of lower motoneurons by descending activating pathways has been lost. Following recovery from central nervous system shock, interneurons and lower motoneurons usually resume activity, although their activity is no longer modulated by upper motoneurons. In many cases, during the months following the upper motoneuron lesion, muscle tone increases as a result of intrinsic changes in the muscles, producing spasticity.

DISORDERS OF LOWER MOTONEURONS

Lower motoneurons innervate muscle fibers. These neurons can be damaged by trauma, infection (poliomyelitis), degenerative or vascular disorders, or tumors. Interruption of lower motoneuron signals to muscle decreases or prevents muscle contraction. If the lower motoneuron cell body or its axon is destroyed, the following signs occur in the affected muscles:

- Loss of reflexes
- Atrophy
- Flaccid paralysis
- Fibrillations

Traumatic injuries to lower motoneurons are discussed in Chapter 11. An example of a lower motoneuron lesion is poliomyelitis, an acute infection of the anterior horn of the spinal cord. Poliovirus invades lower motoneurons and destroys some of the neurons (Fig. 10–1), denervating some muscle fibers. Although some people do not survive the infection, polio survivors recover some muscle strength as surviving neurons sprout new terminal axons and innervate the muscle fibers (Fig. 10–2). Postpolio syndrome occurs years after the acute illness. The syndrome is not due to death of entire motoneurons;

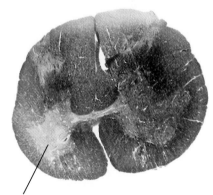

Loss of lower motoneuron cell bodies

FIGURE 10–1

The effect of polio on lower motoneuron cell bodies in the spinal cord. The section has been stained for myelin so the white matter appears dark. Loss of cell bodies is visible in the anterior horn. (Courtesy of Dr. Melvin J. Ball.)

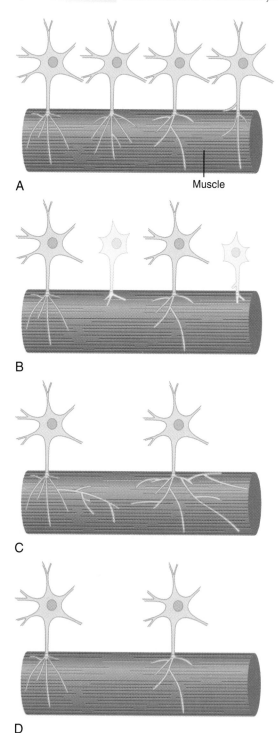

A

Muscle

B

C

D

instead, the overextended surviving neurons cannot support the abnormal number of axonal branches, and the distal branches die. Symptoms include increasing muscle weakness, joint and muscle pain, fatigue, and breathing problems. The muscle weakness may be a normal age-related strength decline, more obvious in people who previously had polio because their muscles were previously weakened.

UPPER MOTONEURON SYNDROME

Neurons whose axons descend from the cerebrum and brain stem to synapse on lower motoneurons (alpha and gamma) are upper motoneurons. Upper motoneuron lesions have four major effects on motor output:

- Abnormal cutaneous reflexes
- Abnormal timing of muscle activation
- Paresis
- Spasticity

Abnormal Cutaneous Reflexes

Changes in cutaneous reflexes include Babinski's sign (Fig. 10–3) and muscle spasms in response to normally innocuous stimuli. Babinski's sign is extension of the great toe, often accompanied by fanning of the other toes. The sign is elicited by firm stroking of the lateral sole of the foot, from the heel to the ball of the foot, then across the ball of the foot. Typically a key or the end of the handle of a reflex hammer is used as the stimulus. Although Babinski's sign* is pathognomonic for corticospinal tract damage in people over 6 months of age, the mechanism is not understood. Muscle spasms commonly occur after re-

*In infants until about 7 months of age, Babinski's sign is normal because the corticospinal tracts are not adequately myelinated.

FIGURE 10–2

Effects of polio on alpha motoneurons. *A,* Healthy motor units with normal innervation. *B,* Acute polio: death of some neurons leading to muscle fiber atrophy. *C,* Recovery: surviving neurons grow new distal branches to reinnervate surviving muscle fibers. *D,* Late post-polio: overextended neurons can no longer support the excessive number of distal branches. The newer distal branches atrophy, leaving some muscle fibers denervated.

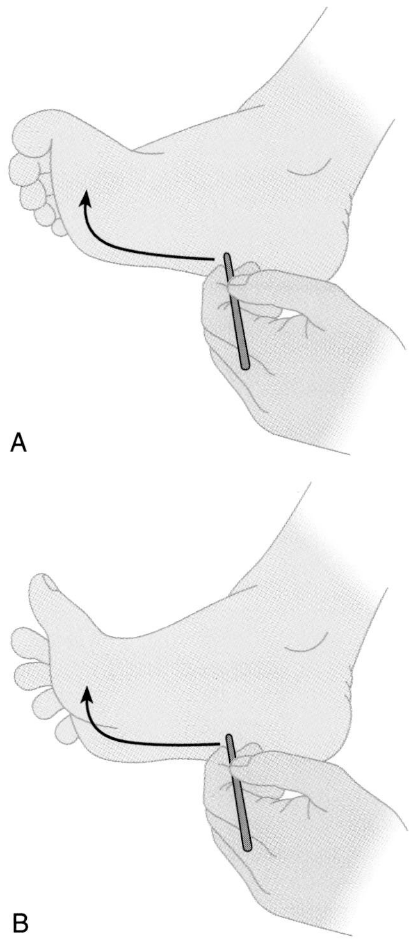

A

B

FIGURE 10–3

A, Stroking from the heel to the ball of the foot along the lateral sole, then across the ball of the foot, normally causes the toes to flex. *B,* Babinski sign in response to the same stimulus. In corticospinal tract lesions, or infants less than six months old, the big toe extends and the other toes fan out.

covery from spinal shock. Mild cutaneous stimulation, such as gentle touch on the foot, may result in abrupt flexion of the lower limb. Occasionally, touch on one lower limb may elicit bilateral flexion. In rare cases, the muscle spasms are severe enough to disturb the person's sitting balance, which can cause the person to fall out of a chair.

Abnormal Timing of Muscle Activation

Disrupted timing of muscle activation also contributes to the movement problems in people with upper motoneuron syndrome. Initiation of movement is delayed, the rate of force development is slowed, muscle contraction time is prolonged, and the timing of activation of antagonists relative to agonists is disrupted in people with upper motoneuron syndrome (Sahrmann and Norton, 1977; Young and Mayer, 1992). Electromyography (EMG), the recording of electrical activity in muscles, can be used to evaluate movements. For example, when a person with upper motoneuron syndrome attempts to flex the elbow, the following differences in biceps activity are found compared to normal: onset of EMG activity is delayed, force develops more slowly, a lower maximal force is achieved, and the biceps continue to contract for a longer period of time than normal.

Paresis

Paresis occurs in upper motoneuron lesions as a consequence of inadequate recruitment of lower motoneurons. For example, paresis is common following stroke (also called cerebrovascular accident, or CVA). Stroke is a sudden onset of neurological deficits due to disruption of the blood supply in the brain. Bobath (1977), the founder of Neurodevelopmental Therapy, contended that people with strokes do not lack muscle power on the affected side. She asserted that apparent weakness in people with strokes was due to antagonist muscle contraction opposing the agonist activity. Subsequent research provides no evidence that antagonist opposition is a limiting factor in movement of people with stroke; instead, paresis and loss of ability to fractionate movement are the major determinants of motor impairment in people with stroke (Gowland et al., 1992; Fellows et al., 1994).

Spasticity

Spasticity is velocity-dependent muscle hypertonia. Velocity dependent means that when a spastic muscle is passively stretched slowly, the resistance to stretch is less than when the same muscle is passively stretched quickly. Hypertonia results from alteration in muscle fiber properties secondary to changes in descending tract activity.

ABNORMAL REFLEXES AND SPASTICITY

Signs sometimes associated with spasticity are the clasp-knife response and clonus. When a spastic muscle is slowly passively stretched, resistance drops at a specific point in the range of motion. This is called the clasp-knife response because of the analogy with opening a pocketknife: the initial strong resistance to opening the knife blade gives way to easier movement. When a therapist is passively stretching a spastic biceps brachii muscle, the resistance to passive movement is strong initially. However, if steady stretch is gradually applied, an abrupt decrease in resistance will occur. Group II afferents, including some joint capsule receptors and cutaneous and subcutaneous touch and pressure receptors, are responsible for eliciting the clasp-knife response.

In some people with damage to descending pathways, lack of descending control allows repeating stretch reflexes to occur on passive dorsiflexion of the foot. This phenomenon is called **clonus**; each relaxation of the soleus allows the muscle to be passively stretched, resulting in another stretch reflex contraction.

Although hypertonia is sometimes associated with hyperactive phasic stretch reflexes, hyperactive phasic stretch reflexes do not contribute to muscle hypertonia. If phasic stretch reflexes did contribute to muscle hypertonia, EMG recordings of muscle activity would be greater in a spastic limb than in a normal neuromuscular system.

COMMON MISCONCEPTIONS ABOUT SPASTICITY

Discussion of spasticity is problematic because the commonly accepted explanation of its etiology is not supported by research. *Stedman's Concise Medical Dictionary* (McDonough, 1994) defines *spasticity* as "a state of increased muscular tone with exaggeration of the tendon reflexes." The dictionary definition implies a relationship between increased muscle tone (hypertonia) and overactive stretch reflexes; however, hypertonia is independent of hyperactive tendon reflexes (Berger et al., 1984).

According to Dietz (1992), a prevalent misconception is that hyperactive stretch reflexes are responsible

TABLE 10–1 DEFINITIONS OF *SPASTICITY*	
Definition supported by research	Velocity-dependent hypertonia due to changes in muscle
Most common definition in the literature (not supported by research)	Hyperreflexia associated with hypertonia
Common clinical use	Entire upper motoneuron syndrome: paresis, hypertonia, cocontraction, hyperreflexia

for muscle hypertonia and for the associated movement disorder.* Although hyperactive phasic stretch reflexes are sometimes associated with spasticity, the normally more powerful polysynaptic (tonic stretch) reflex is diminished or absent.

> Although phasic stretch reflexes may be hyperactive, tonic stretch reflexes are decreased or absent in spasticity.

Another factor that complicates discussion of spasticity is a tendency of clinicians to use the term *spasticity* in a much broader sense than researchers and most authors. Clinicians often use the term to indicate all of the movement abnormalities seen in upper motoneuron syndrome (Campbell et al., 1995). Thus paresis, hypertonia, cocontraction, and hyperreflexia are not distinguished. This leads to confusion in evaluation and treatment. Table 10–1 compares definitions of *spasticity*.

INDEPENDENCE OF PHASIC STRETCH HYPERREFLEXIA AND HYPERTONIA

Dietz and Berger (1983) and Berger et al. (1984) demonstrated that in adults with chronic hemiplegia, force generation in the nonparetic leg correlates with the level of EMG activity (as it does in healthy subjects). However, in the paretic leg, high levels of force

---

*Historical note: Hyperactive stretch reflexes were previously attributed to a hypersensitive spindle (excessive γ output); however, the fusimotor system is normal in hyperreflexia. Hyperactivity of monosynaptic stretch reflexes is probably due to decreased presynaptic inhibition of monosynaptic (group Ia) afferents and a decrease in inhibitory interneuron activity, secondary to the reduced descending inputs to lower motoneurons.

generation are disproportionate to the low level of EMG activity. This indicates that the lower motoneurons are less active than normal in the paretic limb. If hyperreflexia contributed to hypertonia, EMG activity would increase with increased muscle force output.

In adult chronic hemiplegia, the clinically important aspect of human spasticity during movement, hypertonus, is muscle based, not reflex based. Chemical and structural changes occur in muscle fibers subsequent to upper motoneuron lesion, contributing to hypertonus. These changes include atrophy of muscle fibers, fibrosis, and alteration of contractile properties toward tonic muscle characteristics. As Dietz (1992) suggests, these muscle changes may be functionally beneficial, allowing people to support their weight during gait despite the lack of neural control. Thus hypertonia may be a compensation for decreased neural control rather than the cause of disordered movement. However, the muscle changes prevent fast, active movements.

For clarity in evaluation and for evaluating treatment, hyperreflexia of the phasic stretch reflex must be distinguished from hypertonia (changes in muscle and connective tissue). Clinically, hyperreflexia of the phasic stretch reflex is probably less important than abnormal timing, paresis, and muscular/connective tissue changes because people can avoid hyperreflexia by moving slowly.

Given the independence of hypertonia and phasic stretch reflexes, it is not surprising that little correlation has been found between clinical measures of spasticity and reflex testing. (In the following discussion, the term "spasticity" is used according to the study authors' use of the term; that is, the authors' use of the term includes hyperreflexia as a component of spasticity. To distinguish this use of the term from the definition advocated in this textbook, the term "spasticity" is enclosed in quotation marks.) The lack of consistent results when using several measures of "spasticity" on the same individuals was confirmed by Levin and Hui-Chan's (1993) report that the degree of "spasticity" is not reliably indicated by passive reflex measurements. They used three clinical measures of "spasticity" to assess the severity of "spasticity": phasic stretch reflex of the triceps surae, resistance to passive ankle dorsiflexion, and clonus. Note that two of the measures of "spasticity" are reflex measurements. Their passive reflex testing consisted of resting H-reflexes, H-reflexes during tendon vibration, and EMG recording of stretch reflexes. Despite the use of six entirely passive evaluation techniques, the degree

of "spasticity" was not consistently measured. In addition, none of Levin and Hui-Chan's evaluations used active, functional movements (i.e., gait), which alter reflex responses even in people who are neurologically intact.

Evaluation of Upper Motoneuron Lesions

Testing administered while a subject is passive provides little information about how the subject performs actively, regardless of whether the subject is healthy or has a neurological deficit. Clinically pertinent information can be obtained by using surface EMG to differentiate the following factors contributing to movement impairment:

- Paresis
- Hypertonia
- Cocontraction
- Hyperreflexia

Paresis can be defined as 50% or less EMG amplitude in the homologous muscle on the paretic side as compared with the nonparetic side (Crenna et al., 1992). Hypertonia is decreased passive range of motion without increased EMG output. Cocontraction is temporal overlap of EMG activity in antagonist muscles. Cocontraction is only abnormal if it does not contribute to making the movement functional; often healthy people use cocontraction when learning a new movement. Hyperreflexia is defined as EMG activity during muscle stretch, with a positive correlation between EMG amplitude and velocity of muscle stretch.

TYPES OF UPPER MOTONEURON LESIONS

Three types of upper motoneuron lesions are covered in this section: **spinal cord injury, stroke,** and **congenital.** Although head trauma, tumors, and multiple sclerosis can also damage upper motoneurons, the structures affected and thus the clinical outcomes are so variable that a discussion of the motor effects of these conditions is beyond the scope of this text. Because the etiology of each type of upper motoneuron lesion is different, research on movement disorders in one population cannot be generalized to the others; indeed, given the variety in size and location of lesions, research on one type of lesion cannot be indiscriminately applied to individuals. For this reason, accurate assessment of the relative contributions of each component of the movement disorder, as provided by analysis of surface EMG recordings, is important.

Spinal Cord Injury

In a complete spinal cord lesion, all descending neuronal control is lost below the level of the lesion. After recovery from spinal shock, the monosynaptic stretch reflex is of normal amplitude, but activity in the polysynaptic spinal reflex pathways is reduced or absent (Ashby, 1993). Spinal cord injury can be incomplete, with preservation of some ascending and/or descending fibers. Gait recordings from a person with an incomplete spinal cord lesion are shown in Figure 10–5A. A discussion of congenital spinal cord malformation (myelomeningocele) and the variety of deficits following incomplete lesions is deferred to Chapter 12.

Stroke

Stroke damages descending pathways neurons, causing adult-onset upper motoneuron lesions. If the stroke occurs in the most common site for stroke, the middle cerebral artery (see Chaps. 1 and 18), cortical connections with the spinal cord, brain stem, and cerebellum are disrupted. (See the cerebrovascular accident box.) Thus the lesion may alter the output from all supraspinal motor areas, depending on the extent of the lesion. However, with the exception of direct cortical influences, the other supraspinal motor areas continue to exert some control over lower motoneuron activity. The resulting abnormal control of motor activity results in overactivity of antigravity muscles during standing. The antigravity muscles include the extensors in the lower limb and flexors in the upper limb (Fig. 10–4). Abnormal muscle activation occurs because corticospinal input is lost, and the lateral reticulospinal tract is deprived of its normal cortical facilitation, leaving the medial reticulospinal and vestibulospinal tracts to facilitate extension relatively unopposed. The medial reticulospinal and vestibulospinal tracts remain active because their activity is less dependent on cortical facilitation. Activity of the intact rubrospinal tract may contribute to upper limb flexion.

CEREBROVASCULAR ACCIDENT, MIDDLE CEREBRAL ARTERY

Pathology
Interruption of blood supply

Etiology
Occlusion or hemorrhage

Speed of onset
Usually acute

Signs and symptoms

Consciousness
May be temporarily impaired

Communication and memory
May be impaired

Sensory
Usually impaired contralateral to the lesion

Autonomic
May be impaired

Motor
Contralateral to the lesion: paresis, muscle atrophy, loss of fractionation of movement, decreased movement speed and efficiency, impaired postural control, Babinski sign

Region affected
Cerebrum

Demographics
Males and females affected equally; average age at onset approximately 72 years (Alter, 1993)

Prognosis
About 20% die from stroke within the first 30 days; after first 30 days, risk of death approximately double the rate in the general population; after first year, cardiovascular disease the most common cause of death (Dennis, 1993)

Gandevia (1993) summarized his group's research on adults with unilateral CVAs: on the paretic side, weakness in one muscle group was usually associated with weakness in the antagonist muscle group; distal muscles were weaker than proximal muscles in both limbs; and muscles on the nonparetic side (particularly the shoulder) were weak in comparison to the strength of muscles in healthy subjects matched for age and gender. Recordings from a person with hemiplegia during gait are shown in Figure 10–5B.

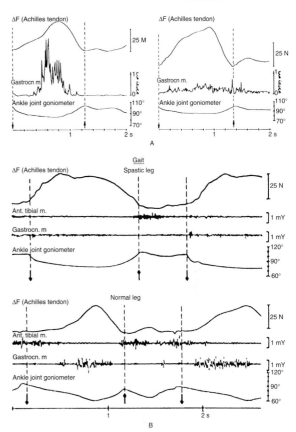

FIGURE 10–4

Typical posture of an adult with a middle cerebral artery CVA.

Spastic Cerebral Palsy

In the congenital syndrome spastic cerebral palsy, abnormal supraspinal influences, failure of normal neuronal selection, and consequent aberrant muscle development lead to movement dysfunction (see Chap. 5). The motor disorders in spastic cerebral palsy include abnormal tonic stretch reflexes both at rest and during movement, reflex irradiation (spread of reflex activity; e.g., tapping biceps tendon causes finger flexor contraction in addition to biceps contraction), lack of postural preparation prior to movement, and abnormal cocontraction of muscles.

Common Characteristics of Upper Motoneuron Lesions

Emotional agitation has similar effects on spasticity among people with CVA, cerebral palsy, and incomplete spinal cord injury, causing increased spasticity via limbic action on motor cortical areas and via the nonspecific activating pathways to lower motoneurons. Pain may also increase spasticity; the neural mechanism is unknown.

FIGURE 10–5

A, Averaged recordings (30 steps) of a step cycle during slow gait of a normal subject (*left*) and subject with paraparesis (*right*) due to a spinal cord lesion. From top to bottom in each recording: Changes in tension recorded from the Achilles tendon, gastrocnemius EMG, and goniometer signal from the ankle joint. Rectified EMG recordings are shown. In the normal subject, an increase in Achilles tendon tension correlates with an increase in gastrocnemius EMG activity. In the paraparetic subject, the increase in Achilles tendon tension does not correlate with an increase in EMG. Instead, the increase in Achilles tendon tension coincides with stretch of the triceps surae during passive dorsiflexion of the foot in stance phase. B, Gait recordings of a step cycle during slow gait of an adult with spastic hemiparesis. The spastic leg is shown above and the normal one below. From top to bottom in each recording: changes in tension recorded from the Achilles tendon, tibialis anterior and gastrocnemius EMG, and goniometer signal from the ankle joint. Vertical lines indicate touch down (↓) and lift off (↑) of the foot. During pushoff (pushoff begins about 0.5 sec), paresis of the gastrocnemius muscle is indicated by the decreased EMG amplitude: In the spastic limb, the gastrocnemius EMG amplitude is less than 50% of the normal side. Hypertonia in the spastic limb during stance phase is indicated by the large, early increase in Achilles tendon force despite very little EMG activity of the gastrocnemius. (From Dietz, V., and Berger, W. [1983]. Normal and impaired regulation of muscle stiffness in gait: A new hypothesis about muscle hypertonia. Exp. Neurology, 79:680–687, with permission.)

In summary, common signs among different types of upper motoneuron lesions include abnormal cutaneous reflexes, abnormal timing of muscle activity, paresis, and spasticity. Two characteristics of upper motoneuron syndromes arising during nervous system development (e.g., cerebral palsy) that do not occur with damage to the mature nervous system (most CVAs and spinal cord injuries) include reflex irradiation and abnormal cocontraction of antagonist muscles. Reciprocal inhibition is preserved in adult CVA and in most spinal cord injuries because damages occurs to a mature nervous system.

TREATMENT OF UPPER MOTONEURON LESIONS

Treatment of Spasticity

Until recently, spasticity was regarded by many therapists as the primary problem in people with upper motoneuron lesions, and "normalizing" muscle tone was considered a worthy goal of therapy. The assumption was that motor control would be normal if the spasticity were successfully reduced. However, this assumption has been thoroughly disproved. Reduction of muscle tone (by selective dorsal rhizotomy and by voluntary control of spasticity) has demonstrated that decreasing tone without other intervention does not improve function (Cahan et al., 1990). Also, subjects with cerebral palsy who have successfully learned to decrease spasticity by decreasing sensitivity of the tonic stretch reflex during rest and active movement have shown no improvement in functional control (Neilson, 1993). This finding reflects the negligible effects of the abnormal tonic stretch reflex when compared with the other effects of upper motoneuron lesions. Wolf and Catlin (1994) report that in people with chronic hemiplegia, practicing elbow extension is as effective in improving range of motion as training that incorporates learning to inhibit biceps activity prior to elbow extension. Dietz (1992) reports that activation of the calf muscles on the paretic side is less than normal in people with CVAs and thus attributes muscle hypertonia to changes in muscle. Dietz et al (1991) report similar findings in upper limb muscles. Paretic elbow flexor and extensor muscles in people with CVAs produce high torque during stretch with less EMG activity than normal. Therefore other factors in addition to

spasticity are involved in the movement problems of people with upper motoneuron lesions. Testing for and remediation of these deficits provide guidance for rehabilitation.

Several pharmacological treatments for spasticity are available. The most commonly used are baclofen and botulinum toxin. Baclofen activates receptors that produce inhibition of neurotransmitter release. Baclofen produces presynaptic inhibition in stretch reflex pathways, both presynaptically (by decreasing calcium influx into the presynaptic terminals of primary afferent fibers) and by stabilizing the postsynaptic membrane. The effectiveness of baclofen in improving function is controversial. The benefits and costs of baclofen have been summarized by Campbell et al. (1995). Their review of the literature reports decreased spasms, pain, and sleep disturbance, along with improved bladder function and increased mobility. However, baclofen may cause a decrease in function if spasticity is used functionally. For example, using spasticity may enable a person who is otherwise unable to sit upright to be stable in sitting. Unlike baclofen, which is used systemically, botulinum toxin in injected directly into spastic muscles. This allows specific targeting of particular muscles without interfering with the function of other muscles. In 7 of 12 people with chronic hemiparesis, injection of botulinum toxin into triceps surae and tibialis posterior resulted in gait improvements, including velocity, stride length, and push off with affected limb (Hesse et al., 1994).

Treatment of hypertonia is beneficial if gains in function or comfort occur. Thus, for some people, reducing hypertonia may allow more comfort during hygiene and dressing activities and ease the work of caregivers. However, some people depend on spasticity to maintain erect posture or to enable gait; in these people, decreasing their spasticity restricts their function (Penn, 1992). For effective evaluation and treatment, the effects of spasticity must be differentiated from the other effects of upper motoneuron lesions.

Improvement of Function in People with Upper Motoneuron Lesions

Function can be promoted in people with upper motoneuron lesions by electrical stimulation of motor points or nerves. No standardized terminology exists for this treatment; the terms *transcutaneous electrical stimulation* (TENS), *neuromuscular electrical stimulation* (NMES), and *functional electrical stimulation* (FES) are used. NMES has been reported to improve function in two children with cerebral palsy (Carmick, 1993). Functional electrical stimulation in subjects with incomplete spinal cord injury has been demonstrated to result in improved gait with decreased physiological cost (Granat et al., 1993).

Treatments effective in improving voluntary movement in adult-onset CVA include TENS (Levin and Hui-Chan, 1992), EMG biofeedback (Schleenbaker and Mainous, 1993), strengthening (improves function without increasing spasticity [Light et al., 1994]) and forced use of the paretic upper limb by restraining the nonparetic upper limb (Taub et al., 1993). Taub et al. based their intervention on the theory that a major contributor to the dysfunction of a paretic limb is learned nonuse. Because the person initially finds using the nonparetic limb easier and more effective than trying to use the paretic limb, there is a tendency to substitute the function of the more easily controlled limb for the role of the paretic limb. As noted by Fisher and Woll (1995), this choice may have long-term consequences, as the changes secondary to disuse (atrophy, contracture) further limit function. In a study comparing the effects of treatments for hand function in hemiparetic adults, Butefisch et al. (1995) report that techniques that focus on spasticity reduction (techniques developed by Bobath [1977]) instead of active movement produce no significant improvement in motor capabilities of the hand. In contrast, training of finger and hand flexion and extension against resistance is shown to result in significantly improved grip strength and other indicators of hand function during training.

VOLUNTARY MOTOR SYSTEM DEGENERATION

A disease restricted to the voluntary motor system is **amyotrophic lateral sclerosis (ALS).** In its most common form, ALS destroys only the lateral activating pathways and anterior horn cells in the spinal cord (Fig. 10–6), resulting in both upper and lower motoneuron signs. Thus paresis, spasticity, hyperreflexia, Babinski's sign, atrophy, fasciculations, and fibrillations are present. Approximately 90% of cases are idiopathic, although the gene responsible for the familial type of ALS has been identified. Death results

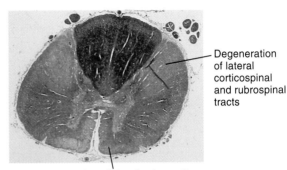

Degeneration of lateral corticospinal and rubrospinal tracts

Degeneration of medial activating pathways

FIGURE 10–6

Spinal cord section, stained for myelin, showing loss of descending activating tract neurons in ALS. The loss is visible dorsolaterally, where the lateral corticospinal and rubrospinal axons should be, and ventromedially, where the medial activating paths should be. (Courtesy of Dr. Melvin J. Ball.)

from respiratory complications. (See the amyotrophic lateral sclerosis box.)

PATHOLOGY OF THE BASAL GANGLIA

Movement disorders in basal ganglia dysfunction range from hypokinetic disorders (too little movement, as in Parkinson's disease) to hyperkinetic (excessive movement, as in Huntington's disease, dystonia, or subtypes of cerebral palsy). The differences in abnormal movements are due to dysfunction in specific parts of the basal ganglia–thalamocortical motor circuit and in the basal ganglia–pedunculopontine nucleus output. The basal ganglia inhibit the motor thalamus and the pedunculopontine nucleus; excessive inhibition results in hypokinetic disorders, and inadequate inhibition results in hyperkinetic disorders.

Parkinson's Disease

The most common basal ganglia motor disorder is Parkinson's disease, characterized by muscular rigidity, shuffling gait, droopy posture, rhythmical muscular tremors, and a masklike facial expression. Parkinson's disease (paralysis agitans) interferes with both voluntary and automatic movements. Despite identification of the pathology—death of dopamine-producing cells in the substantia nigra compacta

AMYOTROPHIC LATERAL SCLEROSIS

Pathology
Degeneration of motoneurons (both upper and lower)

Etiology
Unknown; speculative: excessive levels of glutamate (Lipton, 1994)

Speed of onset
Chronic

Signs and symptoms

Consciousness
Normal

Communication and memory
Normal

Sensory
Normal

Autonomic
Normal

Motor
Paresis, spasticity, clonus, Babinski's sign, fasciculations, fibrillations, muscle atrophy

Region affected
Upper motoneuron in cerebrum, brain stem, and spinal cord; lower motoneuron in spinal and peripheral regions

Demographics
Onset is usually > 50 years old; males outnumber females by 2:1

Prognosis
Progressive; average life span after diagnosis = 3 years; rarely live > 20 years; death usually from respiratory complications

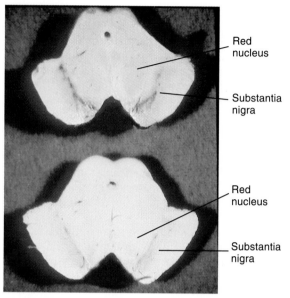

FIGURE 10–7

Horizontal sections of the midbrain. The upper section is normal, with darkly pigmented cells in the substantia nigra. The lower section is from a person with Parkinson's disease, with the characteristic loss of darkly pigmented, dopamine-producing cells. (Courtesy of Dr. Melvin J. Ball.)

(Fig. 10–7) and acetylcholine-producing cells in the pedunculopontine nucleus—the cause of cell death is unknown. Cell death occurs long before clinical signs of Parkinson's disease become evident; about 80% of the dopamine-producing cells die before signs of the disease appear.

The loss of dopamine in the basal ganglia direct pathway reduces activity in the motor areas of the cerebral cortex, decreasing voluntary movements. The loss of pedunculopontine cells, combined with increased inhibition of the pedunculopontine nucleus, disinhibits the reticulospinal and vestibu-lospinal pathways (see Chap. 9), producing excessive contraction of postural muscles.

A person with Parkinson's disease will have difficulty coming to standing from sitting, and the gait will be characterized by a flexed posture, shuffling of the feet, and decreased or absent arm swing. The distinctive signs of Parkinson's disease are rigidity, hypokinesia (decreased movement), resting tremor, and visuoperceptive impairments. Rigidity is increased resistance to passive movement in all muscles. Unlike spasticity, this form of hypertonia is produced via a direct effect on alpha motoneurons (Fig. 10–8). Rigidity, in contrast to spasticity, is not associated with hyperreflexia, velocity dependence, clonus, or clasp-knife response. Hypokinesia is manifested in decreased ranges of active motion and in the lack of automatic movements, including facial expression and normal arm swing during walking. Hypokinesia may be related to decreased ability to control the force output of muscles. Ivry and Keele (1989) demonstrated that people with Parkinson's disease are unable to control force production as well as people without neurological disorders. Resting tremor is movement of the hands as if using the thumb to roll a

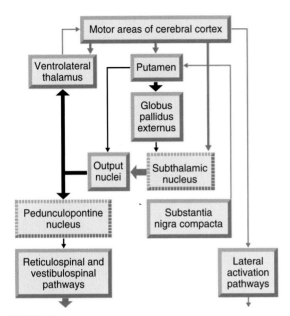

FIGURE 10–8

Changes in neural activity in Parkinson's disease. Compare with normal basal ganglia activity in Fig. 9.16. Red arrows indicate that the connection is excitatory. Black arrows indicate that the connection is inhibitory. Thick arrows represent increased neural activity, and thin arrows represent decreased neural activity. The output nuclei are the globus pallidus internus and the substantia nigra reticularis.

Decreased dopamine from the substantia nigra compacta is the primary change leading to excessive activity of the output nuclei, in turn inhibiting the motor thalamus and thus reducing the output of motor areas of the cerebral cortex. Cells in the pedunculopontine nucleus (PPN) die off. The combination of death of cells in the PPN with excessive inhibition of the motor thalamus results in disinhibition of several of the medial activation pathways.

pill along the fingertips (pill-rolling tremor); the tremor is prominent when the hand is at rest and diminishes during voluntary movement. Visuoperceptive deficits can result in impediments to action. A walker, intended to assist a person with ambulation, often creates a visual block, and movement ceases. A therapist standing near the person may also unwittingly interfere with the person's ability to move. People frequently report difficulties moving past visual movement blocks, like doorways; if a marker is placed on the floor, the person can use the marker as a cue for getting through the doorway. Why the marker does not serve as an additional visual block is unknown. Late in the disease process, dementia may also develop.

Because Parkinson's disease involves loss of dopamine-producing cells in the substantia nigra, drug therapy that replaces dopamine (L-dopa) is initially effective in reducing symptoms. However, tolerance to L-dopa, side effects, and progression of the disease with involvement of other cells and neurotransmitters limit the effectiveness of L-dopa therapy. Even with L-dopa therapy, people with Parkinson's disease may experience periods of near-normal mobility alternating with periods of immobility. This is called the on-off phenomenon. Moreover, motor performance often varies at different times of day regardless of medication.

PARKINSON'S DISEASE

Pathology
Death of dopaminergic neurons in substantia nigra compacta and in pedunculopontine nucleus

Etiology
Unknown

Speed of onset
Chronic

Signs and symptoms

Consciousness
Normal

Communication and memory
Normal; dementia may occur in late stages

Sensory
Normal

Autonomic
Normal

Motor
Hypokinesia; rigidity; stooped posture; shuffling gait; difficulty initiating movements, turning, and stopping; resting tremor; visuoperceptive movement blocks

Region affected
Basal ganglia nuclei in cerebrum and midbrain

Demographics
Age at onset typically between 50 and 65 years of age; men and women affected equally

Prognosis
Progressive; average life span after diagnosis = 13 years (Louis, 1995); death usually by heart disease or infection

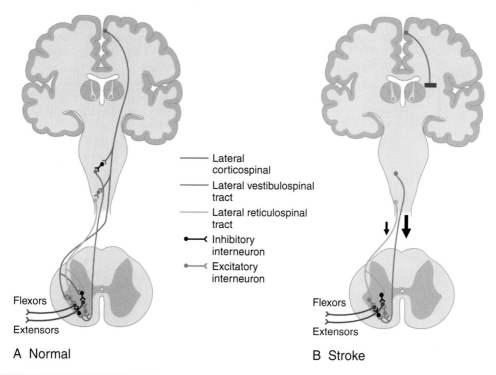

A Normal

B Stroke

	Effect on Flexors	Effect on Extensors
Lateral corticospinal	++	--
Lateral reticulospinal	++	-
Lateral vestibulospinal	-	+

Lateral corticospinal also facilitates lateral reticulospinal and inhibits lateral vestibulospinal tracts.

	Effect on Flexors	Effect on Extensors
Lateral corticospinal		
Lateral reticulospinal	+	-
Lateral vestibulospinal	--	+++

Three changes occur as a result of interruption of the lateral corticospinal tract: loss of its direct input to lower motoneurons, loss of facilitation of lateral reticulospinal tract, and disinhibition of lateral vestibulospinal tract.

FIGURE 10–9

See legend on opposite page

Surgical procedures are an alternative to drug therapy. Short-term benefits have been documented in two surgical procedures: interruption of basal ganglia circuits and transplantation of fetal brain tissue into the basal ganglia (Freed, 1992). Further studies are needed to determine long-term benefits. Physical and occupational therapy generally improves movement and functional abilities in moderately disabled people with Parkinson's disease. Unfortunately, if the exercise activities are abandoned, the gains are lost within 6 months (Comella et al., 1994). Therefore, a moderate level of activity needs to be maintained. People with Parkinson's disease are prone to falls because of their inability to generate adequate muscle force quickly. Their postural corrections may be too slow to be useful. (See the Parkinson's disease box.) Figure 10–9 compares the location and effects of upper motoneuron lesions with the location and effects of Parkinson's disease.

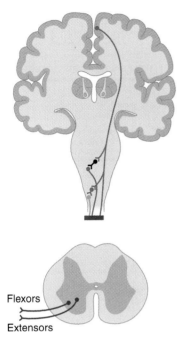

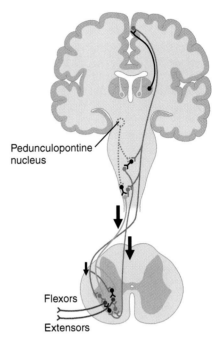

C Complete spinal cord lesion

	Effect on Flexors	Effect on Extensors
Lateral corticospinal		
Lateral reticulospinal		
Lateral vestibulospinal		

All descending tracts are interrupted.

D Parkinson's disease

	Effect on Flexors	Effect on Extensors
Lateral corticospinal	+	−
Lateral reticulospinal	+++	−
Lateral vestibulospinal	−	+++

Decreased activity of motor areas of cortex (due to inhibition of thalamus by globus pallidus) causes decreased output via the lateral corticospinal tract. Disinhibition of the lateral reticulospinal and lateral vestibulospinal tracts result from decreased activity of the pedunculopontine nucleus. These changes lead to muscle rigidity.

FIGURE 10–9
Comparison of the effects of stroke, complete spinal cord injury, and Parkinson's disease on activity in descending activating tracts and the resulting activity levels in skeletal muscles.

Parkinsonism and Parkinsonian Syndrome

The term *parkinsonism* refers to disorders that cause the signs seen in Parkinson's disease, including toxic, infectious, or traumatic causes. Lesions of the lentiform nucleus are associated with parkinsonism. Parkinsonism is often a side effect of drugs that treat psychosis or digestive problems. Drug-induced parkinsonism frequently leads to misdiagnosis and unnecessary treatment for Parkinson's disease in the elderly (Avorn et al., 1995).

HYPERKINETIC DISORDERS

Abnormal involuntary movements are characteristic of Huntington's disease, dystonia, and some types of cerebral palsy.

Huntington's Disease

Huntington's disease is an autosomal dominant hereditary disorder that causes degeneration in many areas of the brain, most prominently in the striatum and cerebral cortex (Fig. 10–10). This causes a decrease in signals from the output nuclei, resulting in disinhibition of the motor thalamus and pedunculopontine nucleus. The result is excessive output from the motor areas of the cerebral cortex (Fig. 10–11). Chorea, consisting of involuntary, jerky, rapid movements, and dementia are the signs. Onset is typically between 40 and 50 years of age, and the disease is progressive, resulting in death about 15 years after signs first appear.

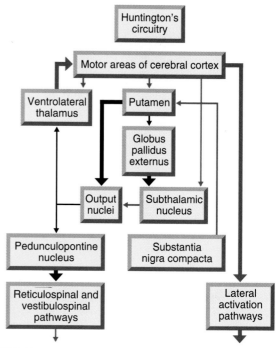

FIGURE 10–11

Changes in neural activity in Huntington's disease. Compare with normal basal ganglia activity in Fig. 9.16. Red arrows indicate that the connection is excitatory. Black arrows indicate that the connection is inhibitory. Thick arrows represent increased neural activity, and thin arrows represent decreased neural activity. The output nuclei are the globus pallidus internus and the substantia nigra reticularis.

Huntington's disease is characterized by excessive direct inhibition of the output nuclei by the putamen. The inhibitory output of the basal ganglia is inadequate, producing disinhibition of the motor thalamus and the pedunculopontine nucleus. This results in excessive activity of the motor areas of the cerebral cortex, producing hyperkinesias, combined with excessive output from the pedunculopontine nucleus, causing insufficient activity in several medial activation pathways.

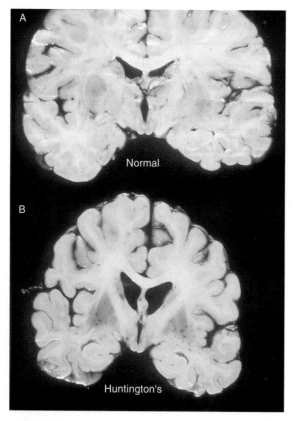

FIGURE 10–10

A, A coronal section of a normal cerebrum. Compare the size of the caudate nucleus and the overall size of the cerebrum to *B,* a cerebrum from a person with Huntington's disease. Atrophy of the caudate nucleus produces enlargement of the lateral ventricles. (Courtesy of Dr. Melvin J. Ball.)

Dystonia

Dystonias are genetic, usually nonprogressive, movement disorders with involuntary sustained muscle contractions causing abnormal postures or twisting, repetitive movements. The cause is currently unknown. The most common are focal dystonias, limited to one part of the body. Focal dystonia of the hand (writers' or musicians' cramp) involves reduced reciprocal inhibition due to decreased descending inhibition (Hallet, 1993). Focal dystonia of the hand is frequently misdiagnosed as carpal tunnel syndrome, tennis elbow, strain, or a psychogenic disorder. Because the muscle contractions are caused by basal ganglia dysfunction, attempts at treating the disorder by stretching the muscles are ineffective. Heat, cold, and exercise may be helpful to relieve pain and/or spasm. (See focal dystonia box.) Severe dystonia can be alleviated by surgical lesion of the motor thalamus or by injection of botulinum toxin into the affected muscles. A very rare disorder, Segawa's dystonia, may mimic the appearance of cerebral palsy; however, Segawa's dystonia progresses slowly and can be effectively treated with medications.

FOCAL DYSTONIA

Pathology
Basal ganglia dysfunction

Etiology
Genetic disorder

Speed of onset
Chronic

Signs and symptoms

Consciousness
Normal

Communications and memory
Normal

Sensory
Normal

Autonomic
Normal

Motor
Involuntary, sustained muscle contractions

Region affected
Cerebrum: basal ganglia

Demographics
Average age at onset 45 years old; males and females affected equally

Prognosis
Normal life span

Choreoathetotic Cerebral Palsy

Abnormal involuntary movements are also observed in people with choreoathetosis, a type of cerebral palsy (the other major types of cerebral palsy are spastic and ataxic [see Chap. 5]). The term **chorea** indicates abrupt, jerky movements, and the term **athetosis** identifies slow, writhing, purposeless movements.

> Basal ganglia disorders interfere with voluntary and automatic movements and produce involuntary movements. Hypokinesia is a decrease in the amount and speed of voluntary and automatic movements, characteristic of Parkinson's disease. However, Parkinson's signs are not purely hypokinetic because resting tremor is an increase in movement compared to normal. Hyperkinesia is abnormal excessive movement, seen in Huntington's disease, dystonia, and choreoathetotic cerebral palsy.

SIGNS AND SYMPTOMS OF CEREBELLAR DISEASE

Unilateral lesions of the cerebellum affect the same side of the body because the output paths of the medial descending pathways remain ipsilateral and because cerebellar efferents project to the contralateral cerebral cortex and red nucleus whose descending pathways cross the midline (see Fig. 9–20). **Ataxia,** a lack of coordination, is the movement disorder common to all lesions of the cerebellum. Ataxia describes voluntary, normal-strength movements, not associated with hypertonia, that are jerky and inaccurate. Midline cerebellar lesions result in truncal ataxia, paravermal lesions result in gait ataxia, and lateral lesions cause limb ataxia.

Lesions involving the vestibulocerebellum cause abnormal eye movements (nystagmus [see Chap. 15]), dysequilibrium, and difficulty maintaining sitting and standing balance (truncal ataxia). Vermal lesions result in **dysarthria** (slurred, poorly articulated speech). Dysfunction of the spinocerebellum results in ataxic gait: a wide-based, staggering gait. In chronic alcoholism, the anterior lobe section of the spinocerebellum is often damaged because of malnutrition, resulting in characteristic ataxic gait.

Cerebrocerebellar lesions result in limb ataxia, with the following manifestations:

- **Dysdiadochokinesia:** inability to rapidly alternate movements; for example, inability to rapidly pronate and supinate the forearm or inability to rapidly alternate toe tapping
- **Dysmetria:** inability to accurately move an intended distance
- **Action tremor:** shaking of the limb during voluntary movement

Action tremor may arise because the onset and offset of muscle activity are delayed. Thus, in a rapid movement, the agonist burst is prolonged, and the onset of braking by the antagonist is delayed, causing overshoot of the target. As correction of the movement is attempted, the same dysfunctions cause repeated overshoot (Vilis and Hore, 1977). Cerebrocerebellar lesions also cause deficits in the ability to perceive time intervals, as in judging the amount of time between two sounds (Ivry and Keele, 1989).

In addition to ataxia, cerebellar lesions may also produce hypotonia, a decreased resistance to passive stretch. Cerebellar hypotonia is usually temporary.

Ataxia is not always due to cerebellar lesions. Interference with the transmission of somatosensory information to the cerebellum, either by lesions of the spinocerebellar tracts or by peripheral disorders, may also produce ataxia. Table 10–2 compares the characteristics of the major types of motor control disorders.

THREE FUNDAMENTAL TYPES OF MOVEMENT

Movements can be classified into three types, depending on the neural substrate for each type:

- Postural
- Ambulatory
- Reaching/grasping

Posture is primarily controlled by brain stem mechanisms, ambulation by brain stem and spinal regions, and reaching/grasping by the cerebral cortex; however, all regions of the nervous system contribute to each type of movement.

TABLE 10–2
CHARACTERISTICS OF MAJOR TYPES OF MOTOR CONTROL DISORDERS

Characteristic	Complete Severance of Peripheral Nerve (Lower Motoneuron Lesions)	Upper Motoneuron Lesions	Parkinson's Disease	Huntington's Disease	Cerebellar Lesions
Muscle strength	Absent (paralysis)	Decreased (paresis)	Normal	Normal	Normal
Muscle bulk	Severe atrophy	Variable atrophy	Normal	Normal	Normal
Involuntary muscle contraction	Fibrillations	Fibrillations	Resting tremor	Chorea	None
Muscle tone*	Decreased	Velocity-dependent increase (spasticity)	Velocity-independent increase (rigidity)	Variable, depending on stage of disease	Hypotonia
Movement speed and efficiency	Absent	Decreased	Decreased	Abnormal	Ataxic
Postural control	Normal	Decreased or normal, depending on location of lesion	Excessive	Abnormal	Depends on location of lesion

*Muscle tone ratings are compared to muscle tone in an alert person. Thus muscle tone in a person with a complete peripheral nerve lesion would be similar to muscle tone in a completely relaxed person with an intact nervous system because only the viscoelastic properties contribute.

Contributions of each region of the central nervous system to an externally imposed movement can be assessed by EMG. For example, a subject is asked to support a light weight and to maintain a constant position of the elbow (flexion at 90°). Then a weight is unexpectedly added, resulting in displacement of the forearm and hand downward. After about half a second, biceps contraction begins to restore the elbow angle to 90°. The EMG response to the displacement shows three distinct increases in activity occurring 30 milliseconds, 50 to 80 milliseconds, and 80 to 120 milliseconds after the imposed biceps stretch.

The shortest latency response, M1, is the monosynaptic stretch reflex; this reflex does not generate enough force to restore the elbow position. The second response, M2, is the long loop response; this response generates enough force to begin to move the hand upward. The subject's intentions influence M2—if the instructions to the subject are to "let go" when the extra weight is added, the M2 response is attenuated or disappears. The final response, voluntary, completes the return of the elbow angle to 90°. The latencies provide clues to the central nervous system region involved in each response. The 30-millisecond latency of M1 is just enough time for the afferent conduction, transmission across one synapse in the spinal cord, and efferent conduction to activate the biceps. Fifty to eighty milliseconds allows enough time for transmission involving brain stem connections. The 120-millisecond response requires synapses in the cerebral cortex.

Postural Control

Postural control provides orientation and balance (equilibrium). Orientation is the adjustment of the body and head to vertical, and balance is the ability to maintain the center of mass relative to the base of support. Postural control is achieved by central commands to lower motoneurons; the central output is adjusted to the environmental context by sensory input.* The central commands are mediated by the tectospinal, medial reticulospinal, vestibulospinal,

and medial corticospinal pathways. Sensory input is used in both feedback and feed-forward mechanisms. Feedback control involves a response to disturbance (e.g., a slip on ice). Feed-forward control involves prediction and anticipation to respond to upcoming hazards to stability (e.g., path turns).

To orient in the world, we use three senses:

- Somatosensation
- Vision
- Vestibular

Somatosensation provides information about weight bearing and the relative positions of body parts. Vision provides information about movement and cues for judging upright. Vestibular input from receptors in the inner ear informs us about head position relative to gravity and about head movement. Visual and somatosensory information can predict destabilization; all three sensations can be used to shape the motor reaction to instability (Fig. 10–12).

Head position in reference to gravity, to the neck, and to the visual world affects muscular activation. Head position in space is signaled by neck proprioception, vestibular, and visual information (visual as-

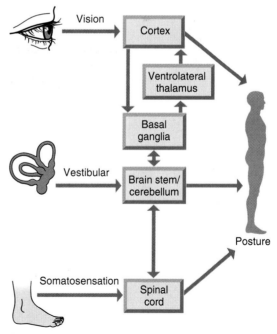

FIGURE 10–12

Sensory influences on postural control.

*Historical note: An earlier theory of postural control posited three separate control mechanisms: reflexive, reactive, and voluntary. The theory proposed that during development, reflexes and reactions were inhibited by the voluntary system; in the mature nervous system, the voluntary system controlled all normal movements. However, as noted in the text, reflexes and reactions help to restore stability before the voluntary system is aware of possible loss of balance and participate extensively in mature motor control.

pects will be covered in Chap. 15). In normal infants and in children and adults with extensive cerebral lesions, neck and vestibular reflexes can be elicited by neck movements or by changing head position. In children and adults with intact nervous systems, the same stimuli do not produce obvious responses. Activity of cervical joint receptors and neck muscle stretch receptors elicits neck reflexes. The symmetrical tonic neck reflex results in flexion of the upper limbs and extension of the lower limbs when the neck is flexed and the opposite pattern in the limbs when the neck is extended. The asymmetrical tonic neck reflex is elicited by head rotation to the right or left; the limbs on the nose side extend, and limbs on the skull side flex (Fig. 10–13). When the head is tilted, information from vestibular gravity receptors is used to right the head by contraction of neck muscles. Vestibular gravity receptors also influence limb muscle activity, in a manner opposite to the neck reflexes; for instance, tilting the head forward causes extension of the upper limbs and flexion of the lower limbs if the position of the head relative to the neck is un-changed (eliminating neck reflexes). Because the vestibular receptors for gravity are in the labyrinthine part of the inner ear, the reflex is called the tonic labyrinthine reflex. Because vestibular and neck reflexes oppose each other, and normally head and neck movements occur together, the reflexes usually counteract each other. By canceling out these two reflexes, our limbs are not compelled to move when we turn or nod or shake our heads.

During stance, we sway continuously; to prevent falling, the postural control system makes adjustments to maintain upright posture. Sometimes inappropriate sensory information is selected, as when we rely on visual information regarding movement and a large object nearby moves. A person standing next to a bus that unexpectedly moves can misinterpret the moving visual information and make inappropriate postural adjustments to compensate for the illusory movement. The person's goal is to maintain upright posture, but reacting to deceptive sensory information leads to instability.

Posturography can be used to determine sensory organization (which sensory stimuli a person relies on preferentially) and muscle coordination (how postural adjustments are coordinated). The subject stands on a force platform, and surface EMGs are recorded from specific muscles. As indicated in Fig. 10–14, six sensory conditions can be tested (Nashner et al., 1982). In the sensory conflict conditions, vision and/or proprioception is rendered inaccurate by having the visual field and/or the support surface move with the subject as the subject sways. Thus subjects have the visual and/or proprioceptive illusion that they are not swaying when in fact they are moving; under these conditions, a person without vestibular function will fall. The results of posturography can be compared to norms for different diagnostic groups. For example, the postural responses of an individual can be compared to those of people with Parkinson's disease and other disorders.

Posturography can also test motor coordination. If the platform moves forward, the subject sways backward; the normal response is contraction of the tibialis anterior, then quadriceps, then abdominal muscles to return the subject to upright (Fig. 10–15). If the platform moves backward, the gastrocnemius, then hamstrings, then back muscles contract to restore the person to upright. In both cases, distal to proximal muscle activation normally occurs.

When the platform moves backward, the subject's forward lean stretches the gastrocnemius muscle. An

FIGURE 10–13

Asymmetrical tonic neck reflex: when the head is rotated right or left, the limbs move into the fencer's position. Obligatory only in infants and people with cerebral damage.

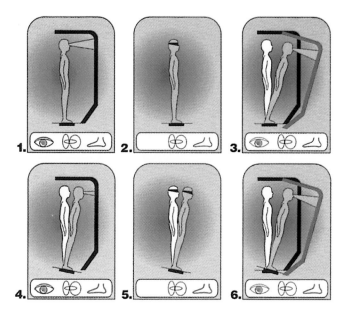

FIGURE 10–14

Posturography. Six sensory conditions are used to evaluate the relative contributions of vision, vestibular, and somato-sensory input in balance function. The icons at the bottom of each figure indicate the eyes, vestibular system, and foot and ankle proprioceptors. If the eye icon is red, this indicates that the visual surround sways with the person's postural sway. In conditions 3 and 6, if the person sways forward, movement of the visual field matches the person's sway; this creates the visual illusion of lack of movement. If the foot icon is red, this indicates that the support surface moves with the person's postural sway, providing inaccurate information about orientation. In conditions 4–6, if the person sways forward, the support surface under their toes tilts downward, so proprioception from the distal lower limbs does not give accurate orientation information. (Courtesy of Neurocom International Inc., Clackamas, OR.)

equivalent stretch of the gastrocnemius can be provided by dorsiflexing the subject's ankles by tilting the platform so that the anterior edge of the platform is higher than the posterior edge. However, in the platform tilt situation, contracting the gastrocnemius further destabilizes the subject. Because the subject's ankles are dorsiflexed, contracting the gastrocnemius pushes the person posteriorly. After a few trials, healthy subjects learn to reduce the response of the gastrocnemius, thus preserving their balance.

Although sophisticated, this type of testing is similar to the clinical practice of pushing a person off balance to test the equilibrium response. Both tests assess the ability to respond to externally imposed displacements; in daily life, these displacements are uncommon. The continual challenge in everyday posture is to anticipate and adjust for voluntary destabilization. For example, prior to a forward reach when a person is standing, the muscles of the opposite leg and back contract to provide stability before the deltoid contracts (Cordo and Nashner, 1982). If a standing person decides to walk, anticipatory adjustments must prepare for the movement of the center of mass. Posturography can also be used to assess these active conditions. A complete postural evaluation includes sensory and motor assessment under three conditions: imposed displacement, active reaching, and ambulation.

Certain nervous system dysfunctions produce recognizable postural abnormalities (see review by Horak [1995]). People with Parkinson's disease have aberrant muscle tone and reduced central control. This combination of deficits results in flexed posture, lack of protective reactions, and weak anticipatory postural adjustments. Postural effects of cerebellar le-

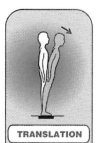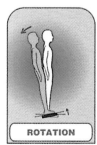

TRANSLATION   ROTATION

FIGURE 10–15

The gastrocnemius can be passively stretched by moving the support surface backward (translation) or by tilting the support surface to raise the toes (rotation). Illustration shows the body position at the end of platform movement in gray and the return to upright in white. If the support surface moves backward, the long latency stretch reflex helps return the body to upright. If platform rotation causes ankle dorsiflexion, the long latency stretch reflex will plantar flex the ankle and result in loss of balance backward. With repeated trials, the long latency reflex decreases as the nervous system learns to maintain posture efficiently when the platform tilts. (Courtesy of Neurocom International Inc., Clackamas, OR.)

sions depend on the cerebellar region involved. As noted earlier, cerebrocerebellar lesions have little effect on posture, spinocerebellar lesions result in gait and stance ataxia, and vestibulocerebellar lesions result in truncal ataxia. Muscle activation patterns are normal in people with spinocerebellar lesions, but the duration and amplitude of postural adjustments are larger than normal, and anticipatory adjustments for predictable conditions are lacking.

People with intact nervous systems respond to movement of the supporting platform differently depending on instructions (Burleigh et al., 1994). If asked to maintain stance when the platform moves, the first response is to sway forward or backward. If asked to step when the platform moves, the first response is to weight shift laterally. This ability to switch motor response to identical stimuli demonstrates the importance of both instructions and the subject's intentions in sculpting motor output.

Ambulation

All regions of the nervous system are required for normal human ambulation. The cerebral cortex provides goal orientation, the basal ganglia govern generation of force, and the cerebellum provides timing, interlimb coordination, and error correction. However, cerebral and basal ganglia neurons are not essential for ambulation in the cat; cats with the cerebral hemispheres isolated from the brain stem and spinal cord (connecting fibers are cut above the superior colliculi) can ambulate if the mesencephalic locomotor region is electrically stimulated (Shik and Orlovsky, 1976). Brain stem descending pathways adjust the strength of muscle contractions by two mechanisms: direct connections and adjusting transmission in spinal reflex pathways. The pattern of muscle activation is controlled by spinal central pattern generators. Sensory information is used to adapt motor output appropriately for environmental conditions.

During normal gait initiation, the swing limb first pushes downward and backward against the support surface. This pattern of force results in the center of mass being moved forward and onto the stance leg in preparation for foot-off and increases the magnitude of the subsequent movement. However, in adults with hemiplegia, the contribution of the paretic leg to weight transfer is much less than normal if the feet are parallel. When the foot of the swing limb is placed

10 to 15 cm behind the stance foot, normal weight transfer occurs, indicating that foot position is an important variable in weight transfer for gait initiation (Malouin et al., 1994). People with Parkinson's disease also fail to adequately move the center of mass prior to attempting stepping (Gantchev et al., 1994), causing inability to initiate gait.

Reaching and Grasping

Vision and somatosensation are essential for normal reaching and grasping. Vision provides information for locating the object in space, as well as assessing the shape and size of the object. Preparation for movement (feed-forward) is the primary role of visual information; if the movement is inaccurate, vision also guides corrections (feedback). The stream of visual information used for movement ("action stream" [Goodale and Milner, 1992]) flows from the visual cortex to the posterior parietal cortex (Fig. 10–16). The posterior parietal cortex contains neurons associated with both sensation and movement; these neurons project to premotor cortical areas controlling reaching, grasping, and eye movements. The premotor cortical areas for each action are somewhat distinct, with the result that reaching, grasping, and eye movements are controlled separately but coordinated by connections among the areas. Information from sensory and planning areas of the cerebral cortex projects to the basal ganglia and cerebellum, then via the

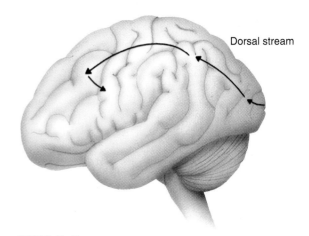

Dorsal stream

FIGURE 10–16

Visual action stream. Visual information traveling from occipital to parietal to premotor cortex helps control movement.

thalamus to the sensorimotor cortex, origin of the corticospinal tract. Proprioception is used similarly to vision, to prepare for movement and to provide information regarding movement errors. Proprioceptive and cutaneous information also trigger changes in movement, as when contact with an object triggers the fingers to close around the object.

Before one can reach accurately, visual grasp (fixing the object in central vision) and proprioceptive information about upper limb position are required. This information allows successful prediction task dynamics and control for the first phase of reaching: fast approach to the object, which is primarily a feedforward process. The second phase of reaching is homing in, a slower, corrective adjustment to achieve contact with the target. Visual guidance (feedback) is necessary for homing in (Jeannerod, 1990). In contrast to classical concepts of proximal to distal muscle activation during reaching, proximal muscles move the hand toward the target; the muscles are controlled by the medial activation system. Simultaneously, distal muscles orient and preshape the hand for grasp; these actions are controlled primarily by feed-forward mechanisms via the lateral activation system. If fractionated movements are used, as in picking up a coin with the index finger and thumb, the neural activity that begins in the prefrontal cortex eventually activates the lateral corticospinal tract (Fig. 10–17).

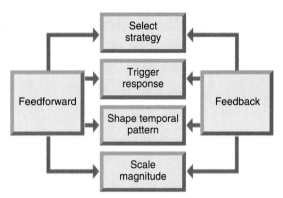

FIGURE 10–18

Interaction of feedforward and feedback in determining movements. (Modified, with permission, from Horak, F. B., presentation "Somatosensory experience in adapting postural coordination," Sensory Mechanisms in Motor Coordination: Implications for Motor Control and Rehabilitation Symposium, Society for Neuroscience, Nov. 13, 1994.)

Grasping is coordinated with activity of the eyes, head, proximal upper limb, and trunk; orientation and postural preparation are integral to the movement. When the object is contacted, grip force adjusts quickly, indicating feed-forward control. After the object is grasped, somatosensory information corrects any error in grip force. Somatosensory information is also used to trigger shifts in movement, for example, to switch from touch to grasp or from grasp to lift (Johansson and Edin, 1992).

During development, normal infants grasp objects before they can control their posture, and the ability of infants to manipulate objects in their hands does not depend on proximal control (vonHofsten, 1992). In fact, early manual activity may be important in developing normal proximal control (Bradley, 1992).

Adults with parietal lobe damage have abnormal timing of reaching and grasp, lack anticipatory adjustments of the fingers, and use a palmar, rather than pincer, grasp (Jeannerod, 1984). Pincer grasp requires corticospinal control, which explains the use of palmar grasp by infants prior to myelination of the corticospinal tracts as well as the palmar grasp of adults with parietal lobe damage.

In normal actions, feed-forward and feedback interact to create movement. Preparation for movement (feed-forward) is based on prediction, and the movement is adjusted according to the resulting sensory information (feedback) (Fig. 10–18). The source of the signal to initiate movements is unknown.

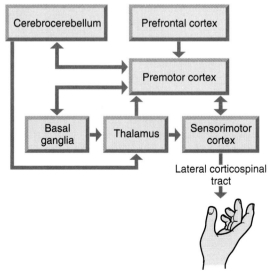

FIGURE 10–17

The motor circuit for fractionated finger movements.

TESTING THE MOTOR SYSTEM

An evaluation of the motor system assesses strength, muscle bulk, muscle tone, reflexes, movement efficiency and speed, postural control, and abnormal movements (Table 10–3).

Electrodiagnostic Studies

Frequently the purpose of nerve conduction studies in motor disorders is to differentiate among three possible sites of dysfunction: nerve, neuromuscular junction, and muscle. In nerve conduction velocity studies of motor nerve, the skin over a nerve is electrically stimulated, and potentials are recorded from the skin over an innervated muscle. Diagnostic EMG, using a needle electrode inserted into muscle, is commonly used to distinguish between denervated muscle and myopathy. Myopathy is an abnormality or disease intrinsic to muscle tissue. The electrical activity of a muscle is recorded using an oscilloscope and loudspeaker.

NERVE CONDUCTION STUDIES

The function of motor fibers in the median nerve can be tested by electrically stimulating the median nerve at the wrist while recording from electrodes over the abductor pollicis brevis muscle and then stimulating at the elbow while recording from the same site. The depolarization of the muscle is recorded as a compound muscle action potential (CMAP). The nerve conduction velocity equals the distance between the proximal and distal stimulation sites, divided by the difference between the latencies. CMAP amplitude

---

TABLE 10–3
EVALUATION OF THE MOTOR SYSTEM

Evaluation	Procedure
Strength	
Quick screening	Resist abduction, adduction, flexion, and extension at the shoulder, elbow, wrist, fingers, hip, knee, and ankle.
Manual muscle test	Position person appropriately (see Kendall and McCreary [1983] or Daniels and Worthingham [1986]); apply manual resistance to patient's movement.
Muscle bulk	Visually inspect for disparity in muscle size. Measure if a difference is suspected.
Muscle tone	
Passive range of motion	Passively flex and extend the elbow, wrist, knee, ankle, and neck; note resistance to movement. Hypotonia may indicate a lower motoneuron or cerebellar lesion; hypertonia may be a sign of upper motoneuron or basal ganglia lesion.
Reflexes	
Phasic stretch reflex (also called deep tendon reflex, muscle stretch reflex, and tendon jerk)	Use a reflex hammer to tap the tendon of a relaxed muscle; muscle should contract. Indicates whether the reflex loop (spindle, afferents, spinal segment, efferent, and muscle) is functioning. Biceps, triceps, quadriceps, and triceps surae tendons are most commonly tested.
Babinski's sign	Stroke the outer edge of patient's foot. Babinki's sign is present if great toe extends; the other toes may spread apart. Absent if no response or all toes curl.
H-reflex*	Electrically stimulate skin over large fiber afferents; a reflexive contraction of muscle should occur. Indicates excitability of alpha motoneurons.
Movement efficiency and speed	
Rapid alternating movements	Patient taps both index fingers or both feet, then pronates and supinates hands; note speed, smoothness, symmetry, and rhythm of movements. If patient has difficulties with these movements in the absence of weakness, indicates cerebellar or proprioceptive dysfunction.

(continued)

TABLE 10–3
EVALUATION OF THE MOTOR SYSTEM *Continued*

Evaluation	Procedure
Accuracy and smoothness of movement	Patient performs the following movements several times; abduct arm, touch own nose; abduct arm, touch examiner's finger; walk heel-to-toe. Observe for jerky movements (ataxia), tremor worsening with movement (action tremor), and inability to move the precise distance required (dysmetria). Difficulty indicates cerebellar or proprioceptive dysfunction.
Transcranial magnetic stimulation*	Currently experimental, not used clinically. Investigates possible upper motoneuron lesions by inducing magnetic activation of cortical neurons, then recording muscle activity.
Movement analysis by surface EMG tests	
During active movement	
Paresis	Compare EMG output of involved muscle to EMG output to contralateral homologous muscle: paresis is 50% or less EMG without antagonist cocontraction.
Hyperreflexia	Increase in EMG activity during active or passive muscle lengthening.
Cocontraction	Measure the amount of time antagonist muscles are contracting simultaneously.
During muscle stretch	
Hypertonia	If EMG does not increase during stretch despite limitation in range motion, restriction is due to non-neural factors, i.e., intrinsic changes in muscle. If EMG increases, neural factors are contributing to decreased range of motion.
Postural control	
Romberg's test	Patient stands, arms folded across chest. Time how long patient can maintain balance with eyes open, then with eyes closed. Stop test if time exceeds 1 minute.
Sharpened Romberg	Same as Romberg's test, but with one foot in front of other foot. Stop test if time exceeds 1 minute.
Tinetti balance scale (Tinetti, 1986)	Balance in sitting, standing, moving from sit to stand and stand to sit, turning 360°, and in response to push on sternum is assessed.
Clinical test for sensory interaction in balance	Three visual conditions—eyes open, eyes closed, and vision altered by a dome covering the face—are combined with standing on firm surface or on foam. See Shumway-Cook and Horak (1986) for interpretation.
Tiltboard	Patient stands on a board repeatedly tipped by the examiner.
Functional reach	Patient stands with feet together, shoulder flexed 90°; then patient reaches as far forward as possible (Duncan et al., 1990).
Computerized balance testing	Stationary posturography: Patient stands on force platform; variations in force exerted are recorded. May incorporate EMG recording. May vary foot position and eyes open, closed. Moving platform posturography: As above, with movement of platform and different sensory conditions (see text).
Postural muscle EMG	During three conditions: external displacement (push on shoulder), prior to voluntary upper limb movement, and prior to walking.

Abnormal involuntary movements	Patient sits quietly; note any of the following:
	   Athetosis
	   Chorea
	   Dystonia
	These involuntary movements indicate basal ganglia disorders. A more sensitive test is to have patient stand with eyes closed, arms outstretched and note any involuntary movements.

*Indicates a test used by researchers but not routinely used by therapists.

indicates the function of the neuromuscular junction and muscle fibers in addition to the conduction ability of the alpha motoneurons (Fig. 10–19).

ELECTROMYOGRAPHY

Diagnostic EMG requires inserting an electrode directly into muscle because surface electrodes cannot detect spontaneous (involuntary) muscle activity. Muscle electrical activity is recorded during four conditions: on insertion of the needle into the muscle (insertional activity), during rest, during minimal voluntary contraction, and during maximal voluntary contraction. Normal response to needle insertion is a brief interval of depolarization, due to mechanical irritation of muscle fibers. At rest, normal muscle is electrically silent. Minimal voluntary contraction elic-

its single motor unit action potentials. Increasing voluntary contraction generates a full interference pattern, created by the asynchronous discharge of many muscle fibers.

If a muscle membrane is unstable, due to denervation, trauma, electrolyte imbalance, or upper motoneuron lesion, its fibers become hypersensitive to acetylcholine, and fibrillation (random, spontaneous contraction of individual muscle fibers) ensues. Fibrillation occurs when the subject intends for the muscle to be at rest. If muscle is reinnervated, larger than normal amplitude muscle potentials are recorded due to axons innervating a greater than normal number of muscle fibers. In primary disease of muscle (myopathy), axons innervate fewer than normal numbers of muscle fibers, resulting in a small-amplitude CMAP. Myopathy is indicated by short-duration, low-

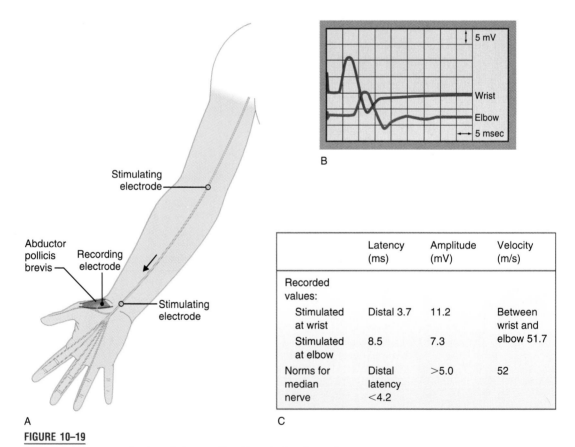

	Latency (ms)	Amplitude (mV)	Velocity (m/s)
Recorded values:			
Stimulated at wrist	Distal 3.7	11.2	Between wrist and elbow 51.7
Stimulated at elbow	8.5	7.3	
Norms for median nerve	Distal latency <4.2	>5.0	52

FIGURE 10–19

Normal medial nerve conduction velocity recording. Normal values for median nerve conduction are listed below the recorded values.

amplitude potentials during voluntary contraction, lack of spontaneous muscle activity (fasciculations, fibrillations), and absence of sensory involvement. No spontaneous activity (fasciculations or fibrillations) occurs in myopathy.

> Nerve conduction studies differentiate among nerve, neuromuscular junction, and muscle disorders. Diagnostic EMG distinguishes between denervated muscle and myopathy.

 CLINICAL NOTES

CASE 1

M.V. is a 62-year-old man. While having breakfast, he suddenly lost control of the left side of his body and face. He fell to the floor but did not lose consciousness. Now, 2 weeks later, he is examined in the hospital. The results are as follows:

- He has complete loss of sensation and voluntary movement of his left side.

- He requires assistance to move from supine to sitting and from sitting to standing.

- He cannot sit or stand independently.

- He has difficulty speaking because of lack of sensation and reduced control of the oral and pharyngeal muscles on the left side. The nursing staff reports that he also has difficulty eating.

- Babinski's sign is present on the left side.

Question

What is the location of the lesion and probable etiology?

CASE 2

K.L. is a 73-year-old woman recently admitted to long-term care because she is unable to care for herself. Ten years ago she developed tremor in both hands; the tremor has gotten progressively worse. Her current status is as follows:

- Consciousness, mentation, sensation, and autonomic functions are intact.

- The tremor is worse when her hands are resting and improves when she is using her hands.

- She tends to remain in the position the staff leaves her in. She does not move spontaneously.

- She cannot independently change from sitting to standing, nor can she initiate walking. However, if someone assists her by leaning her forward, she can walk with a shuffling gait.

- During passive range of motion, all muscles show increased resistance compared to normal.

Question

What is the location of the lesion and probable etiology?

(continued)

CASE 3

A.F. is a 15-year-old girl who was thrown from a horse. She sustained fractures of the humerus and the C5 vertebra. Her coma lasted 2 days.

- Mentation and consciousness are normal.
- Except for the distal right upper limb, sensation, autonomic function, and movement are normal.
- Sensation and sweating are absent in the little finger, medial half of the ring finger, and the adjacent palm of the right hand. The skin is warm and red in the same distribution.
- Wrist flexion on the ulnar side and ulnar deviation are impaired. She cannot flex the distal interphalangeal joints in the fourth and fifth digits.

Question

What is the location of the lesion and probable etiology?

CASE 4

P.A. is a 39-year-old woman, 1 month postinjury from a 30-foot fall while mountain climbing. She sustained multiple injuries, most prominently fractures of the right femur, right fibula, and the T10 vertebra.

- The right lower limb is in a cast, restricting evaluation, but sensation is absent in the L1 dermatome above the cast and in the toes. No voluntary movement of the right quadriceps or toes can be elicited.
- Sensation is absent throughout the left lower limb and bilaterally in the sacral region.
- No voluntary movement can be elicited in the left lower limb.
- The Achilles tendon reflex and Babinski's sign are present bilaterally.

Question

What is the location of the lesion and probable etiology?

CASE 5

R.J. is a 71-year-old man concerned about regaining his strength. Prior to 4 months ago, he had considered himself healthy and strong. He competed regularly in master's swimming events and walked several miles daily. Gradually he has been becoming weaker; although he continues swimming, his times are not competitive, and he can walk only half a mile.

- Mentation, consciousness, sensation, and autonomic functions are normal. When asked, he mentions that he has noticed muscle twitching.
- Throughout the body, skeletal muscles are visibly atrophied.
- Babinski's signs are present in both lower limbs.
- On passive movement, the therapist notes that faster movements meet with more resistance than slower movements. If R.J.'s limbs are moved slowly, the initially strong resistance gives way to easier movement.
- Diagnostic EMG studies reveal fasciculations and fibrillations in the muscles tested.

Question

What is the location of the lesion and probable etiology?

(continued)

CASE 6

B.F. is a 52-year-old accountant. His only complaint is right forearm muscle pain and cramping when playing tennis and when writing.

- Right upper limb cutaneous sensation and autonomic function are normal.
- When the therapist palpates the muscles of the forearm, B.F. reports no pain in wrist extensor muscles and deep, aching pain on palpation of the wrist and finger flexor muscles.
- Active range of motion is normal throughout the right upper limb.
- When B.F. is asked to write, after 2 minutes his forearm and finger flexor muscles cramp. To continue writing, B.F. must use an abnormal upper limb position with the shoulder abducted 60°.

Question

What is the location of the lesion and probable etiology?

REVIEW QUESTIONS

1. What is hemiplegia?
2. Which of the following signs always indicate pathology: muscle spasms, cramps, fasciculations, fibrillations, abnormal movements?
3. What is hypertonia? What is the difference between the two types of hypertonia? How is hypertonia produced?
4. What is spinal shock?
5. If a person has loss of reflexes, muscle atrophy, flaccid paralysis, and fibrillations, what is the location of the lesion?
6. What does Babinski's sign in an adult indicate?
7. What is the location of a lesion that produces abnormal cutaneous reflexes, abnormal timing of muscle activation, paresis, and spasticity?
8. What is clonus?
9. Do hyperactive stretch reflexes produce spasticity?
10. How can paresis, cocontraction, hypertonia, and hyperflexia each be quantified using surface EMG?
11. What is the difference in effects on lower motoneurons between a complete spinal cord lesion and a typical cerebrovascular accident?
12. What motor signs occur in spastic cerebral palsy that are not also seen in adult-onset upper motoneuron lesions?
13. What is learned nonuse?
14. Amyotrophic lateral sclerosis destroys what parts of the nervous system?
15. What disease is characterized by rigidity, hypokinesia, resting tremor, and visuoperceptive impairments? Cell death in which nuclei produces this disease?
16. List three disorders that cause hyperkinesia.
17. Involuntary abnormal postures or repetitive twisting movements are signs of what disorder?
18. Slurred, poorly articulated speech and ataxic gait are signs of damage to what part of the cerebellum?
19. List and define the signs of cerebrocerebellar lesions.
20. What is a long loop response?
21. What is an asymmetrical tonic neck reflex?
22. What information can posturography provide?
23. Give an example of a preparatory postural adjustment.
24. Give an example of identical stimuli eliciting different responses depending on instructions given to a person.
25. If a person with hemiplegia is having difficulty initiating gait, what simple adjustment might make gait initiation easier?

26. What are the two phases of reaching?

27. What is diagnostic EMG used for?

28. What is the difference in purpose between motor nerve conduction velocity studies and movement analysis by surface EMG tests?

References

Alter, M., Friday, G., Sobel, E., et al. (1993). The Lehigh Valley Recurrent Stroke Study: Description of design and methods. Neuroepidemiology 12(4):241–248.

Ashby, P. D. (1993). The neurophysiology of human spinal spasticity. In: Science and practice in clinical neurology. Cambridge, England: Cambridge University Press, pp. 106–129.

Avorn, J., Bohn, R. L., Mogun, H., et al. (1995). Neuroleptic drug exposure and treatment of parkinsonism in the elderly: A case-control study. Am. J. Med. 99(1):48–54.

Berger, W., Horstmann, G., and Dietz, V. (1984). Tension development and muscle activation in the leg during gait in spastic hemiparesis: Independence of muscle hypertonia and exaggerated stretch reflexes. J. Neurol. Neurosurg. Psychiatry 27: 1029–1033.

Bobath, B. (1977). Treatment of adult hemiplegia. Physiother. Can. 63:310–313.

Bradley, N. S. (1992). What are the principles of motor development? In: Forssberg, H., and Hirschfeld, H. (Eds.). Movement disorders in children. Basel: Karger, pp. 41–49.

Burke, D., and Gandevia, S. C. (1993). Muscle spindles, muscle tone and the fusimotor system. In: Gandevia, S. C., Burke, D., and Anthony, M. (Eds.). Science and practice in clinical neurology. Cambridge, England: Cambridge University Press, pp. 89–105.

Burleigh, A. L., Horak, F. B., and Malouin, F. (1994). Modification of postural responses and step initiation: Evidence for goal directed postural interactions. J. Neurophysiol. 72(6):2892–2902.

Butefisch, C., Hummelsheim, H., Denzler, P., et al. (1995). Repetitive training of isolated movements improves the outcome of motor rehabilitation of the centrally paretic hand. J. Neurol. Sci. 130(1):59–68.

Cahan, L., Adams, J., Perry, J., et al. (1990). Instrumented gait analysis after selective dorsal rhizotomy. Dev. Med. Child Neurol. 32:1037–1043.

Campbell, S., Almeida, G., Penn, R. D., et al. (1995). The effects of intrathecally administered baclofen on function in patients with spasticity. Phys. Ther. 75:352–362.

Carmick, J. (1993). Clinical use of neuromuscular electrical stimulation for children with cerebral palsy: Part I. Lower extremity. Part II. Upper extremity. Phys. Ther. 73(8):505–527.

Comella, C. L., Stebbins, G. T., Brown-Toms, N., et al. (1994). Physical therapy and Parkinson's disease: A controlled clinical trial. Neurology 44:376–378.

Cordo, P. J., and Nashner, L. M. (1982). Properties of postural adjustments associated with rapid arm movements. J. Neurophysiol. 47:287–302.

Crenna, P., Inverno, M., Frigo, C., et al. (1992). Pathophysiological profile of gait in children with cerebral palsy. In: Movement disorders in children. Basel: Karger, pp. 186–198.

Daniels, L., and Worthingham, C. (1986). Muscle testing: Techniques of manual examination. Philadelphia: W. B. Saunders.

Dennis M. S., and Burn, J. P. (1993). Long-term survival after first-ever stroke: The Oxfordshire Community Stroke Project. Stroke 24(6):796–800.

Dietz, V. (1992). Spasticity: Exaggerated reflexes or movement disorder? In: Forssberg, H., and Hirshfeld, H. (Eds.). Movement disorders in children. Basel: Karger, pp. 225–233.

Dietz, V., and Berger, W. (1983). Normal and impaired regulation of muscle stiffness in gait: A new hypothesis about muscle hypertonia. Exp. Neurol. 79:680–687.

Dietz, V., Trippel, M., and Berger, W. (1991). Reflex activity and muscle tone during elbow movements in patients with spastic paresis. Ann. Neurol. 30:767–779.

Duncan, P., Weiner, D. K., Chandler, J., et al. (1990). Functional reach: A new clinical measure of balance. J. Gerontol. 45(6):192–197.

Fellows, S., Kaus, C., Thilman, A. F., et al. (1994). Voluntary movement at the elbow in spastic hemiparesis. Ann. Neurol. 36(3):397–407.

Fisher, B., and Woll, S. (1995). Considerations in the restoration of motor control. In: Montgomery, J. (Ed.). Physical therapy for traumatic brain injury. New York: Churchill Livingstone, pp. 55–78.

Freed, C. R. (1992). Survival of implanted fetal dopamine cells and neurologic improvement 12 to 46 months after transplantation for Parkinson's disease. N. Engl. J. Med. 327(22):1549–1555.

Gandevia, S. C. (1993). Assessment of corticofugal output: Strength testing and transcranial stimulation of the motor cortex. In: Gandevia, S. C., Burke, D., and Anthony, M. (Eds.). Science and practice in clinical neurology. Cambridge, England: Cambridge University Press, pp. 76–88.

Gantchev, N., Viallet, F., Aurenty, R., et al. (1996). Impairment of posturo-kinetic coordination during initiation of forward oriented step in parkinsonian patients. EEG and Clinical Neurophys. 101(2):110–120.

Goodale, M. A., and Milner, A. D. (1992). Separate visual pathways for perception and action. Trends Neurosci. 15:20–25.

Gowland, C., deBruin, J., Basmajian, J. V., et al. (1992). Agonist and antagonist activity during voluntary upper-limb movement in patients with stroke. Phys. Ther. 72:624–633.

Granat, M. H., Ferguson, A. C., Andrews, B. J., et al. (1993). The role of functional electrical stimulation in the rehabilitation of patients with incomplete spinal cord injury—observed benefits during gait studies. Paraplegia 31(4):207–215.

Hallett, M. (1993). Physiology of basal ganglia disorders: An overview. Can. J. Neurol. Sci. 20:177–183.

Hesse, S., Lucke, D., Malezic, M., et al. (1994). Botulinum toxin treatment for lower limb extensor spasticity in chronic hemiparetic patients. J. Neurol. Neurosurg. Psychiatry 57(11):1321–1324.

Horak, F. B., and MacPherson, J. M. (1995). Postural orientation and equilibrium. In: Rowell, L. B., and Shepherd, J. T. (Eds.).

Handbook of physiology: Section 12, Integration of motor, circulatory, respiratory and metabolic control during exercise. New York: Oxford University Press, pp. 255–292.

Ivry, R, and Keele, S. (1989). Timing functions of the cerebellum. J. Cognitive Neurosci. 1(2):136–152.

Jeannerod, M. (1990). The neural and behavioral organization of goal-directed movements. Oxford: Clarendon Press.

Jeannerod, M. (1984). The contribution of open-loop and closed-loop modes in prehension movements. Preparatory states and processes. Hillsdale, N. J.: Lawrence Erlbaum.

Johansson, R. S., and Edin, B. B. (1992). Neural control of manipulation and grasping. In: Forssberg, H., and Hirschfeld, H. (Eds.). Movement disorders in children. Basel: Karger, pp. 107–112.

Kendall, F. P., and McCreary, E. K. (1983). Muscles, testing and function. Baltimore: Williams & Wilkins.

Levin, M., and Hui-Chan, C. (1992). Relief of hemiparetic spasticity by TENS is associated with improvement in reflex and voluntary motor functions. Electroencephalogr. Clin. Neurophysiol. 85:131–142.

Levin, M. F., and Hui-Chan, C. (1993). Are H and stretch reflexes in hemiparesis reproducible and correlated with spasticity? J. Neurol. 240:63–71.

Light, K. E., Rehm, S., et al. (1994). Does heavy resistive exercise promote muscular cocontraction and loss of reciprocal movement in brain-injured subjects? Soc. Neurosci. Abstr. 20:844.

Lipton, S. A., and Rosenberg, P. A. (1994). Excitatory amino acids as a final common pathway for neurologic disorders. New Eng. J. Med. 330(9):613–622.

Louis, E. D., Goldman, J. E., Powers, J. M., et al (1995). Parkinsonian features of eight pathologically diagnosed cases of diffuse Lewy body disease. Movement Disorders 10(2):188–194.

Malouin, F., Menier, C., et al. (1994). Dynamic weight transfer during gait initiation in hemiparetic adults and effect of foot position. Soc. Neurosci. Abstr. 20:571.

McDonough, J. T. (1994). Stedman's concise medical dictionary. Baltimore: Williams & Wilkins.

Nashner, L. M., Black, F. O., and Wall, C. (1982). Adaptation to altered support and visual conditions during stance: Patients with vestibular deficits. J. Neurosci. 2:536–544.

Neilson, P. D. (1993). Tonic stretch reflex in normal subjects and in cerebral palsy. In: Ganderia, S. C., Burke, D., and Anthony, M. (Eds.). Science and practice in clinical neurology. Cambridge, England: Cambridge University Press, pp. 169–190.

Penn, R. (1992). Intrathecal baclofen for spasticity of spinal origin: Seven years of experience. J. Neurosurg. 77:236–240.

Sahrmann, S., and Norton, B. (1977). The relationship of voluntary movement to spasticity in the upper motor neuron syndrome. Ann. Neurol. 2:460–465.

Schleenbaker, R. E., and Mainous, A. G. (1993). Electromyographic biofeedback for neuromuscular reeducation in the hemiplegic stroke patient: A meta-analysis. Arch. Phys. Med. Rehabil. 74:1301–1304.

Shik, M. L., and Orlovsky, G. N. (1976). Neurophysiology of locomotor automatism. Physiol. Rev. 56:465–501.

Shumway-Cook, A., and Horak, F. B. (1986). Assessing the influence of sensory interaction on balance. Phys. Ther. 66: 1548–1550.

Taub, E., Miller, N. E., Novack, T. A., et al. (1993). Technique to improve chronic motor deficit after stroke. Arch. Phys. Med. Rehabil. 74:347–354.

Tinetti, M. (1986). Performance oriented assessment of mobility problems in elderly patients. J. Am. Geriatr. Soc. 41:479.

Vilis, T., and Hore, J. (1977). Effects of changes in mechanical state of limb on cerebellar intention tremor. J. Neurophysiol. 40: 1214–1224.

vonHofsten, C. (1992). Development of manual actions from a perceptual perspective. In: Forssberg, H., and Hirschfeld, H. (Eds.). Movement disorders in children. Basel: Karger, pp. 113–123.

Wolf, S. L., and Catlin, P. A. (1994). Overcoming limitations in elbow movement in the presence of antagonist hyperactivity. Phys. Ther. 74:826–835.

Young, J. L., and Mayer, R. F. (1992). Physiological alterations of motor units in hemiplegia. J. Neurol. Sci. 54:401–412.

Suggested Readings

Cordo, P., and Harnad, S. (1994). Movement control. Cambridge, England: Cambridge University Press.

Rothwell, J. (1994). Control of human voluntary movement (2nd ed.). London: Chapman and Hall.

Umphred, D. A. (1995). Neurological rehabilitation (3rd ed.). St. Louis: Mosby.

11

Peripheral Nervous System

I was a 32-year-old woman working a 40-hour week as a chef's assistant. I first noticed pain in my wrist and hand while working. After work, my hand would be numb and have tingling sensations. As the problem progressed, it became difficult to grip a knife or cleaver.

When I first went to the doctor, the problem was diagnosed as tendinitis, and I was advised to use my other hand more. I kept working, and the condition worsened; on returning to the doctor, I received pain medication and a wrist brace. When these did not help, I was referred to an orthopedic specialist. Nerve conduction studies were performed by a physical therapist. The condition was diagnosed as carpal tunnel syndrome. I went to a physical therapist two times a week for heat treatments and exercises for about 3 months. I was told I could no longer continue my line of work. I ended up with a full cast for 6 weeks to prevent me from using my left arm and hand.

The pain in my wrist and hand continued to be intense, much worse at night. I could only sleep with my arm propped up on a pillow above my head. I ended therapy after having two cortisone shots into my wrist, which did not have any effect.

Today if I garden or use my left hand too long typing or playing tennis, I will have pain and know I need to lighten up.

—Genevieve Kelly

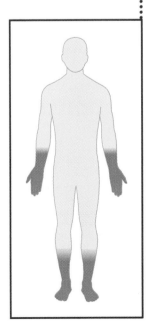

INTRODUCTION

The peripheral nervous system includes all neural structures distal to the spinal nerves. Thus, axons of sensory, motor, and autonomic neurons, along with specialized sensory endings and entire postganglionic autonomic neurons, form the peripheral nervous system. Examples of peripheral nerves are the median, ulnar, and tibial nerves. Although cranial nerves are also peripheral nerves, cranial nerves will be covered in Chapters 13 and 14 because their function can best be understood in the context of brain stem function. In this textbook, all nervous system structures enclosed by bone are considered parts of the central nervous system; nerve roots, dorsal root ganglia, and spinal nerves are therefore within spinal region (see Fig. 1–4). Distal to the spinal nerve, the groups of axons split into posterior and anterior rami. The posterior rami innervate structures along the posterior midline of the body, and the anterior rami supply structures in the lateral and anterior body.

Classifying nerve roots, dorsal root ganglia, and spinal nerves as spinal and the remaining axons as peripheral allows the clinical difference between spinal and peripheral lesions to be easily distinguished: sensory, autonomic, and motor deficits in spinal lesions show a myotomal and/or dermatomal distribution; sensory, autonomic, and motor deficits in peripheral lesions show a peripheral nerve distribution (see Fig. 6–4). Signs of peripheral neuron lesions are paresis or paralysis, sensory loss, abnormal sensations, muscle atrophy, and reduced or absent deep tendon reflexes.

> Peripheral nerve lesions produce signs and symptoms in a peripheral nerve distribution. Spinal cord lesions produce signs and symptoms in a myotomal and/or dermatomal distribution.

ANATOMY OF PERIPHERAL NERVES

Peripheral nerves consist of parallel bundles of axons. The axons are surrounded by three connective tissue sheaths:

- Endoneurium
- Perineurium
- Epineurium

Endoneurium separates individual axons, perineurium surrounds bundles of axons, and epineurium encloses the entire nerve trunk (Fig. 11–1). Peripheral nerves receive blood supply via arterial branches that enter the nerve trunk. Within the nerve, axons are electrically insulated from each other by endoneurium and by a myelin sheath. The myelin sheath is provided by Schwann cells, which either partially surround a group of small-diameter axons or

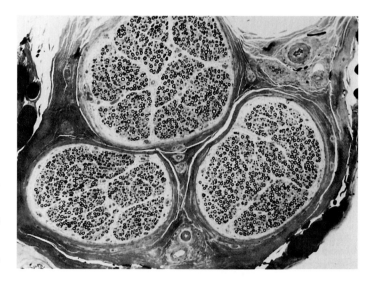

FIGURE 11–1

Cross section of a normal peripheral nerve, showing three fascicles. Within each fascicle are many darkly stained myelin sheaths, appearing as small oval structures enclosing the axons, which appear white. Endoneurium surrounds each axon. Perineurium surrounds each fascicle. Epineurium surrounds the three fascicles. (Reproduced by permission from Richardson, E. P., Jr., and DeGirolami, U. [1995]. Pathology of the peripheral nerve. Philadelphia: W. B. Saunders, p. 3.)

completely envelop a section of a single large axon. The small-diameter axons that share Schwann cells are called unmyelinated (although "partially myelinated" would be a more accurate term), while the large-diameter axons that are fully wrapped by individual Schwann cells are designated myelinated.

Peripheral nerves supply either viscera or somatic structures. The visceral supply, via splanchnic nerves, is discussed in Chapter 8.

Somatic peripheral nerves are usually mixed, consisting of sensory, autonomic, and motor axons. Cutaneous branches supply the skin and subcutaneous tissues; muscular branches supply muscle, tendons, and joints. Cutaneous branches are not purely sensory because they deliver the sympathetic efferent axons to sweat glands and arterioles. Muscular branches are not purely motor because they contain sensory axons from proprioceptive structures.

Peripheral axons are classified into groups according to their speed of conduction and their diameter (Table 11–1). Two classification systems for peripheral axons are in common use. The letter classification system (A, B, C) applies to both afferent and efferent axons; the Roman numeral system applies only to afferent axons.

Nerve Plexuses

Four nerve plexuses are formed by the junctions of anterior rami. The cervical plexus arises from anterior rami of C1 to C4 and lies deep to the sternocleido-mastoid muscle. The cervical plexus provides cutaneous sensory information from the posterior scalp to the clavicle and innervates the anterior neck muscles and the diaphragm. The phrenic nerve, whose cell bodies are in the cervical spinal cord (C3 to C5), is the most important single branch from the cervical plexus because the phrenic nerve is the only motor supply and the main sensory nerve for the diaphragm.

The brachial plexus is formed by anterior rami of C5 to T1. The plexus emerges between the anterior and middle scalene muscles, passes deep to the clavicle, and enters the axilla. In the distal axilla, axons from the plexus become the radial, axillary, ulnar, median, and musculocutaneous nerves. The entire upper limb is innervated by brachial plexus branches (Fig. 11–2).

The lumbar plexus is formed by anterior rami of L1 to L4; the plexus forms in the psoas major muscle. Branches of the lumbar plexus innervate skin and muscles of the anterior and medial thigh (Fig. 11–3). A cutaneous branch from the plexus, the saphenous nerve, continues into the leg to innervate the medial leg and foot. Branches of the cervical, brachial, and lumbar plexuses provide sympathetic innervation via connections with the sympathetic chain.

The sacral plexus innervates the posterior thigh and most of the leg and foot. Unlike the other plexuses, containing sympathetic axons, the sacral plexus contains parasympathetic axons.

TABLE 11–1 PERIPHERAL AXONS			Efferent Axons		Afferent Axons	
Axon	Conduction Speed (m/s)	Axon Diameter (μm)	Group	Innervates	Group	Innervates
Large myelinated	7–130	7–22	Aα	Extrafusal muscle fibers	Ia, Ib, II	Spindles, Golgi tendon organs, touch and pressure receptors
Medium myelinated	2–10	2–15	Aγ	Intrafusal muscle fibers		
Small myelinated	12–45	2–10			Aδ	Pain, temperature, visceral receptors
	4–25	1–5	B	Presynaptic autonomic		
Unmyelinated	0.2–2.0	0.2–.05	C	Postsynaptic autonomic	C	Pain, temperature, visceral receptors

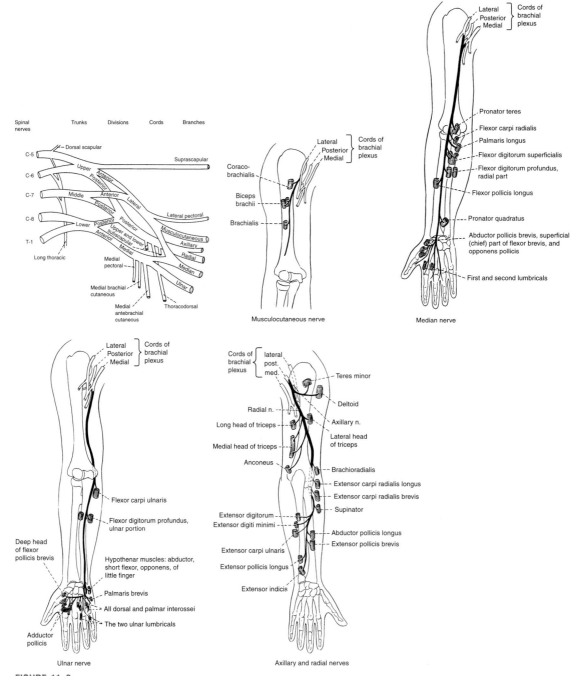

FIGURE 11-2

Brachial plexus and the distribution of the branches of the brachial plexus. (Reproduced by permission from Jenkins, D. B. [1991]. Hollinshead's functional anatomy of the limbs and back [6th ed.]. Philadelphia: W. B. Saunders, pp. 74, 116, 134, 135, 145.)

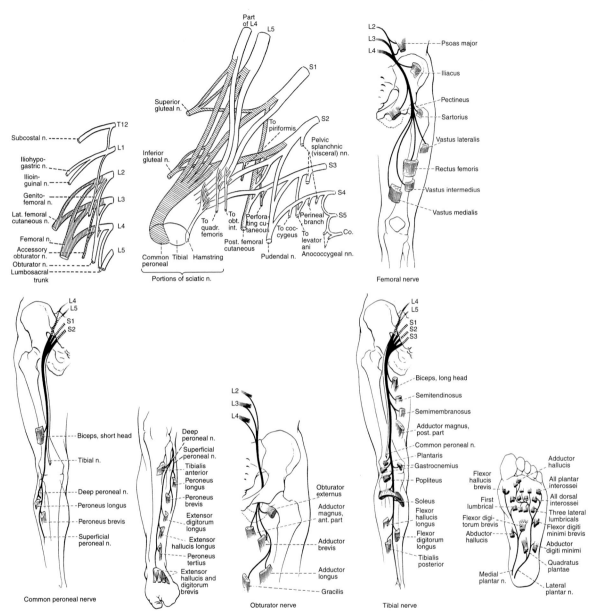

FIGURE 11–3

Lumbar and sacral plexuses and the distribution of their branches. (Reproduced by permission from Jenkins, D. B. [1991]. *Functional anatomy of the limbs and back* [6th ed.]. Philadelphia: W. B. Saunders, pp. 239, 253, 256, 300, 304.)

NEUROMUSCULAR JUNCTION

Motor axons synapse with muscle fibers at neuromuscular junctions. This nerve-muscle synapse requires only depolarization of the motor axon, releasing acetylcholine, which diffuses across the synaptic cleft and binds with receptors to cause depolarization of the muscle membrane. Unlike neuron-neuron synapses, no summation of action potentials is required to depolarize the postsynaptic membrane. No inhibition is possible because only one branch of an axon synapses with a muscle fiber and the action of the neurotransmitter is always excitatory. In a normal motor unit, every depolarization of the motor axon releases sufficient acetylcholine to initiate action potentials in the innervated muscle fibers. Even when a lower motoneuron is inactive (no action potentials are occurring), it spontaneously releases minute amounts of acetylcholine. Binding of the small quantity of acetylcholine to the receptors on muscle membrane causes miniature end plate potentials. These potentials, although not sufficient to initiate the process of muscle contraction, are believed to supply factors necessary to maintain muscle health. Without miniature end plate potentials, muscles atrophy.

DYSFUNCTION OF PERIPHERAL NERVES

The signs of peripheral nerve damage include sensory, autonomic, and motor changes. All signs are in a peripheral nerve distribution.

Sensory Changes

Sensory changes include decreased or lost sensation and/or abnormal sensations: hyperalgesia, dysesthesia, and paresthesia (see Chap. 7).

Autonomic Changes

Autonomic signs depend on the pattern of axonal dysfunction. If a single nerve is damaged, autonomic signs are usually only observed if the nerve is completely severed. These signs include lack of sweating and loss of sympathetic control of smooth muscle fibers in arterial walls. The latter may contribute to edema in an affected limb. If many nerves are involved, autonomic problems may include difficulty regulating blood pressure, heart rate, sweating, bowel and bladder functions, and impotence.

Motor Changes

Motor signs of peripheral nerve damage include paresis (weakness) or paralysis. If muscle is denervated, electromyography (EMG) recordings show no activity for about 1 week following injury. Muscle atrophy progresses rapidly. Then muscle fibers begin to develop generalized sensitivity to acetylcholine along the entire muscle membrane, and fibrillation ensues. Fibrillation is spontaneous contraction of individual muscle fibers. Fibrillation is only observable with needle EMG. Unlike fasciculation, fibrillation cannot be observed on the skin surface. Fibrillation is not diagnostic of any specific lesion; historically, fibrillation was considered diagnostic of muscle denervation (Johnson, 1980).

Denervation: Trophic Changes

When nerve supply is interrupted, trophic changes begin in the denervated tissues. Muscles atrophy, skin becomes shiny, the nails become brittle, and subcutaneous tissues thicken. Ulceration of cutaneous and subcutaneous tissues, poor healing of wounds and infections, and neurogenic joint damage are common, secondary to blood supply changes, loss of sensation, and lack of movement.

CLASSIFICATION OF NEUROPATHIES

Peripheral neuropathy can involve a single nerve (mononeuropathy), several nerves (multiple mononeuropathy), or many nerves (polyneuropathy). Mononeuropathy is focal dysfunction, and multiple mononeuropathy is multifocal. Multiple mononeuropathy presents as asymmetrical involvement of individual nerves. Polyneuropathy is a generalized disorder that typically presents distally and symmetrically (Schaumberg and Spencer, 1983). Dysfunction can be due to damage to the axon, myelin sheath, or both. Table 11–2 summarizes the types, pathology, and prognosis of peripheral neuropathies.

TABLE 11–2

PERIPHERAL NEUROPATHIES

Neuropathy	Usual Cause	Pathology	Typical Recovery
Mononeuropathy	Trauma		
	Class I	Demyelination	Complete and rapid, by remyelination
	Class II	Axonal damage	Slow, by regrowth of axons, but good recovery because Schwann cell and connective tissue sheaths intact
	Class III	Axon and myelin degeneration	Slow, with poor results, due to inappropriate reinnervation and traumatic neuroma
Multiple mononeuropathy	Complication of diabetes or blood vessel inflammation	Ischemia of neuron	Slow, by regrowth of axons, usually good recovery
Polyneuropathy	Complication of diabetes or autoimmune disorder (Guillain-Barré syndrome)	Metabolic or inflammatory	Diabetic may be stable, progressive, or improve with better blood sugar control; Guillain-Barré syndrome usually improves gradually

Mononeuropathies

Depending on the severity of damage, traumatic injuries to peripheral nerves are classified into three classes (Schaumberg and Spencer, 1983). Class I results from focal compression due to entrapment or pressure. Entrapment is most common in the following nerves: median (carpal tunnel), ulnar (ulnar groove), radial (spiral groove), and peroneal (fibular head). Prolonged pressure from casts, crutches, or sustained positions (e.g., sitting with knees crossed) may also compress nerves. Compression temporarily interferes with blood supply or, in the case of prolonged compression, may cause local demyelination. Local demyelination slows or prevents nerve conduction at the demyelinated site (Fig. 11–4). Signs of Class I injury are decreased or lost function of large-diameter axons (motor, touch and proprioceptive, lost phasic stretch reflex), intact autonomic function, and lack of damage to the axon. Recovery tends to be complete because remyelination can occur rapidly, before irreversible damage occurs in the target tissues. See Dawson et al. (1990) for a review of compression injuries.

Carpal tunnel syndrome is a common compression injury of the median nerve in the space between the carpal bones and the flexor retinaculum. (See the carpal tunnel syndrome box.) Initially, pain and numbness are noted at night. Later, these symptoms persist throughout the day, and sensation is decreased or lost in the lateral three and one-half digits and adjacent palm of the hand. On the dorsum of the hand, the distal half of the same digits are involved. Paresis and atrophy of the thumb intrinsic muscles may follow. Pain from carpal tunnel syndrome may radiate into the forearm and occasionally to the shoulder (Kasdan et al., 1993). Provocative tests include Tinel's sign, tapping on the wrist to evoke paresthesia in the median nerve distribution. Direct pressure applied over the carpal tunnel to evoke symptoms has better specificity, that is, elicits fewer false-positive results, than Tinel's sign (Durkan, 1991).

Carpal tunnel syndrome is more prevalent in people whose occupations require repetitive hand movements or the gripping of vibrating tools than in the general population. Thus food service, factory, data entry, and carpentry workers are at high risk. For mild cases, often 1 month of rest, splinting, and anti-inflammatory medication followed by exercises designed to promote gliding of the tendons in the carpal tunnel are sufficient treatment. For more severe cases, changing to a different occupation and surgical release by severing the transverse carpal ligament may be necessary.

Class II injuries usually arise from crushing the nerve. This type of damage affects all sizes of axons, so reflexes are markedly reduced or absent. The connective tissue sheaths and myelin sheaths remain uninterrupted, but the axons are disrupted, and wallerian degeneration occurs distal to the lesion.

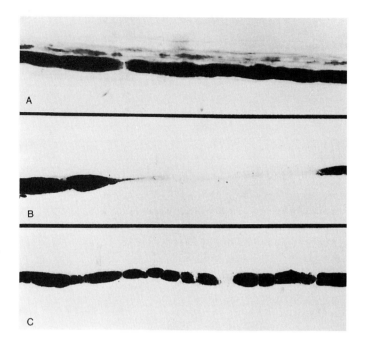

FIGURE 11–4

Nerve fiber with myelin stained to appear black. *A,* Normal myelin. *B,* Segmental demyelination severe enough to cause secondary axonal degeneration. *C,* Remyelination, with abnormally short distance between nodes of Ranvier. (Reproduced by permission from Richardson, E. P., Jr., and DeGirolami, U. [1995]. Pathology of the peripheral nerve. Philadelphia: W. B. Saunders, p. 18.)

Subsequently, muscle atrophy ensues. Because the Schwann cell basement membrane and the connective tissues are intact, regenerating axons are able to reinnervate appropriate targets. Axon regrowth typically proceeds at a rate of 1 mm/day. Recovery from Class II injuries is generally good, resulting in return of nerve conduction in the regenerating axon.

Class III injuries occur when nerves are physically severed, by excessive stretch or laceration. The nerve and connective tissue are completely interrupted, causing immediate loss of sensation and/or muscle paralysis in the area supplied. Wallerian degeneration begins distal to the lesion 3 to 5 days later. Then axons in the proximal stumps begin to sprout. If the proximal and distal nerve stumps are apposed and scarring doesn't interfere, some sprouts enter the distal stump and are guided to their target tissue in the periphery. Other sprouts meet obstacles, and their growth goes awry, with axons growing in random directions from the proximal stump. In a mixed peripheral nerve, the lack of guidance from connective tissue and Schwann cells may allow the axon sprouts to reach inappropriate end-organs, resulting in poor recovery. For example, a motor axon may innervate a Golgi tendon organ; although the motor neuron could fire, the tendon organ would not respond, so the connection would be nonfunctional. If the stumps are displaced or scar tissue intervenes between the stumps, sprouts may grow into a tangled mass of nerve fibers, forming a traumatic neuroma (tumor of axons and Schwann cells). The course of Class III injuries is initially similar to Class II; however, nerve conduction distal to the injury may never return because of poor regeneration.

Multiple Mononeuropathy

Involvement of two or more nerves in different parts of the body occurs most commonly from ischemia of the nerves, either from diabetes or inflammation of the blood vessels. In multiple mononeuropathy, individual nerves are affected, producing a random, asymmetrical presentation of signs.

Polyneuropathy

Symmetrical involvement of sensory, motor, and autonomic fibers, often progressing from distal to proximal, is the hallmark of polyneuropathy. The symptoms typically begin in the feet and then appear in the hands, areas of the body supplied by the longest axons. Degeneration of the distal part of long axons

CARPAL TUNNEL SYNDROME

Pathology
Compression of median nerve in carpal tunnel

Etiology
Repetitive finger movements, gripping vibrating tools

Speed of onset
Chronic

Signs and symptoms

Consciousness
Normal

Communication and memory
Normal

Sensory
Numbness, tingling, burning sensation in median nerve distribution

Autonomic
Lack sweating in median nerve distribution

Motor
Paresis and atrophy of thenar muscles

Region affected
Peripheral

Demographics
Most common in people over 30 years of age; women more often affected than men

Prognosis
Variable; according to the National Center for Health Statistics, "carpal tunnel syndrome required the longest recuperation period of all conditions resulting in lost workdays, with a median 30 days away from work" (Rosenstock, 1996).

may occur because of inadequate axonal transport to keep the distal axons viable. Demyelination is also likely to produce distal symptoms first because the longer axons have more myelin along their length and thus have a greater chance of being affected by the random destruction of myelin.

In contrast to mononeuropathies, polyneuropathies are not due to trauma or ischemia. The etiology can be toxic, metabolic, or autoimmune. The most common causes of polyneuropathies are diabetes, nutritional deficiencies secondary to alcoholism, and autoimmune diseases. A variety of therapeutic drugs, industrial and agricultural toxins, and nutritional disorders (including

malnutrition secondary to alcoholism) can cause polyneuropathy. In severe polyneuropathy, trophic changes (poor healing, ulceration of skin, neurogenic joint damage) often occur; these changes probably occur because the person is unaware of injuries to the part, due to lack of sensation (Sabin and Swift, 1975). Thus, education regarding monitoring and care of insensitive areas is vital. Physical therapists are likely to treat people with diabetic (metabolic) and Guillain-Barré (autoimmune) polyneuropathies.

In diabetic polyneuropathy (Greene et al., 1990), axons and myelin are damaged. Usually sensation is affected most severely, often in a stocking/glove distribution (Fig. 11–5). All sizes of sensory axons are damaged (Fig. 11–6), resulting in decreased sensations and pain, paresthesias, and dysesthesias. Loss of pain sensation often leads to damaged joints in the

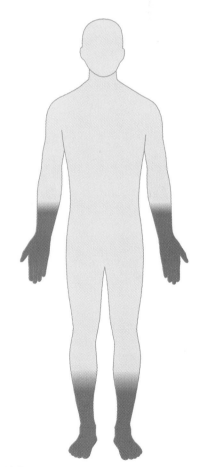

FIGURE 11–5
Stocking/glove distribution of sensory impairment in diabetic neuropathy.

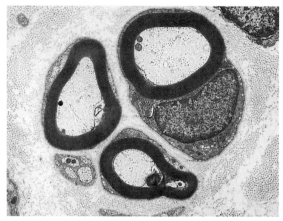

A

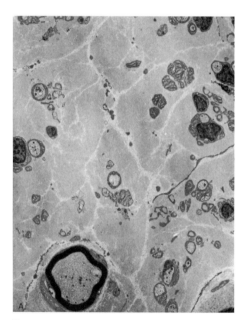

B

FIGURE 11–6

Sural nerve biopsies. The sural nerve is often used for biopsies because it is a purely sensory nerve, and thus the removal of a small section does not cause motor loss. *A,* Cross section of normal nerve, with three myelinated axons (surrounded by darkly stained rings of myelin) and a small group of unmyelinated axons to the left of the bottom myelinated axon. *B,* Cross section of a nerve with damage by diabetic neuropathy. All sizes of axons have been lost, only one myelinated fiber is present, and many axons have been replaced by collagen. (Reproduced by permission from Richardson, E. P., Jr., and DeGirolami, U. [1995]. Pathology of the peripheral nerve. Philadelphia: W. B. Saunders, pp. 5 and 80.)

feet (Charcot's joints) and to foot ulcers. Proper diabetic foot care, including regular sensory testing with monofilaments (see Chap. 7), wearing of appropriate shoes, regular self-inspection of the feet, and proper care of the skin and toenails, may prevent or forestall limb amputations in people with diabetes. Later in the disease process muscle weakness and atrophy also tend to occur distally. All autonomic functions are susceptible to diabetic neuropathy; cardiovascular,

DIABETIC POLYNEUROPATHY

Pathology
Demyelination and axon damage

Etiology
Metabolic

Speed of onset
Chronic

Signs and symptoms
Distal more involved than proximal

Consciousness
Normal

Communication and memory
Normal

Sensory
Numbness, pain, paresthesias (tingling, pins and needles), dysesthesias (burning, aching)

Autonomic
Orthostatic hypotension; impaired sweating, bowel, bladder, digestive, genital, pupil, and lacrimal function

Motor
Balance and coordination problems (secondary to sensory deficits); weakness

Cranial nerves
Usually normal; occasionally CN III is involved, producing drooping of upper eyelid and paresis of four extraocular muscles.

Region affected
Peripheral

Demographics
Affects all ages; no gender predominance

Prognosis
Stable or progressive; occasionally better control of blood sugar levels leads to improvement

gastrointestinal, genitourinary, and sweating dysfunction (lack of sweating distally, excessive compensatory sweating proximally) are common. (See the diabetic polyneuropathy box.)

The polyneuropathy in Guillain-Barré syndrome is characterized by more severe effects on the motor than sensory system (see Guillain-Barré syndrome box in Chap. 2). Contrary to the pattern in most polyneuropathies, paresis may be worst proximally.

DYSFUNCTIONS OF THE NEUROMUSCULAR JUNCTION

Two typical problems at the neuromuscular junction have similar effects. In myasthenia gravis, an autoimmune disease that damages acetylcholine receptors at the neuromuscular junction, repeated use of a muscle leads to increasing weakness. In botulism, ingesting the botulinum toxin from improperly stored foods causes interference with the release of acetylcholine from the motor axon. This produces acute, progressive weakness, with loss of stretch reflexes. Sensation remains intact. Botulinum toxin is used therapeutically in people with movement dysfunctions such as spasticity and dystonia to weaken overactive muscles by directly injecting the toxin into muscles. This treatment frequently improves function.

MYOPATHY

Myopathies are disorders intrinsic to muscle. An example is muscular dystrophy; random muscle fibers degenerate, leaving motor units with fewer muscle fibers than normal. Activating such a motor unit produces less force than a healthy motor unit. Because the nervous system is not affected by myopathy, sensation and autonomic function remain intact. Coordination, muscle tone, and reflexes are unaffected until muscle atrophy becomes so severe that muscle activity cannot be elicited.

ELECTRODIAGNOSTIC STUDIES

Dysfunction of peripheral nerves and the muscles they innervate can be evaluated by electrodiagnostic studies. Recording the electrical activity from nerves and muscles by nerve conduction velocity (NCV) and

EMG studies (see Chaps. 7 and 10) reveals the location of pathology and is often diagnostic. Nerve conduction velocity studies can be used to differentiate between the following:

- Processes that are primarily demyelinating (myelinopathy) and those that primarily damage axons (axonopathy). Myelinopathies produce marked slowing of NCV. Axonopathies produce decreases in the amplitude of recorded potentials and may produce slowing of conduction velocity.
- Upper motoneuron and lower motoneuron paresis. Upper motoneuron lesions have no effect on NCV, so NCV is normal. Lower motoneuron lesions produce abnormal NCV.
- Local conduction block and wallerian degeneration. Local conduction block interferes with NCV only at one site, while wallerian degeneration affects the entire axon distal to the lesion.

Electromyography differentiates between nerve and muscle disorders, thus distinguishing neuropathy from myopathy (see Chap. 10).

The effect of Class I injury on NCV is to slow or stop conduction across the site of damage, with normal conduction in the axon segments proximal to and distal to the injury (Fig. 11–7). In Class II injuries, axons lose their ability to conduct action potentials across the damaged site at the time of injury. Thus amplitude of the evoked potential is decreased (Fig. 11–8). Nerve conduction velocity in the section of nerve distal to the injury gradually decreases over several days, eventually ceasing as a result of wallerian degeneration distal to the lesion. In Class III injuries, nerve conduction may never return distal to the injury.

Generalized neuropathies (i.e., polyneuropathies) are characterized by slowed NCV throughout the affected nerves and by decreased amplitude, particularly with increased distance between stimulation and recording sites. In myopathy, NCV is normal, but the amplitude of the potential recorded from muscle is decreased.

CLINICAL APPLICATION

Evaluation

Sensory, autonomic, and motor functions are evaluated as indicated in their respective chapters (see Chaps. 7, 8, and 10). Signs of peripheral nervous sys-

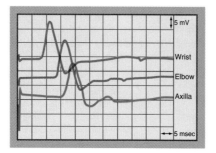

	Latency (ms)	Amplitude (mV)	Distance (mm)	Velocity (m/s)
Recorded values:				
Stimulated at wrist	8.0 (distal)	13.0		
Stimulated at elbow	12.8	13.2	240	49.6
Stimulated at axilla	15.2	11.5	145	62.1
Normal values for median nerve	Distal latency <4.2	>5.0		>50

A

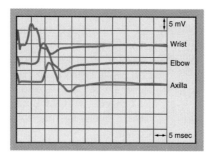

	Latency (ms)	Amplitude (mV)	Distance (mm)	Velocity (m/s)
Recorded values:				
Stimulated at wrist	3.0 (distal)	8.0		
Stimulated at elbow	6.7	7.5	205	55.9
Stimulated at axilla	8.8	7.8	150	69.2
Normal values for ulnar nerve	Distal latency <3.4	>5.0		>49.5

B

FIGURE 11–7

Motor nerve conduction velocity. *A,* Severe demyelination of the median nerve at the wrist. This is indicated by the 8.0 distal latency and the slow forearm conduction velocity, combined with normal amplitude of the recorded potential. *B,* Normal, ipsilateral ulnar NCV in the same person. (Courtesy of Robert A. Sellin, PT.)

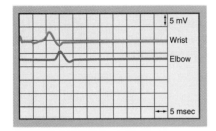

	Latency (ms)	Amplitude (mV)	Distance (mm)	Velocity (m/s)
Recorded values:				
Stimulated at wrist	9.2	3.3		
Stimulated at elbow	13.2	3.2	230	57.5
Normal values for median nerve	Distal latency <4.2	>5.0		>50

FIGURE 11–8

Median motor nerve conduction velocity recording, showing severely prolonged distal latency and a marked decrease in amplitude compared to normal. Conduction velocity between the wrist and elbow is normal. (Courtesy of Robert A. Sellin, PT.)

tem damage result from hypoactivity or hyperactivity of neurons. Neuronal hypoactivity is due to the decrease or loss of neuronal activity, for example, loss of proprioception. Neuronal hyperactivity causes light touch to elicit a painful sensation. Table 11–3 lists the signs and symptoms of mononeuropathy. In polyneuropathy, the same characteristics are found but in a symmetrical distribution, sometimes with additional autonomic signs: postural hypotension, bowel or bladder incontinence, and inability to have a sexual erection.

Clinically, making the distinction between peripheral neuropathy and central nervous system dysfunction is vital. Table 11–4 indicates factors that differentiate peripheral from central nervous system lesions.

Treatment

Results of sensory, manual muscle, and, if indicated, electrodiagnostic testing guide treatment decisions. Education is necessary to prevent complications from damage due to lack of sensation, disuse, or overuse. The person with peripheral neuropathy that affects sensation should be taught to visually inspect the involved areas daily, using mirrors if necessary, to monitor for wounds and for reddening of the skin that

persists more than a few minutes. If the feet are involved, proper foot care should be taught. Interventions for edema include elevation of the limb, compression bandaging with an elastic wrap, and

TABLE 11–3		
SIGNS AND SYMPTOMS OF MONONEUROPATHY		
	Neuronal Hypoactivity	**Neuronal Hyperactivity**
Sensory	Decrease or lack of sensation (touch, pressure, proprioception, or pain)	Pain, dysesthesia, hyperesthesia
Autonomic	Flushing of skin, edema Lack of sweating	Vasoconstriction: cold skin, pallor, cyanosis (dark blue color of skin) Perpetuation of pain (see Chap. 7) Excessive sweating
Motor	Paresis, paralysis, hypotonia, muscle atrophy	Spasms, fasciculations, fibrillations in the muscles
Reflexes	Decreased or absent	Normal

TABLE 11–4

DISTINGUISHING PERIPHERAL FROM CENTRAL NERVOUS SYSTEM DYSFUNCTION

	Peripheral Nervous System	Central Nervous System
Distribution of signs and symptoms	Peripheral nerve pattern	Dermatomal or myotomal pattern
Nerve conduction velocity	Slowed or blocked conduction; decreased amplitude of recorded potentials	Normal
Muscle tone	If lower motoneuron involvement, hypotonia	If upper motoneuron involvement, hypertonia
Muscle atrophy	Rapid muscle atrophy indicates denervation	Muscle atrophy progresses slowly
Phasic stretch reflexes	Reduced or absent	Hyperactive or normal
Paraspinal sensation and/or paraspinal muscles	Normal	Involved

electrical stimulation. Contractures may be prevented by prolonged stretching or by daily activities.

Exercise beginning the day following a peripheral nerve crush injury has been demonstrated to enhance both sensory and motor recovery (vanMeeteren et al., 1997). Exercises should emphasize gradual strengthening and functional use of individual muscles and muscle groups. Orthoses (braces) are frequently used to stabilize weight-bearing joints, thus preventing sprains and strains, and to prevent drop-ping of the forefoot during gait in cases of paresis or paralysis of the tibialis anterior muscle. Orthoses are also used to prevent deformities that can result from paresis, paralysis, and lack of sensation. The use of electrical stimulation to prevent atrophy of denervated muscles by evoking muscle contractions is controversial because controlled studies have not been performed in humans and the results from animal studies are inconsistent (Stillwell and Thorsteinsson, 1993).

CLINICAL NOTES

CASE 1

R.V. is a 35-year-old man, brought to the emergency room by a friend 2 days ago. At that time, R.V. complained of aching, burning pain in his thighs and a feeling of weakness that began 2 days prior to admission. He has no history of trauma. His current condition is as follows:

- He is unable to communicate, so sensation and cognitive functions cannot be tested.

- He is subject to abnormal variations in blood pressure and heart rate.

- He is completely paralyzed. His breathing is maintained by a respirator.

- Nerve conduction velocity is markedly slowed bilaterally in the tested nerves, the median and tibial nerves. Amplitude of recorded potentials is normal.

Question

What is the location of the lesion(s) and probable etiology?

(continued)

CASE 2

A 16-year-old woman was injured 2 days ago when a load of lumber fell from a shelf, pinning her left forearm. The following signs and symptoms are noted on the left:

- She does not feel pinprick, touch, temperature differences, or vibration on the medial hand, little finger, and medial half of the ring finger.

- Sweating is absent in the same distribution as the sensory loss.

- Radial wrist extension and flexion and finger extension are normal strength on manual muscle tests.

- She is unable to flex the middle and distal phalanges of the fourth and fifth digits, abduct or adduct her fingers, or adduct the hand.

Questions

1. What is the location of the lesion(s)?
2. How could the probable rate of recovery be predicted?

CASE 3

A 7-year-old boy has progressive proximal muscle weakness. Clinical examination and electrodiagnostic tests reveal the following:

- Sensation and coordination are within normal limits.

- He falls twice when walking 100 feet.

- He has difficulty coming to standing and climbing stairs.

- Lumbar lordosis is increased.

- Manual muscle tests indicate that shoulder girdle and hip muscles are approximately 50% of normal strength, knee and elbow muscles are about 75% of normal strength, and distal muscles have near-normal strength.

- Nerve conduction velocity is normal.

- Electromyographic potentials recorded from hip girdle muscles are of small amplitude.

Question

What is the location of the lesion(s) and probable etiology?

REVIEW QUESTIONS

1. Why are cutaneous nerve branches not purely sensory?

2. What signs are produced by complete severance of a peripheral nerve?

3. List the trophic changes that occur in denervated tissues.

4. Give an example of a Class I mononeuropathy. What part of the nerve is damaged in Class I mononeuropathy?

5. Describe a Class II mononeuropathy.

6. Why is the prognosis for Class III mononeuropathy poor?

7. What is multiple mononeuropathy?

8. What are the most common causes of polyneuropathy?

9. Why do the signs and symptoms of polyneuropathy usually appear distally first?

10. How can myelinopathy be distinguished from axonopathy?

11. How can neuropathy be distinguished from myopathy?

12. Classify each of the following signs as resulting from neuronal hyperactivity or hypoactivity: absent reflexes, muscle spasms, lack of sweating, paralysis, muscle fasciculations, and cyanosis of the skin.

13. Sensory loss in a dermatomal pattern, normal NCV, muscle hypertonia, and slowly progressive muscle atrophy that includes the paraspinal muscles are indicative of lesions in what region of the nervous system?

References

Dawson, D. M., Hallet, M., and Millender, L. H. (1990). Entrapment neuropathies. Boston: Little, Brown.

Durkan, J. A. (1991). A new diagnostic test for carpal tunnel syndrome. J. Bone Joint Surg. Am. 73A:535–538.

Greene, D. A., Simal, A. A. F., Pfeifer, M. A., et al. (1990). Diabetic neuropathy. Annu. Rev. Med. 41:303–317.

Johnson, E. W. (1980). Practical electromyography. Baltimore: Williams & Wilkins.

Kasdan, M., Lane, C., Merritt, W. H., et al. (1993). Carpal tunnel syndrome: The workup. Patient Care 27(7):97–102, 104–108.

Kasdan, M., Lane, C., Merritt, W. H., et al. (1993). Carpal tunnel syndrome: Management techniques. Patient Care 27(7):111–112, 115, 123–126.

Rosenstock, L. (1996). National Occupational Research Agenda (NORA) Priority Research Areas. National Center for Health Statistics, Hyattsville, MD.

Sabin, T. D., and Swift, T. R. (1975). Leprosy. In: Dyck, P. J., Thomas, P. K., Griffin, J. W., et al. (Eds.). Peripheral neuropathy. Philadelphia: W. B. Saunders, vol. 2, pp. 1166–1198.

Schaumberg, H. H., Spencer, P. S., Thomas, P. K., et al. (1983). Disorders of peripheral nerves. Philadelphia: F. A. Davis.

Stillwell, G. K., and Thorsteinsson, G. (1993). Rehabilitation procedures. In: Dyck, P. J., Thomas, P. K., Griffin, J. W., et al. (Eds.). Peripheral neuropathy. Philadelphia: W. B. Saunders, vol. 2, pp. 1692–1708.

vanMeeteren, N. L., Brakkee, J. H., Hamers, F. P., et al. (1997). Exercise training improves functional recovery and motor nerve conduction velocity after sciatic nerve crush lesion in the rat. Arch. Phys. Med. Rehabil. 78(1):70–77.

Suggested Readings

Adams, R. D., and Victor, M. (1993). Laboratory aids in the diagnosis of spinal cord and neuromuscular disease; Diseases of the peripheral nerves. In: Principles of Neurology. New York, McGraw-Hill, pp. 1059–1077, 1117–1169.

Asbury, A. K., and Bird, S. J. (1992). Disorders of peripheral nerve. In: Asbury, A. K., McKhann, G. M., and McDonald, W. I. (Eds.). Diseases of the nervous system: Clinical neurobiology. Philadelphia: W. B. Saunders, vol. 1, pp. 252–269.

Dyck, P. J. (1993). Quantitating severity of neuropathy. In: Dyck, P. J., Thomas, P. K., Griffith, J. W., et al. (Eds.). Peripheral neuropathy. Philadelphia: W. B. Saunders, vol. 1, pp. 686–697.

Richardson, E. P., and Girolami, U. D. (1995). General reactions of peripheral nerve to disease; Metabolic and nutritional neuropathies. In: Pathology of the peripheral nerve. Philadelphia: W. B. Saunders, pp. 8–21, 78–93.

12

Spinal Region

Five years ago, I had an accident. I recall the doctor saying afterwards: "You have a spinal cord injury, a thoracic 7 lesion, but you can manage yourself in the future."

The last part was most important, since I have two children. What the doctor didn't tell me was how to achieve independence and how to return to a normal life. I am a physical therapist, and my specialty was in treating the neurological problems of children. I am a pioneer in this field in the Netherlands, and for the past 25 years I have worked with handicapped children in their daily situations.

I left the rehabilitation center after 9 months of therapy and training. It could have been earlier, but my home was not ready for my return. Some things needed to be adapted and made accessible to me from my wheelchair. I have a car that my work paid to have adapted for hand control. I can organize all the daily things in life for me and my children. We are a good team.

Now, I had to work for a new life for myself. Because of my profession and my specialty, I was able to return to my job after only about 6 months. Part of my job involved my own physical therapy practice, and the other part was working as an instructor/senior tutor for children with cerebral palsy. Due to my injury, I sold my physical therapy practice and began teaching, from my wheelchair, at a physical therapy school. In this surrounding, nobody noticed the wheelchair; I was just myself.

Now it has been 5 years and sometimes I think to myself: "What is different?" I can do all the things I want and enjoy. I cannot walk, and sometimes I have a lot of pain. Once I spilled hot tea on my stomach and burned myself quite severely without realizing it until later. Because I lack sensation in my abdomen and legs, I did not become aware of the burn until I saw blisters on my skin. But I am happy in my wheelchair and I am happy with my son (18) and my daughter (16). The

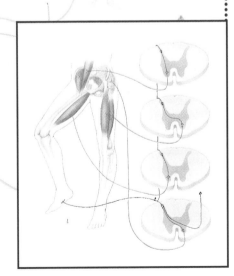

doctor was right. It is a hard and long way to come, but it is possible.

Last year, while visiting the United States, I learned how to catheterize my bladder while remaining sitting in my wheelchair. This was very important to my independence. Now I can go anywhere and not need special equipment. This year I went to the United States for my work and was driving a car on the interstate. I thought to myself, "It really is true, you can do almost anything if you have friends and your own desires." When I use the terms impairment, disability, and handicap, I can say that I am not handicapped.

I use no medications. I can deal with the spasticity very well, since for 25 years in my profession I worked with spasticity in other people. I control the spasticity by using slow stretch, correct foot and leg positioning, prolonged positions, and making sure to empty my bladder on schedule. My professional knowledge helped me a lot, but on the other hand I am now a patient and sometimes need the guidance of professionals.

—Tineke Dirks

ANATOMY OF THE SPINAL REGION

The spinal region includes all neural structures contained within the vertebrae: spinal cord, dorsal and ventral roots, spinal nerves, and meninges (see Fig. 1–4). The spinal cord is continuous with the medulla and extends to the L1–2 intervertebral space in adults. Because the spinal cord is not present below the L1 vertebral level, long roots are required for axons from the termination of the cord to exit the lumbosacral vertebral column. These long roots form the cauda equina within the lower vertebral canal (see Fig. 5–5). Lateral enlargements of the cord at the cervical and lumbosacral levels accommodate the neurons for upper and lower limb innervation.

Vertical grooves mark the external spinal cord. The anterior cord has a deep median fissure, and the posterior cord has a shallow median sulcus. The anterior cord also has two anterolateral sulci, where nerve rootlets emerge from the cord. The posterior cord has two posterolateral sulci, where nerve rootlets enter the cord.

Ventral and Dorsal Roots

Axons sending information to the periphery (motor) leave the anterolateral cord in small groups, called rootlets. Ventral rootlets from a single segment coalesce to form a **ventral root**. The **dorsal root** contains sensory axons bringing information into the spinal cord and enters the posterolateral spinal cord via rootlets. Unlike the ventral roots, each dorsal root has a **dorsal root ganglion** located outside the spinal cord. The dorsal root ganglion contains the cell bodies of sensory neurons. Where sensory axons enter the spinal cord, the large-diameter fibers, transmitting proprioceptive and touch information, are located medially, and the small-diameter fibers, transmitting pain and temperature information, are located laterally (Fig. 12–1).

The dorsal and ventral roots join briefly to form a **spinal nerve**. The spinal nerve is a mixed nerve because it contains both sensory and motor axons. Spinal nerves are located in the intervertebral foramen.

> The ventral root contains motor axons. The dorsal root contains sensory neurons. The somas of sensory neurons are found in the dorsal root ganglion. The spinal nerve consists of all sensory and motor axons connecting with a single segment of the cord.

Segments of the Spinal Cord

A striking and significant feature of the spinal cord is **segmental organization**. Each segment of the cord is connected to a specific region of the body by axons traveling through a pair of spinal nerves. The connections of nerve rootlets to the exterior of the cord indicate the segments (Fig. 12–2). Segments are identified by the same designation as their corresponding spinal nerves. For example, the term *L4 spinal segment* refers to the section of the cord whose spinal nerve traverses the L4 intervertebral foramen. However, within the cord, the distinct segments are not evident because the cord consists of continuous vertical columns extending from the brain to the cord termination.

Spinal Nerves and Primary Rami

Spinal nerves are unique in carrying all of the motor and sensory axons of a single spinal segment. In the cervical region, spinal nerves are found above the corresponding vertebra, except for the eighth spinal nerve, which emerges between the C8 and T1 vertebrae. In the remainder of the cord, spinal nerves lie below the corresponding vertebra. After a brief transit

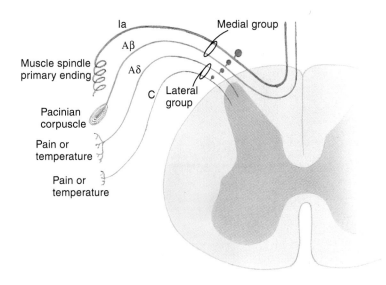

FIGURE 12–1

In the dorsal root entry zone, axons conveying information from touch and proprioceptive receptors enter the cord medially, while axons carrying information about painful stimuli and temperature enter the cord laterally.

through the intervertebral foramen, the spinal nerve splits into two primary rami; this division marks the end of the spinal region and the beginning of the peripheral nervous system. The **dorsal primary rami** innervate the paravertebral muscles, posterior parts of the vertebrae, and overlying cutaneous areas. The **ventral primary rami** innervate the skeletal, muscular, and cutaneous areas of the limbs and of the anterior and lateral trunk. Both primary rami are mixed nerves.

> The dorsal primary rami innervate the paravertebral muscles and the adjacent skin and posterior vertebrae. The ventral primary rami innervate the limbs and anterolateral trunk. A segment of the spinal cord is connected to a specific region of the body by a pair of spinal nerves.

Internal Structure of the Spinal Cord

The internal structure of the spinal cord can be observed in horizontal sections. Throughout the spinal cord, white matter surrounds the gray matter. White matter contains the axons connecting various levels of the cord and linking the cord with the brain. Axons that begin and end within the spinal cord are called **propriospinal.** The propriospinal axons are adjacent to the gray matter. Cells with long axons connecting the spinal cord with the brain are **tract cells.** The dorsal and lateral columns of white matter contain axons of tract cells, transmitting sensory information upward to the brain. The lateral and anterior white matter contains axons of upper motoneurons, conveying information descending from the brain to interneurons and lower motoneurons. Specific tracts have been discussed in Chapters 6 to 10. The propriospinal axons and tracts in the spinal cord are illustrated in Fig. 12–3.

The central part of the cord is marked by a distinctive H-shaped pattern of gray matter (Fig. 12–4). Lat-

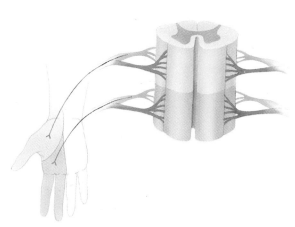

FIGURE 12–2

Two segments of the spinal cord. The axons traveling through the rootlets, roots, and spinal nerves connect a spinal segment with a specific part of the body. The axons shown are sensory axons, conveying information from the C6 and C7 dermatomes through the dorsal root into the C6 and C7 spinal cord segments.

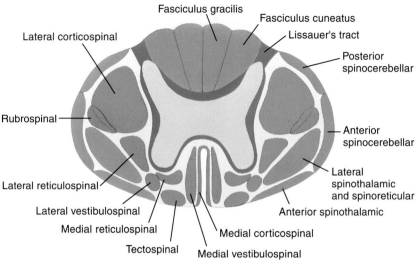

FIGURE 12–3

White matter of the spinal cord. The propriospinal fibers are indicated in purple, the sensory tracts in blue, and the motor tracts in red.

eral sections of spinal gray matter are divided into three regions, called horns:

- Dorsal horn
- Lateral horn
- Ventral horn

The **dorsal horn** is primarily sensory, containing endings and collaterals of first-order sensory neurons,

interneurons, and dendrites and somas of tract cells. For example, the somas of second-order neurons in the spinothalamic pathway are in the dorsal horn (see Chap. 6). The **lateral horn** (present only at T1 to L2 spinal segments) contains the cell bodies of preganglionic sympathetic neurons. A region analogous to the lateral horn in the S2 to S4 spinal segments contains the preganglionic parasympathetic cell bodies. Preganglionic autonomic neurons are efferent neu-

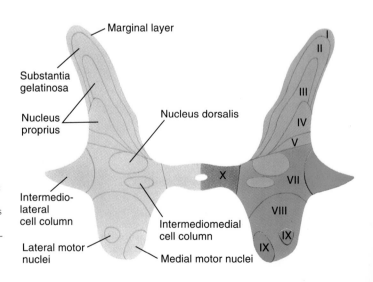

FIGURE 12–4

Gray matter of lower thoracic spinal cord. Named regions are indicated on the left, and Rexed's laminae are indicated on the right. Lamina VI is present in the segments of the spinal cord that innervate the limbs and is not present between T4 and L2. The correspondence between Rexed's laminae and the named areas is inconsistent. For example, some authors include laminae III to VI in the nucleus proprius.

rons. Both sympathetic and parasympathetic preganglionic neurons exit the cord via the ventral root. The **ventral horn** contains cell bodies of lower motoneurons whose axons exit the spinal cord via the ventral root.

> The dorsal horn processes sensory information, the lateral horn processes autonomic information, and the ventral horn processes motor information.
>
> Much of the gray matter is composed of spinal interneurons, cells with their somas in the gray matter that act upon other cells within the cord. Spinal interneurons include cells that remain entirely within the gray matter and also cells whose axons travel in white matter to different levels of the cord.

Rexed's Laminae

Spinal gray matter has been classified into 10 histologic regions called Rexed's laminae; these regions are illustrated in Fig. 12–4. In the dorsal horn, laminae I to VI are numbered from dorsal to ventral. Lamina VII includes the intermediolateral horn and part of the ventral horn. In the ventral horn, locations of laminae VII to IX vary, depending on the level of the cord. Lamina X is the central region of gray matter.

Laminae I and II, called the marginal layer and substantia gelatinosa respectively, process information about noxious stimuli. Laminae III and IV, together known as the nucleus proprius, process proprioceptive and two-point discrimination information* Lamina V cells process information about noxious stimuli and information from the viscera. Cells in lamina VI process proprioceptive information. Lamina VII includes the nucleus dorsalis, or Clarke's column, extending from T1 to L3 and the intermediolateral horn. The nucleus dorsalis receives proprioceptive information, and its axons relay unconscious proprioceptive information to the cerebellum. The intermediolateral horn contains somas of autonomic efferents. Lamina VIII cells connect with the contralateral cord and the brain. Lamina IX contains cell bodies of lower motoneurons, whose axons travel through a ventral

root, a spinal nerve, and then a peripheral nerve to innervate skeletal muscles. Lamina X consists of neurons whose axons cross to the opposite side of the cord.

> Rexed's laminae are 10 histological and functionally specific regions in the spinal cord gray matter.

Meninges

The meninges, layers of connective tissue surrounding the spinal cord, are continuous with the meninges surrounding the brain. Pia matter closely adheres to the spinal cord surface, the arachnoid is separated from the pia by cerebrospinal fluid in the subarachnoid space, and the dura is the tough, outer layer. Between the arachnoid and dura is the subdural space, and the epidural space separates the dura from the vertebrae.

FUNCTIONS OF THE SPINAL CORD

Segments of the spinal cord exchange information with other spinal cord segments, with peripheral nerves, and with the brain. Tracts convey this information, yet spinal cord functions are far more complex than a simple conduit. Only for one type of information does the spinal cord serve as a simple conduit: axons carrying touch and proprioceptive information enter the dorsal column and project to the medulla without synapsing. All other tracts conveying information in the spinal cord synapse in the cord, and thus their information is subject to processing and modification within the cord.

For example, after one hammers a thumb, the pain signals can be modified by rubbing the thumb and/or by activity of the descending pain inhibition pathways (see Chap. 7). The pain information is modified within the spinal cord by signals from large-diameter sensory afferents and by signals in the descending tracts, both of which decrease the frequency of signals in slow pain pathways. Similarly, the information conveyed by an axon in a descending tract to a lower motoneuron is only one of many influences on that lower motoneuron (see Chap. 9). The origins and functions of the tracts in the spinal cord are listed in Table 12–1.

---

*The correspondence between Rexed's laminae and some of the named spinal cord areas is controversial. For example, some authors include laminae III to VI in the nucleus proprius.

TABLE 12-1
ORIGINS AND FUNCTIONS OF TRACTS OF SPINAL CORD

Tract	Origin	Function
Dorsal column/ medial lemniscus	Peripheral receptors; first-order neuron synapses in medulla	Conveys information about discriminative touch and conscious proprioception
Spinothalamic	Dorsal horn of spinal cord	Conveys information about pain and temperature
Spinocerebellar	High-fidelity paths originate in peripheral receptors; first-order neurons synapse in nucleus dorsalis or medulla	Conveys unconscious proprioceptive information
	Internal feedback paths originate in the dorsal horn of the spinal cord	Conveys information about activity in descending activating pathways and spinal interneurons
Lateral corticospinal	Supplementary motor, premotor, and primary motor cerebral cortex	Fractionation of movement, particularly of hand movements
Medial corticospinal	Supplementary motor, premotor, and primary motor cerebral cortex	Control of neck, shoulder, and trunk muscles
Tectospinal	Superior colliculus of midbrain	Reflexive movement of head toward sounds or visual moving objects
Rubrospinal	Red nucleus of midbrain	Facilitates contralateral upper limb flexors
Medial reticulospinal	Pontine reticular formation	Facilitates postural muscles and limb extensors
Lateral reticulospinal	Medullary reticular formation	Facilitates flexor muscle motoneurons and inhibits extensor motoneurons
Medial vestibulospinal	Vestibular nuclei in pons	Adjusts activity in neck and upper back muscles
Lateral vestibulospinal	Vestibular nuclei in medulla and pons	Ipsilaterally facilitates lower motoneurons to extensors; inhibits lower motoneurons to flexors
Ceruleospinal	Locus ceruleus in the brain stem	Enhances the activity of interneurons and motoneurons in the spinal cord
Raphespinal	Raphe nucleus in the brain stem	Same as ceruleospinal

Classification of Spinal Interneurons

In most textbooks, spinal interneurons are considered only in the context of reflexes. To study interneurons, experimenters have often disconnected the spinal cord from the brain, stimulated only one type of afferent neuron, and then recorded from interneurons. These experiments led to the concept of reflexes as an invariant coupling of input and output, with discrete spinal circuits dedicated to each reflex. Voluntary movement was considered to be entirely separate from reflexes. Although reductionism may be required to simplify the system for experiments, interneurons do not normally function with isolated inputs. Subsequent research has demonstrated the following:

- Natural stimuli simultaneously excite a variety of receptor types. For example, flexing a joint stimulates muscle spindles, Golgi tendon organs, joint stretch and pressure receptors, and cutaneous stretch and pressure receptors.

- Afferent and descending information converges on the same spinal interneurons.
- Reflexes and voluntary control act together to produce goal-oriented movements.

By integrating volleys of peripheral, ascending, and descending inputs, spinal circuitry provides the following:

- Modulation of sensory information
- Coordination of movement patterns
- Autonomic regulation

In this text, interneurons are categorized by function.* Modulation of sensory information was cov-

*Historical note: Until recently, spinal interneurons have been categorized according to the earliest discovery of associated afferents. Thus, interneurons activated by Ia spindle afferents are often called Ia inhibitory interneurons, despite subsequent findings that these interneurons are also strongly influenced by other afferents and by descending tracts. Interneurons that previously were called Ib are now called nonreciprocal inhibition neurons.

ered in Chapter 7 and will not be considered here. The other mechanisms will be discussed individually for simplicity; however, recall that none of these mechanisms act in isolation.

SPINAL CORD MOTOR COORDINATION

Interneuronal circuits integrate the activity from all sources and then adjust the output of lower motoneurons (Davidoff and Hackman, 1991). Thus interneurons coordinate activity in all the muscles when a limb moves.

What determines whether a single alpha motoneuron will fire? Activity at 20,000 to 50,000 synapses, providing information from the following:

- Ia, Ib, and II afferents
- Interneurons
- Descending tracts, including the medial, lateral, and nonspecific activation pathways

The following reflexive and inhibitory circuits are examples of connections that use interneuron activity to shape motor output.

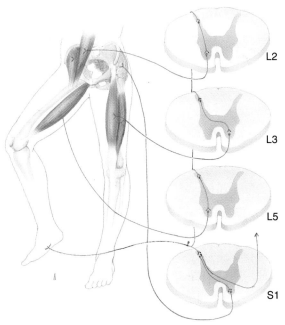

FIGURE 12–5

Withdrawal reflex of the right leg and crossed extension reflex in the left leg. The interaction of several spinal cord segments is required to produce the coordinated muscle action.

Reflexes

Except for the monosynaptic phasic stretch reflex, spinal reflexes involve interneurons. The phasic and tonic stretch reflexes, reciprocal inhibition, and withdrawal reflexes were introduced as spinal region reflexes in Chapter 9. In this chapter, the focus is on the capacity of interneuronal circuits to generate complex movements. This is demonstrated by the **withdrawal reflex.** Afferent information from skin, muscles, and/or joints can elicit a variety of withdrawal movements. Each withdrawal movement is specific for most effectively removing the stimulated area away from the provocation. For example, if one steps on a tack, the involved lower limb flexes to remove the foot from the stimulus. However, if a bee stings the inside of one's calf, the lower limb abducts. The specificity of the movement pattern is referred to as local sign, indicating that the response depends on the site of stimulation. Because the muscles removing the part from the stimulation are usually not innervated by the same cord segment that received the afferent input, the information is relayed to other cord segments by collaterals of the primary afferent and by interneurons. In an intact nervous system, the stimulation must be quite strong to evoke a powerful withdrawal reflex. If one is standing when one lower limb is abruptly withdrawn, another interneuronal circuit quickly adjusts the muscle activity in the stance limb to prevent falling; this is the **crossed extension reflex.** The withdrawal and the associated crossed extension reflexes are illustrated in Figure 12–5.

Inhibitory Circuits

Interneurons in inhibitory circuits also contribute to spinal cord motor coordination. Inhibitory interneurons provide the following:

- Reciprocal inhibition
- Recurrent inhibition
- Nonreciprocal inhibition

Reciprocal inhibition decreases activity in an antagonist when an agonist is active, allowing the agonist to act unopposed. When agonists are voluntarily recruited, the reciprocal inhibition interneurons prevent unwanted activity in the antagonists (McCrea, 1994), (Fig. 12–6). Thus, reciprocal inhibition separates muscles into agonists and antagonists. For efficient motor control, collaterals of descending pathways activate reciprocal inhibitory interneurons simultaneously with excitation of selected lower motoneurons.

Input to reciprocal inhibition interneurons is also provided by Ia, cutaneous, and joint afferents, by other interneurons, and by cortico-, rubro-, and vestibulospinal tracts (see review in McCrea, 1994). Reciprocal inhibition occurs with afferent input, as well as during voluntary movement. For example, during a quadriceps stretch reflex, reciprocal inhibition interneurons inhibit the hamstrings. Occasion-

ally, reciprocal inhibition is suppressed to allow co-contraction of antagonists. This occurs in people with intact nervous systems when they are anxious, anticipate unpredictable movement disturbances, or are learning new movements.

Recurrent inhibition has effects opposite to reciprocal inhibition: inhibition of agonists and synergists, with disinhibition of antagonists (Fig. 12–7). **Renshaw cells,** interneurons that produce recurrent inhibition, are stimulated by a recurrent collateral branch from the alpha motoneuron. A recurrent collateral branch is a side branch of an axon that turns back toward its own cell body. Renshaw cells inhibit the same alpha motoneuron that gives rise to the collateral and also inhibit alpha motoneurons of synergists. Renshaw cells focus motor activity, thus isolating desired motor activity from gross activation (Veale and Rees, 1973). Loss of descending influence on Renshaw cell activity may cause difficulty in achieving fine motor control.

Nonreciprocal inhibition is provided by interneurons that selectively inhibit agonist, synergist, and antagonist muscles. Nonreciprocal inhibitory neurons receive afferent information from muscle spindles and Golgi tendon organs of numerous muscles throughout an entire limb. In addition, nonreciprocal inhibitory interneurons receive input from cortico-, rubro-,

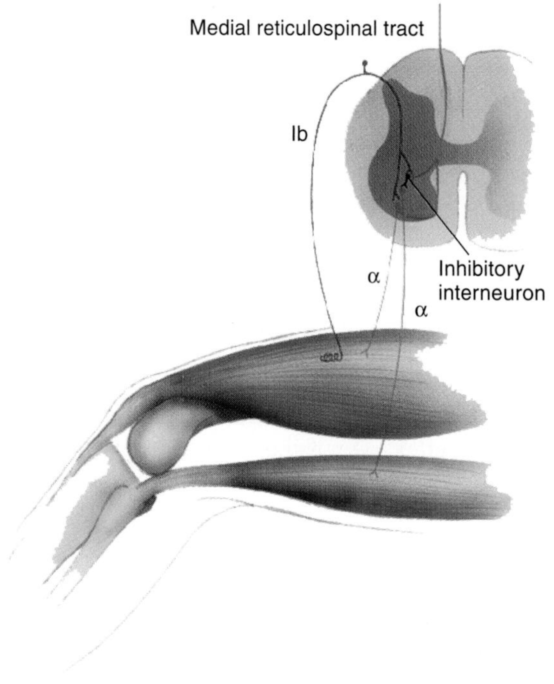

Medial reticulospinal tract

Ib

α

Inhibitory interneuron

α

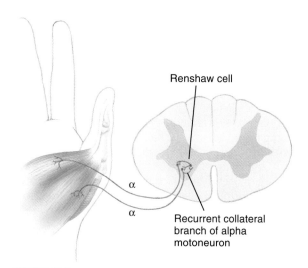

Renshaw cell

α

α

Recurrent collateral branch of alpha motoneuron

FIGURE 12–6

Reciprocal inhibition. For simplicity, only the medial reticulospinal input to an alpha motoneuron activating fibers in the quadriceps and to a reciprocal inhibition interneuron inhibiting an alpha motoneuron to fibers in the semitendinosus muscle are shown.

FIGURE 12–7

Recurrent inhibition. The recurrent collateral branch of the alpha motoneuron stimulates the Renshaw cell. The Renshaw cell inhibits agonists and synergists and facilitates antagonists. For simplicity, the antagonist facilitation is not shown.

and reticulospinal tracts (see review in Davidoff, 1992). Input to nonreciprocal inhibitory interneurons is shown in Figure 12–8. Nonreciprocal inhibitory interneurons coordinate the actions of various muscles during limb movements.

> Reciprocal inhibition decreases antagonist opposition to the action of agonist muscles. Recurrent inhibition focuses motor activity. Nonreciprocal inhibition sculpts the motor output of large groups of muscles.

In normal movement, motor activity elicited by descending commands can be modified by afferent input. The contribution of interneurons to this modification is illustrated in Figure 12–9.

Alternatively, descending commands can also modify the motor activity elicited by afferent input. Jendrassik's maneuver provides a demonstration of the effect of descending influences on alpha motoneurons. The maneuver consists of voluntary contraction of certain muscles during reflex testing of other muscles. For example, subjects hook their flexed fingers together, then pull isometrically against their own resistance; this activity facilitates the quadriceps deep tendon reflex by producing a generalized increase in spinal interneuron activity. In Jendrassik's maneuver, activity in descending activating pathways contributes to increasing the general level of excitation in the cord.

SPINAL CONTROL OF PELVIC ORGAN FUNCTION

The sacral spinal cord contains centers for the control of urination, bowel function, and sexual function. In a normal infant, when the bladder is empty, the sympathetic efferents from T11 to L2 levels inhibit contraction of the bladder wall and maintain contraction of the internal sphincter. When the bladder fills, stretching of the bladder wall is sensed by proprioceptors, impulses regarding fullness of the bladder are transmitted to the reflex center in the sacral cord, and efferent impulses initiate voiding. Parasympathetic efferent impulses stimulate contraction of the bladder wall and somatic efferents (S2 to S4) open the external sphincter. Thus **reflexive bladder function**, which is normal in infants, requires the following:

- Afferents
- T11 to L2 and S2 to S4 cord levels
- Somatic, sympathetic, and parasympathetic efferents

For voluntary control of voiding, information is sent from the reflex center for urination to the brain; if circumstances are appropriate, the brain initiates voiding by corticospinal inhibition of lower motoneurons that innervate the external sphincter and by brain stem pathways to the autonomic efferents. Fig-

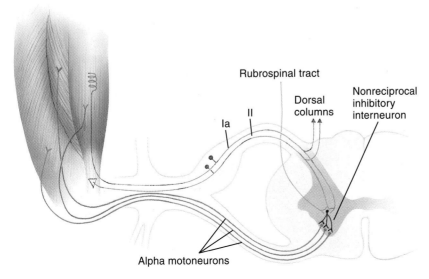

FIGURE 12–8

Nonreciprocal inhibition. Nonreciprocal inhibitory interneurons inhibit agonists, synergists, and antagonists.

Rubrospinal tract

Dorsal columns

Nonreciprocal inhibitory interneuron

Ia

II

Alpha motoneurons

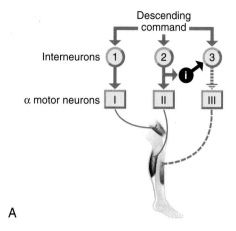

A

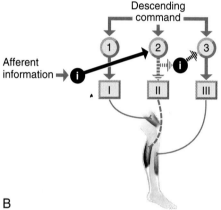

B

FIGURE 12–9

Modification of the action of descending commands by afferent information. At the bottom of both *A* and *B* are three muscles in the lower limb. Solid lines indicate active axons. Dotted lines indicate inactive axons. *A,* Descending commands stimulate all three interneurons (1, 2, and 3). A collateral of interneuron 2 excites an interneuron (black) that inhibits interneuron 3. As a result, alpha motoneurons I and II fire, and III is silent. *B,* Afferent input excites an interneuron (black) that inhibits interneuron 2. As a result, alpha motoneurons I and III fire, and II is silent. (Modified by permission from McCrea, D. A. [1994]. Can sense be made of spinal interneuron circuits? In: Cordo, P., and Harnad, S. [Eds.]. Movement control. Cambridge, England: Cambridge University Press, pp. 31–41.)

ure 12–10 illustrates the neural control of the bladder.

Bowel control is similar to bladder control. The signal to empty the bowels is stimulation of stretch receptors in the wall of the rectum. Afferent fibers transmit the information to the lumbar and sacral cord, the information is conveyed to the brain, and if appropriate, the efferent signal is sent to relax the sphincters.

The lower spinal cord is also vital for sexual function. Erection of the penis or clitoris is controlled by parasympathetic fibers from S2 to S4; ejaculation is controlled by sympathetic nerves at the L1 to L2 levels and the pudendal nerve, S2 to S4.

Reflexive functions of the bladder, bowels, and male sexual organ require intact afferents, lumbar and sacral cord segments, and somatic and autonomic efferents. Voluntary control of these functions requires intact neural pathways between the organ and the cerebral cortex.

EFFECTS OF SEGMENTAL AND TRACT LESIONS IN THE SPINAL REGION

A lesion in the spinal region may interfere with the following:

- Segmental function
- Vertical tract function
- Both segmental and vertical tract function

Segmental Function

Segmental function is the function of a spinal cord segment. Segmental lesions interfere with neural function only at the level of the lesion. For example, complete severance of the C5 dorsal root (roots are considered within the spinal region, although not in the spinal cord) would prevent sensory information from the C5 dermatome, myotome, and sclerotome from reaching the spinal cord.

Vertical Tract Function

Vertical tracts convey ascending and descending information. Lesions interrupting the vertical tracts result in a loss of function below the level of the lesion. A complete lesion prevents sensory information from below the lesion from ascending to higher levels of the central nervous system and interrupts descending commands from reaching levels of the spinal cord below the lesion.

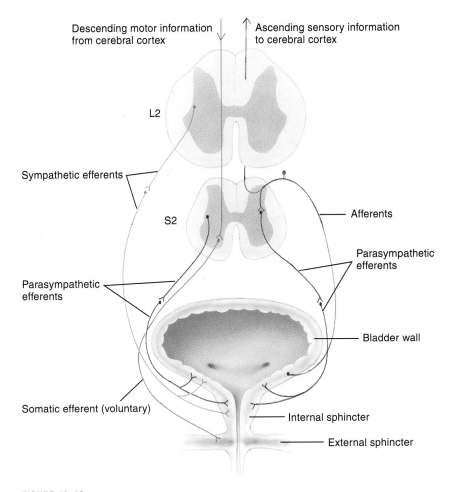

Descending motor information
from cerebral cortex

Ascending sensory information
to cerebral cortex

L2

Sympathetic efferents

S2

Afferents

Parasympathetic
efferents

Parasympathetic
efferents

Bladder wall

Somatic efferent (voluntary)

Internal sphincter

External sphincter

FIGURE 12–10

Neural control of the bladder. Descending activating neurons and efferents are indicated on the left
side of the illustration. Ascending sensory neurons, afferents, and a reflexive connection between a
sensory afferent and parasympathetic efferents are shown on the right side.

Segmental and Vertical Tract Function

Lesions may cause both segmental and tract signs. A
lesion at the C5 level on the right that involves the
right dorsal quadrant would prevent discriminative
touch and conscious proprioception from the right
side of the body below C5 from reaching the brain
(tract signs), and the sensory information from the C5
dermatome, myotome, and sclerotome would be lost
(segmental signs).

Signs of Segmental Dysfunction

A focal lesion involving a single level of the spinal
cord, the dorsal or ventral roots, or a spinal nerve results in segmental signs due to interruption of pathways. At the level of the lesion, sensory, motor,
and/or reflexive changes occur. In Figure 12–11, the
effects of a C5 spinal nerve lesion are contrasted with
the effects of a C5 hemisection of the spinal cord. Autonomic signs are difficult to demonstrate with a lesion at a single level because of the overlapping dis-

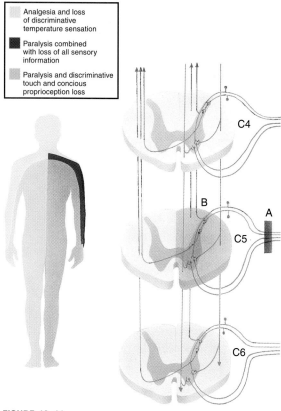

| Analgesia and loss of discriminative temperature sensation |
| Paralysis combined with loss of all sensory information |
| Paralysis and discriminative touch and concious proprioception loss |

FIGURE 12–11

Spinal region lesions: segmental signs versus vertical tract signs. *A*, This lesion interrupts all axons in the left C5 spinal nerve. This produces loss of sensation from the C5 spinal nerve and weakness of the biceps and brachioradialis, innervated by the C5 spinal nerve. The biceps and brachioradialis are not paralyzed because C6 also supplies these muscles. Thus the losses are limited to only part of the left arm. The entire remainder of the nervous system functions normally. *B*, In contrast, the hemisection of the cord at C5 produces the following conditions below the C5 level: paralysis on the left side, loss of discriminative touch and conscious proprioceptive information from the left side, and analgesia and loss of discriminative temperature sensation from the right side. In addition, the segmental losses are the same as in lesion A.

tribution of autonomic fibers from adjacent cord segments.

A lesion of the dorsal root, spinal nerve, or dorsal horn interferes with sensory function in a spinal segment, causing abnormal sensations or loss of sensation in a dermatomal distribution. For example, a dorsal root can be avulsed from the cervical spinal cord by extreme traction on the upper limb. If avulsion occurs at C5, the spinal cord is deprived of sen-

sory information from the C5 dermatome, myotome (proprioceptive and muscle pain information), and sclerotome innervated by that dorsal root.

A lesion of the ventral horn, ventral root, or spinal nerve interferes with lower motoneuron function. Signs of lower motoneuron dysfunction include flaccid weakness, atrophy, fibrillation, and fasciculation. If lower motoneuron signs occur in a myotomal pattern (see Chap. 9), the lesion is in the spinal region. A myotome includes paraspinal muscles, so signs of paraspinal involvement help differentiate spinal region from peripheral nerve lesions. Reflexes are absent if either the motor or sensory fibers contributing to the reflex circuit are damaged.

> Segmental signs include abnormal or lost sensation in a dermatomal distribution and/or lower motoneuron signs in a myotomal distribution.

Signs of Vertical Tract Dysfunction

Lesions interrupting the vertical tracts result in loss of communication to and/or from the spinal levels below the lesion. Therefore, all signs of damage to the vertical tracts occur below the level of the lesion. Ascending tract (sensory information) signs are ipsilateral if the dorsal column is interrupted and contralateral if the spinothalamic tracts are involved because the dorsal columns remain ipsilateral throughout the cord, while the spinothalamic tracts cross the midline within a few levels of where the information enters the cord. Autonomic signs may include problems with regulation of blood pressure, sweating, and bladder and bowel control.

Descending tract (upper motoneuron) signs include paralysis, hyperreflexia, and muscle hypertonia; if the lateral corticospinal tract is interrupted, Babinski's sign (see Chap. 10) is present. Deep tendon reflex testing (biceps, triceps, patellar, and tendo calcaneus [see Chap. 7]) may help to distinguish between upper motoneuron and lower motoneuron involvement: hyperreflexia indicates upper motoneuron, and hyporeflexia or areflexia may indicate lower motoneuron. However, hyporeflexia or areflexia may also occur with damage to Ia afferents.

An incomplete bilateral lesion at the C5 level limited to the dorsal columns would prevent ascending conscious proprioceptive and discriminative touch information from reaching the brain. Thus a person

TABLE 12–2

DIFFERENTIATING BETWEEN SPINAL REGION AND PERIPHERAL REGION LESIONS

	Spinal Region Lesion	Peripheral Region Lesion
Segmental signs and symptoms		
Sensory (abnormal sensation, loss of sensation)	Dermatomal distribution. Skin over paraspinal muscles involved	Peripheral nerve distribution. Skin overlying paraspinal muscles spared
Motor (flaccid paresis or paralysis, atrophy, fibrillation, fasciculation)	Myotomal distribution. Paraspinal muscles involved	Peripheral nerve distribution. Paraspinal muscles spared
Vertical tract involvement		
Sensory	Abnormal sensation or loss of sensation below the level of the lesion	None
Autonomic	Loss of control of blood pressure, pelvic viscera, thermoregulation	None
Motor	Muscle paresis or paralysis with hyperreflexia and/or hypertonia Babinski's sign if lateral corticospinal tract involved	None

with a spinal cord tumor that damaged the dorsal columns at C5 would not be aware of the location of light touch or passive joint movement below the C5 level but would be able to distinguish between sharp and dull stimuli, locations of pinprick, and among different temperatures. Information in the descending activating pathways would also be intact, although coordination would be somewhat impaired because of the lack of conscious proprioceptive information.

Differentiating Spinal Region from Peripheral Region Lesions

Table 12–2 summarizes patterns of signs and symptoms that differentiate spinal region from peripheral region lesions. The major distinctions are a lack of paraspinal muscle and/or overlying skin involvement and the lack of vertical tract signs in peripheral region lesions.

SPINAL REGION SYNDROMES

Syndromes are a collection of signs and symptoms that do not indicate a specific etiology. The following syndromes usually result from tumors or trauma (Fig. 12–12):

- **Anterior cord syndrome** (Fig. 12–12A) interrupts ascending spinothalamic tracts and descending

motor tracts and damages the somas of lower motoneurons. Thus anterior cord syndrome interferes with pain and temperature sensation and with motor control. Because tracts conveying proprioception and discriminative touch information are located in the posterior cord, these functions are spared.
- **Central cord syndrome** (Fig. 12–12B) usually occurs at the cervical level. If the lesion is small, loss of pain and temperature information occurs at the level of the lesion because spinothalamic fibers crossing the midline are interrupted. Larger lesions additionally impair upper limb motor function, due to the medial location of upper limb fibers in the lateral corticospinal tracts.
- **Brown-Séquard syndrome** (Fig. 12–12C) results from a hemisection of the cord. Segmental losses are ipsilateral and include loss of lower motoneurons and all sensations. Below the level of the lesion, voluntary motor control, conscious proprioception, and discriminative touch are lost ipsilaterally; pain and temperature sensation is lost contralaterally. This syndrome is also illustrated and explained in Figure 12–11B.
- **Cauda equina syndrome** (Fig. 12–12D) indicates damage to the lumbar and or sacral nerve roots, causing sensory impairment and flaccid paralysis of lower limb muscles, bladder, and bowels. Muscle hypertonia and hyperreflexia do not occur because the upper motoneurons are intact.

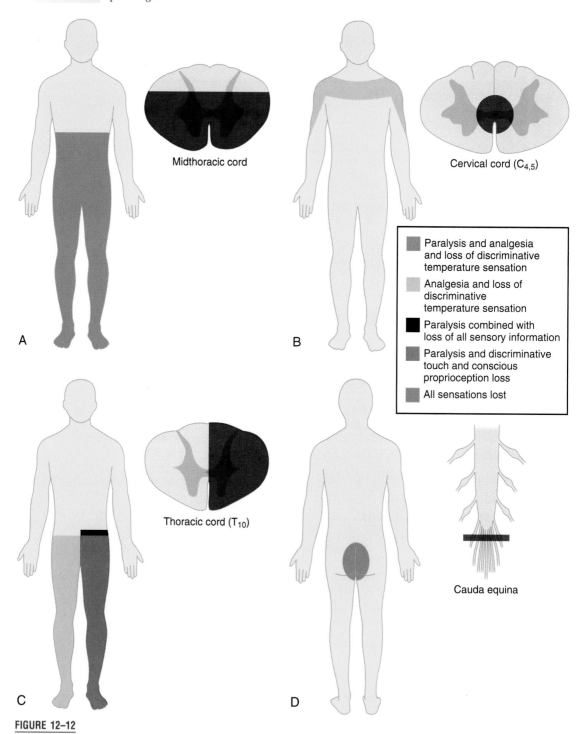

Paralysis and analgesia and loss of discriminative temperature sensation

Analgesia and loss of discriminative temperature sensation

Paralysis combined with loss of all sensory information

Paralysis and discriminative touch and conscious proprioception loss

All sensations lost

FIGURE 12–12

Spinal cord syndromes: *A,* Anterior. *B,* Central. *C,* Brown-Séquard. *D,* Cauda equina. The cauda equina syndrome shown affects only S3 through S5, so no skeletal muscles are paralyzed, but the bladder and anal sphincters are paretic. The sphincters are not paralyzed because S2 also innervates the sphincters.

Syndromes are a collection of signs and symptoms that occur together. Spinal cord syndromes indicate the location of a lesion but do not signify etiology. Thus an anterior cord syndrome could be caused by trauma, loss of blood supply, or other pathology.

EFFECTS OF SPINAL REGION DYSFUNCTION ON PELVIC ORGAN FUNCTION

Control of bladder, bowel, and sexual function depends on the level of cord damage. Lesions above the sacral level of the cord produce signs similar to upper motoneuron lesions. Lesions in the S2 to S4 spinal cord levels or the parasympathetic afferents and/or efferents produce signs similar to lower motoneuron lesions.

Complete lesions that damage any part of the reflexive bladder emptying circuit, that is, levels S2 to S4, or the parasympathetic afferents and/or efferents produce a flaccid, paralyzed bladder (Fig. 12–13A). The flaccid, paralyzed bladder overfills with urine, and when the bladder cannot stretch any further, urine dribbles out.

In contrast, complete lesions above the sacral cord interrupt descending axons that normally control bladder function but do not interrupt sacral level reflexive control of the bladder. This results in a hypertonic, hyperreflexive bladder, with reduced bladder capacity (Fig. 12–13B). Because the reflex circuit for bladder emptying is intact, reflexive emptying may occur automatically whenever the bladder is

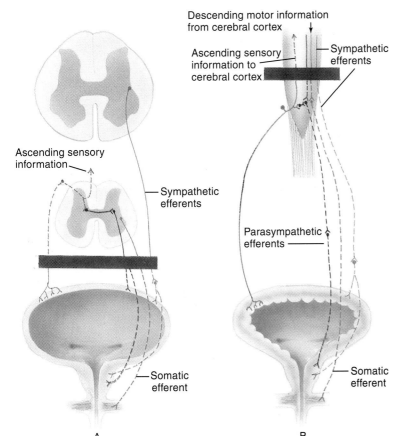

Descending motor information from cerebral cortex

Ascending sensory information to cerebral cortex

Sympathetic efferents

Ascending sensory information

Sympathetic efferents

Parasympathetic efferents

Somatic efferent

Somatic efferent

A   B

FIGURE 12–13

Bladder dysfunction after spinal cord injury. Dotted lines indicate neural pathways that have been interrupted and do not convey information. A, Flaccid bladder due to a complete lesion of the cauda equina. All neural connections with the bladder are severed. A complete lesion of spinal cord levels S2 to S4 would also produce a flaccid bladder, due to interruption of the reflexive bladder emptying circuit. B, Hypertonic bladder caused by a complete lesion above the S2 level. Communication between the brain and the sacral level parasympathetic neurons controlling the bladder are interrupted, preventing voluntary control. The reflexive connections between the bladder and spinal cord are intact, so reflexive emptying of the bladder can occur.

stretched, or if the sphincter is also hypertonic, flow of urine is functionally obstructed and the kidneys can be damaged (Madersbacher, 1990).

Bowel control and sexual organ function are similarly affected by spinal cord lesions because the parasympathetic reflexive connections for these organs are also located at levels S2 to S4. The person with a spinal cord lesion above the sacral cord is unaware of rectal stretch and has no voluntary control of sphincters, yet rectal stretch can elicit reflexive emptying of the lower bowel because the reflexive lower bowel emptying circuit is intact. If the bowel emptying reflex circuit is interrupted by a lesion of S2 to S4 or the parasympathetic connections with S2 to S4, the parasympathetic influence on peristalsis and reflex emptying of the bowels is lost.

Reflexive sexual erection can occur if the sacral cord is intact. In some men with complete spinal lesions above the lumbar level in whom the lumbosacral cord is intact, ejaculation can be elicited reflexively because sympathetic axons from the L1 and L2 levels and somatic nerves from the S2 to S4 levels control ejaculation. Fertile women with spinal cord lesions can conceive, often have a normal pregnancy, but frequently require cesarean delivery.

Complete lesions above the sacral cord interfere with the transmission of sensory information from the pelvic organs to the brain and with descending control of pelvic organ function. Complete sacral spinal cord or parasympathetic lesions interfere with reflexive control of the pelvic organs.

TRAUMATIC SPINAL CORD INJURY

Traumatic injuries to the spinal cord are usually caused by motor vehicle accidents, sports injuries, falls, or penetrating wounds. The first three types of injuries typically do not sever the cord. Instead, damage is due to crush, hemorrhage, edema, and infarction. Penetrating wounds, by a knife or a bullet, directly sever neurons in the cord.

Immediately after a traumatic injury to the spinal cord, cord functions below the lesion are depressed or lost. This condition, known as **spinal shock**, is due to interruption of descending tracts that supply

tonic facilitation to the spinal cord neurons. During spinal shock, the following are lost or impaired:

- Somatic reflexes, including stretch reflexes, withdrawal reflexes, and crossed extension reflexes are lost.
- Autonomic reflexes, including loss of smooth muscle tone and reflexive emptying of the bladder and bowels, are lost or impaired.
- Autonomic regulation of blood pressure is impaired, resulting in hypotension.
- Control of sweating and piloerection is lost.

Several weeks after the injury, most people experience some recovery of function in the cord, leading to return of reflex activity below the lesion. In some people, spinal neurons become excessively excitable, resulting in hyperreflexia (see Chap. 10).

Damage of the cervical cord results in **tetraplegia** (quadriplegia), with impairment of arm, trunk, lower limb, and pelvic organ function. People with lesions above the C4 level cannot breathe independently, because the diaphragm is innervated by the phrenic nerve (C3 to C5) and the intercostal and abdominal muscles are innervated by thoracic nerves. **Paraplegia** results from damage to the cord below the cervical level, sparing arm function. Function of the trunk, lower limbs, and pelvic organs in paraplegia depends on the level of the lesion. Table 12–3 lists the motor capabilities, reflexes, and sensations mediated by each spinal cord level.

Abnormal Interneuron Activity in Chronic Spinal Cord Injury

Chronic spinal cord injury is the period after recovery from spinal shock when the neurological deficit is stable, neither progressing nor improving. (See the chronic spinal cord injury box.) This period can last for decades. In chronic spinal cord injury, three abnormalities occur in interneuron activity below the level of the lesion:

- Inhibitory interneuron response to Ia afferent activity is diminished.
- Nonreciprocal inhibition is reduced.
- Transmission from cutaneous afferents to motoneurons is facilitated.

The first change correlates with hyperreflexia, and the second correlates with muscle hypertonia, providing evidence that the two characteristics of spasticity

TABLE 12-3

FUNCTIONAL ABILITIES ASSOCIATED WITH COMPLETE SPINAL CORD LESIONS AT VARIOUS LEVELS

Level of Lesion	Motor Capability*	Reflexes Present	Intact Sensation
Above C4	Facial, pharyngeal, laryngeal movements	—	Neck and head (cranial nerves from face; C2: posterior head, upper neck; C3: lower neck)
C4	Scapular elevation, adduction	—	
C5	Deltoids, elbow flexion (biceps is innervated by C5 and C6)	Biceps	Lateral upper arm
C6	Pectoralis major, radial wrist extensors, serratus, anterior	Brachioradialis	Lateral forearm and lateral hand
C7	Triceps (C7, C8), latissimus dorsi	Triceps	Middle finger
C8	Flexor digitorum muscles	—	Medial hand
T1	Finger abduction	—	Medial forearm
T1 to T6	Erector spinae above the injury	—	T2: medial upper arm T3 to T6: torso
T7 to T12	Abdominal muscles above the injury	Abdominal	T7 to T12: torso (T10: level of umbilicus)
L1	Psoas	—	Anterior upper thigh
L2	Iliacus	—	Anterior thigh, below L1
L3	Quadriceps (L3, L4)	Quadriceps	Anterior knee
L4	Tibialis anterior	—	Medial leg
L5	Extensor hallucis longus	—	Lateral leg, dorsum of foot
S1	Peroneus longus & brevis, triceps surae, hamstrings, gluteus maximus	Tendo calcaneus	Posterior calf and lateral foot
S2	—	—	Posterior thigh
S3	—	Bladder	Ring surrounding S4 to S5 (see below)
S4 to S5	Voluntary anal contraction	Bladder	Ring surrounding anus

*Each additional level adds functions to the capabilities of the higher levels. Muscles listed may be only partially innervated at the level indicated. Thus the quadriceps usually has some voluntary activity if the L3 level is intact; however, the action is weak unless the L4 level is also intact.

are separate (Delwaide and Pennisi, 1994). The third change occurs because of the loss of descending inhibition by the reticulospinal tracts. Interneurons that produce the withdrawal reflex are normally inhibited by the reticulospinal tracts. Without this inhibition, an exaggerated withdrawal reflex occurs in response to normally innocuous stimuli in some people with spinal cord injuries (Ashby, 1993). For example, light touch on the thigh may trigger a withdrawal reflex of the entire lower limb. Additional changes secondary to spinal cord injury include loss of lower motoneurons and changes in mechanical properties of muscle fibers: atrophy of muscle fibers, fibrosis, and alteration of contractile properties toward tonic muscle characteristics.

Classification of Spinal Cord Injuries

Spinal cord injuries are classified according to two criteria (American Spinal Cord Injury Association [Ditunno et al., 1994]):

- Whether the injury is complete or incomplete
- The neurological level of injury

A **complete injury** is defined as lack of sensory and motor function in the lowest sacral segment. An **incomplete injury** is defined as preservation of sensory and/or motor function in the lowest sacral segment.

The **neurological level** is the most caudal level with normal sensory and motor function bilaterally.

CHRONIC SPINAL CORD INJURY

Pathology
Crush, hemorrhage, edema, and/or infarction

Etiology
Trauma

Speed of onset
Acute

Signs and symptoms

Consciousness
Normal

Communication and memory
Normal

Sensory
Depends on what part of the spinal cord is damaged; in a complete spinal cord lesion, all sensation is lost below the level of the lesion

Autonomic
Depends on what part of the spinal cord is damaged; in a complete spinal cord lesion, all autonomic control is lost below the level of the lesion, including voluntary bladder and bowel control and, if the lesion is above T6, autonomic dysreflexia, poor thermoregulation, and orthostatic hypotension may occur

Motor
Depends on what part of the spinal cord is damaged; in a complete spinal cord lesion, all voluntary motor control is lost below the level of the lesion

Region affected
Spinal region

Demographics
4:1 ratio of males to females; 80% of injuries occur between the ages of 16 and 45 years

Prognosis
Currently no functional regeneration of neurons in the central nervous system occurs in humans. Once the lesion is stable (no more bleeding, infarction, edema), the neurological deficit does not change. People with spinal cord injuries may live a normal life span

However, motor function may be impaired at a level different from sensory function, and the losses may be asymmetrical. In these cases, up to four different neurological segments may be described in a single patient: right sensory, left sensory, right motor, and left motor.

Determination of Neurological Levels

The American Spinal Injury Association (ASIA) has developed a standardized assessment for evaluating neurological level in spinal cord injury. The ASIA classification form is presented in Figure 12–14. Key sensory points (28 bilateral points) are tested with a safety pin to determine the person's ability to distinguish sharp from dull and with light touches with cotton to determine the ability to localize light touch. In addition, testing of deep pressure and of position sense in the index fingers and great toes is recommended. Key muscles are tested on the right and left

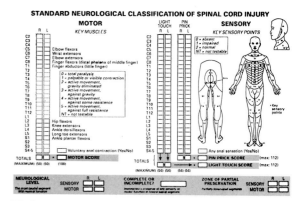

FIGURE 12–14

American Spinal Injury Association classification of spinal cord injury. Motor scores are recorded on the left half of the form. The scoring criteria are listed in the large box. The two columns, headed by the letters *R* (right) and *L* (left), are for recording the scores of the listed muscles. To the left of the columns is a list of the segments of the spinal cord. Sensory scores are recorded on the right half of the form. The scoring criteria are in the small box. The areas of impaired or absent sensation can be indicated on the dermatome diagrams. At the bottom of the motor and sensory sections are small boxes for totaling the motor and sensory scores. The neurological level is recorded at the bottom of the form, according to the criteria listed there. (Courtesy of American Spinal Injury Association International, Atlanta.)

sides of the body. Scoring criteria for each test are listed on the form.

Autonomic Dysfunction in Spinal Cord Injury

During spinal shock, neural control of the pelvic organs is depressed. Therefore, the bladder and bowels walls are atonic, allowing overfilling of these viscera, and overflow leaking occurs (see Fig. 12–13A). Overfilling and overflow leaking can be avoided by establishing a regular bladder and bowel emptying routine. After recovery from spinal shock, a complete lesion above the sacral level usually allows some reflexive functioning of the pelvic organs, but voluntary control is not possible, and the person is deprived of conscious awareness of the state of the pelvic organs.

Complete lesions at higher levels of the spinal cord cause more serious abnormalities of autonomic regulation because more segments of the cord are free from descending sympathetic control. Loss of descending sympathetic control due to lesions above T6 results in three dysfunctions:

- Autonomic dysreflexia
- Poor thermoregulation (body temperature regulation)
- Orthostatic hypotension

Autonomic dysreflexia (also called mass reflex) is excessive activity of the sympathetic nervous system, elicited by noxious stimuli below the lesion. Often the precipitating stimulus is overstretching of the bladder or rectum. The response is characterized by an abrupt increase in heart rate and blood pressure, flushing of the skin, profuse sweating, emptying of the bladder, and a pounding headache. The sudden spike in blood pressure may be life-threatening.

Normally, **body temperature regulation** is achieved by descending sympathetic innervation. In spinal cord injury, reflexive sweating below the lesion may be intact; however, interruption of sympathetic pathways prevents thermoregulatory sweating (response to increased ambient temperature) below the level of injury. To compensate, excessive sweating may occur above the level of the lesion. People with complete lesions above the T6 level should avoid exposure to high ambient temperatures because of the risk of heat stroke. The signs of heat stroke are high

body temperature, rapid pulse, and dry, flushed skin. These signs indicate a medical emergency because untreated heat stroke can cause permanent brain damage or convulsions and death. In cold weather, hypothermia is a risk because the person with a complete lesion above T6 has lost descending control of blood vessels and the ability to shiver below the lesion. The signs of hypothermia include irritability, mental confusion, hallucinations, lethargy, clumsiness, slow respiration, and slowing of the heartbeat.

Orthostatic hypotension is an extreme fall in blood pressure on assuming an upright position. In people with spinal cord injury, this results from the loss of sympathetic vasoconstriction combined with loss of muscle pumping action for blood return. Figure 12–15 summarizes the autonomic dysfunctions associated with various levels of spinal cord injury.

Prognosis and Treatment in Spinal Cord Injury

Unlike axons in the peripheral nervous system, severed axons in the adult spinal cord fail to functionally regenerate. The barriers to regeneration include inhibitory molecules on oligodendrocytes, impenetrable glial scars, and decreased rate of growth (compared to embryonic neurons) in mature neurons (Fawcett, 1992). However, some of the functional losses after spinal cord injury are not due to the original trauma but are instead due to secondary changes: bleeding, edema, ischemia, pain, and inflammation. Recent pharmacological advances alter the outcome of acute spinal cord injury by preventing the secondary changes. High doses of the steroid methylprednisolone, given within 8 hours after injury, significantly decrease the secondary changes in the spinal cord. Bracken (1993) reported significantly greater improvement in motor function, touch sensation, and pinprick sensation in people given methylprednisolone than in people given a placebo; the differences in recovery were sustained 1 year after injury. Further, the drug reduced the severity of spinal cord lesions. Drug treatment more than doubled the conversion rate from complete to partial motor and complete to partial sensory loss (Young and Bracken, 1992).

People with incomplete paraplegia have the highest rate of recovery during the first 3 months postinjury,

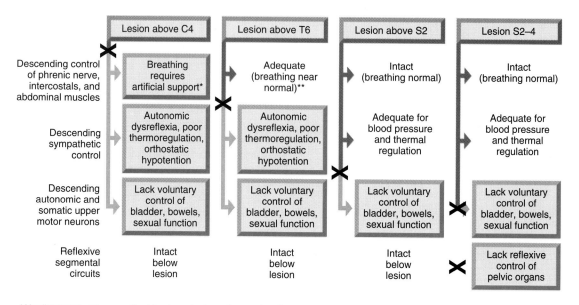

* Ventilator or phrenic nerve stimulator dependent; may learn to breathe using glossopharyngeal technique for short periods
** Abdominal muscles and lower intercostal muscles do not receive descending control.

FIGURE 12–15

Autonomic dysfunctions associated with various levels of spinal cord injury.

TABLE 12–4		
PROGNOSIS FOR AMBULATION IN PEOPLE WITH PARAPLEGIA		
Criteria	**Group**	**Ambulation Outcome**
Zero score on the ASIA lower limb motor scale 1 month postinjury (Waters et al., 1994)	Complete paraplegia	Fewer than 1% were community ambulators a year later
	Incomplete paraplegia	33% using reciprocal gait pattern in the community; all people with incomplete paraplegia who had a lower limb motor score greater than 10 points were successful community ambulators 1 year later
Initially motor complete (Crozier et al., 1991)	Only touch sensation intact below the injury	Unlikely to ambulate
	Both touch and pinprick sensation intact below the injury	Likely to become ambulatory
Motor incomplete lesions (Crozier and Cheng, 1992)	<3/5 quadriceps strength 2 months postinjury	Less likely to recover walking
	>3/5 quadriceps strength 2 months postinjury	Excellent chance of becoming functional ambulators

with relatively small gains after 3 months. The contrast between functional recovery in complete versus incomplete paraplegia at 1 year postinjury is striking. Table 12–4 summarizes ambulation prognosis for people with paraplegia. In tetraplegic subjects, recovery rates of arm muscle function have been reported by Ditunno et al. (1992).

Atrice et al. (1995) describe the expected functional outcomes (mobility, self-care, communication, and home management), prevention of complications, and rehabilitation techniques appropriate for the various levels of spinal cord injury. Typical complications the second year after spinal cord injury (in order of prevalence) include urinary tract infection, spasticity, chills and fever, decubiti, autonomic dysreflexia, contractures, heterotropic ossification, and pneumonia (Ditunno and Formal, 1994). Upright posture can provide some protection against urinary tract infection and pneumonia; mobility can help avoid contractures and decubiti. Currently strengthening and range of motion exercises, mobility and activities of daily living training, adaptive equipment, and environmental modifications are commonly used in spinal cord injury rehabilitation.

Functional electrical stimulation (FES) to restore movement in people with spinal cord injuries is currently primarily an experimental technique, but advances in technology including combining FES with orthotic systems offer promise for future treatment. Stein et al. (1993) reported the effects of FES for ambulation in people with incomplete spinal cord injury. Prior to muscle stimulation, participants were either unable to walk or exhibited an exceptionally slow gait. Stimulation was applied to the common peroneal nerve. In all people selected for the trials, walking and/or gains in speed were achieved. According to Stein, the stimulation has three effects: direct stimulation of dorsiflexor muscles; temporary reduction of plantar flexor activity; and with more intense stimulation, a withdrawal reflex with flexion of the hip and knee.

SPECIFIC DISORDERS AFFECTING SPINAL REGION FUNCTION

Other disorders in addition to traumatic spinal cord injury interfere with spinal region function. These include developmental disorders, lesions of dorsal and ventral nerve roots, multiple sclerosis, and lesions that cause compression in the spinal cord.

Developmental Disorders

MENINGOMYELOCELE

Outcomes of meningomyelocele, a developmental defect arising from failure of the inferior neuropore to close (see Chap. 5), are roughly equivalent to spinal cord injury in later life. If the lesion is in the lower lumbar spinal cord, anterior thigh muscles may be functional and sensation intact in the overlying skin, with the remainder of the lower limbs nonmoving and insensitive to sensory stimulation and with no voluntary or reflexive control of the pelvic organs. If the lesion is below S1, bladder and bowel reflexive (as well as voluntary) control is absent because the sacral cord contains the connections for these reflexes, but skeletal muscle control is intact throughout the body.

SPASTIC CEREBRAL PALSY

Cerebral palsy is a motor disorder that develops in utero or during infancy. Spastic cerebral palsy is characterized by muscle hypertonia and phasic stretch hyperreflexia. The hyperreflexia resulting from overexcitation of local reflex circuits can be inhibited by surgically cutting selected dorsal rootlets, thus decreasing the sensory input to the reflex circuit. Thus, to alleviate lower limb hyperreflexia in children with spastic cerebral palsy, selected dorsal rootlets are sometimes surgically severed (**dorsal rhizotomy**). Dorsal rhizotomy reduces hyperreflexia by interrupting the afferent limb of the stretch reflex. Each rootlet is electrically stimulated, and only rootlets that contribute to abnormal muscle activity are cut. Dorsal rhizotomy is typically performed at the L2 to L5 level. The goals of the surgery are to improve motor function or to make bathing, positioning, and dressing easier. However, motor control remains abnormal following the surgery, and intensive physical therapy is required to achieve benefits from the surgery (see review by Giuliani, 1991). Careful evaluation of the child's potential for improved function is vital prior to the surgery.

Lesions of Dorsal and Ventral Nerve Roots

Lesion of a nerve root is termed **radiculopathy;** however, this term is also often used clinically to refer to damage of a spinal nerve. Mechanical irritation or infection of a dorsal root produces pain in the innervated dermatome and also in the muscles innervated by the spinal cord segment. Mechanical irritation can be produced by a herniated intervertebral disk, a tumor, or a dislocated fracture. However, herniated vertebral disks do not always cause symptoms; Teresi et al. (1987) reported that 24 of 42 people over age 64 without symptoms referable to the cervical spine had herniated disks in the cervical region. When a dorsal root is irritated, coughing or sneezing often aggravates the pain.

Other conditions affecting the spinal nerve roots include infections, avulsions, and severance. A common infection of the somas in the dorsal root is varicella zoster, also called herpes zoster and shingles (see Chap. 7). Avulsion or complete severance of the dorsal root causes loss of sensation in the dermatome. A complete severance of a ventral root deprives the muscles in its myotome of motor innervation, resulting in muscle atrophy and fibrillation.

Multiple Sclerosis

Multiple sclerosis is characterized by random, multifocal demyelination limited to the central nervous system (see Chap. 2). Signs and symptoms of multiple sclerosis are exceptionally variable because the demyelination can occur in a wide variety of locations and the extent of each lesion also varies. Sensory complaints may include numbness, paresthesias, and Lhermitte's sign. Lhermitte's sign is the radiation of a sensation like electrical shock down the back or limbs, elicited by neck flexion. Frequently multiple sclerosis of the spinal cord produces asymmetrical weakness due to plaques interfering with the descending motor tracts and ataxia of the lower limbs due to interruption of conduction in the dorsal columns.

Compression in the Spinal Region

Pressure in the spinal region or restriction of blood flow due to compression can cause any of the following symptoms: pain (usually constant), sensory changes, weakness, paralysis, hypertonia, ataxia, and impaired bladder and/or bowel function. The clinical presentation depends on the location of the lesion. Gradual onset, progressive worsening, no history of trauma, and the combination of segmental and verti-

cal tract signs indicate the possibility of a spinal region tumor, cervical spondylosis, or syringomyelia.

SPINAL REGION TUMORS

Tumors outside the dura mater or in the subarachnoid space may compress the spinal cord, nerve roots, spinal nerve, or their blood supply. Tumors can also occur within the spinal cord, resulting in pressure on neurons and vascular supply from within the cord. Tumors can produce segmental and/or vertical tract signs, depending on their location.

MYELOPATHY DUE TO CERVICAL SPONDYLOSIS

Myelopathy is a disorder of the spinal cord. Cervical spondylosis is degeneration of the cervical vertebrae and disks that produces narrowing of the vertebral canal and intervertebral foramina. This narrowing progressively impinges on the enclosed spinal cord and nerve roots. The compression results in the following:

- Neck and upper-limb pain and stiffness
- In some cases, numbness and paresthesias of the feet
- Lower-limb paresis and hypertonia
- Babinski's sign

The neck and upper-limb pain are segmental signs, caused by irritation of nerve roots. The lower-limb signs are vertical tract signs produced by compression of ascending and descending tracts in their course through the lower cervical vertebral canal.

SYRINGOMYELIA

Syringomyelia is a rare, progressive disorder, most frequently occurring in people 35 to 45 years of age. A syrinx, or fluid-filled cavity, develops in the spinal cord, almost always in the cervical region. Syringomyelia is usually congenital but can be secondary to trauma or tumors. The accumulation of cerebrospinal fluid in the syrinx causes increased pressure inside the spinal cord, expanding the cavity and compressing the adjacent nerve fibers. Syringomyelia is often associated with trauma to the spinal cord or with congenital malformations of the cord. Segmental signs occur in the upper limbs, including loss of sensitivity to pain and temperature stimuli, due to interruption of axons crossing the midline in the anterior white commissure; paresis; and muscle atrophy. The sensory loss is often distributed like a cape draped over the shoulders (see Fig. 12–13B). Vertical tract signs in the lower limbs include paresis, muscle hypertonicity, phasic stretch hyperreflexia, and loss of bowel and bladder control.

Summary

Lesions of the spinal cord produce segmental and/or vertical tract signs. Segmental signs include the following:

- Sensory changes: impaired sensations, paresthesias, dysesthesias, in a dermatomal distribution.
- Lower motoneuron signs (paresis or paralysis, atrophy, cramps) in a myotomal distribution.
- If dorsal nerve roots are involved, increasing intra-abdominal pressure by straining, sneezing, or coughing may produce sharp, radiating pain.

Common vertical tract signs include the following:

- Sensory changes: decreased or lost sensation below the level of the lesion
- Autonomic signs: decreased or lost voluntary control of pelvic organs, autonomic dysreflexia, poor thermoregulation, and/or orthostatic hypotension
- Upper motoneuron lesion signs: muscle hypertonia, paresis, phasic stretch hyperreflexia, Babinski's sign

CLINICAL NOTES

CASE 1

P.E. is a 17-year-old woman. She fractured the C7 vertebra in a diving accident 2 months ago. The fracture is stable. Current findings are as follows:

- Sensation is intact (pinprick, temperature, conscious proprioception, and discriminative touch) in her head, neck, and lateral upper limbs.
- She has no sensation in the medial upper limbs, the trunk below the sternal angle, and the lower limbs.
- All head and shoulder movements are normal strength except shoulder extension.
- Elbow flexion and radial wrist extensors are normal strength.
- The remaining upper limb, trunk, and lower limb muscles have no trace of voluntary movement.
- Babinski's signs are present bilaterally.

Without adaptive equipment, P.E. is unable to care for herself. Using adaptive equipment, she is able to eat, dress, and groom independently. She uses a wheelchair. She cannot voluntarily control her bladder or bowels.

Questions

1. Is the lesion in the dorsal or ventral root or in the spinal cord?

2. What neurological level is the lesion? Note: the neurological level in a spinal cord injury is the most caudal level with *normal* sensory and motor function bilaterally. Refer to Table 12–3 to determine the neurological level. Is the lesion complete or incomplete?

(continued)

CASE 2

B.D. is a 16-year-old man. He sustained a spinal cord injury 2 months ago in a fall from a bicycle. Current findings are as follows:

- Pinprick and temperature sensation are impaired as indicated in Figure 12–16. All other sensations are fully intact.

- Manual muscle test scores are also indicated in Figure 12–16.

- Babinski's signs are present bilaterally.

- He is independent in all activities. He is able to walk 30 meters using an ankle-foot orthosis on his left leg and a cane.

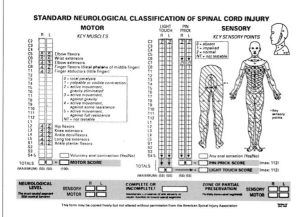

FIGURE 12–16

Motor and sensory test results for Case 2. (Form courtesy of American Spinal Injury Association International, Atlanta.)

Questions

1. What level is the cord lesion? Is the lesion complete, or does the pattern indicate a spinal cord syndrome?

2. Why is this patient independent, while the patient in Case 1 requires adaptive equipment, a wheelchair, and maximal assistance on stairs?

CASE 3

V.K. is a 30-year-old man. He plays recreational sports 4 days a week and is a highly competitive soccer player. Two years ago, he noted temporary weakness in his left lower leg, which gradually resolved without consultation or treatment. His primary complaint now is inability to control his right foot. He first noted poor kicking skills 3 weeks ago. Sensation and motor control are normal except in the right lower limb. The following deficits are observed in the right lower limb:

- Discriminative touch, vibration sense, and position sense are impaired throughout.

- Pain and temperature sensations are intact.

- Movement is ataxic, including gait: dragging of toes on the ground during swing phase of walking (foot drop), poor placement of the foot on the ground, weight bearing on the right lower limb only half the time spent weight bearing on the left lower limb.

- Gluteals, hamstrings, and all muscles originating below the knee are weak, less than half the strength of the homologous muscles on the left.

(continued)

Questions

1. Why are pain and temperature sensations intact bilaterally?

2. Where is the lesion?

3. What is the probable etiology?

CASE 4

E.V. is a 62-year-old woman. She reports constant burning pain radiating down the back of her left leg into her foot. When she coughs or sneezes, sharp, stabbing pains become excruciating. The pain began as a backache 3 months ago. Pain intensity has been consistently increasing. Following are the results of testing:

- Sensation is intact in the right lower limb.

- Sensory testing results for the left lower limb are shown in Table 12–5.

- Strength is within the normal limits in all limbs.

- Ankle deep tendon reflex is absent on the left.

TABLE 12–5
SENSORY TESTING RESULTS FOR LEFT LOWER LIMB FOR CASE 4

Spinal Level	Discriminative Touch	Joint Kinesthesia	Pinprick	Warm	Cold
L4	2	Knee 2	2	2	2
L5	2	Ankle 1	2	2	2
S1	0	Ankle 1	0	0	0
S2	0	—	1	1	1
S3	0	—	1	1	1
S4	2	—	2	2	2
S5	2	—	2	2	2

Scoring: 2 = intact, 1 = impaired, 0 = absent.

Questions

1. Where is the lesion?

2. What is the probable etiology?

3. Why is discriminative touch more affected than pain and temperature sensations?

REVIEW QUESTIONS

1. What is a spinal nerve?

2. What is the difference between a ventral root and a ventral primary ramus?

3. What is a spinal segment?

4. What is the function of the dorsal horn?

5. Which of Rexed's laminae is also known as the substantia gelatinosa?

6. Are reflexes and voluntary motor control entirely separate systems?

7. What is the function of nonreciprocal inhibition?

8. How is voluntary voiding of urine controlled?

9. What are the differences in signs between segmental and vertical tract lesions?

10. List the four spinal cord syndromes, and draw spinal cord cross sections that illustrate the location of the lesion in each syndrome.

11. Why are cord functions below the lesion depressed or lost immediately after a spinal cord injury?

12. Why do some people with spinal cord injuries have exaggerated withdrawal reflexes?

13. What is an incomplete spinal cord injury? Give two examples of syndromes that may result from incomplete spinal cord injury.

14. List the three conditions that arise when the spinal cord below the T6 level is deprived of descending sympathetic innervation.

References

Ashby, P. (1993). The neurophysiology of human spinal spasticity. In: Gandevia, S. C., Burke, D., Anthoney, M., et al. (Eds.). Science and practice in clinical neurology. Cambridge, England: Cambridge University Press, pp. 106–129.

Atrice, M. B., Gonter, M., Griffin, D. A., et al. (1995). Traumatic spinal cord injury. In: Umphred, D. A. (Ed.). Neurological rehabilitation (3rd ed.). St. Louis: Mosby, pp. 484–534.

Bracken, M. B. (1993). Pharmacological treatment of acute spinal cord injury: Current status and future projects. J. Emerg. Med. 11 (Suppl. 1):43–48.

Crozier, K. S., and Cheng, L. L. (1992). Spinal cord injury: Prognosis for ambulation based on quadriceps recovery. Paraplegia 30(11):762–767.

Crozier, K. S., Graziani, V., Ditunno, J. F., et al. (1991). Spinal cord injury: Prognosis for ambulation based on sensory examination in patients who are initially motor complete. Archives Phys. Med. Rehabil. 72(2):119–121.

Davidoff, R. A. (1992). Skeletal muscle tone and the misunderstood stretch reflex. Neurology 42:951–963.

Davidoff, R. A., and Hackman, J. C. (1991). Aspects of spinal cord structure and reflex function. Neurol. Clin. 9(3):533–550.

Delwaide, P. J., and Pennisi, G. (1994). Tizanidine and electrophysiologic analysis of spinal control mechanisms in humans with spasticity. Neurology 44(Suppl. 9):S21–S28.

Ditunno, J. F., and Formal, C. S. (1994). Chronic spinal cord injury. N. Engl. J. Med. 330(8):550–556.

Ditunno, J. F., Stover, S. L., Freed, M. M., et al. (1992). Motor recovery of the upper extremities in traumatic quadriplegia: A multicenter study. Arch. Phys. Med. Rehabil. 73:431–436.

Ditunno, J. F., Young, W., Donovan, W. H., et al. (1994). The international standards booklet for neurological and functional classification of spinal cord injury: American Spinal Injury Association. Paraplegia 32(2):70–80.

Fawcett, J. W. (1992). Spinal cord repair: Future directions. Paraplegia 30(2):83–85.

Giuliani, C. A. (1991). Dorsal rhizotomy for children with cerebral palsy: Support for concepts of motor control. Phys. Ther. 71:248–259.

Madersbacher, H. (1990). The various types of neurogenic bladder dysfunction: An update of current therapeutic concepts. Paraplegia 28:217–229.

McCrea, D. A. (1994). Can sense be made of spinal interneuron circuits? In: Cordo, P., and Harnad, S. (Eds.). Movement control. Cambridge, England: Cambridge University Press, pp. 31–41.

Stein, R. B., Yang, J. F., Belanger, M., et al. (1993). Modification of reflexes in normal and abnormal movements. Prog. Brain Res. 97:189–196.

Teresi, L. M., Lufkin, R. B., Riecher, M. A., et al. (1987). Asymptomatic degenerative disk disease and spondylosis of the cervical spine: MR imaging. Radiology 164(1):83–88.

Veale, J. L., and Rees, S. (1973). Renshaw cell activity in man. J. Neurol. Neurosurg. Psychiatry 36(4):674–683.

Waters, R. L., Adkins, R. H., et al. (1994). Motor and sensory recovery following incomplete paraplegia. Arch. Phys. Med. Rehabil. 75(1):67–72.

Young, W., and Bracken, M. B. (1992, March 9). The second national acute spinal cord injury study. J. Neurotrauma Suppl. 1:S397–405.

13

Cranial Nerves

I'm 25 years old. On Saturday I swam in very cold water. On Sunday the right side of my tongue and mouth had a coated feeling, and by that evening my lips were twitching ever so slightly. Monday, my right eyelids were occasionally twitching uncontrollably. Tuesday morning I had only 25% to 50% control over my right eyelid and facial muscles; I could only get three-quarters of a smile. Wednesday morning I had 0% to 5% control of the right facial muscles. I couldn't close my right eye. I felt like I had Novocain in the right side of my face, except that I had sensation in the affected area. It was scary. The physician performed a nerve conduction velocity test, eye blink reflex tests, and needle electromyography to determine the status of the nerve. The disorder was diagnosed as Bell's palsy.

—Darren Larson

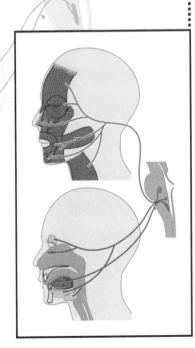

INTRODUCTION

Cranial nerves exchange information between the periphery and the brain. Twelve pairs of cranial nerves emerge from the surface of the brain to innervate structures of the head and neck. Cranial nerve X, the vagus, innervates thoracic and abdominal viscera in addition to structures in the head and neck. The axons and receptors of the cranial nerves that are outside the skull are part of the peripheral nervous system.

Like peripheral nerves connected to the spinal cord, cranial nerves serve sensory, motor, and autonomic functions. Generally, cell bodies for cranial nerves are located similarly to the cell bodies that contribute axons to peripheral nerves that connect with the spinal cord. Cell bodies of sensory neurons in cranial nerves are usually in ganglia outside of the brain stem (the exception, neurons that convey proprioceptive information from the face, have cell bodies inside the brain stem). Motor cell bodies are located in nuclei inside the brain stem.

Cranial nerves differ from spinal nerves in specialization. Some cranial nerves are only motor, others only sensory, and some are both sensory and motor. The cranial nerve fibers that innervate muscles of the head and neck are lower motoneurons. As in the spinal cord, these lower motoneurons are influenced by input from upper motoneurons and sensory afferent fibers. Several cranial nerves have unique functions not shared by any other nerves, such as conveying visual, auditory, or vestibular information.

Cranial nerves have three functions:

- They supply motor innervation to muscles of the face, eyes, tongue, jaw, and two neck muscles (sternocleidomastoid and trapezius).
- They transmit somatosensory information from the skin and muscles of the face and the temporomandibular joint and special sensory information related to visual, auditory, vestibular, gustatory, olfactory, and visceral sensations.
- They provide parasympathetic regulation of heart rate, blood pressure, breathing, and digestion.

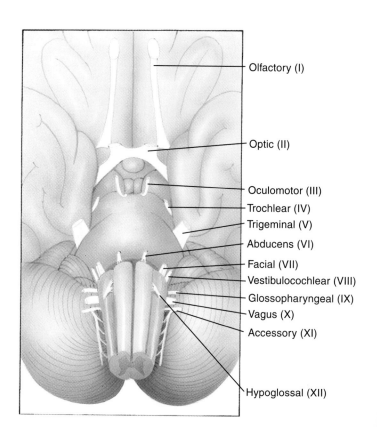

FIGURE 13–1

Inferior view of brain, showing cranial nerve connections.

TABLE 13–1
CRANIAL NERVES

Number	Name	Function	Connection to Brain
I	Olfactory	Smell	Inferior frontal lobe
II	Optic	Vision	Diencephalon
III	Oculomotor	Moves eye up, down, medially; raises upper eyelid; constricts pupil	Midbrain (anterior)
IV	Trochlear	Moves eye medially and down	Midbrain (posterior)
V	Trigeminal	Facial sensation, chewing, sensation from temporomandibular joint	Pons (lateral)
VI	Abducens	Abducts eye	Between pons and medulla
VII	Facial	Facial expression, closes eye, tears, salivation, taste	Between pons and medulla
VIII	Vestibulocochlear	Sensation of head position relative to gravity and head movement; hearing	Between pons and medulla
IX	Glossopharyngeal	Swallowing, salivation, taste	Medulla
X	Vagus	Regulates viscera, swallowing, speech, taste	Medulla
XI	Accessory	Elevates shoulders, turns head	Spinal cord and medulla
XII	Hypoglossal	Moves tongue	Medulla

All cranial nerve connections to the brain are visible on the inferior brain (Fig. 13–1), except cranial nerve IV, which emerges from the posterior midbrain. Cranial nerve names, primary functions, and connections to the brain are listed in Table 13–1.

CRANIAL NERVE I: OLFACTORY

The **olfactory nerve** is sensory, conducting information from nasal chemoreceptors to the olfactory bulb. Signals from the olfactory bulb travel in the olfactory tract to the medial temporal lobe of the cerebrum. The sense of smell is dependent on olfactory nerve function. Much of the information attributed to taste is olfactory because information from taste buds is limited to chemoreceptors for salty, sweet, sour, and bitter tastes.

CRANIAL NERVE II: OPTIC

The **optic nerve** is sensory, transmitting visual information from the retina to the lateral geniculate body of the thalamus and to nuclei in the midbrain (Fig. 13–2). The retina is the inner layer of the posterior eye,

formed by photosensitive cells. Light striking the retina is converted into neural signals by the photosensitive cells. Axons from neurons in the retina travel in the optic nerve, through the optic chiasm, and in the optic tract before synapsing in the lateral geniculate body. The lateral geniculate body is a relay along a pathway to the primary visual cortex. This pathway projects to areas involved in the analysis and conscious awareness of visual information. The central processing of visual signals will be discussed in Chapter 15.

The visual signals sent to the midbrain are involved in reflexive responses of the pupil, awareness of light and dark, and orienting the head and eyes. Reflexes involving cranial nerves are listed in Table 13–2 (on p. 253).

CRANIAL NERVES III, IV, AND VI: OCULOMOTOR, TROCHLEAR, AND ABDUCENS

The **oculomotor, trochlear,** and **abducens nerves** are primarily motor, containing lower motoneuron axons innervating the six extraocular muscles that move the eye (Fig. 13–3 on p. 253) and control reflexive constriction of the pupil.

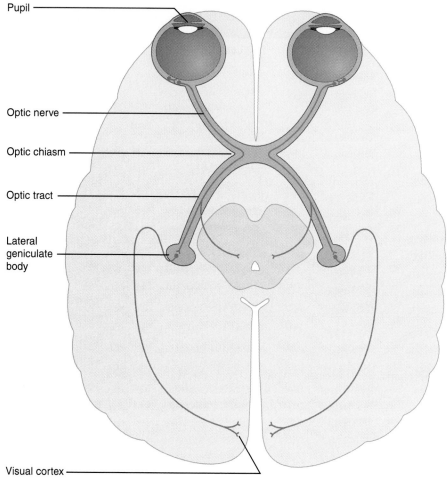

Pupil

Optic nerve

Optic chiasm

Optic tract

Lateral
geniculate
body

Visual cortex

FIGURE 13–2

The optic nerve projects from the retina to the midbrain and to the lateral geniculate. Reflex connections in the midbrain control the constriction of the pupil and reflexive eye movements. Visual information relayed by the lateral geniculate to the visual cortex provides conscious vision.

Control of Eye Movement

The extraocular muscles include four straight (rectus) muscles and two oblique muscles. The rectus muscles attach to the anterior half of the eyeball. The lateral rectus moves the eye laterally, and the medial rectus moves the eye medially; thus these muscles form a pair controlling horizontal eye movement. With the eyes looking straight forward, the actions of the superior and inferior rectus are primarily elevation and depression of the eye, respectively. The two oblique muscles attach to the posterior half of the eyeball. If the eye is abducted, the obliques primarily rotate the

eye. When the eye is adducted, the superior oblique depresses the eye, and the inferior oblique elevates it (see Fig. 13–3). Cranial nerve supply to the extraocular muscles is shown in Figure 13–4 (on p. 254).

Cranial nerve III, the oculomotor nerve, has cell bodies located in the oculomotor nucleus that control the superior, inferior, and medial rectus, the inferior oblique, and the levator palpebrae superioris muscles. These muscles move the eye upward, downward, and medially; rotate the eye around the axis of the pupil; and assist in elevating the upper eyelid.

Cranial nerve IV, the trochlear nerve, has cell bodies located in the trochlear nucleus in the midbrain

TABLE 13–2
CRANIAL NERVE REFLEXES

Reflex	Description of Reflex	Afferent Neurons	Efferent Neurons
Pupillary	Pupil of eye constricts when light shined into eye	Optic	Oculomotor
Consensual	Pupil of eye constricts when light shined into other eye	Optic	Oculomotor
Accommodation	Lens of eye adjusts to focus light on the retina, pupil constricts, and pupils move medially when viewing an object at close range	Optic	Oculomotor
Masseter	When masseter is tapped with a reflex hammer, the muscle contracts	Trigeminal	Trigeminal
Corneal (blink)	When the cornea is touched, the eyelids close	Trigeminal	Facial
Gag	Touching of pharynx elicits contraction of pharyngeal muscles	Glossopharyngeal	Vagus
Swallowing	Food touching entrance of pharynx elicits movement of the soft palate and contraction of muscles of the pharynx	Glossopharyngeal	Vagus

and is the only cranial nerve to emerge from the dorsal brain stem, below the inferior colliculus. The trochlear nerve controls the superior oblique muscle, which rotates the eye or, if the eye is adducted, depresses the eye.

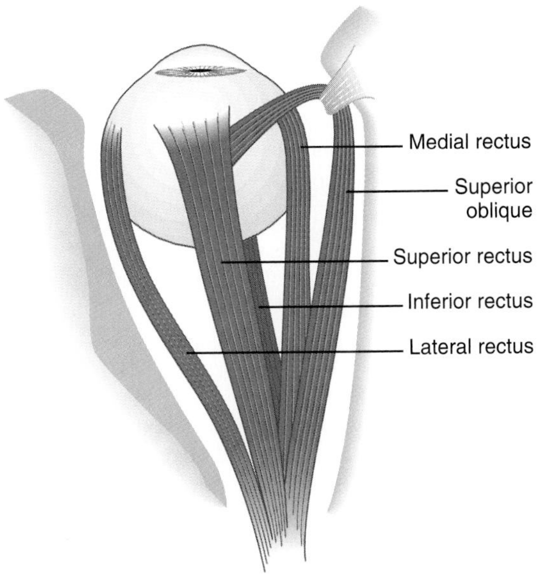

FIGURE 13–3

Superior view of the extraocular muscles. When the eye is adducted, the superior oblique moves the pupil down, and inferior oblique moves the pupil up. When the eye is abducted, the obliques rotate the eye. These actions occur because of the angle of muscle pull and because the obliques attach to the posterior half of the eyeball.

Labels in figure:
- Medial rectus
- Superior oblique
- Superior rectus
- Inferior rectus
- Lateral rectus

Cranial nerve VI, the abducens nerve has cell bodies located in the abducens nucleus in the pontine tegmentum. The abducens nerve controls the lateral rectus muscle that moves the eye laterally.

Coordination of Eye Movements

Coordination of the two eyes is maintained via synergistic action of the eye muscles. For instance, to look toward the right, the abducens nerve activates the lateral rectus to move the right eye laterally, while the oculomotor nerve activates the medial rectus to move the left eye medially. This coordination requires connections among the cranial nerve nuclei that control eye movements. A brain stem tract, the **medial longitudinal fasciculus,** coordinates head and eye movements by providing bilateral connections among vestibular, oculomotor, and spinal accessory nerve nuclei in the brain stem. The medial longitudinal fasciculus is discussed further in Chapter 15.

> Eye and head movements are coordinated by the medial longitudinal fasciculus.

Parasympathetic Fibers of Cranial Nerve III

In addition to the voluntary control of eye movements, cranial nerve III is involved in reflexive constriction of the pupil and the muscles controlling the

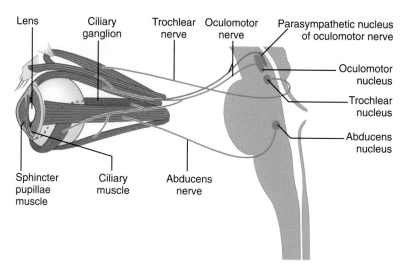

Cranial Nerve	Muscle	Movement
III: oculomotor	Levator palpebrae superioris	Lifts eyelid
	Superior rectus	Pupil up
	Medial rectus	Pupil medial
	Inferior rectus	Pupil down
	Inferior oblique	If eye adducted, pupil up; if eye abducted, rotates eye
IV: trochlear	Superior oblique	If eye adducted, pupil down and in; if eye abducted, rotates eye
VI: abducens	Lateral rectus	Pupil lateral

FIGURE 13–4

The innervation of extraocular and intraocular eye muscles. CN, cranial nerve.

lens of the eye. The oculomotor nerve contains parasympathetic fibers with their cell bodies in the parasympathetic nucleus of the oculomotor nerve (also called the Edinger-Westphal nucleus). The parasympathetic fibers are preganglionic and synapse with postganglionic fibers behind the eyeball in the ciliary ganglion. The parasympathetic connections innervate the intrinsic muscles of the eye: the pupillary sphincter and the ciliary muscle. When the pupillary sphincter constricts, the amount of light reaching the retina is decreased. When viewing objects closer than 20 centimeters, the ciliary muscle contracts, increasing the curvature of the lens. This action, called accommodation, increases refraction of light rays so that the focal point will be maintained on the retina.

Pupillary, Consensual, and Accommodation Reflexes

The pupillary, consensual, and accommodation reflexes involve the optic and oculomotor nerves (Figs. 13–5 and 13–6). The optic nerve is the afferent (i.e., sensory) limb of these reflexes, while the oculomotor nerve provides the efferent (i.e., motor) limb. The pupillary and consensual reflexes are elicited by the same stimulus: shining a bright light into one eye. The pupillary reflex is pupil constriction in the eye directly stimulated by the bright light. Consensual reflex is constriction of the pupil in the other eye.

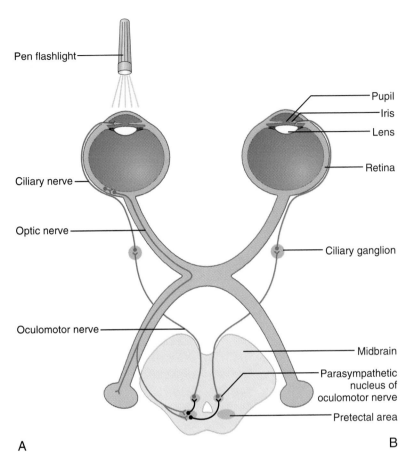

Pen flashlight

Pupil
Iris
Lens
Retina

Ciliary nerve

Ciliary ganglion

Optic nerve

Oculomotor nerve

Midbrain

Parasympathetic
nucleus of
oculomotor nerve

Pretectal area

A

B

FIGURE 13–5

Eye reflexes. Both the pupillary and consensual reflexes are responses to bright light shown into one eye. Light shown into the left eye elicits reflexive constriction of both pupils. The optic nerve conveys information from the retina to the pretectal area. Interneurons from the pretectal area synapse in the parasympathetic nucleus of the oculomotor nerve. Efferents travel in the oculomotor and then the ciliary nerves. *A,* The pupillary reflex is produced by ipsilateral neural connections. *B,* The consensual reflex, constriction of the opposite pupil, is elicited by the neuron connecting the left pretectal area with the right parasympathetic nucleus of the oculomotor nerve.

The pathways for the pupillary and consensual reflexes consist of neurons that sequentially connect the following:

• The retina to the pretectal nucleus in the midbrain
• The pretectal nucleus to the parasympathetic nuclei of the oculomotor nerve
• The parasympathetic nuclei of the oculomotor nerve to the ciliary ganglion
• The ciliary ganglion to the pupillary sphincter

The size of the pupil and the shape of the lens of the eye are reflexively controlled by afferents in the optic nerve and parasympathetic efferents in the oculomotor nerve.

Accommodation consists of adjustments to view a near object: the pupils constrict, eyes converge (adduct), and the lens becomes more convex.

CRANIAL NERVE V: TRIGEMINAL

The trigeminal nerve is a mixed nerve containing both sensory and motor fibers. The sensory fibers transmit information from the face and temporomandibular joint. The motor fibers innervate the muscles of mastication. The trigeminal nerve is named for its three branches: ophthalmic, maxillary, and mandibular (Fig. 13–7A). All three branches convey somatosensory signals; the **mandibular branch** also contains lower motoneuron axons to the muscles used in chewing. Pathways carrying information from the trigeminal nerve are illustrated in Figure 13–7B.

Cell bodies of the neurons carrying sensory information for **discriminative touch** are found in the trigeminal ganglion. The central axons synapse in the **main sensory nucleus** in the pons. Second-order neurons cross the midline and project to the ventral

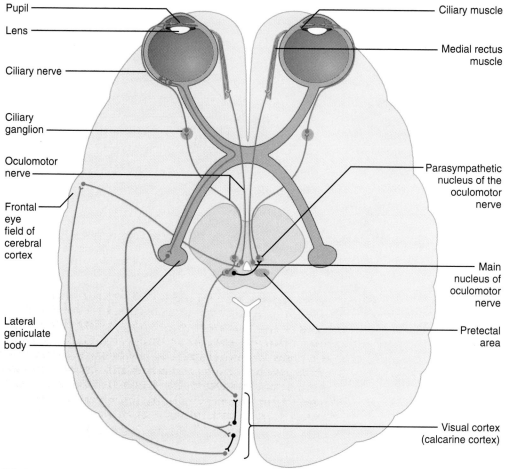

Pupil

Lens

Ciliary nerve

Ciliary ganglion

Oculomotor nerve

Frontal eye field of cerebral cortex

Lateral geniculate body

Ciliary muscle

Medial rectus muscle

Parasympathetic nucleus of the oculomotor nerve

Main nucleus of oculomotor nerve

Pretectal area

Visual cortex (calcarine cortex)

FIGURE 13–6

Accommodation is a change in curvature of the lens, contraction of the pupil, and position of the eyes in response to viewing a near object. The afferent limb is the retino-geniculo-calcarine pathway. The efferent limb to control the curvature of the lens and to contract the pupil is from the visual cortex to nuclei in the midbrain, then via parasympathetic neurons to the ciliary muscle. The efferent limb to move the pupils toward the midline is from the visual cortex to the frontal eye fields, then to the main oculomotor nucleus, which controls contraction of the medial rectus muscles.

posteromedial nucleus of the thalamus. Third-order neurons then project to the somatosensory cortex, where discriminative touch signals are consciously recognized.

Proprioceptive information from the muscles of mastication and extraocular muscles is transmitted ipsilaterally by axons of cranial nerve V to the **mesencephalic nucleus** in the midbrain. The primary neuron cell bodies are found inside the brain stem, in the

mesencephalic nucleus rather than the trigeminal ganglion. This location for sensory cell bodies is atypical because the usual location of primary neuron cell bodies is cranial nerve or dorsal root ganglia outside the brain stem or spinal cord. Central branches of mesencephalic tract neurons project to the reticular formation. The pathways from the reticular formation to conscious awareness are not known. Collaterals of proprioceptive fibers project to the trigeminal motor

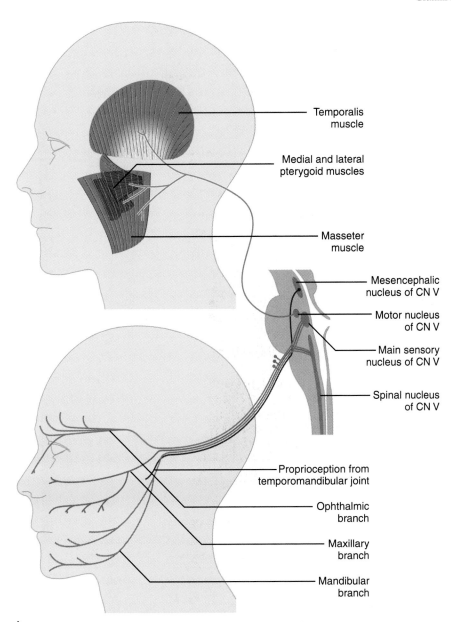

Temporalis
muscle

Medial and lateral
pterygoid muscles

Masseter
muscle

Mesencephalic
nucleus of CN V

Motor nucleus
of CN V

Main sensory
nucleus of CN V

Spinal nucleus
of CN V

Proprioception from
temporomandibular joint

Ophthalmic
branch

Maxillary
branch

Mandibular
branch

A

FIGURE 13–7

Trigeminal nerve. *A,* Distribution to skin of the face, temporomandibular joint, and muscles of mastication.

Illustration continued on following page

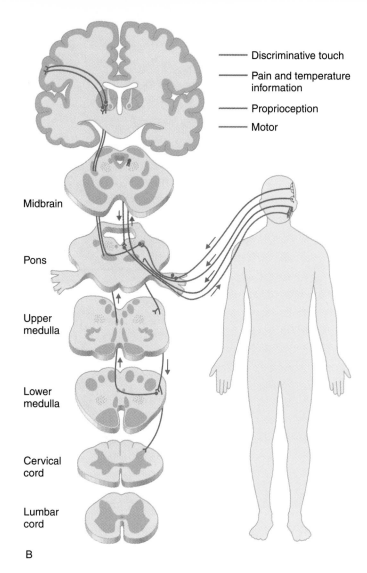

	Discriminative touch
	Pain and temperature information
	Proprioception
	Motor

Midbrain

Pons

Upper medulla

Lower medulla

Cervical cord

Lumbar cord

B

Sensation	Primary Neuron Cell Body	1st Synapse	2nd Synapse	Termination
Discriminative touch	Trigeminal ganglion	Main sensory nucleus	Ventral posteromedial nucleus of thalamus	Somatosensory cortex
Proprioception	Mesencephalic nucleus	Reticular formation	Unknown	Unknown
Fast pain	Trigeminal ganglion	Spinal trigeminal nucleus	Ventral posteromedial nucleus of thalamus	Somatosensory cortex
Slow pain	Trigeminal ganglion	Spinal trigeminal nucleus	Reticular formation and intralaminar nuclei	Limbic system and throughout cortex

C

FIGURE 13–7

Continued B, Pathways conveying somatosensory information from the face and motor signals to the muscles involved in chewing.

nucleus (reflex connections) and to the cerebellum (motor coordination).

The cell bodies of nociceptive, Aδ and C, fibers are in the trigeminal ganglion. The central axons enter the pons, then descend as the spinal tract of the trigeminal nerve into the cervical spinal cord. These neurons synapse with second-order neurons in the **spinal trigeminal nucleus.** Axons of second-order neurons transmitting fast pain information cross the midline and ascend to the ventral posteromedial nucleus of the thalamus. Third-order neurons arise in the ventral posteromedial nucleus and project to the somatosensory cortex. Slow pain information travels in the **trigeminoreticulothalamic pathway.** C fibers from the trigeminal nerve synapse in the reticular formation. Projection neurons end in the intralaminar nuclei. The projections from the intralaminar nuclei are similar to the paleospinothalamic pathways, with projections to many areas of cortex.

Reflex actions are also mediated by the trigeminal nerve. Ophthalmic fibers of the trigeminal nerve provide the afferent limb of the corneal (blink) reflex. When the cornea is touched, information is relayed to the spinal trigeminal nucleus via the trigeminal nerve. From the spinal trigeminal nucleus, interneurons convey the information bilaterally to the facial nerve (VII) nuclei. The facial nerve then reflexively activates muscles to close eyelids of both eyes. Another reflex, the masseter reflex, relies entirely on trigeminal nerve connections. When a light, downward tap is delivered to the chin, a monosynaptic stretch reflex closing the jaw occurs. The afferent information from the muscle spindles and the efferent signals to the muscles both travel in the trigeminal nerve. The synapse is in the motor trigeminal nucleus.

> Somatosensory information from the face is conveyed by the trigeminal nerve and distributed to the three trigeminal nuclei: mesencephalic (proprioceptive), main sensory (discriminative touch), and spinal (fast pain and temperature). Slow pain information projects to the reticular formation.

CRANIAL NERVE VII: FACIAL

The facial nerve (Fig. 13–8 on p. 260) is a mixed nerve containing both sensory and motor fibers. The sensory fibers transmit touch, pain, and pressure information from the tongue and pharynx and informa-

tion from the chemoreceptors located in the taste buds of the anterior tongue to the solitary nucleus.

Motor innervation by the facial nerve includes the muscles that close the eyes, move the lips, and produce facial expressions. The facial nerve provides the efferent limb of the corneal reflex. The trigeminal nerve provides the afferent information from the cornea, and the facial nerve activates eyelid closure. Cell bodies for the motor fibers are in the motor nucleus of the facial nerve. The facial nerve also innervates salivary, nasal, and lacrimal (tear-producing) glands. Cell bodies for the preganglionic parasympathetic neurons that innervate the glands are all located in the superior salivary nucleus of the medulla.

> The facial nerve innervates the muscles of facial expression and provides sensory information from the oral region.

CRANIAL NERVE VIII: VESTIBULOCOCHLEAR

Cranial nerve VIII, **vestibulocochlear,** is a sensory nerve with two distinct branches. The vestibular branch transmits information related to head position and head movement. The cochlear branch transmits information related to hearing. The peripheral receptors for these functions are located in the inner ear, in a structure called the labyrinth. The **labyrinth** consists of the vestibular apparatus and the cochlea (Fig. 13–9 on p. 261).

Vestibular Apparatus

The vestibular organs detect position and movement of the head by sensing the motion of fluid inside a fluid-filled vestibular apparatus consisting of three hollow rings (semicircular canals) and two sacs, the utricle and the saccule. The **semicircular canals** are perpendicular to each other (Fig. 13–10A on p. 262). Each semicircular canal has a swelling, called the ampulla, containing sensory hair cells. The hairs are embedded in a gelatinous mass, the cupula. If the head begins to turn, inertia causes the fluid in the canal to lag behind, resulting in bending of the cupula and the ends of the hair cells (Fig. 13–10B). Bending of the hairs results in excitation or inhibition of the vestibular

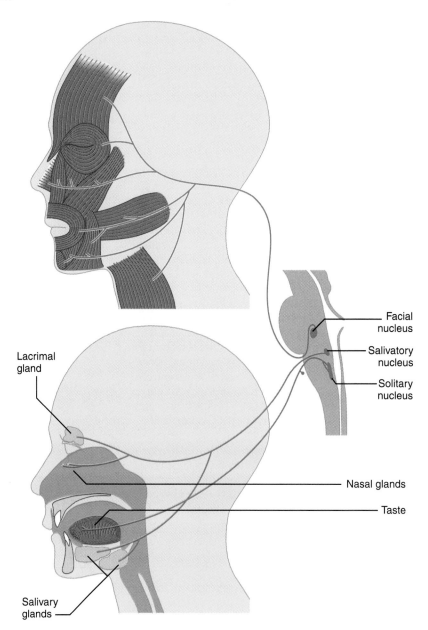

Facial
nucleus

Salivatory
nucleus

Solitary
nucleus

Lacrimal
gland

Nasal glands

Taste

FIGURE 13–8

Facial nerve, supplying innervation to
the muscles of facial expression and
most glands in the head. Facial nerve
also transmits sensory information from
the tongue and pharynx.

Salivary
glands

nerve endings, depending on the direction of bend.
The receptors in semicircular canals are only sensitive
to rotational acceleration or deceleration (i.e., speed-
ing up or slowing down rotation of the head).

The **utricle** and **saccule**, also called the **otolithic
organs,** are not sensitive to rotation but instead respond to head position relative to gravity and to lin-

ear acceleration and deceleration. In each of these
sacs is a macula, consisting of hair cells enclosed by
gelatinous mass topped by calcium salt crystals (Fig.
13–10C). The crystals, called otoliths or otoconia, are
more dense than the surrounding fluid and their
gelatinous support. Changing head position tilts the
macula, and the weight of the otoconia displaces the

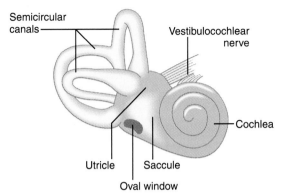

FIGURE 13–9

Vestibulocochlear nerve and the labyrinth of the inner ear.

Labels on figure: Semicircular canals; Vestibulocochlear nerve; Cochlea; Utricle; Saccule; Oval window

gelatinous mass, bending the embedded hair cells. Again, bending the hairs stimulates the hair cells, and this stimulates the vestibular nerve. In addition to head position, the maculae also respond to linear acceleration and deceleration. As the head moves forward, the otoconia in the utricular macula fall back, and the resulting impulses signal head acceleration.

The information from the vestibular apparatus is transmitted by the vestibular nerve to the vestibular nuclei in the medulla and pons and to the cerebellum. The central connections of the vestibular nerve are discussed in Chapter 15.

> Receptors in semicircular canals respond to angular acceleration or deceleration of the head. Receptors in the utricle and saccule encode information about linear acceleration or deceleration and head position relative to gravity.

Cochlea

The **cochlea** is a snail shell–shaped organ, formed by a spiraling, fluid-filled tube (Fig. 13–11A on p. 263). A basilar membrane extends almost the full length of the cochlea, dividing the cochlea into upper and lower chambers. The upper chamber (scala vestibuli) is further divided by a membrane that separates the **cochlear duct** from the remainder of the upper chamber. Within the cochlear duct, resting on the basilar membrane, is the **organ of Corti**, the organ of hearing. The organ of Corti is composed of receptor cells (hair cells), supporting cells, a tectorial membrane, and the

terminals of the cochlear branch of cranial nerve VIII (Fig. 13–11B). The tips of the hairs of the hair cells are embedded in the overlying tectorial membrane.

CONVERTING SOUND TO NEURAL SIGNALS

Sound is converted to neural signals by a sequence of mechanical actions. The tympanic membrane (eardrum), small bones called ossicles, and a membrane at the opening of the upper chamber of the cochlea are connected in series. When sound waves enter the external ear, the vibration of the tympanic membrane moves the ossicles. The ossicles in turn vibrate the membrane at the opening of the upper chamber, moving the fluid contained in the upper chamber. This moves the fluid inside the cochlea, vibrating the basilar membrane and its attached hair cells. Because the tips of the hair cells are embedded in the tectorial membrane, movement of the hair cells bends the hairs. This bending results in excitation of the hair cell and stimulation of the cochlear nerve endings (Fig. 13–11C). The neural signals travel in the cochlear nerve to the cochlear nuclei, located at the junction of the medulla and the pons. The central connections of the auditory system are discussed in Chapter 15.

> The organ of Corti converts mechanical energy into neural signals conveyed by the auditory nerve.

The shape of the basilar membrane is important in coding the frequency of sounds. Because the basilar membrane is narrowest near the middle ear and widest at the free end, the fibers at the free end of the basilar membrane are longer than the fibers at the attached end. The longer fibers vibrate at a lower frequency than the shorter fibers. A low-frequency (low-pitched) sound will cause the longer fibers at the free end to vibrate more than the fibers at the attached end of the membrane. When the free end of the basilar membrane vibrates, the resulting neural signals are eventually perceived as low-pitched sounds.

CRANIAL NERVE IX: GLOSSOPHARYNGEAL

The glossopharyngeal nerve is a mixed nerve containing both sensory and motor fibers. The sensory fibers transmit somatosensation from the soft palate and

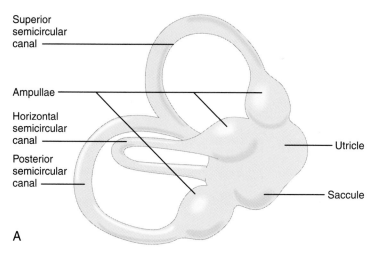

Superior semicircular canal

Ampullae

Horizontal semicircular canal

Posterior semicircular canal

Utricle

Saccule

A

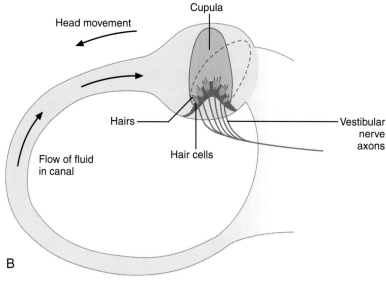

Cupula

Head movement

Hairs

Vestibular nerve axons

Hair cells

Flow of fluid in canal

B

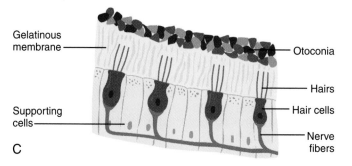

Gelatinous membrane

Otoconia

Hairs

Hair cells

Supporting cells

Nerve fibers

C

FIGURE 13–10

A, The vestibular apparatus consists of the utricle, saccule, and semicircular canals. The three semicircular canals are at right angles to each other. Each semicircular canal has a swelling, the ampulla, that contains a receptor mechanism, the crista. B, A section through a semicircular canal showing the crista inside the ampulla. The flow of fluid in the canal, indicated by the arrow, moves the cupula and in turn bends the hair cells. Bending of the hair cells changes the pattern of firing in the vestibular neurons. C, Inside the utricle and saccule is a receptor called the macula. In the macula, hairs projecting from hair cells are embedded in a gelatinous material. Atop the gelatinous material are small, heavy, sandlike crystals. When the macula is moved into different positions, the weight of the crystals (otoconia) bends the hairs, stimulating the hair cells and changing the pattern of vestibular neuron firing.

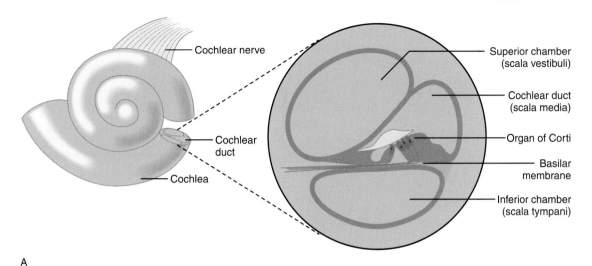

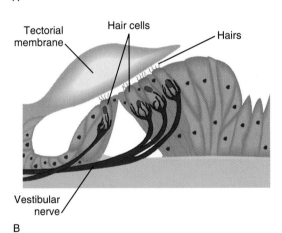

FIGURE 13–11

A, Cochlea with a small section cut away and enlarged to show the fluid-filled spaces inside and the organ of Corti. *B,* Organ of Corti.

Illustration continued on following page

pharynx and information from chemoreceptors in the posterior tongue (Fig. 13–12 on p. 265). The motor component innervates a pharyngeal muscle and the parotid salivary gland.

Glossopharyngeal sensory fibers contribute the afferent limb of the gag reflex, which can be activated by touching the pharynx with a cotton-tipped swab. The information is conveyed to the solitary nucleus located in the dorsal medulla, then by interneurons to

the nucleus ambiguus located in the lateral medulla. Cranial nerve X (see subsequent section) then provides the efferent fibers, causing the pharyngeal muscles to contract.

> The primary function of the glossopharyngeal nerve is to convey somatosensory information from the soft palate and pharynx.

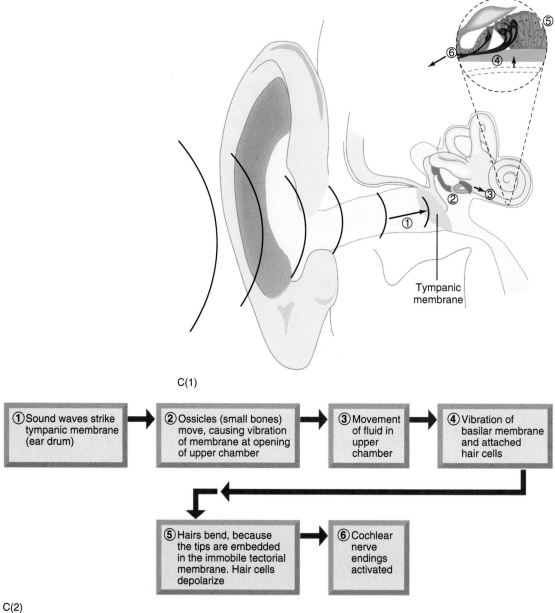

Tympanic
membrane

C(1)

① Sound waves strike tympanic membrane (ear drum)	→	② Ossicles (small bones) move, causing vibration of membrane at opening of upper chamber	→	③ Movement of fluid in upper chamber	→	④ Vibration of basilar membrane and attached hair cells

⑤ Hairs bend, because the tips are embedded in the immobile tectorial membrane. Hair cells depolarize	→	⑥ Cochlear nerve endings activated

C(2)

FIGURE 13–11

Continued C, The conversion of sound waves into neural signals.

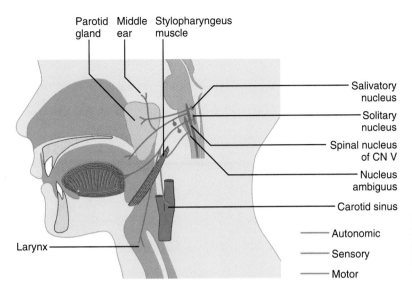

Parotid gland · Middle ear · Stylopharyngeus muscle

Salivatory nucleus

Solitary nucleus

Spinal nucleus of CN V

Nucleus ambiguus

Carotid sinus

Larynx

—— Autonomic

—— Sensory

—— Motor

FIGURE 13–12

The glossopharyngeal nerve provides the afferent limb of the gag and swallowing reflexes, supplies taste information, and innervates a salivary gland.

CRANIAL NERVE X: VAGUS

The vagus nerve provides sensory and motor innervation of the larynx, pharynx, and viscera (Fig. 13–13). Cell bodies of the visceral afferent fibers are located in the inferior nucleus of the vagus, outside the brain stem. Cell bodies of the motor fibers are in the nucleus ambiguus and the dorsal motor nucleus of the vagus, both in the medulla.

Vagal parasympathetic fibers, both sensory and motor, are extensively distributed to the larynx, pharynx, trachea, lungs, heart, gastrointestinal tract (except the lower large intestine), pancreas, gallbladder, and liver. These far-reaching connections allow the vagus to decrease heart rate, constrict the bronchi, affect speech production, and increase digestive activity. The motor function of the vagus nerve can be tested by eliciting the gag reflex, discussed in the previous section.

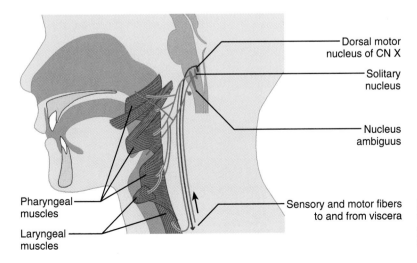

Dorsal motor nucleus of CN X

Solitary nucleus

Nucleus ambiguus

Pharyngeal muscles

Laryngeal muscles

Sensory and motor fibers to and from viscera

FIGURE 13–13

The vagus nerve regulates viscera, swallowing, and speech and supplies taste information. CN, cranial nerve.

CRANIAL NERVE XI: ACCESSORY

The accessory nerve is motor, providing innervation to the trapezius and sternocleidomastoid muscles. The accessory nerve (Fig. 13–14) originates in the upper cervical cord, travels upward through the foramen magnum, then leaves the skull through another foramen. The cell bodies are in the ventral horn at levels C1 to C4.

CRANIAL NERVE XII: HYPOGLOSSAL

The hypoglossal nerve is motor, providing innervation to the intrinsic and extrinsic muscles of the ipsilateral tongue (Fig. 13–15). Cell bodies are located in the hypoglossal nucleus of the medulla. Activity of the hypoglossal nerve is controlled by both voluntary and reflexive neural circuits.

CRANIAL NERVES INVOLVED IN SWALLOWING AND SPEAKING

Swallowing

Swallowing involves three stages: **oral, pharyngeal/laryngeal**, and **esophageal**. Table 13–3 indicates the participation of the cranial nerves in each stage.

Speaking

Speaking requires cortical control, which will be discussed in Chapter 16. At the cranial nerve level, sounds generated by the larynx (cranial nerve X) are articulated by the soft palate (cranial nerve X), lips (cranial nerve VII), jaws (cranial nerve V), and tongue (cranial nerve XII).

SYSTEMS CONTROLLING CRANIAL NERVE LOWER MOTONEURONS

Cranial nerves III to VII and IX to XII contain lower motoneuron fibers. The activity of the motor fibers in cranial nerves is controlled via descending inputs from voluntary and limbic structures of the brain stem and cerebrum and also via local reflex mechanisms.

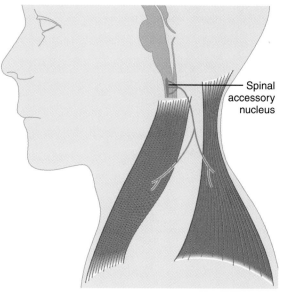

FIGURE 13–14

The accessory nerve innervates the sternocleidomastoid and trapezius muscles.

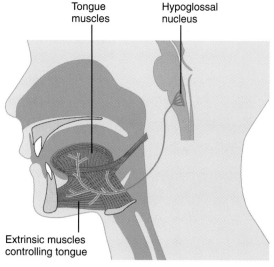

FIGURE 13–15

The hypoglossal nerve innervates muscles of the tongue.

TABLE 13–3
PHASES OF SWALLOWING

Stage	Description	Cranial Nerve
Oral	Food in mouth, lips close	VII
	Jaw, cheek, and tongue movements manipulate food	V, VII, XII
	Tongue moves food to pharynx entrance	XII
	Larynx closes	X
	Swallow reflex triggered	IX
Pharyngeal/laryngeal	Food moves into pharynx	IX
	Soft palate rises to block food from nasal cavity	X
	Epiglottis covers trachea to prevent food from entering lungs	X
	Peristalsis moves food to entrance of esophagus, sphincter opens, food moves into esophagus	X
Esophageal phase	Peristalsis moves food into stomach	X

Descending Control of Motor Cranial Nerves

Like the lower motoneurons in the spinal cord, the cranial nerve efferents receive descending regulation by the corticofugal tracts and the limbic system. Thus their activity can be affected by voluntary, emotional, or as mentioned previously for individual nerves, by reflexive pathways. Descending limbic pathways are separate from the corticobulbar tracts (Holstege, 1991).

An example of the dissociation of limbic and voluntary controlled movements is the facial nerve activity that produces a spontaneous smile, a result of limbic innervation and an expression of true emotion, versus an insincere smile that is produced voluntarily and can usually be detected. The facial expressions associated with powerful emotions are difficult to suppress voluntarily, but the same expressions may be difficult to produce intentionally.

Similarly, eye movements can be voluntarily controlled, or the eyes may be automatically drawn toward or avoid disturbing sights. Speaking is mainly voluntary but can occur automatically in highly emotional contexts. In some instances where brain damage interferes with voluntary speech, the ability of the limbic system to produce emotionally charged words, such as profanity, may be preserved. Extreme emotions, by activating limbic pathways that influence motor activity, can interfere with the ability to eat and speak.

DISORDERS OF CRANIAL NERVES

Olfactory Nerve

Lesions of the olfactory nerve can result in an inability to smell. However, smoking or excessive nasal mucus may also interfere with the function of the olfactory nerve.

Optic Nerve

A complete interruption of the optic nerve results in ipsilateral blindness and loss of the pupillary light reflex. The pupillary light reflex is pupil constriction in response to a light shining into the client's eye. Loss of the pupillary light reflex could also occur with a lesion of cranial nerve III because the oculomotor nerve is the efferent limb of the reflex. Lesions at other sites in the visual pathway can also cause blindness (see Chap. 15).

Oculomotor Nerve

A complete lesion of the oculomotor nerve causes the following deficits:

- The ipsilateral eye looks outward and down because the actions of the lateral rectus and the superior oblique muscles are unopposed.

- Double vision, due to the difference in position of the eyes. Because the eyes do not look in the same direction, the light rays from objects do not fall on corresponding areas of both retinas, producing double vision.
- Deficits in moving the ipsilateral eye medially, downward, and upward.
- Drooping of the eyelid, due to paralysis of the voluntary muscle fibers that elevate the eyelid. The autonomic muscle fibers are able to keep the eyelid partially elevated.
- Loss of pupillary reflex and consensual response to light.
- Loss of constriction of the pupil in response to focusing on a near object.

The eye movement deficits seen in an oculomotor nerve lesion must be differentiated from the asymmetrical eye movements that occur with upper motoneuron lesions or medial longitudinal fasciculus lesions. The reflexes producing pupillary constriction will be spared in upper motoneuron or medial longitudinal fasciculus lesions. Either of these disorders will be accompanied by more extensive brain stem or cerebral signs and symptoms (see Chap. 14).

Trochlear Nerve

A lesion of the trochlear nerve prevents activation of the superior oblique muscle, so the ipsilateral eye cannot look downward and inward. People with lesions of the trochlear nerve complain of double vision, difficulty reading, and visual problems when descending stairs. Other possible causes of eye movement asymmetry must be ruled out, as discussed for the oculomotor nerve.

Trigeminal Nerve

A complete severance of a branch of the trigeminal nerve results in anesthesia of the area supplied by the ophthalmic, maxillary, or mandibular branch. If the ophthalmic division is affected, the afferent limb of the blink reflex will be interrupted, preventing blinking in response to touch stimulation of the cornea. If the mandibular branch is completely severed, the jaw will deviate toward the involved side when the mouth is opened, and the masseter reflex will be lost.

TRIGEMINAL NEURALGIA

Trigeminal neuralgia (also known as tic douloureux) is a dysfunction of the trigeminal nerve, producing severe, sharp, stabbing pain in the distribution of one or more branches of the trigeminal nerve. (See the trigeminal neuralgia box.) Several etiologies have been proposed:

- A peripheral lesion leading to segmental demyelination of the nerve, ephaptic transmission, and subsequent secondary changes in the spinal trigeminal nucleus (Young, 1990)
- Hyperexcitability of damaged fibers in the trigeminal root ganglia, leading to spontaneous neuronal firing (Rappaport and Devor, 1994)
- Pressure of a blood vessel on the nerve

The pain is triggered by stimuli that are normally not noxious, such as eating, talking, or touching the face. The pain begins and ends abruptly, lasts less than 2 minutes, and is not associated with sensory loss. Trigeminal neuralgia can be treated effectively by drugs or surgery (Mauskop, 1993).

Abducens Nerve

A complete lesion of the abducens nerve will cause the eye to look inward because the paralysis of the lateral rectus muscle leaves the pull of the medial rectus muscle unopposed. A person with this lesion will be unable to voluntarily abduct the eye and will have double vision. Other causes of asymmetrical eye movements must be ruled out, as discussed regarding the oculomotor nerve.

Facial Nerve

A lesion of the facial nerve causes paralysis or paresis of the ipsilateral muscles of facial expression. This causes one side of the face to droop and prevents the person from being able to completely close the ipsilateral eye. Unilateral facial palsy can result from a lesion of the cranial nerve VII nucleus or from a lesion of the axons of cranial nerve VII. If the lesion involves the axons, the disorder is called **Bell's palsy**. (See the Bell's palsy box.)

TRIGEMINAL NEURALGIA (TIC DOULOUREUX)

Pathology
Unknown

Etiology
Unknown; hypotheses include segmental demyelination with subsequent changes in the spinal trigeminal nucleus, damage of trigeminal root ganglia, and compression of the nerve branch by a blood vessel

Speed of onset
Abrupt

Signs and symptoms

Consciousness
Normal

Communication and memory
Normal

Sensory
Normal except for sharp, severe pains that last less than 2 minutes, usually only in one branch of the trigeminal nerve distribution and typically triggered by chewing, talking, brushing the teeth, or shaving

Autonomic
Normal

Motor
Normal

Region affected
Peripheral part of cranial nerve V; may also involve spinal nucleus of cranial nerve V

Demographics
Women are 1.5 times more likely than men to have trigeminal neuralgia; mean age at onset is 55 years

Prognosis
Variable; may resolve spontaneously after a few bouts, may recur, or may require medication or surgery to decompress the nerve

BELL'S PALSY

Pathology
Paralysis of the muscles innervated by the facial nerve (cranial nerve VII) on one side of the face

Etiology
Unknown; hypotheses include viral infection or immune disorder causing swelling of facial nerve within the temporal bone, resulting in compression and ischemia of the nerve

Speed of onset
Acute

Signs and symptoms

Consciousness
Normal

Communication and memory
Normal

Sensory
Normal

Autonomic
In severe cases, salivation and production of tears may be affected

Motor
Paresis or paralysis of entire half of face, including frontalis and orbicularis oculi muscles; in severe cases, the ipsilateral eye cannot be closed

Region affected
Peripheral part of cranial nerve VII

Demographics
Men and women affected equally; usually affects older adults

Prognosis
80% recover neural control of facial muscles within 2 months; recovery depends on severity of damage, which can be assessed by nerve conduction velocity and electromyography; paresis is typically followed by complete recovery; outcome after complete paralysis varies from complete recovery to permanent paralysis

Vestibulocochlear Nerve

Bilateral lesions of the vestibular branch of cranial nerve VIII interfere with reflexive eye movements in response to head movement. People with a vestibular nerve lesion initially complain of the world seeming to bounce up and down as they walk because normal reflexive adjustments for head movement are decreased. Over time, the nervous system adapts to the change, and people report less difficulty with disorienting movements of the visual field. A complete lesion of the cochlear branch of cranial nerve VIII causes unilateral deafness. Vestibular and auditory system dysfunctions are discussed more fully in Chapter 15.

Glossopharyngeal Nerve

A complete lesion of cranial nerve IX interrupts the afferent limb of both the gag reflex and the swallowing reflex (cranial nerve X provides the efferent limb for both reflexes). Salivation is also decreased.

Vagus Nerve

A complete lesion of the vagus nerve results in difficulty speaking, swallowing, poor digestion due to decreased digestive enzymes and decreased peristalsis, asymmetrical elevation of the palate, and hoarseness.

Accessory Nerve

A complete lesion of the accessory nerve paralyzes the ipsilateral sternocleidomastoid and trapezius muscles. Upper motoneuron lesions, in contrast, cause paresis rather than paralysis because cortical innervation is bilateral, and the muscles become hypertonic rather than hypotonic.

Hypoglossal Nerve

A complete lesion of the hypoglossal nerve causes atrophy of the ipsilateral tongue. When a person with this lesion is asked to stick out the tongue, the tongue protrudes ipsilaterally rather than in the midline. The problems with tongue control result in difficulty speaking and swallowing.

Swallowing Dysfunctions

Frequent choking, lack of awareness of food in one side of the mouth, or food coming out of the nose may indicate dysfunctions of cranial nerves V, VII, IX, X, and/or XII. Upper motoneuron lesions may also cause swallowing dysfunctions.

Dysarthria

Poor control of the speech muscles is **dysarthria**. In dysarthria, only vocal speech is affected, that is, motor production of sounds. People with dysarthria can understand spoken language, write, and read. Lower motoneuron involvement of cranial nerves V, VII, X, or XII can cause dysarthria. Dysarthria can also result from upper motoneuron lesions or muscle dysfunction.

TESTING CRANIAL NERVES

Standard tests are used to assess cranial nerve function. Table 13–4 lists the cranial nerve tests and the effects of lesions of individual cranial nerves.

| TABLE 13-4 |
| CRANIAL NERVE TESTS |

Nerve	Test	Normal Response	Cranial Nerve Lesion	Differentiate From
Olfactory	Patient closes eyes, closes one nostril, then smells coffee or cloves.	Patient identifies substance.	Lack of ability to smell; however, mucus or smoking may interfere.	
Optic	With one eye covered, patient looks in examiner's eye. Examiner moves a finger in from outside the patient's visual field (above, below, left, and right) until patient reports seeing the finger.	Patient reports seeing finger.	If optic nerve is completely interrupted, patient is ipsilaterally blind.	Lesions at other sites in the visual pathway also interfere with vision (see Chap. 15).
	Shine light into subject's eye. Observe pupil (pupillary reflex).	Pupil constricts (cranial nerve III provides the reflex efferent limb).	Response is slow or absent.	
Oculomotor	Observe position of patient's eyes with patient looking forward.	Both eyes appear to look in same direction; no nystagmus.*	Ipsilateral eye looks outward and down (pulled by unopposed lateral rectus and superior oblique). Patient reports double vision.	
	Observe size of pupil in room light.	Moderate size.	Dilated pupil.	Extremely small pupil: sympathetic dysfunction (Horner's syndrome [see Chap. 8]).
	Patient's eyes follow examiner's finger, moving eyes up, down, and in (testing superior, medial, and inferior rectus muscles).	Eyes move symmetrically, smoothly.	Deficits in adduction, depression, or elevation of the eye.	Asymmetrical eye movements; due to upper motoneuron or medial longitudinal fasciculus lesion.
	Patient's eye follows examiner's finger to about 50° adduction, then up (testing inferior oblique).	Eye follows finger movement.	Unable to adduct and elevate eye.	Asymmetrical eye movements; due to upper motoneuron or medial longitudinal fasciculus lesion.
	Shine light into subject's eye (pupillary reflex and consensual response to light).	Constriction of pupils (requires cranial nerve II for afferent limb of reflex).	Pupil unchanged.	
	Patient looks at examiner's finger, then examiner's nose. Observe pupillary response to near and far objects.	Near object: constriction. Far object: dilation.	Pupil unchanged.	
Trochlear	Patient's eye follows examiner's finger to about 50° adduction, then down (testing superior oblique).	Eye moves in, then down.	Deficit in looking inferomedially. Patient reports double vision, difficulty reading, descending stairs.	Asymmetrical eye movements; due to upper motoneuron or medial longitudinal fasciculus lesion.

(continued)

271

TABLE 13–4
CRANIAL NERVE TESTS *Continued*

Nerve	Test	Normal Response	Cranial Nerve Lesion	Differentiate From
Trigeminal	Use light touch and pin to assess facial sensation in three areas: forehead, cheek, chin.	Distinguishes between sharp and dull and can localize stimulus.	Anesthesia in affected area; or patient reports severe pain in trigeminal branch distribution (trigeminal neuralgia, a severe neuropathic pain).	
	Touch outer cornea with a wisp of cotton (corneal reflex).	Eye blinks (requires efferent limb via cranial nerve VII).	Eye does not close (because response is bilateral, can stimulate other eye to determine if absence of reflex is due to problem with afferent or efferent limb).	
	Manual muscle test: jaw opening strength.	Jaw opens strongly and symmetrically.	Unilateral damage: jaw deviates toward weak side.	
	Tap downward on patient's chin with reflex hammer.	Masseter contracts, elevating chin.	Lost or decreased reflex.	Hyperreflexia; due to upper motoneuron lesion.
Abducens	Observe position of eyes with patient looking forward.	Both eyes appear to look in same direction; no nystagmus.	One eye looks inward (pulled by unopposed medial rectus); patient reports double vision.	
	Patient follows examiner's finger to look laterally.	Eye moves laterally.	Deficit of abduction.	Asymmetrical eye movements; due to upper motoneuron or medial longitudinal fasciculus lesion.
Facial	Facial movements: smile, puff cheeks, close eyes, wrinkle forehead.	Able to perform requested movements.	Paralysis or paresis, with upper and lower face equally involved: cranial nerve VII nucleus or Bell's palsy. Bell's palsy affects axons of facial nerve, preventing patient from completely closing ipsilateral eye.	Lower face paralyzed, forehead spared: upper motoneuron lesion.
Vestibulocochlear	Vestibular function: not tested unless patient reports vertigo or dysequilibrium. Slowly turn patient's head side to side or up and down (vestibulo-ocular reflex).	Eyes move in direction opposite the head movement.	Absence of adequate compensatory eye movement.	

Cranial Nerve	Test	Normal	Abnormal	Interpretation
	Past pointing: patient touches examiner's finger, closes eyes, raises arm overhead, touches examiner's finger again.	Contacts examiner's finger.	Misses examiner's finger.	Incoordination due to cerebellar or spinocerebellar tract dysfunction
	Caloric stimulation of vestibular system (not typically performed by therapists): cold or warm water (30° or 40°C) poured into ear. Allows unilateral testing of vestibular function.	Nystagmus, due to cooling or warming of fluid in semicircular canal causing fluid flow followed by vestibulo-ocular reflex	Absence of nystagmus.	
	Auditory function: examiner rubs fingers together near patient's ear; performance can be compared with examiner's own hearing.	Patient reports hearing the stimulus equally in each ear.	Difference in acuity of patient's ears or in patient's and examiner's ability to hear should be investigated further.	
	Auditory function: hold vibrating tuning fork on mastoid bone; when patient no longer hears it, move tuning fork into the air about 1 inch from ear canal.	Patient hears through air after cannot hear through bone.	Patient hears through air after cannot hear through bone, but volume is reduced for both air and bone conduction (sensorineural hearing loss).	Hearing longer through bone indicates conduction loss (lesion in auditory canal or middle ear).
Glossopharyngeal	Touch soft palate with cotton swab (gag reflex).	Gagging and symmetrical elevation of soft palate; requires cranial nerve X efferents.	Lack of gag reflex, or asymmetrical elevation of palate.	
Vagus	Patient opens mouth and says "ah." Examiner observes soft palate.	Elevation of soft palate.	Asymmetrical elevation of palate; hoarseness.	
Accessory	Manual muscle test: sternocleidomastoids and upper trapezius.	Normal strength.	Paralysis or paresis.	Upper motoneuron lesion: paresis with hypertonia; bilateral cortical innervation prevents complete paralysis.
Hypoglossal	Patient protrudes tongue.	Tongue protrudes in midline.	Protruded tongue deviates to the side of the lesion, and ipsilateral tongue atrophies.	

*Involuntary back-and-forth movement of the eyes.

CASE 1

R.F. is a 62-year-old man who was involved in a car accident 5 days ago. He sustained fractures of the skull, both femurs, and the right tibia. He complains of double vision.

- Consciousness, cognition, language, memory, and somatosensation are normal.

- All autonomic functions, including pupillary and accommodation reflexes, are normal.

- Motor function is normal except for an inability to look downward and inward with the right eye. All other eye movements, including the ability to look medially and laterally with the right eye, are normal. When asked, he says he has been having trouble reading since the accident.

Question

What is the most likely location of the lesion?

CASE 2

A.K., a 46-year-old engineer, is complaining of double vision. She cannot read or drive unless she closes one eye. The following are findings on the cranial nerve examination:

- Olfaction, vision, facial sensation, control of the muscles of facial expression, mastication, hearing, equilibrium, gag and swallowing reflexes, and contraction of the sternocleidomastoid, trapezius, and tongue muscles are normal.

- When A.K. is instructed to look straight ahead, her right eye looks outward and down.

- A.K. cannot look in, down, or up with her right eye.

- A.K. can only open the right eyelid halfway.

- No pupillary reflex or consensual response occurs when a flashlight is shown into her right eye, nor does the pupil constrict when she focuses on an object 6 inches from her right eye.

Questions

1. Is this an upper motoneuron lesion? Why or why not?

2. Where is the lesion?

CASE 3

M.R. awoke with inability to move the left side of his face. His neurological examination was normal, including facial sensation and control of the muscles of mastication and the tongue, except for the following:

- Drooping of the left side of the face.

- Complete lack of movement of the muscles of facial expression on the left. Examples include M.R.'s inability to voluntarily smile, pucker his lips, or raise his eyebrow on the left side. Emotional facial expressions were also absent; when he smiled with the right side of his lips, the left half of his lips did not move.

- Inability to completely close his left eye.

Question

What is the most likely location of the lesion?

REVIEW QUESTIONS

1. How are the eyes voluntarily moved to follow an examiner's finger to the right and then look up? Include both muscle and cranial nerve activity in the answer.

2. Which cranial nerve provides efferents for the pupillary reflex?

3. Which cranial nerve provides efferents to the tongue muscles?

4. Which cranial nerve provides for afferents for the gag reflex?

5. Which cranial nerve provides control of the muscles of facial expression?

6. Which cranial nerve provides somatosensation from the face?

7. Diagram the accommodation reflex.

8. How does rotational acceleration of the head stimulate endings of the vestibular nerve?

9. Describe the mechanism for converting head position relative to gravity into neural signals.

10. What is the function of the organ of Corti?

11. Why is there a difference between an authentic smile and a false smile?

12. Which cranial nerves are required for swallowing?

13. Lesions of what structures could cause double vision?

References

Holstege, G. (1991). Descending motor pathways and the spinal motor system: Limbic and non-limbic components. Prog. Brain Res. 87:307–421.

Mauskop, A. (1993). Trigeminal neuralgia (tic douloureaux). J. Pain Symptom Manage. 8(3):148–154.

Rappaport, A. H., and Devor, M. (1994). Trigeminal neuralgia: The role of self-sustaining discharge in the trigeminal ganglion. Pain 56:127–138.

Young, R. F. (1990). The trigeminal nerve and its central pathways. In: Rovit, R. L., Muali, R., and Janneta, P. J. (Eds.). Trigeminal neuralgia. Baltimore: Williams & Wilkins, pp. 27–51.

14

Brain Stem Region

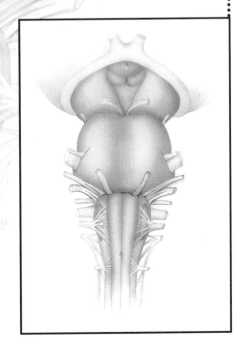

INTRODUCTION

The brain stem is superior to the spinal cord and inferior to the cerebrum, with the cerebellum appended posteriorly. From inferior to superior, the parts of the brain stem are the medulla, pons, and midbrain (Fig. 14–1). The locations of cranial nerve nuclei in the brain stem are shown in Figure 14–2.

ANATOMY OF THE BRAIN STEM

Vertical Tracts in the Brain Stem

Sensory, autonomic, and motor vertical tracts travel through the brain stem, just as in the spinal cord. The sensory tracts conveying information from the spinal cord to the brain, and the motor tracts conveying signals from the cortex to the brain stem and spinal

cord, have been discussed in Chapters 6 through 10. Some of these tracts continue through the brain stem without alteration. For these tracts, the brain stem acts as a conduit. Other vertical tracts leave the brain stem or synapse in brain stem nuclei. Modifications of the vertical tracts in the brain stem are summarized in Table 14–1 and illustrated in Figure 14–3.

Vertical tracts that originate in the brain stem and project to the spinal cord are the tecto-, rubro-, reticulo-, vestibulo-, ceruleo-, and raphespinal. The origins and functions of these tracts are discussed in Chapter 9.

Longitudinal Sections of the Brain Stem

The brain stem is divided longitudinally into two sections: the basilar section and the tegmentum. Throughout the brain stem, the **basilar section** is

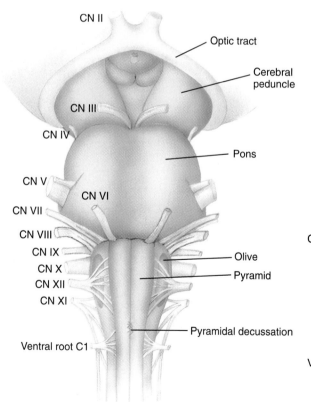

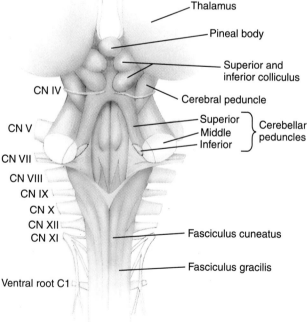

FIGURE 14–1

Anterior and posterior views of the brain stem.

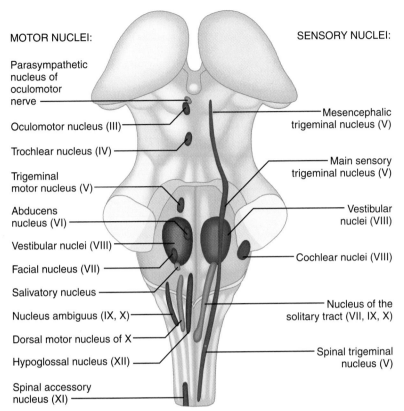

MOTOR NUCLEI:

Parasympathetic nucleus of oculomotor nerve

Oculomotor nucleus (III)

Trochlear nucleus (IV)

Trigeminal motor nucleus (V)

Abducens nucleus (VI)

Vestibular nuclei (VIII)

Facial nucleus (VII)

Salivatory nucleus

Nucleus ambiguus (IX, X)

Dorsal motor nucleus of X

Hypoglossal nucleus (XII)

Spinal accessory nucleus (XI)

SENSORY NUCLEI:

Mesencephalic trigeminal nucleus (V)

Main sensory trigeminal nucleus (V)

Vestibular nuclei (VIII)

Cochlear nuclei (VIII)

Nucleus of the solitary tract (VII, IX, X)

Spinal trigeminal nucleus (V)

FIGURE 14–2

The locations of cranial nerve nuclei inside the brain stem. Motor nuclei are indicated on the left and sensory nuclei on the right.

TABLE 14–1	
VERTICAL TRACTS IN THE BRAIN STEM	
Vertical Tract	**Modification of Tract in Brain Stem**
Sensory (ascending) tracts	
Neospinothalamic	Not modified (tract passes through brain stem without alteration)
Dorsal column	Axons synapse in nucleus gracilis or cuneatus; second-order neurons cross midline to form medial lemniscus
Spinocerebellar	Axons leave the brain stem via the inferior and superior cerebellar peduncles to enter the cerebellum
Autonomic (descending) tracts	
Sympathetic	Not modified (tract passes through brain stem without alteration)
Parasympathetic	Axons synapse with brain stem parasympathetic nuclei or continue through brain stem and cord to the sacral level of spinal cord
Motor (descending) tracts	
Corticospinal	Not modified (tract passes through brain stem without alteration)
Corticobulbar	Axons synapse with cranial nerve nuclei in brain stem
Corticopontine	Axons synapse with nuclei in the pons
Corticoreticular	Axons synapse within the reticular formation

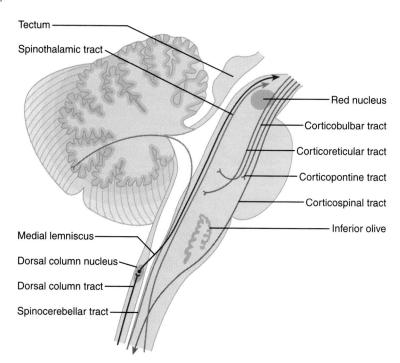

Tectum

Spinothalamic tract

Red nucleus

Corticobulbar tract

Corticoreticular tract

Corticopontine tract

Corticospinal tract

Inferior olive

Medial lemniscus

Dorsal column nucleus

Dorsal column tract

Spinocerebellar tract

FIGURE 14–3

Vertical pathways in the brain stem. For simplicity, the autonomic pathways are omitted.

located anteriorly and contains predominantly motor system structures (see Chap. 10):

- Descending axons from the cerebral cortex: corticospinal, corticobulbar, corticopontine, and corticoreticular tracts
- Motor nuclei: the substantia nigra, pontine nuclei, and inferior olive
- Pontocerebellar axons

The **tegmentum,** located posteriorly, includes the following:

- The reticular formation, which adjusts the general level of activity throughout the nervous system
- Sensory nuclei and ascending sensory tracts (see Chaps. 6 and 7)
- Cranial nerve nuclei (discussed later in this chapter)
- The medial longitudinal fasciculus, a tract that coordinates eye and head movements

In addition to basilar and tegmentum sections, the midbrain has an additional longitudinal section, posterior to the tegmentum, called the **tectum.** The tectum includes structures involved in reflexive control

of intrinsic and extrinsic eye muscles and in movements of the head:

- Pretectal area
- Superior and inferior colliculi

> The longitudinal sections of the brain stem are the basilar, tegmentum, and in the midbrain, the tectum. The basilar section is primarily motor. The tegmentum is involved in adjusting the general level of neural activity, integrating sensory information, and cranial nerve functions. The tectum regulates eye reflexes and reflexive head movements.

The structures listed above are discussed in the context of their location in the medulla, pons, or midbrain. Because the reticular formation extends vertically throughout the brain stem, it is discussed next.

RETICULAR FORMATION

The reticular formation is a complex neural network including the reticular nuclei, their connections, and ascending and descending reticular pathways. The reticular formation can be divided into three zones: lateral, medial, and midline (Fig. 14–4), each with distinct functions.

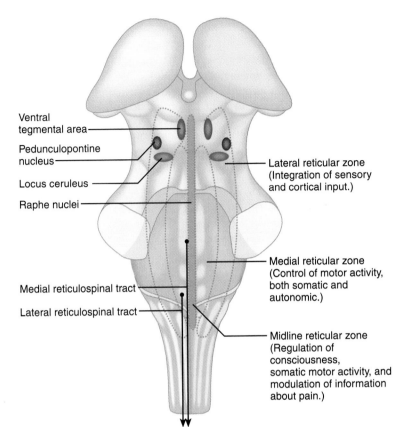

Ventral tegmental area

Pedunculopontine nucleus

Locus ceruleus

Raphe nuclei

Lateral reticular zone
(Integration of sensory
and cortical input.)

Medial reticulospinal tract

Lateral reticulospinal tract

Medial reticular zone
(Control of motor activity,
both somatic and
autonomic.)

Midline reticular zone
(Regulation of
consciousness,
somatic motor activity, and
modulation of information
about pain.)

FIGURE 14–4

Zones and nuclei of the reticular formation.
The projections of the reticular nuclei are illustrated in Figure 14–5.

Lateral Zone: Sensory and Cortical Input, Generalized Arousal

All sensory information (somatosensation, vision, hearing, vestibular, taste, and smell) and cortical input into the reticular formation are integrated by the lateral zone. This information is then conveyed to the medial and midline regions. In addition, axons from nuclei in the lateral reticular zone project to the cerebrum, cerebellum, and spinal cord, adjusting neural activity throughout the central nervous system.

Medial Zone: Vital Functions, Somatic Motor Activity

Vital functions required to sustain life are regulated by numerous autonomic nuclei in the medial reticular formation. Networks of brain stem neurons regulate cardiovascular function, respiration, swallowing, and vomiting. These networks operate by projections to cranial nerve nuclei, to spinal lower motoneurons, and to preganglionic sympathetic neurons. Two reticulospinal tracts also originate in the medial zone: the medial and lateral reticulospinal tracts. The medial reticulospinal tract is important in postural control and facilitates lower motoneurons to limb extensors. The lateral reticulospinal tract also contributes to movement, but the effects of activation are dependent on the movement. In general, lateral tract activity facilitates flexion and inhibits extension.

Midline Zone: Analgesia, Somatic Motor Activity, Awareness

The flow of information about pain, the general level of motoneuron activity, and the level of awareness are regulated by groups of neurons located in the midline of the brain stem. Activity of various nuclei in the midline zone results in the following:

- Analgesia by activating inhibitory neurons in the dorsal horn, or
- Increased background activity in the spinal cord (via the raphespinal tract, enhancing the activity of interneurons and motoneurons in the spinal cord), or
- Drowsiness, followed by initiation of sleep

Because the raphe nuclei are associated with causing drowsiness, lesions of these nuclei can cause insomnia.

> The lateral zone of the reticular formation integrates sensory and cortical input and produces generalized arousal. The medial zone regulates vital functions, somatic motor activity, and attention. The midline zone adjusts the transmission of pain information, somatic motor activity, and consciousness levels.

RETICULAR NUCLEI AND THEIR NEUROTRANSMITTERS

Reticular nuclei regulate neural activity throughout the central nervous system. Neurons in each nucleus produce a different neuromodulator. Neuromodulators are chemicals that alter neurotransmitter release or the response of receptors to neurotransmitters. These neuromodulators markedly influence activity in other parts of the brain stem and in the cerebrum and cerebellum. Several also influence neural activity in the spinal cord.

Although the reticular nuclei are confined to small regions in the brain stem, their axons project to widespread areas of the brain and, in some cases, to the spinal cord. The major reticular nuclei are as follows:

- Ventral tegmental area
- Pedunculopontine nucleus
- Raphe nuclei
- Locus ceruleus and medial reticular area

Ventral Tegmental Area: Dopamine

Most neurons that produce dopamine are located in the midbrain. Of the two midbrain areas that produce dopamine, only one, the ventral tegmental area, is part of the reticular formation. The other dopamine-producing area is the substantia nigra, discussed in Chapter 9 as part of the basal ganglia circuit that supplies dopamine to the caudate and putamen. The ventral tegmental area provides dopamine to cerebral areas important in motivation and in decision making (Fig. 14–5A). The powerful effect of ventral tegmental area activity is demonstrated in addiction to amphetamines and cocaine. Both drugs activate the ventral tegmental area dopamine system. Morphine is habit forming because it inhibits inhibitory inputs to the ventral tegmental area, thus increasing dopamine release. Dopamine antagonists can cause addicts to cease self-administering amphetamines, cocaine, and heroin. Excessive ventral tegmental area activity has been hypothesized to explain certain aspects of schizophrenia, because drugs that block a particular type of dopamine receptor (D_2) have antipsychotic effects. Schizophrenia is a disorder of perception and thought processes, characterized by withdrawal from the outside world.

Pedunculopontine Nucleus: Acetylcholine

The pedunculopontine nucleus is located in the caudal midbrain (Fig. 14–5B). Ascending axons from the pedunculopontine nucleus project to the inferior part of the frontal cerebral cortex and the intralaminar nuclei of the thalamus. The pedunculopontine nucleus influences movement via connections with the following:

- Globus pallidus and subthalamic nucleus
- Vestibular nuclei
- Reticular areas that give rise to the reticulospinal tracts

In cats that have a lesion separating the brain stem from the cerebrum, electrical stimulation to the pedunculopontine nucleus can induce walking despite the lack of cerebral connection with the spinal cord.

Raphe Nuclei: Serotonin

Most cells that produce serotonin are found along the midline of the brain stem, in the raphe nuclei (Fig. 14–5C). The midbrain raphe nuclei project throughout the cerebrum. Activity in these serotonergic projections is believed to play a role in the onset of sleep.

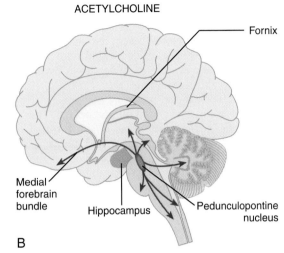

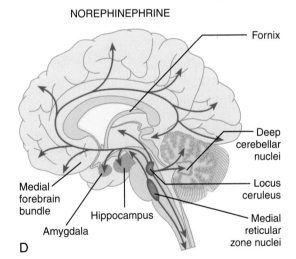

FIGURE 14–5

Neuromodulators are produced in the brain stem by reticular nuclei. The ascending fibers from the reticular nuclei form the reticular activating system, which regulates activity in the cerebral cortex. Descending fibers adjust the general level of activity in the spinal cord. *A,* The ventral tegmental area supplies dopamine to the frontal cortex and limbic areas. *B,* The pedunculopontine nucleus provides acetylcholine to the thalamus, frontal cerebral cortex, the brain stem and cerebellum, and the spinal cord (ceruleospinal tract). *C,* The raphe nuclei supply serotonin to the thalamus, midbrain tectum, striatum, amygdala, hippocampus, cerebellum, throughout the cerebral cortex, and to the spinal cord (raphespinal tract). *D,* The locus ceruleus and medial reticular zone nuclei provide norepinephrine in a wide distribution similar to the pattern of serotonin distribution. The tracts descending into the spinal cord are the reticulospinal tracts.

Activity in the serotonergic projections then lessens during sleep and disappears during rapid eye movement (REM) sleep. Serotonin levels also have profound effects on mood. The antidepressant fluoxetine (Prozac) prolongs the availability of serotonin by inhibiting the reuptake of serotonin.

The pontine raphe nuclei modulate neural activity throughout the brain stem and in the cerebellum. The medullary raphe nuclei send axons into the spinal cord to modulate sensory, autonomic, and motor activity. Some medullary raphe nuclei are part of the fast-acting neuronal pathway for descending pain inhibition (see Fig. 7–8). Ascending pain information stimulates both the periaqueductal gray and the medullary raphe nuclei. In response, axons from the medullary raphe nuclei release serotonin onto interneurons in the dorsal horn that inhibit the transmission of pain information (see Chap. 7). Raphespinal endings in the lateral horn influence the cardiovascular system. Raphespinal endings in the anterior horn provide nonspecific activation of interneurons and lower motoneurons (see Chap. 9).

Locus Ceruleus and Medial Reticular Zone: Norepinephrine

The locus ceruleus and medial reticular zone are the source of most norepinephrine in the central nervous system (Fig. 14–5D). Axons from the locus ceruleus project throughout the brain and spinal cord. The locus ceruleus is inactive during sleep and most active in situations where watchfulness is essential. Activity of the ascending axons from the locus ceruleus provides the ability to direct attention. Descending axons from the locus ceruleus form the ceruleospinal tract, providing nonspecific activation of interneurons and motoneurons in the spinal cord. Ceruleospinal endings in the dorsal horn provide direct inhibition of spinothalamic neurons conveying pain information.

The medial reticular zone produces both norepinephrine and epinephrine. It regulates autonomic functions—respiratory, visceral, and cardiovascular—by projections to the hypothalamus, brain stem nuclei, and lateral horn of the spinal cord.

> Arousal levels in the cerebrum are influenced by the raphe nuclei, and attention is directed by the locus ceruleus. Axons from the locus ceruleus and the raphe nuclei determine the general level of neuronal activity in the spinal cord.

Regulation of Consciousness by the Ascending Reticular Activating System

Consciousness is the awareness of self and surroundings. The **consciousness system** governs alertness, sleep, and attention. Brain stem components of the consciousness system are the reticular formation and its ascending reticular activating system. The axons of the ascending reticular activating system project to the cerebral components of the consciousness system: basal forebrain (anterior to the hypothalamus), thalamus, and cerebral cortex. For normal sleep-wake cycles and the ability to direct attention while awake, all brain stem and cerebral components of the consciousness system must be functional.

Sleep, a periodic loss of consciousness, is actively induced by activity of areas within the ascending reticular activating system. The function of sleep is controversial. Current speculation on the role of sleep includes consolidation of memory, particularly memory for motor skills (Wilson and McNaughton, 1994), and adjusting immune activity (Moldofsky, 1994).

MEDULLA

The medulla is the inferior part of the brain stem, continuous with the spinal cord inferiorly and the pons superiorly.

External Anatomy of the Medulla

Anteriorly, the medulla has two vertical bulges, called the **pyramids.** Lateral to the pyramids are two small oval lumps, called the **olives** (see Fig. 14–1). Cranial nerve XII connects with the medulla between the pyramid and the olive. In a vertical groove lateral to the olive, cranial nerves IX, X, and XI attach to the medulla. The most prominent features of the posterior medulla are the inferior cerebellar peduncle and the widening of the central canal to become a larger space, the fourth ventricle.

Inferior Medulla

The inferior half of the medulla contains a central canal, continuous with the central canal of the spinal cord. Anteriorly, the pyramids are formed by the descending axons of the corticospinal tract. Most (85%) of the corticospinal axons cross the midline in the pyramidal decussation at the inferior border of the medulla. The spinothalamic tracts maintain an anterolateral position, similar to their location in the cord (Fig. 14–6A). The dorsal column tracts synapse in their associated nuclei, the nucleus gracilis and cuneatus. The second-order fibers cross the midline in the decussation of the medial lemniscus, attaining a position posterior to the pyramids before ascending.

In addition to the connections between the spinal cord and cerebrum, the lower medulla also contains cranial nerve structures. The spinal tract and nucleus of the trigeminal nerve are located anterolateral to the nucleus cuneatus and convey pain and temperature information from the face. The medial longitudinal fasciculus, located near the center of the inferior medulla, coordinates eye and head movements via connections between the vestibular nuclei and the nuclei that control eye movements (see Fig. 15–5).

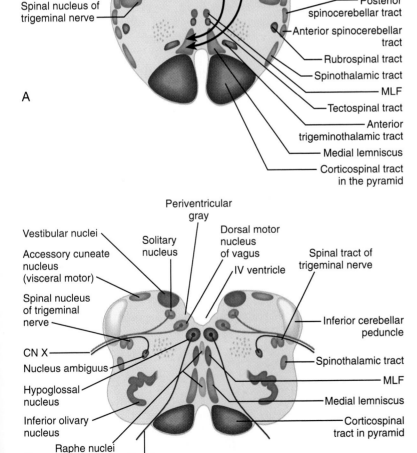

FIGURE 14–6

Horizontal sections of the medulla. *A,* Inferior medulla. *B,* Upper medulla. MLF, medial longitudinal fasciculus; CN, cranial nerve. Stippled areas are the reticular formation.

> The corticospinal and dorsal column/medial lemniscus pathways cross the midline in the caudal medulla. Thus these tracts connect the spinal cord with the opposite cerebral cortex. Cranial nerve V fibers conveying pain and temperature synapse in the caudal medulla.

Upper Medulla

In the upper half of the medulla, the central canal widens to form part of the fourth ventricle. Tracts in the rostral medulla maintain approximately the same positions as in the caudal medulla, except that the medial longitudinal fasciculus is located more posteriorly (Fig. 14–6B). Most cranial nerve nuclei in the rostral medulla are clustered in the dorsal section; from medial to lateral, the nuclei are the hypoglossal (cranial nerve XII), dorsal motor nucleus of the vagus (cranial nerve X), the solitary nucleus (visceral sensory from cranial nerves VII, IX, and X), and the vestibular and cochlear nuclei (cranial nerve VIII). The solitary nucleus receives visceral and taste afferent information. The nucleus ambiguus is the only cranial nerve nucleus in the medulla that is separate from the dorsally located group. The nucleus ambiguus is located more anteriorly and contributes motor fibers to striated muscles in the pharynx, larynx, and upper esophagus via cranial nerves IX and X. Corticobulbar tracts provide cortical input to the nucleus ambiguus and the hypoglossal nucleus. The corticobulbar projections are usually bilateral; however, occasionally the projections to the hypoglossal nucleus are contralateral.

At the junction of the medulla and the pons are the **cochlear and vestibular nuclei,** which receive auditory and vestibular information via cranial nerve VIII. Auditory information from the cochlea of the inner ear is transmitted to the cochlear nuclei by the cochlear nerve. Head movement and head position relative to gravity are signaled by receptors in the labyrinths of the inner ear (see Chap. 13); this information is relayed to the vestibular nuclei by the vestibular nerve. The medial and lateral vestibulospinal tracts (see Chap. 10) that arise from the vestibular nuclei contribute to the control of postural muscle activity.

Deep to the olive is the **inferior olivary nucleus** (see Fig. 14–6B). Shaped like a wrinkled paper bag, this nucleus receives input from most motor areas of the brain and spinal cord. Axons from the inferior olivary nucleus project to the contralateral cerebellar hemisphere via the olivocerebellar tract. Current theory on the role of the inferior olivary nucleus is that these neurons signal the cerebellum when a movement deviates from the planned movement. For example, if someone is picking up a book and the movement occurs as planned, no signals are sent via the olivocerebellar tract. However, if the person's arm is bumped during the movement so that a movement correction is required, a signal is sent via the olivocerebellar tract to alert the cerebellum to the error.

The medulla sends many fibers (spino-, olivo-, vestibulo-, and reticulocerebellar) to the cerebellum via the inferior cerebellar peduncle. Only one fiber tract, the cerebellovestibular tract, sends information from the cerebellum into the medulla.

> The rostral medulla contains nuclei for cranial nerves VII through X and XII. Most of the cranial nerve nuclei are located dorsally. The inferior olive, a movement error detector, notifies the cerebellum when a movement is not carried out as intended. Vestibular nuclei help regulate head and eye movements and postural activity.

Functions of the Medulla

Medullary neuronal networks coordinate cardiovascular control, breathing, head movement, and swallowing. These activities are partially executed by cranial nerves with nuclei in the medulla: VII through X and XII. The medullary neuronal networks regulating these functions are normally influenced by cerebral activity. For example, the tonic neck reflexes seen in infants less than 6 months old require reflex circuits in the medulla (see Chap. 10). As the cerebral cortex matures, information from the cortex modulates the activity of the reflex circuit, modifying the reflexive activity.

> The medulla contributes to the control of eye and head movements, coordinates swallowing, and helps regulate cardiovascular, respiratory, and visceral activity.

PONS

The pons is located between the midbrain and medulla. The posterior pons borders on the fourth ventricle. Most vertical tracts continue unchanged through the pons (Fig. 14–7A). Only the corticopontine tracts and some corticobulbar tracts synapse in the pons. The corticopontine tracts synapse on pontine nuclei; then the postsynaptic axons, called pontocerebellar fibers, leave the pons to enter the cerebellum via the middle cerebellar peduncle. The corticobulbar tracts synapse with neurons in the trigeminal motor nucleus and the facial nucleus.

The basilar (anterior section) of the pons contains descending tracts (corticospinal, corticobulbar, and corticopontine axons), pontine nuclei, and ponto-cerebellar axons. The posterior section of the pons, the tegmentum, contains sensory tracts, reticular formation, autonomic pathways, medial longitudinal fasciculus, and nuclei for cranial nerves V through VII. These cranial nerves are involved in the following:

- Processing sensation from the face (cranial nerve V)
- Controlling lateral movement of the eye (cranial nerve VI)

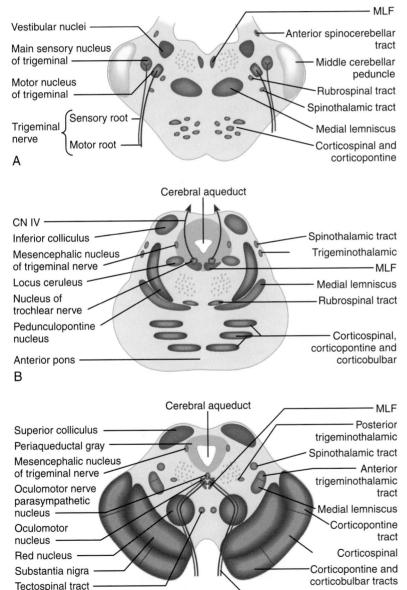

FIGURE 14-7

Horizontal section of *A*, pons; *B*, junction of pons and midbrain; and *C*, upper midbrain. MLF, medial longitudinal fasciculus; CN, cranial nerve. Stippled areas are the reticular formation.

• Control of facial and chewing muscles (cranial nerves VII and V respectively)

> The pons processes motor information from the cerebral cortex and forwards the information to the cerebellum. Pontine cranial nerve nuclei process sensory information from the face (cranial nerve V) and control contraction of muscles involved in facial expression (cranial nerve VII), lateral movement of the eye (cranial nerve VI), and chewing (cranial nerve V).

MIDBRAIN

The uppermost part of the brain stem, the midbrain, connects the diencephalon and the pons. The cerebral aqueduct, a small canal through the midbrain, joins the third and fourth ventricles. The midbrain can be divided into three regions, from anterior to posterior: basis pedunculi, tegmentum, and tectum.

Basis Pedunculi

Anteriorly, the basis pedunculi is formed by the **cerebral peduncles** (composed of descending tracts from the cerebral cortex) and an adjacent nucleus, the **substantia nigra** (Fig. 14–7C). Substantia nigra is one of the nuclei in the basal ganglia circuit (see Chap. 10). The other basal ganglia nuclei are the caudate, putamen, globus pallidus, pedunculopontine nucleus and subthalamic nucleus.

Midbrain Tegmentum

The middle region of the midbrain, tegmentum, contains vertical sensory tracts, the superior cerebellar peduncle, the red nucleus, the pedunculopontine nucleus (Fig. 14–7B), and the nuclei of cranial nerves III and IV. Most vertical tracts occupy similar positions as in the pons, except that the spinothalamic tract and medial lemniscus are located more laterally in the midbrain. The superior cerebellar peduncle connects the midbrain with the cerebellum, transmitting primarily efferent information from the cerebellum.

The **red nucleus** is a sphere of gray matter that receives information from the cerebellum and cerebral cortex and projects to the cerebellum, spinal cord (via rubrospinal tract), and reticular formation. Activity in the rubrospinal tract contributes to upper limb flex-

ion. **Pedunculopontine nucleus** neurons are part of the basal ganglia circuit and are involved in the initiation and termination of locomotor activity (Garcia-Rill and Skinner, 1988).

Anterior to the cerebral aqueduct are the **oculomotor complex** (nuclei of cranial nerve III) and the nucleus of the trochlear nerve (cranial nerve IV). The oculomotor complex consists of the oculomotor nucleus, supplying efferent somatic fibers to the extraocular muscles innervated by the oculomotor nerve, and the oculomotor parasympathetic (Edinger-Westphal) nucleus, supplying parasympathetic control of the pupillary sphincter and the ciliary muscle. The oculomotor complex is superior to the trochlear nucleus. The trochlear nerve innervates the superior oblique muscle that moves the eye.

Surrounding the cerebral aqueduct is the **periaqueductal gray.** Involvement of the periaqueductal gray in pain suppression was discussed in Chapter 7. The periaqueductal gray also coordinates somatic and autonomic reactions to pain, threats, and emotions. Activity of the periaqueductal gray results in fight-or-flight reaction (Bandler et al., 1991) and in vocalization during laughing and crying (Holstege, 1991).

Midbrain Tectum

The posterior region of the midbrain, the tectum, contains the pretectal area and the colliculi. The pretectal area is involved in the pupillary, consensual, and accommodation reflexes of the eye (see Chap. 13). The inferior colliculi relay auditory information from the cochlear nuclei to the superior colliculus and to the medial geniculate body of the thalamus (see Chap. 15). The superior colliculi are involved in reflexive eye and head movements (see Chap. 15).

CEREBELLUM

The cerebellum is discussed briefly in this chapter because cerebellar function is entirely dependent on input and output connections with the brain stem. Furthermore, the cerebellum and brain stem share the tightly confined space of the posterior fossa, bringing them into a close anatomical relationship. The following list summarizes cerebellar functions:

• Coordination of movement, including fine finger movements, limb and head movements, postural control, and eye movements

- Motor planning
- Cognitive functions, including rapid shifts of attention

Specific motor functions and the role in motor planning of the cerebellum are discussed in Chapter 10. The cerebellum is also vital for voluntarily shifting attention between auditory and visual stimuli (Akshoomoff and Courchesne, 1992). The attention-shifting role of the cerebellum is believed to be particularly important in social and communication situations (Courchesne et al., 1994). Fibers from the cerebellum synapse with fibers in the reticular formation to achieve their role in directing attention.

> In addition to its roles in motor control and motor planning, the cerebellum also contributes to voluntary shifting of attention.

DISORDERS IN THE BRAIN STEM REGION

Lesions within the brain stem can be localized by evaluating the function of cranial nerves (see Chap. 13) and vertical tracts. A single brain stem lesion may cause a mix of ipsilateral and contralateral signs (see Fig. 7–5). The mix of ipsilateral and contralateral signs occurs because cranial nerves supply the ipsilateral face and neck, while many of the vertical tracts cross the midline in the brain stem to supply the contralateral body. Aside from the outcomes of vertical tract and cranial nerve damage, lesions in the brain stem may also interfere with vital functions and consciousness.

Vertical Tract Signs

The lateral corticospinal, dorsal column/medial lemniscus, and spinothalamic tracts connect the spinal cord with the contralateral cerebrum. Lesions of the lateral corticospinal and dorsal column tracts in the brain stem usually cause contralateral signs because these tracts cross the midline in the inferior medulla. The only location where a brain stem lesion would cause ipsilateral lateral corticospinal or dorsal column/medial lemniscus signs would be a lesion of the corticospinal tract or dorsal column nuclei in the inferior medulla. The spinothalamic tract crosses the

midline in the spinal cord, so any brain stem lesion that damages the spinothalamic tract causes contralateral signs.

CORTICOBULBAR LESIONS

The corticobulbar tracts convey motor signals from the cerebral cortex to cranial nerve nuclei in the brain stem. Thus neurons with axons in the corticobulbar tract serve as upper motoneurons to the lower motoneurons in cranial nerves V, VII, IX, X, XI, and XII. Although both upper and lower motoneuron lesions cause paresis or paralysis, upper motoneuron lesions are associated with hyperreflexia and muscle hypertonia, while lower motoneuron lesions are associated with hyporeflexia and muscle flaccidity. Corticobulbar projections are bilateral, except to lower motoneurons innervating the muscles of the lower face and sometimes to the hypoglossal nucleus.

FACIAL NERVE VERSUS CORTICOBULBAR TRACT LESIONS

A complete lower motoneuron lesion of the facial nerve, cranial nerve VII, prevents commands from reaching all ipsilateral facial muscles. The result is flaccid paralysis of the muscles in the ipsilateral face. A person with a complete facial nerve lesion is completely unable to contract the muscles of facial expression and cannot close the ipsilateral eye. In contrast, unilateral upper motoneuron lesions interrupt voluntary control of contralateral facial muscles in the lower half of the face. The muscles in the upper half of the face are spared because both the right and left cerebral cortex have projections to lower motoneurons innervating muscles of the forehead and surrounding the eye. Thus an upper motoneuron lesion that prevents corticobulbar information from the left cerebral cortex from reaching the facial nerve nuclei causes paresis or paralysis of the right lower face, but cerebral control of muscles of the upper face is relatively unaffected. The difference between a lesion of the facial nerve (lower motoneuron lesion) and a corticobulbar lesion (upper motoneuron lesion) is illustrated in Figure 14–8. People with upper motoneuron lesions preventing voluntary control of the contralateral lower face are able to laugh and cry normally because the pathway involved in emotional vocalization is separate from the corticobulbar tract for the same activity (Holstege, 1991).

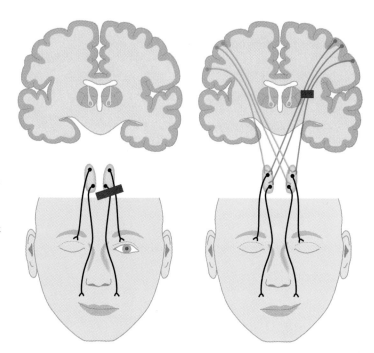

FIGURE 14–8

Lower motoneuron lesion versus upper motoneuron lesion (corticobulbar) affecting the facial nerve. In both *A* and *B*, the person has been requested to close the eyes and smile. *A*, With a facial nerve lesion, the lower motoneurons are interrupted, preventing control of the ipsilateral muscles of facial expression. Therefore the person cannot close the eye or contract the muscles that move the lips on the left. *B*, An upper motoneuron lesion prevents information from the left cortex from reaching the facial nerve nuclei. Because the muscles of the lower face are controlled by the contralateral cortex, the person is unable to generate a smile on the right side. However, because the upper face is innervated bilaterally, the person with this upper motoneuron lesion can close both eyes.

In the brain stem, lesions cause contralateral vertical tract signs unless the lesion affects the corticospinal tracts or dorsal column nuclei in the inferior medulla. Complete lesions of the facial cranial nerve cause ipsilateral paralysis of the facial muscles, while lesions of the corticobulbar axons to the facial nucleus cause paralysis of the contralateral lower face with sparing of control of the upper face.

CONTRALATERAL AND IPSILATERAL SIGNS

A single lesion in the upper anteromedial medulla (Fig. 14–9A) on the left side can cause paralysis of the right hand and foot, loss of discriminative touch and proprioceptive information on the right side of the body, and paralysis of the left side of the tongue. The right hand and foot paralysis is due to interruption of lateral corticospinal tract in the pyramids, superior to the pyramidal decussation where the tract crosses the midline. The contralateral sensory loss occurs because the dorsal column/medial lemniscus tract crosses the midline in the lower medulla. The left tongue paralysis occurs following damage to the left hypoglossal nerve. Pain and temperature information is spared because the lateral medulla is undamaged. Additional examples of brain stem lesions are provided in Figure 14–9.

Disorders of Vital Functions

Disruption of vital functions secondary to brain stem damage may cause the heart to stop beating, blood pressure to fluctuate, and/or breathing to cease. Centers in the medulla and pons regulate vital functions.

Disorders of Consciousness

States of altered consciousness may occur with lesions to either the brain stem or the cerebrum because structures in both regions are required for consciousness. Brain stem damage that affects the reticular formation and/or the axons of the ascending reticular activating system interferes with consciousness. Damage to the cerebrum that interferes with hypothalamic/thalamic activating areas or with the function of the entire cerebral cortex may also impair consciousness. States of altered consciousness are defined in Table 14–2 (on p. 294).

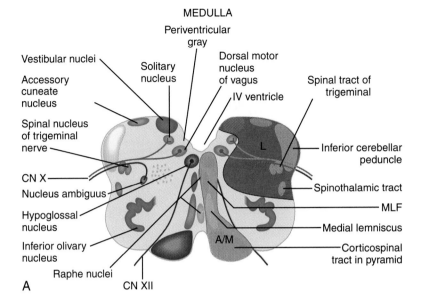

MEDULLA

ANTEROMEDIAL LESIONS

	Structures involved	Lesion interferes with
Either medulla or pons	Corticospinal tract	Fractionated movements
	Medial lemniscus	Discriminative touch and conscious proprioception
Medulla only	Hypoglossal nerve	Tongue movement
Pons only	Medial longitudinal fasciculus	Adduction of eye past midline during lateral gaze

FIGURE 14–9

Lesions are indicated on the brain stem sections. The charts summarize the structures damaged by the lesion and the results of the damage. Because the corticospinal tract and medial lemniscus are located in the anteromedial medulla and pons, an anteromedial lesion in either the medulla or the pons will damage those tracts. Because the spinothalamic tract, spinal tract and nucleus of cranial nerve V, and vestibular nuclei are located laterally in the medulla and pons, a lateral lesion in either the medulla or pons will damage those structures. A/M, anteromedial lesion; CN, cranial nerve; L, lateral lesion; MLF, medial longitudinal fasciculus.

Illustration continued on following page

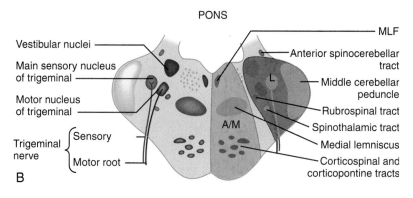

PONS

Vestibular nuclei

Main sensory nucleus
of trigeminal

Motor nucleus
of trigeminal

Trigeminal
nerve { Sensory

Motor root

B

MLF

Anterior spinocerebellar
tract

Middle cerebellar
peduncle

Rubrospinal tract

Spinothalamic tract

Medial lemniscus

Corticospinal and
corticopontine tracts

	LATERAL LESIONS	
	Structures involved	Lesion interferes with
Either medulla or pons	Spinothalamic tract	Pain and temperature sensation from the body
	Spinal tract and nucleus of CN V	Pain and temperature sensation from the face
	Vestibular nuclei (in upper medulla, lower pons)	Control of posture, head position, and eye movement
Medulla only	Nucleus ambiguus	Swallowing, vocalization
	Inf. cerebellar peduncle	Smoothness of movement
	Descending sympathetic pathway	Sympathetic control of face; causes Horner's syndrome
	Vagus nerve	Digestion, ability to slow heart rate
Pons only	Middle cerebellar peduncle	Smoothness of movement
	Main sensory nucleus of CN V	Corneal reflex and discriminative touch information from face
	Facial nerve	Control of muscles of facial expression

FIGURE 14–9

Continued. Refer to part *A* for lesions in medulla.

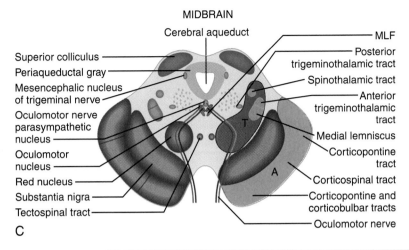

MIDBRAIN

Cerebral aqueduct — MLF

Superior colliculus — — Posterior trigeminothalamic tract

Periaqueductal gray — — Spinothalamic tract

Mesencephalic nucleus of trigeminal nerve — — Anterior trigeminothalamic tract

Oculomotor nerve parasympathetic nucleus — — Medial lemniscus

Oculomotor nucleus — — Corticopontine tract

Red nucleus — — Corticospinal tract

Substantia nigra — — Corticopontine and corticobulbar tracts

Tectospinal tract — — Oculomotor nerve

C

ANTERIOR LESIONS		
	Structures involved	Lesion interferes with
Midbrain	Corticospinal tract	Control of fractionated movement; face may be involved
	Frontopontine tracts	Deficit should cause ataxia, but deficit not visible because hemiparesis or hemiplegia prevents movement
	Oculomotor nerve	Ability to move eye medially, downward, and upward; also causes drooping upper eyelid, dilated pupil

TEGMENTAL LESIONS		
	Structures involved	Lesion interferes with
Midbrain	Oculomotor nerve	Same functions as above
	Medial lemniscus, trigeminothalamic and spinothalamic tract	Discriminative tactile, conscious proprioception, temperature, and pain sensation from face and body
	Superior cerebellar peduncle and red nucleus	Smoothness of movement; causes ataxia, dysdiadochokinesis, and hypotonia

FIGURE 14–9

Continued. A, anterior lesion; T, tegmental lesion.

TABLE 14–2
STATES OF ALTERED CONSCIOUSNESS

Coma	Unarousable; no response to strong stimuli such as strong pinching of the Achilles tendon
Stupor	Arousable only by strong stimuli, such as strong pinching of the Achilles tendon
Obtunded	Sleeping more than awake, drowsy and confused when awake
Vegetative state	Complete loss of consciousness, without alteration of vital functions*
Syncope (fainting)	Brief loss of consciousness due to a drop in blood pressure[†]
Delirium	Reduced attention, orientation, and perception, associated with confused ideas and agitation

*Vegetative state is distinguished from coma by the following signs: spontaneous eye opening, regular sleep-wake cycles, and normal respiratory patterns.
[†]Benign syncope results from overactivity of the vagus nerve (vasovagal syncope). Orthostatic hypotension (decreased blood pressure in the upright position) may cause syncope in patients with spinal cord injury or who have experienced prolonged bed rest.

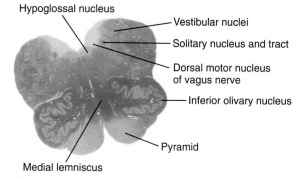

FIGURE 14–10

Section of the medulla, myelin darkly stained, illustrating degeneration of the medullary pyramids. The destruction of the corticospinal and other descending pathways produced locked-in syndrome.

A disconnection syndrome, called locked-in syndrome, may mimic the signs of impaired consciousness. In locked-in syndrome, consciousness is intact, but the person is completely unable to move because of damage to descending activating pathways. In some cases, the person is able to voluntarily control eye movements and can communicate by coded eye movements. Figure 14–10 is a section of medulla from a person with locked-in syndrome.

The integrity of brain stem function can be assessed with auditory evoked potentials. As in somatosensory evoked potentials, a sense organ is stimulated, and resulting electrical activity is recorded from electrodes on the scalp. For auditory evoked potentials, a brief burst of tone is presented, and the brain stem response is recorded. Auditory evoked potentials are most commonly used to assess brain stem function in comatose patients. Auditory evoked potentials can also be used to evaluate whether the cochlea, cochlear nerve, and auditory nuclei in the brain stem are functioning.

Compression in the Brain Stem Region

Damage caused by a benign tumor may be extensive because the confines of unyielding bone and dura prevent brain tissue from moving away from the pressure. For example, an acoustic neuroma is a benign tumor of the Schwann cells surrounding the vestibulocochlear nerve. If the nerve was not confined by the surrounding bones, the acoustic neuroma could enlarge without compromising function. Unfortunately, bony restriction causes the enlarging tumor to compress the vestibulocochlear nerve, resulting in tinnitus and eventual deafness. As the tumor grows, more and more structures are compressed. The trigeminal and facial nerves will be the next structures compressed by the enlarging tumor, causing loss of sensation and paresis of facial muscles. Next, cerebellar signs, including ipsilateral limb ataxia, intention tremor, and nystagmus, appear as the pressure builds on the cerebellum. Eventually, brain stem compression interferes with vertical tracts and nuclear functions.

Summary

The brain stem contains the origin of most descending activating tracts; axons transmitting somatosensory information; and nuclei for cranial nerves III–X and XII and the reticular formation. The reticular formation is essential for modulation of neural activity throughout the central nervous system.

CLINICAL NOTES

CASE 1

P.C. is a 32-year-old man found unconscious at home 4 days ago. He regained consciousness today. The therapist's evaluation reveals the following:

- Lack of pain and temperature sensation from the right side of the body
- Lack of somatosensation from the left side of the face
- Ataxia on the left side of the body
- Paralysis of muscles of facial expression on the left
- Loss of corneal reflex on the left side

The therapist also notes nystagmus, vertigo, oscillopsia, nausea, and vomiting when P.C. turns his head.

Questions

1. List the structure associated with each loss.
2. Where is the lesion?

CASE 2

L.D., a 78-year-old woman, awoke with inability to voluntarily move the muscles of facial expression in her right lower face. In the clinic, the following signs are noted:

- Sensation is intact throughout the body and face and movements of the limbs and trunk are normal.
- Movement of the upper face and the left lower face are normal. She is able to completely close both eyes on request. When she is asked to smile or frown, muscles in the right lower face do not contract. However, when she frowns due to frustration, muscles in the right lower face contract.
- Test results for all cranial nerves other than cranial nerve VII are normal.

Question
Where is the lesion?

CASE 3

M.Z. is 17 years old. He suffered a severe head injury in a car accident 2 months ago. Following a month-long hospitalization, M.Z. has been in a long-term care facility for 4 weeks. Notes in his chart indicate that M.Z. is in a vegetative state and is not expected to recover. M.Z. is completely immobile except for eye movements. M.Z.'s family believes that he is aware and able to communicate with them via eye movements. When the therapist asks him to blink three times, M.Z. complies. When the therapist asks him to look toward his right, he does. However, M.Z. does not move any other part of his body on request.

Questions

1. Is M.Z.'s behavior consistent with a vegetative state?
2. If not, what is the condition?

(continued)

CASE 4

R.V., a 58-year-old man, was in a meeting when suddenly he lost control of the right side of his body, including his face. He slumped in his chair, and the right side of his face appeared to sag, but he did not lose consciousness. R.V. complains of double vision. The clinical findings are as follows:

- Somatosensation is intact.

- Movement and strength on the left side of his body are normal. He is able to sit unassisted in a chair with arm and back support but cannot sit unassisted without support. R.V. is able to voluntarily move his right upper limb at the shoulder and his right lower limb at the hip, but strength is less than half that of the left side. He cannot move any other joints in his limbs on the right.

- All cranial nerves are intact except for the following:

  He is unable to voluntarily move his right lower face.

  He cannot move his left eye medially, downward, or upward.

  He cannot fully open his left eye (left eyelid droops).

  The left pupil is dilated and does not contract in response to light shined into the eye.

Questions

1. List the structures associated with the functional losses.

2. Where is the lesion?

REVIEW QUESTIONS

1. List the vertical tracts that are modified in the brain stem.

2. What are the functions of each of the zones of the reticular formation?

3. List the major reticular nuclei and the neuromodulators produced by these nuclei.

4. Which neuromodulator produced in the brain stem is important in the cerebral processes of motivation and decision making?

5. How does the pedunculopontine nucleus affect movement?

6. Which medullary nuclei are part of a system that inhibits the transmission of pain information?

7. What is the role of the ascending fibers from the locus ceruleus?

8. For each of the following functions, list the cranial nerve nucleus responsible for the function and the part of the brain stem where the nucleus is located (lower medulla, upper medulla, junction of the medulla and pons, pons, junction of the pons and midbrain, midbrain):
   - Control of voluntary muscles in the pharynx and larynx
   - Integration and transmission of pain information from the face
   - Control of tongue muscles
   - Processing of information about sounds
   - Signals the cerebellum when movement errors occur
   - Control of muscles of mastication
   - Contraction of the pupillary sphincter and change in curvature of the lens to focus on near objects

9. Which nuclei in the midbrain are part of the basal ganglia circuit?

10. What are the functions of the cerebellum?

11. Why do brain stem lesions above the inferior medulla cause contralateral loss of discriminative touch information from the body?

12. What midbrain region coordinates somatic and autonomic reactions to pain, threats, and emotions?

13. What tracts convey motor signals from the cerebral cortex to cranial nerve motor nuclei?

14. How can a complete lesion of the facial nerve be differentiated from a lesion affecting the corticobulbar tracts that convey information from the cerebral cortex to the facial nerve nucleus?

15. Would a person with a complete facial nerve lesion on the left side be able to smile involuntarily on the left side of the face? Why or why not?

16. Disorders of consciousness can occur with damage to what brain stem structures?

17. A person who has a complete loss of consciousness combined with normal vital functions is in what state of altered consciousness?

18. Why are space-occupying lesions in the brain stem region, such as benign tumors, so disruptive of brain stem function?

References

Akshoomoff, N. A., and Courchesne, E. (1992). A new role for the cerebellum in cognitive operations. Behav. Neurosci. 106(5): 731–738.

Bandler, R., Carrive, P., Zhang, S. P., et al. (1991). Integration of somatic and autonomic reactions within the midbrain periaqueductal grey: Viscerotopic, somatotopic and functional organization. Prog. Brain Res. 87:269–305.

Courchesne, E., Townsend, J., Akshoomoff, N. A., et al. (1994). Impairment in shifting attention in autistic and cerebellar patients. Behav. Neurosci. 108(5):848–865.

Garcia-Rill, E., and Skinner, R. D. (1988). Modulation of rhythmic function in the posterior midbrain. Neuroscience 27:639–654.

Holstege, G. (1991). Descending motor pathways and the spinal motor system: Limbic and non-limbic components. Prog. Brain Res. 87:307–421.

Moldofsky, H. (1994). Central nervous system and peripheral immune functions and the sleep-wake system. J Psychiatry Neurosci. 19(5):368–374.

Wilson, M. A., and McNaughton, B. L. (1994). Reactivation of hippocampal ensemble memories during sleep. Science 265(5172): 676–679.

15

Auditory,
Vestibular, and
Visual Systems

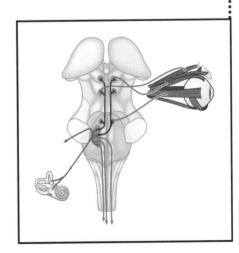

INTRODUCTION

The auditory, vestibular, and visual systems have specialized receptors and cranial nerve axons in the periphery, specific nuclei in the brain stem, and areas of the cerebral cortex dedicated to their function. The auditory system usually functions independently of the other two systems, although loud or unexpected sounds can trigger reflexive movements of the eyes and head to direct vision toward the source of the sound. These reflexive movements require visual and vestibular system activity. The function of the visual system is partially dependent on the vestibular system because vestibular information contributes to compensatory eye movements that maintain stability of the visual world when the head moves. Vision makes a minor contribution to understanding spoken language because people understand spoken language better if they can see the movements of the speaker's lips.

AUDITORY SYSTEM

Auditory information

- Orients the head and eyes toward sounds
- Increases the activity level throughout the central nervous system
- Provides conscious awareness and recognition of sounds

The receptors for converting sound into neural signals are located in the organ of Corti, inside the cochlea of the inner ear (see Chap. 13). Signals from the organ of Corti are conveyed to the cochlear nuclei by the cochlear nerve (cranial nerve VIII). The cochlear nuclei, devoted to processing auditory information, are located at the junction of the medulla and pons. From the cochlear nuclei, auditory information is transmitted to three structures (Fig. 15–1):

- Reticular formation
- Inferior colliculus
- Medial geniculate body

The reticular formation connections account for the activating effect of sounds on the entire central nervous system. For example, loud sounds can rouse a person from sleep. The inferior colliculus integrates auditory information from both ears to de-

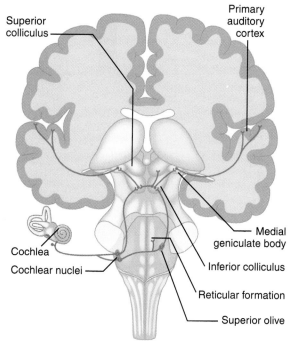

FIGURE 15–1

Pathway for auditory information from the cochlea to the cochlear nuclei, then to the reticular formation, inferior colliculus, and medial geniculate. Information from the medial geniculate projects to the primary auditory cortex.

tect the location of sounds. When the location information is conveyed to the superior colliculus, neural activity in the superior colliculus elicits movement of the eyes and face toward the sound. The medial geniculate body serves as a thalamic relay station for auditory information to the primary auditory cortex, where sounds reach conscious awareness. The routing of auditory information is illustrated in Figure 15–2.

Three cortical areas are dedicated to processing auditory information. The primary auditory cortex is the site of conscious awareness of the intensity of sounds. An adjacent cortical area, the auditory association cortex, compares sounds with memories of other sounds, then categorizes the sounds as language, music, or noise. Comprehension of spoken language occurs in yet another cortical area, called Wernicke's area. These cortical areas are discussed further in Chapters 16 and 17.

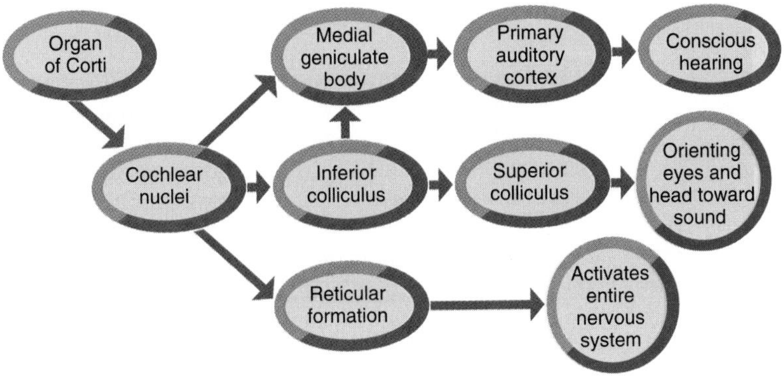

FIGURE 15–2

Flow of signals from the hearing apparatus (organ of Corti) to the outcomes of hearing: conscious hearing, orientation toward sound, and increased general arousal level.

VESTIBULAR SYSTEM

The vestibular system provides the following:

- Sensory information about head movement and head position relative to gravity
- Gaze stabilization (control of eye movements when the head moves)
- Postural adjustments
- Effects on autonomic function and consciousness

Receptors

The receptors for converting motion and position of the head into neural signals are located in the semicircular canals and the otolithic organs of the inner ear (see Chap. 13). Signals from the vestibular apparatus are conveyed to the vestibular nuclei by the vestibular nerve (cranial nerve VIII).

Semicircular Canals

Maximum fluid flow in each semicircular canal and thus maximal change in the frequency of signals generated by bending of the hairs embedded in the cupula occur when the head turns on the canal's axis of rotation (Fig. 15–3). Two semicircular canals that have maximal fluid flow during rotation in a single plane form a pair. For example, when the head is flexed 30°, the horizontal canals are parallel to the ground. Rotation of the head in 30° flexion around the vertical axis maximizes fluid flow in both horizontal canals. The horizontal canals are classified as a pair because they are both maximally stimulated by movement in a single plane.

The anatomical arrangement of the canals, with the semicircular canals oriented at 90° angles to each other, ensures that acceleration or deceleration in a plane of movement that causes maximal fluid flow in a pair of semicircular canals does not stimulate the other semicircular canals. The anterior and posterior semicircular canals are oriented vertically at a 45° angle to the midline. Turning the head 45° to the left and then doing somersaults causes maximal fluid flow in the right anterior canal. This somersault causes no fluid flow in the left anterior canal because the left anterior canal is moving perpendicular to its axis. Because the anterior canals are 90° to each other, there is no plane of movement in which the fluid flow in the anterior canals can be maximized simultaneously. However, the same somersault causes maximal fluid flow in the left posterior canal. Therefore, the right anterior and the left posterior canals are classified as a pair, since fluid flow in both is maximized by movement in a single plane. Similarly, the left anterior and the right posterior canals are a pair.

Each of the canals in a pair produces reciprocal signals; that is, increased signals from one canal occur simultaneously with decreased signals from its partner. These reciprocal signals are essential for normal vestibular function. If the signals from a pair of semicircular canals are not reciprocal, difficulty with control of posture, eye movements, and nausea result.

Otolithic Organs

Unlike the semicircular canals, the otolithic organs are not sensitive to rotation. The otolithic organs respond to head position relative to gravity and to linear acceleration and deceleration. So, as one moves

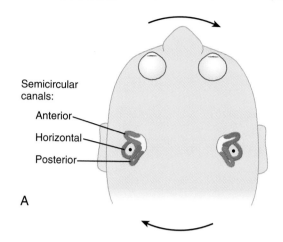

Semicircular
canals:

Anterior

Horizontal

Posterior

A

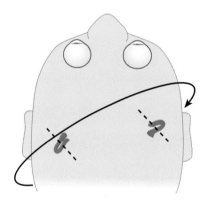

B

FIGURE 15–3

Axis of rotation of the semicircular canals. The three pairs of canals—horizontal, the right anterior with left posterior, and the left anterior with right posterior—are indicated by colors. *A,* The axes of the horizontal canals are indicated by a dot in the center of each horizontal canal. Rotating the head as indicated by the arrows causes maximum fluid flow in both horizontal canals. *B,* Only the left posterior and right anterior canals are shown. The axes are indicated by dotted lines. Because their axes are parallel, rotation in one plane (indicated by the arrow) simultaneously maximally stimulates both canals in the pair.

from standing to lying down or begins or stops walking, the otolithic organs signal the changes in head position or movement. Motion sickness, the nausea sometimes experienced in moving vehicles, is usually caused by a conflict between different types of sensory information. For example, when one reads in a moving car, the visual information is not moving, yet the vestibular system is sensing movement.

Vestibular Role in Motor Control

In addition to providing sensory information about head movement and position, the vestibular system has two roles in motor control: gaze stabilization and postural adjustments. Gaze stabilization operates by the vestibulo-ocular reflex, discussed later in this chapter. Postural adjustments are achieved by reciprocal connections between the vestibular nuclei and the spinal cord, reticular formation, superior colliculus, nucleus of cranial nerve XI, and the cerebellum (Fig. 15–4).

Connections of the Vestibular Nuclei

Rapidly rotating the head, by simply spinning around or by riding a spinning amusement park ride, activates the semicircular canal connections, eliciting the following:

• Altered postural control (leading to leaning or falling)
• Head orientation adjustment
• Eye movement reflexes
• Autonomic changes (nausea, vomiting)
• Changes in consciousness (light-headedness)

In addition to being the source of the vestibulospinal tracts, the vestibular nuclei are linked with areas that affect the corticospinal, reticulospinal, and tectospinal descending tracts. By these connections, the vestibular nuclei strongly influence the posture of the head and body.

Cerebellar connections with the vestibular apparatus, vestibular nuclei, spinal cord, and inferior olive enable the cerebellum to adjust the gain of the postural adjustments and vestibulo-ocular reflex. Thus the magnitude of the reflex responses to changes in position and movement (of the head, body, and/or external objects) depends on cerebellar processing of vestibular and visual information.

The **medial longitudinal fasciculus** consists of axons that bilaterally connect cranial nerve nuclei III, IV, VI, VIII, and XI and the superior colliculus (Fig. 15–5 on p. 304). This linkage is essential for coordinated movements of the eyes and head. For example, the right lateral rectus (cranial nerve VI) and the left medial rectus (cranial nerve III) muscles must contract simultaneously to turn the eyes to the right. Coordination of this movement is achieved by fibers of the medial longitudinal fasciculus.

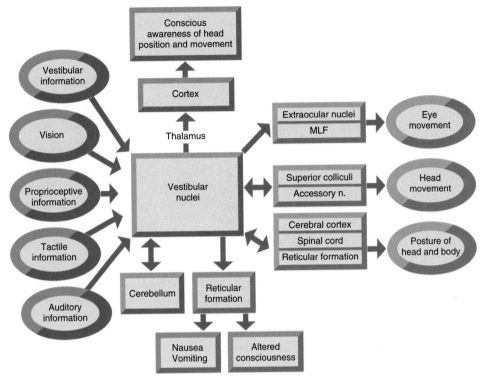

FIGURE 15–4

Connections of the vestibular nuclei. Sensory inputs are shown on the left (in blue), the motor output on the right (in red). Note the wide variety of sensory information feeding into the vestibular nuclei. The vestibular nuclei integrate all types of sensory information that can be used for orientation, not only information from the vestibular receptors. MLF, medial longitudinal fasciculus; n., nerve.

The vestibular connections with the reticular formation, in addition to affecting the reticulospinal tracts, affect the autonomic nervous system. Excessive activity of the circuits linking vestibular nuclei and the reticular formation may result in nausea, vomiting, and changes in consciousness. A pathway to the thalamus and then to the cerebral cortex provides conscious awareness of head orientation and movement.

> Vestibular connections influence body, head, and eye movements, autonomic functions, and consciousness.

VISUAL SYSTEM

The visual system provides the following:

- Sight, for the recognition and location of objects
- Eye movement control

- Information used in postural and limb movement control (see Chap. 9)

Sight

INFORMATION FROM RETINA TO CORTEX

The visual pathway begins with cells in the retina that convert light into neural signals. The signals are processed within the retina and conveyed to the retinal output cells. Retinal output is conveyed by the axons that travel in the optic nerve, optic chiasm, and optic tract, then synapse in the lateral geniculate of the thalamus. The optic nerve is the bundle of axons from the retina to the optic chiasm. The optic nerves merge at the optic chiasm, and some axons cross the midline in the chiasm. The optic tract conveys visual information from the chiasm to the lateral geniculate.

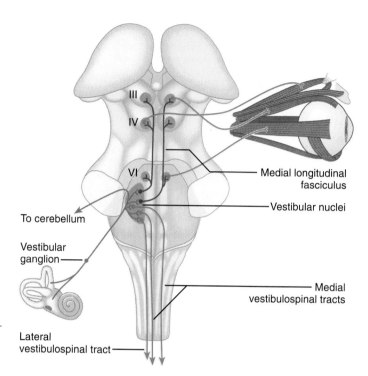

III

IV

VI

To cerebellum

Vestibular
ganglion

Medial longitudinal
fasciculus

Vestibular nuclei

Medial
vestibulospinal tracts

Lateral
vestibulospinal tract

FIGURE 15–5

The medial longitudinal fasciculus connects the vestibular nuclei (shown on the left only) with nuclei that control eye movements. The medial and lateral vestibulospinal tracts are also shown.

Postsynaptic neurons travel from the lateral geniculate in the geniculocalcarine tract (optic radiations) to
the primary visual cortex. As the optic radiations emerge
from the lateral geniculate, they travel in the posterior
part of the internal capsule. The primary visual cortex is
the region of cortex that receives direct projections of
visual information. Thus, to reach conscious awareness,
the neural signals travel to the visual cortex via the
retinogeniculocalcarine pathway (Fig. 15–6).

The cortical destination of visual information depends on which half of the retina processes the visual
information: the nasal retina, nearest the nose, or the
temporal retina, nearest the temporal bone. Information from the nasal half of each retina crosses the
midline in the optic chiasm and projects to the contralateral visual cortex. Information from the temporal half of each retina continues ipsilaterally through
the optic chiasm and projects to the ipsilateral cortex.

The outcome of the fiber rearrangement in the chiasm is to deliver all visual information from one visual
field to the opposite visual cortex. For example, the
right visual field is the part of the environment that a
person sees to the right of their own midline when
looking straight ahead. Light from the right visual field

strikes the left half of each retina. The left half of the
left retina is temporal and projects to the ipsilateral visual cortex. The left half of the right retina is nasal,
and its projections cross the midline in the chiasm.
Thus the axons leaving the chiasm in the left optic
tract all carry information from the right visual field.
Axons of the left optic tract synapse in the left lateral
geniculate, and then the information is relayed to the
left visual cortex via the geniculocalcarine tract. This
results in projection of the right visual field information to the left visual cortex. Similarly, left visual field
information projects to the right visual cortex.

> The retinogeniculocalcarine pathway conveys visual in-
> formation that reaches conscious awareness. Informa-
> tion from a visual field is conveyed to the contralateral
> visual cortex.

PROCESSING OF VISUAL INFORMATION

Visual information reaching the primary visual cortex
stimulates neurons that discriminate the shape, size,
or texture of objects. Information conveyed to the ad-

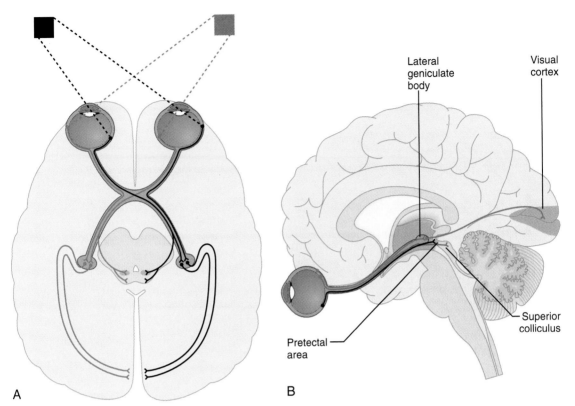

FIGURE 15–6

Visual pathways. *A,* Visual information from the blue box in the right visual field activates neurons in the left half of the retina of both eyes. Axons from the temporal half of the retina project ipsilaterally to the lateral geniculate body, while axons from the nasal half of the retina cross the midline in the optic chiasm to project to the contralateral lateral geniculate body. Thus all visual information from the right visual field projects to the left lateral geniculate, then through the optic radiations to the left visual cortex. Collaterals from axons in the optic nerve to the pretectal area and to the superior colliculus are also shown. *B,* A lateral view of the projections from the retina to the superior colliculus, pretectal area, and lateral geniculate/visual cortex.

jacent cortical areas, called the visual association cortex, is analyzed for colors and motion. From the visual association cortex, the information flows to other areas of the cerebral cortex where the visual information is used to adjust movements or to visually identify objects (see Chap. 16). The stream of visual information that flows dorsally is called the action stream because this information is used to direct movement, and the stream of visual information that flows ventrally is called the perception stream because this information is used to recognize visual objects (Fig. 15–7).

Two areas processing unconscious visual information are discussed in Chapter 13: the superior colliculus and the pretectal area. Projections from the retina to the brain stem and visual cortex are illustrated in Figure 15–6. The conscious and unconscious pathways transmitting visual information are summarized in Figure 15–7.

Eye Movement System

Normal eye movements require the synthesis of information about the following:

- Head movements (vestibular information)
- Visual objects (vision)

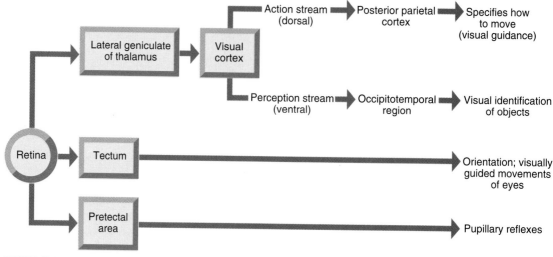

FIGURE 15–7

Flow of visual signals from the retina to the visual cortex, tectum, and pretectal area. Signals arriving in the visual cortex are analyzed and then sent to other areas of cerebral cortex where directions for movement are created and where objects are recognized visually. Signals arriving in the tectum are used for orientation and eye movement control. Signals arriving in the pretectal area produce pupillary reflexes.

- Eye movement and position (proprioceptive information)
- Selection of a visual target (brain stem and cortical areas)

Precise control of eye position is vital for vision because the best visual acuity is available only in a small region of the retina (the fovea) and because binocular perception of an object as a single object requires that the image be received by corresponding points on both retinas. This exquisite control of eye position is achieved by the medial longitudinal fasciculus, reflexes, and cerebral centers.

The medial longitudinal fasciculus is the neural connection among the vestibular nuclei, the nuclei that control eye movements, the spinal accessory nucleus, and the superior colliculus. The superior colliculus coordinates reflexive orienting movements of the eyes and head via the medial longitudinal fasciculus and the tectospinal tract.

VESTIBULO-OCULAR REFLEXES

Vestibulo-ocular reflexes stabilize visual images during head and body movements. This stabilizing prevents the visual world from appearing to bounce or jump around when the head moves, especially during walking. Lack of visual image stability can be seen in videotapes when the videographer walks with the camera, as the videotaped objects appear to bounce. Even more disconcerting to the viewer are abrupt swings of the video image, causing the visual objects to jump. Although these visual effects can be entertaining in giant-screen movies of airplanes swooping over canyons, in daily life lack of image stability can be disabling because the ability to use vision for orientation is lost.

Normally, when the head turns to the right, afferent information from the right horizontal semicircular canal is relayed to the vestibular nuclei for coordination of visual stabilization. Information is sent from the vestibular nuclei to the nuclei of cranial nerves III and VI, activating the rectus muscles that move the eyes to the left and inhibiting the rectus muscles that move the eyes to the right (Fig. 15–8).

Similarly, vertical vestibulo-ocular reflexes can be elicited by flexion of the head and by extension of the head. All vestibulo-ocular reflexes move the eyes in the direction opposite to the head movement to maintain stability of the visual field and visual fixation on objects.

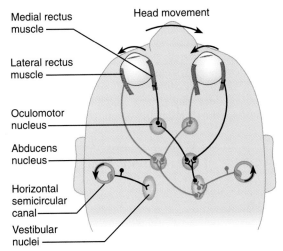

Medial rectus
muscle

Head movement

Lateral rectus
muscle

Oculomotor
nucleus

Abducens
nucleus

Horizontal
semicircular
canal

Vestibular
nuclei

FIGURE 15–8

Vestibulo-ocular reflex. When the head is turned to the right, inertia causes the fluid in the horizontal semicircular canals to lag behind the head movement. This bends the cupula in the right semicircular canal in a direction that increases firing in the right vestibular nerve. The cupula in the left semicircular canal bends in a direction that decreases the tonic activity in the left vestibular nerve. Neurons whose activity level increases with this movement are indicated in red. Neurons whose activity level decreases are black. For simplicity, the connections of the left vestibular nuclei are not shown. Via connections in the vestibular nuclei and the nuclei of cranial nerves III and IV, both eyes move in the direction opposite to the head turn.

Testing Vestibulo-Ocular Reflexes The vestibulo-ocular reflexes can be tested by passively moving an individual's head and observing the associated eye movements. Alternatively, the vestibulo-ocular reflex can be tested with the individual seated in a rotating chair. With the head in neutral position, while the individual is rotated to the left, the eyes will move slowly to the right, as if to maintain fixation on an object in the visual field. These slow-speed eye movements are called pursuit movements and are used to visually follow an object. When the eyes reach the extreme right, they shift quickly to the left, then resume moving to the right. The high-speed eye movements, called saccades, bring new objects into central vision, where details of images are seen. When the head is rotated to the left, the pursuit eye movements (slow movements) are toward the right, and the saccades (rapid movements) are toward the left.

If a person rotates quickly several times to the left, then abruptly stops rotating, the direction of slow and fast eye movements reverses; that is, the eyes repeatedly move slowly to the left and then quickly to the right. This reversal of eye movement is due to the inertia of the fluid continuing to flow in the horizontal canals after the head stops moving. The fluid movement bends the cupulas in the opposite direction to their bend during acceleration, producing the reversal of eye movements.

Nystagmus Involuntary back and forth movements of the eyes, as occur during or after rotation, are called **nystagmus**. Physiological nystagmus is a normal response that can be elicited in an intact nervous system by rotational or temperature stimulation of the semicircular canals (see Table 13–4) or by moving the eyes to the extreme horizontal position. Pathological nystagmus, a sign of nervous system abnormality, is discussed with disorders of the eye movement system.

VISUALLY GUIDED EYE MOVEMENTS (VISUO-OCULAR SYSTEM)

Visually guided eye movements are orchestrated by the superior colliculus. Maps of visual and auditory space, the body surface, and motor representations converge within the superior colliculus. As a result of connections among these maps, the superior colliculus directs saccades to areas of interest. For example, a fast-moving object in peripheral vision elicits reflexive movements of the eyes and head toward the stimulus. The motor output from the visuo-ocular system includes eye movement, head and body orientation, and postural adjustments. The superior colliculus also provides involuntary pursuit eye movements, with the eyes following moving visual targets reflexively.

The different actions of the visuo-ocular system can be demonstrated by the following task. Extend your arm, and place your index finger about 1.5 feet in front of you. Compare the visual clarity when you move your finger back and forth rapidly versus when the finger is held steady and you move your head rapidly from side to side. Explain the difference in ability to see details.*

---

*The difference in clarity is due to the ability of the nervous system to adjust eye movements based on anticipated head location versus the slower process of adjusting after visual information indicates loss of the target.

CORTICAL CONTROL OF EYE MOVEMENTS

Cortical centers influencing eye movements include the frontal eye fields and the occipitotemporal eye fields (Fig. 15–9). The frontal eye fields provide voluntary control of eye movements when a decision is made to look at a particular object. Saccades quickly switch vision from one object to another. Saccades are controlled by an area in the pontine reticular formation that receives information from the superior colliculus, frontal eye fields, and vestibular nuclei. The occipitotemporal eye fields contribute to the ability to visually pursue moving objects (pursuit eye movements).

> Eye movements are influenced by the following:
> - Auditory information (via superior colliculus)
> - Vestibulo-ocular reflex
> - Visuo-ocular system
> - Sensory information from extraocular muscles
> - Limbic system (see Chaps. 16 and 17) and voluntary control

DISORDERS OF AUDITORY, VESTIBULAR, AND VISUAL SYSTEMS

In each of the systems, disorders may affect the peripheral receptors, the cranial nerves, brain stem nuclei, tracts within the central nervous system, or the associated cortical areas. A complete lesion in the peripheral region, involving the receptors or cranial nerve dedicated to that system, will cause total, ipsilateral loss of sensory input for that system. Thus a complete peripheral lesion causes ipsilateral deafness in the auditory system, loss of vestibular information in the vestibular system, and blindness in the visual system. Lesions of these systems within the central nervous system cause more variable outcomes, depending on the location and extent of the lesion.

Disorders of the Auditory System

Deafness usually results from disorders affecting peripheral structures of the auditory system: the cochlea, the organ of Corti within the cochlea, or the cochlear branch of the vestibulocochlear nerve. Deafness due to peripheral disorders is classified as either conductive or sensorineural deafness.

Conductive deafness occurs when transmission of vibrations is prevented in the outer or middle ear. The common causes of conductive deafness are excessive wax in the outer ear canal or otitis media, inflammation in the middle ear. In otitis media, movement of the ossicles is restricted by thick fluid in the middle ear.

Sensorineural deafness, due to damage of the receptor cells or the cochlear nerve, is less common than conductive deafness. The usual causes are ototoxic drugs, Meniere's disease (see discussion of vestibular disorders), and acoustic neuroma. Ototoxic drugs have a poisoning effect on auditory structures, damaging cranial nerve VIII and/or the hearing and vestibular organs. An acoustic neuroma is a benign tumor of myelin cells surrounding cranial nerve VIII within the cranium.

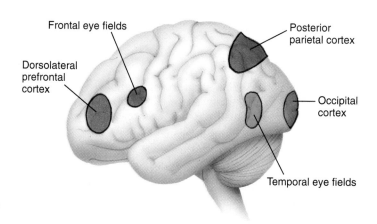

FIGURE 15–9

Cerebral cortex areas that direct eye movements. The frontal eye fields control voluntary eye movements. Occipitotemporal regions control pursuit eye movements. Posterior parietal cortex provides spatial information for eye movements.

Disorders within the central nervous system rarely cause deafness because auditory information projects bilaterally in the brain stem and cerebrum. Thus, small lesions in the brain stem typically do not interfere with the ability to hear. In the cerebral cortex, each primary auditory cortex receives auditory information from both ears, so hearing remains fairly normal when one primary auditory cortex is damaged. If the primary auditory cortex is destroyed on one side, the only loss is the ability to consciously identify the location of sounds because conscious location of sound is accomplished by comparing the time lag between auditory information reaching the cortex on one side versus the time required for auditory information to reach the opposite cortex.

Disorders of the Vestibular System

The most common symptom of vestibular system dysfunction is **vertigo,** an illusion of motion. People may falsely perceive movement of themselves or their surroundings. Vertigo results from either central or peripheral disorders.

CENTRAL VESTIBULAR DISORDERS

Central vestibular disorders result from brain stem ischemia, multiple sclerosis, or tumors in the brain stem/cerebellar region. Vertigo persisting more than 3 days with mild nausea and vomiting usually indicates a central nervous system dysfunction (Warner et al., 1992).

COMMON PERIPHERAL VESTIBULAR DISORDERS

Common peripheral vestibular disorders include benign paroxysmal positional vertigo, vestibular neuritis, Meniere's disease, and traumatic injury (Table 15–1).

Benign paroxysmal positional vertigo (BPPV) is a condition of acute onset of vertigo. The term *benign* indicates not malignant, *paroxysmal* means a sudden onset of a symptom or a disease, and *positional* denotes the provoking stimulus. In BPPV, a rapid change of head position results in vertigo that subsides in less than 2 minutes, even if the provoking head position is sustained. Displacement of otoconia from the macula into the semicircular canals, secondary to trauma, may be the cause of BPPV.

Vestibular neuritis is inflammation of the vestibular nerve, usually caused by a virus. Dysequilibrium, spontaneous nystagmus, nausea, and severe vertigo persist up to 3 days. Caloric testing (see Chap. 13) shows decreased or absent response on the involved side.

Meniere's disease causes a sensation of fullness in the ear, tinnitus (ringing in the ear), severe acute vertigo, nausea, vomiting, and hearing loss. Meniere's disease is associated with abnormal fluid pressure in the inner ear, but whether this is a cause of the disease or an effect is unknown. In extreme cases, the vestibulocochlear nerve may be surgically severed to relieve symptoms.

Traumatic injury of the head may cause concussion of the inner ear, fractures of the bone surrounding the vestibular apparatus and nerve, or pressure changes in the inner ear. Any of these injuries can compromise vestibular function.

TABLE 15–1

COMPARISON OF PERIPHERAL VESTIBULAR DISORDERS

	Benign Paroxysmal Positional Vertigo	Vestibular Neuritis	Meniere's Disease
Etiology	May be otoconia in semicircular canals	Infection	Unknown
Speed of onset	Acute	Acute	Chronic
Duration of typical incident	<2 minutes	2–3 days	0.5–24 hours
Prognosis	Improves in weeks or months	Improves after 3–4 days	Progressive worsening
Unique signs	Elicited by change of head position	None	Associated with hearing loss, tinnitus, and feeling of fullness in the ear

UNILATERAL VESTIBULAR LOSS

Unilateral vestibular loss causes problems with posture and nausea because information from the damaged side is not the reciprocal of information from the intact side. Some people with vestibular problems may have to consciously control their balance, decreasing their ability to concentrate on other tasks.

BILATERAL VESTIBULAR LOSS

Bilateral vestibular loss causes failure of the vestibulo-ocular reflex. When the person walks, the world appears to bounce up and down. This lack of visual stabilization is **oscillopsia.** People with chronic vestibular dysfunction often have stiffness of the neck and shoulders. This stiffness may be an attempt to stabilize the head, to lessen vertigo or oscillopsia (Horak and Shupert, 1994).

PHYSICAL THERAPY IN VESTIBULAR DISORDERS

Postural tests can be used to assess vestibular system function (see Table 10–3). However, none of these tests identify the etiology of equilibrium problems. The postural tests are either static or dynamic. Static tests include Romberg's test and stationary posturography. These tests do not evaluate the ability of the subject to prepare for or adapt to challenges to equilibrium. Dynamic tests include tilt boards and dynamic posturography. As discussed in Chapter 10, these tests only assess reactions to externally imposed displacements. Postural tests may help identify whether the person relies excessively on visual or proprioceptive sensory information and thus aid in devising compensatory strategies.

Physical therapy does not directly affect central dysfunctions of the vestibular system and is ineffective for active Meniere's disease. Physical therapy is effective for BPPV, unilateral vestibular loss or dysfunction, bilateral vestibular loss, and central vestibular disorders that benefit from movement retraining (Cohen et al., 1992). Exercises are designed to promote movement retraining, habituation, or substitution to improve function. Movement retraining consists of practicing and modifying movements as appropriate. Habituation is exposure to positions or movements that produce symptoms, followed by relaxation until the symptoms abate. The provocative stimuli are repeated frequently until the nervous system adapts to the stimuli. Habituation is effective for BPPV and unilateral vestibular loss.

Substitution (also called compensation) consists of using alternative sensory inputs or motor responses or using predictive/anticipatory strategies. To compensate for bilateral vestibular loss, people learn to substitute visual and somatosensory cues and anticipation for absent or unreliable vestibular information. See Herdman (1994) for a comprehensive treatment of vestibular rehabilitation.

DIFFERENTIATION OF DIZZINESS COMPLAINTS

People reporting dizziness are often describing quite different experiences. The therapist must distinguish among the following:

- Vertigo (illusion of movement)
- Near syncope (feeling of impending faint)
- Dysequilibrium (loss of balance)
- Light-headedness (inability to concentrate)

The differential diagnosis of these conditions is important because the etiology and treatment of each are different (Samuels, 1991). Vertigo indicates vestibular etiology, while the other symptoms typically do not indicate vestibular disorders. Near syncope is commonly caused by cardiovascular disorders. Dysequilibrium results from somatosensory deficits, basal ganglia disorders, cerebellar dysfunction, drug use, complete loss of vestibular function, or tumors in the brain stem/cerebellar region. Light-headedness is associated with psychological disorders, including affective and anxiety disorders and hyperventilation syndrome.

Disorders of the Visual System

The consequences of damage along the retinogeniculo-cortical pathway vary according to the location of the lesion (Fig. 15–10). Clinically, visual losses are described by referring to the visual field deficit. Interruption of the optic nerve results in total loss of vision in the ipsilateral eye. Damage to fibers in the center of the optic chiasm interrupts the fibers from the nasal half of each retina, resulting in loss of information from both temporal visual fields, called **bitemporal hemianopsia.** A complete lesion of the pathway anywhere posterior to the optic chiasm, in the optic tract, lateral geniculate, or optic radiations, results in loss of information from the contralateral visual field because all visual information posterior to the chiasm is from the contralat-

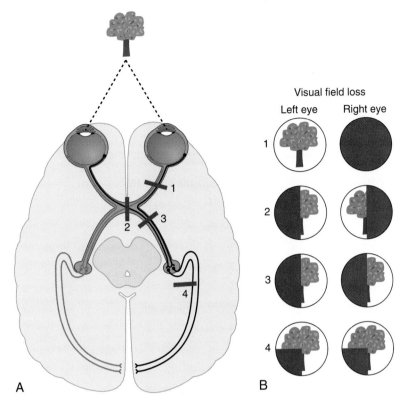

Visual field loss

FIGURE 15-10

Results of lesions at various locations in the visual system. *A,* Locations of the lesions. *B,* Visual field loss with each lesion. A lesion at location 1, optic nerve, causes loss of vision from the right eye. A lesion at location 2, the middle of the optic chiasm, causes bitemporal hemianopsia, loss of the temporal visual field from both eyes. Any lesion that completely interrupts tracts posterior to the optic chiasm, such as the lesion at location 3, optic tract, causes loss of vision from the contralateral visual field of both eyes. An incomplete lesion of tracts posterior to the optic chiasm, as shown at location 4, causes partial loss of vision from the contralateral visual field.

eral visual field. This loss of visual information from one hemifield is called **homonymous hemianopsia.**

Following complete, bilateral loss of visual cortex function, some people retain the ability to orient their head position to objects, despite being cortically blind. Cortically blind means the person has no awareness of any visual information. The ability for a cortically blind individual to orient to visual objects is called blind sight. Blind sight is possible because the ability to perceive light and dark vaguely is retained in the visual system. Blind sight is contingent on intact function of the retina and pathways from the retina to the superior colliculus.

Disorders of the Eye Movement System

Abnormalities of eye movement occur with damage to the following:

- Cranial nerves that control extraocular muscles
- Medial longitudinal fasciculus
- Vestibular system
- Cerebellum
- Eye fields in the cerebral cortex

Eye movement disorders that result from cranial nerve lesions are discussed in Chapter 13. If a lesion affects the medial longitudinal fasciculus, the movements of the eyes will not be coordinated with each other or with the movements of the head. Damage to the vestibular system or to the cerebellum can cause **pathological nystagmus,** abnormal oscillating eye movements that occur with or without external stimulation. Damage to a frontal eye field results in temporary ipsilateral gaze deviation; that is, the eyes look toward the damaged side. Recovery occurs because frontal eye field control of eye movement is controlled bilaterally. Damage to a parieto-occipital eye field causes inadequate pursuit eye movements. Although the lag of eye movements behind a moving target cannot be seen by an examiner, the disorder is visible because of the compensatory saccades that are required to catch up with a moving object.

CLINICAL NOTES

CASE 1

A.J. is a 57-year-old construction worker. In a fall from a scaffolding 1 week ago, he fractured his right temporal bone. He complains of difficulty maintaining his balance, neck and shoulder stiffness, blurred vision, nausea, and a spinning sensation. Clinical observation reveals the following:

- Walking is slow and unsteady, requiring contact with walls or other objects to avoid falling.
- A.J. avoids moving his head as much as possible, resulting in a rigid linkage between his trunk and head.
- Nystagmus is continuous, even when his head is stationary. Hearing is impaired on the right side.
- Muscle strength and somatosensation are normal. Vision and eye movements are intact.

Questions

1. Where is the lesion?
2. How can each of A.J.'s symptoms be explained?

CASE 2

B.F., a 37-year-old woman presents with the following signs and symptoms on the right:

- Loss of sensation from the face
- Loss of voluntary movement of the face
- Ataxia of the limbs
- Inability to move the right eye toward the right
- Deafness

In addition, pain and temperature sensations are impaired from the left side of the body, and she has vertigo, nystagmus, oscillopsia, and vomiting. The onset of symptoms has been gradual over the last 6 months, but unremitting.

Questions

1. Where is the lesion?
2. What is the most likely etiology?

CASE 3

K.W. is a 27-year-old man who abruptly sustained the following losses 2 days ago:

- The ability to localize pain, temperature, touch, and proprioceptive information from the right side of his body
- Voluntary motor control of the right side of his body
- The ability to see objects in his right visual field

Questions

1. What are possible locations of lesions that would explain the visual loss?
2. Given the combination of visual, motor, and somatosensory loss, where is the most likely location of the lesion?

REVIEW QUESTIONS

1. What midbrain structure is important in orienting the eyes and head toward the source of a sound?

2. How does the vestibular system contribute to gaze stabilization?

3. What does "each pair of semicircular canals produces reciprocal signals" mean?

4. Name the tracts that use information from the vestibular nuclei to control posture.

5. Explain how information from the left visual field reaches the right visual cortex.

6. What tract coordinates eye and head movements?

7. What is the difference between physiological and pathological nystagmus?

8. Name the sources of information used to direct eye movements.

9. What is BPPV?

10. What is oscillopsia?

References

Cohen, H., Rubin, A. M., Gombash, L., et al. (1992). The team approach to treatment of the dizzy patient. Arch. Phys. Med. Rehabil. 73(8):703–708.

Herdman, S. J. (Ed.) (1994). Vestibular rehabilitation. Philadelphia: F. A. Davis.

Horak, F. B., and Shupert, C. L. (1994). Role of the vestibular system in postural control. In: Herdman, S. J. (Ed.). Vestibular rehabilitation. Philadelphia: F. A. Davis, pp. 22–46.

Samuels, M. A. (1991). Manual of neurology: Diagnosis and therapy (4th ed.). Boston: Little, Brown.

Warner, E. A., Wallach, P. M., Adelman, M. M., et al. (1992). Dizziness in primary care patients. J. Gen. Intern. Med. 7(4):454–463.

16

Cerebrum

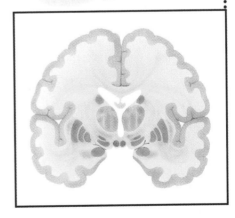

INTRODUCTION

Perception, moving voluntarily, using language and nonverbal communication, understanding spatial relationships, using visual information, making decisions, consciousness, emotions, mind-body interactions, and remembering all rely on systems in the cerebrum. These complex activities require extensive networks of neural connections, some involving brain stem circuits.

The cerebrum consists of the diencephalon and the cerebral hemispheres.* The diencephalon is in the center of the cerebrum, superior to the brain stem, and almost entirely enveloped by the cerebral hemispheres (Fig. 16–1). In the intact adult brain, only a small part of the diencephalon is visible: on the inferior surface, the region between the optic chiasm and the cerebral peduncles, marked by the mamillary bodies. The cerebral hemispheres include both subcortical structures and the cerebral cortex. The subcortical structures include the subcortical white matter, basal ganglia, and amygdala. The cerebral cortex is the gray matter on the external surface of the hemispheres. A collection of diencephalic, subcortical, and cortical structures involved with emotional and some memory functions is the limbic system.

DIENCEPHALON

The diencephalon includes all structures with the term *thalamus* in their names. The thalamus proper, the largest subdivision of the diencephalon, receives information from the basal ganglia, cerebellum, and all sensory systems except olfactory. The thalamus processes the information and then relays the information to specific areas of cerebral cortex. The other areas in the diencephalon are named for their locations relative to the thalamus, not similarities of function. Thus the hypothalamus is inferior and anterior to the thalamus, the epithalamus is superior and posterior to the thalamus, and the subthalamus is directly inferior to the thalamus.

Thalamus

The thalamus is a large, egg-shaped collection of nuclei located bilaterally above the brain stem. A Y-shaped sheet of white matter (intramedullary lam-

*Some authors consider the diencephalon to be part of the brain stem.

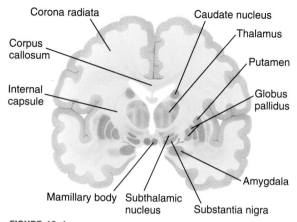

FIGURE 16–1

Cerebrum: diencephalon and cerebral hemispheres.

ina) divides the nuclei of each thalamus into three groups: anterior, medial, and lateral. The lateral group is further subdivided into dorsal and ventral tiers. All nuclei in these groups are named for their location. For example, the ventral anterior nucleus is the most anterior nucleus of the ventral tier.

Additional thalamic nuclei—intralaminar, reticular, and midline—are not included in the three major groups. Intralaminar nuclei are found within the white matter of the thalamus. The reticular and midline nuclei form thin layers of cells on the lateral and medial surfaces of the thalamus (Fig. 16–2).

The thalamus acts as an executive assistant for the cerebral cortex, directing attention to important information by regulating the flow of information to the cortex. Thus, overall, the thalamus regulates the activity level of cortical neurons. Individual thalamic nuclei can be classified into three main functional groups:

- Relay nuclei convey information from the sensory systems (except olfactory), the basal ganglia, or the cerebellum to the cerebral cortex.
- Association nuclei process emotional and some memory information or integrate different types of sensation, that is, touch and visual information.
- Nonspecific nuclei regulate consciousness, arousal, and attention.*

Relay nuclei receive specific information and serve as relay stations by sending the information directly

*Several classifications of thalamic nuclei are available, and the terminology is inconsistent. See Anthoney (1994) for a discussion of the variety of categories employed by various authors.

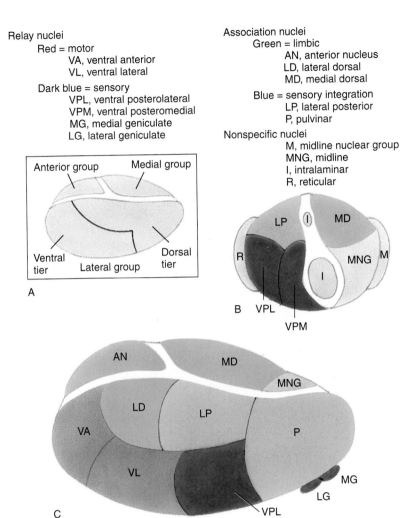

Relay nuclei
  Red = motor
    VA, ventral anterior
    VL, ventral lateral
  Dark blue = sensory
    VPL, ventral posterolateral
    VPM, ventral posteromedial
    MG, medial geniculate
    LG, lateral geniculate

Association nuclei
  Green = limbic
    AN, anterior nucleus
    LD, lateral dorsal
    MD, medial dorsal
  Blue = sensory integration
    LP, lateral posterior
    P, pulvinar
Nonspecific nuclei
    M, midline nuclear group
    MNG, midline
    I, intralaminar
    R, reticular

FIGURE 16–2

Thalamus. *A,* Three major groups of nuclei. *B,* Coronal section through thalamus. *C,* Nuclei of thalamus.

to localized areas of cerebral cortex. For example, the ventral posteromedial nucleus receives somatosensory information from the face and relays the information to the somatosensory cortex. All relay nuclei are found in the ventral tier of the lateral nuclear group.

Association nuclei connect reciprocally to large areas of cortex; that is, axons from association nuclei project to the cerebral cortex, and axons from the same cerebral cortical regions project to the association nuclei. Examples include the anterior nucleus, with reciprocal connections to areas of the cortex involved in emotions, and the pulvinar nucleus, reciprocally connecting with parietal, temporal, and occipital cortices. Association nuclei are found in the anterior thalamus, medial thalamus, and dorsal tier of the lateral thalamus.

Nonspecific nuclei receive multiple types of input and project to widespread areas of cortex. This functional group includes the reticular, midline, and intralaminar nuclei, important in consciousness and arousal. Table 16–1 lists the functions and connections of the thalamic nuclei.

Hypothalamus

The hypothalamus is essential for individual and species survival because integration of behaviors with visceral functions by the hypothalamus is required for continued existence. For example, small areas in the hypothalamus coordinate eating behavior with digestive activity. Electrical stimulation of these hypothala-

TABLE 16-1
THALAMIC NUCLEI

Functional Classification	Nuclei	Function	Afferents	Efferents
Relay nuclei	Ventral anterior	Motor	Globus pallidus	Motor planning areas
	Ventral lateral	Motor	Dentate	Motor cortex, motor planning areas
	Ventral posterolateral	Somatic sensation from body	Spinothalamic and medial lemniscus paths	Somatosensory cortex
	Ventral posteromedial	Somatic sensation from face	Sensory nucleus trigeminal nerve	Somatosensory cortex
	Medial geniculate	Hearing	Inferior colliculus	Auditory cortex
	Lateral geniculate	Vision	Optic tract	Visual cortex
Association nuclei	Anterior	Limbic	Reciprocal with limbic cortex	
	Medial dorsal	Limbic	Reciprocal with limbic cortex	
	Lateral dorsal	Limbic	Reciprocal with limbic cortex	
	Lateral posterior	Sensory integration	Reciprocal with parietal cortex	
	Pulvinar	Sensory integration	Reciprocal with parietal, occipital, and temporal cortex	
Nonspecific nuclei	Midline	Limbic	Viscera	Hypothalamus, amygdala, cerebral cortex
	Intralaminar	Limbic, arousal	Ascending reticular system	Widespread areas of cortex
	Reticular	Adjusts thalamic activity	Interconnections with other thalamic nuclei	

mic areas causes an animal to search for and ingest food as long as the stimulation is applied. At the same time, peristalsis and blood flow increase throughout the intestine. Bilateral destruction of the areas associated with eating behaviors results in refusal of food, causing starvation even when food is readily available. The following functions are orchestrated by the hypothalamus:

- Maintaining homeostasis: adjustment of body temperature, metabolic rate, blood pressure, water intake and excretion, and digestion
- Eating, reproductive, and defensive behaviors
- Emotional expression of pleasure, rage, fear, and aversion
- Regulation of circadian (daily) rhythms, such as sleep-wake cycles, in concert with other brain regions

- Endocrine regulation of growth, metabolism, and reproductive organs

These functions are carried out by hypothalamic regulation of pituitary gland secretions (hormones) and by efferent neural connections with the cortex (via thalamus), limbic system, brain stem, and spinal cord (Fig. 16–3).

Epithalamus

The major structure of the epithalamus is the pineal gland, an endocrine gland innervated by sympathetic fibers. The pineal gland is believed to help regulate circadian (daily) rhythms and influence the secretions of the pituitary gland, adrenals, parathyroids, and islets of Langerhans.

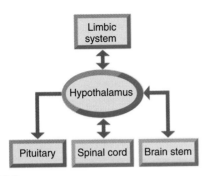

FIGURE 16-3

Interactions of the hypothalamus.

Subthalamus

Subthalamus is located superior to the substantia nigra of the midbrain. Functionally, the subthalamus is part of the basal ganglia circuit, involved in regulating movement. The subthalamus facilitates basal ganglia output nuclei.

SUBCORTICAL STRUCTURES

Subcortical White Matter

All white matter, whether located in the spinal cord or brain, consists of myelinated axons. In the cerebrum, the white matter is deep to the cortex and thus called subcortical. Subcortical white matter fibers are classified into three categories, depending on their connections:

• Projection
• Commissural
• Association

PROJECTION FIBERS

Projection fibers extend from subcortical structures to the cerebral cortex and from the cerebral cortex to the spinal cord, brain stem, basal ganglia, and thalamus. Almost all projection fibers travel through the internal capsule, a section of white matter bordered by the thalamus medially and the basal ganglia laterally (Fig. 16–4). Like the stems of a bouquet of flowers, the axons of projection neurons are gathered into a small bundle, the internal capsule. Above the internal cap-

sule, the axons spread apart to form the corona radiata, connecting with all areas of the cerebral cortex.

Regions of the internal capsule are the anterior limb, genu (from the Latin for "knee," indicating a bend), and posterior limb. The **anterior limb,** lateral to the head of the caudate, contains corticopontine fibers and fibers interconnecting thalamic and cortical limbic areas. The most medial part of the internal capsule, the **genu,** contains cortical fibers that project to cranial nerve motor nuclei and to the reticular formation. The **posterior limb** is located between the thalamus and lenticular nucleus, with additional fibers traveling posterior and inferior to the lenticular nucleus (retrolenticular and sublenticular fibers). The posterior limb contains corticopontine, corticospinal, and thalamocortical projections. The thalamocortical projections relay somatosensory, visual, auditory, and motor information to the cerebral cortex. Because axons from so many areas are together in the internal capsule, small lesions in the internal capsule have consequences disproportionate to their size.

COMMISSURAL FIBERS

Unlike projection fibers connecting cortical and subcortical structures, **commissural fibers** connect homologous areas of the cerebral hemispheres (Fig. 16–5 on p. 321). The largest group of commissural fibers is the corpus callosum, linking many areas of the right and left hemispheres. Fibers of the other two commissures, anterior and posterior, link the right and left temporal lobes.

ASSOCIATION FIBERS

Association fibers connect cortical regions within one hemisphere. The short association fibers connect adjacent gyri, while the long association fibers connect lobes within one hemisphere. For example, the cingulum connects frontal, parietal, and temporal lobe cortices. Additional long association fiber bundles are listed in Table 16–2 (on p. 321). Figure 16–5 illustrates the three types of white matter fibers.

Basal Ganglia

As noted in Chapter 9, the basal ganglia are vital for normal motor function. The basal ganglia sequence movements, regulate muscle tone and muscle force, and select and inhibit specific motor synergies. In

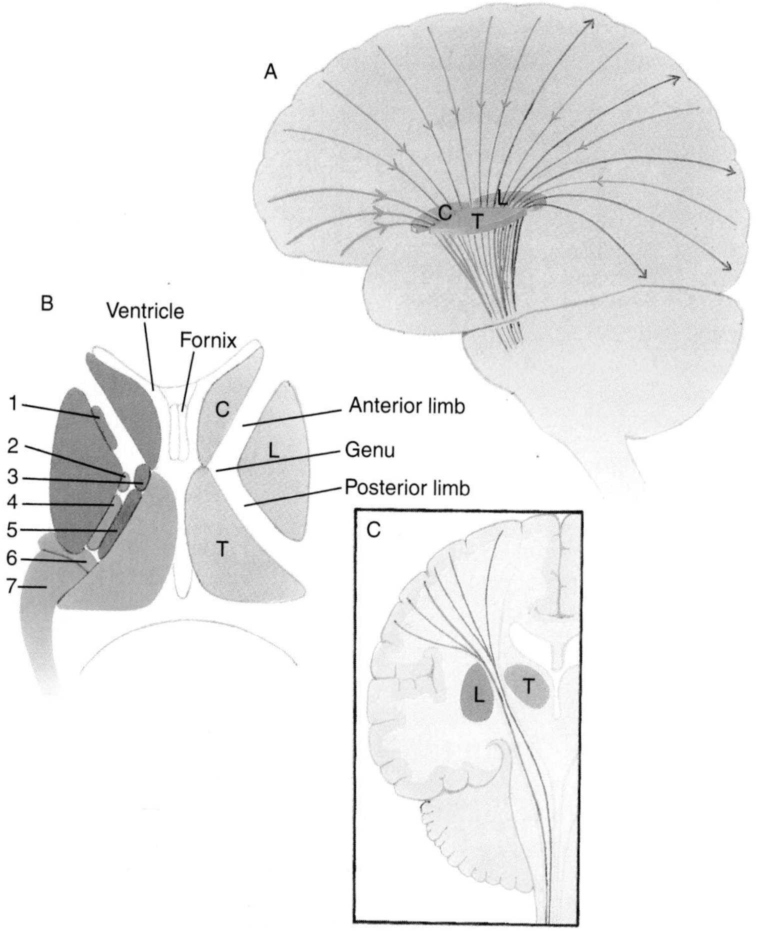

FIGURE 16–4

Internal capsule. Only the fibers projecting beyond the cerebrum are illustrated. Green, frontopontine fibers; red, motor fibers; blue, sensory fibers. *A*, Schematic view of the left internal capsule. The superior parts of the caudate (C) and lenticular (L) nuclei and the thalamus (T) have been removed. *B*, Horizontal section through the internal capsule. The limbs of the capsule are indicated on the right; the fiber tracts passing through are indicated on the left. The white areas of the internal capsule on the left contain thalamocortical fibers. Fiber tracts: (1) frontopontine, (2) corticorubral, (3) corticobulbar, (4) ascending sensory, (5) corticospinal, (6) auditory radiation, and (7) optic radiation. *C,* Coronal section of internal capsule. (Modified with permission from Moore, J. C. [1983]. Association fibers of cerebrum and commissural fibers. University of Puget Sound: A Look at the Nervous System from Five Perspectives. Handout.)

addition to their motor functions, the basal ganglia are involved in cognitive functions (see Alexander et al. [1990] for review), including the following:

- Awareness of body orientation in space
- Memory for location of objects
- Ability to change behavior as task requirements change
- Motivation

CEREBRAL CORTEX

The cerebral cortex is a vast collection of cell bodies, axons, and dendrites covering the surface of the cerebral hemispheres. The most common types of cortical neurons are pyramidal, fusiform, and stellate cells.

Pyramidal cells have an apical dendrite that extends toward the surface of the cortex, several basal dendrites extending laterally from the base of the soma, and one axon (Fig. 16–6 on p. 322). Although some pyramidal cells have short axons that synapse without leaving the cortex, almost all pyramidal cell axons travel through white matter as projection, commissural, or association fibers. Thus most pyramidal cells are output cells for the cerebral cortex. Fusiform cells are spindle-shaped and are also output cells, projecting mainly to the thalamus. Stellate (granule) cells are smaller than pyramidal cells, remain within the cortex, and serve as interneurons.

The cerebral cortex contains layers, differentiated by the size and connectivity of the constituent cells. In the olfactory and medial temporal cortex, there are only three layers of cells. In the remainder of the cere-

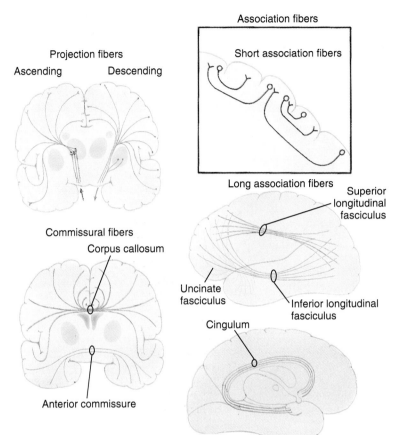

FIGURE 16–5

Types of white matter fibers. (Modified with permission from Moore, J. C. [1983]. Association fibers of cerebrum and commissural fibers. University of Puget Sound: A Look at the Nervous System from Five Perspectives. Handout.)

TABLE 16–2
SUBCORTICAL WHITE MATTER

Type of Fibers	Examples
Projection	Thalamocortical
	Corticospinal
	Corticobulbar
Commissural	Corpus callosum
	Anterior commissure
Association	Short association fibers (connect adjacent gyri)
	Cingulum (connects frontal, parietal, and temporal lobe cortices)
	Uncinate fasciculus (connects frontal and temporal lobe cortices)
	Superior longitudinal fasciculus (connects cortex of all lobes)
	Inferior longitudinal fasciculus (connects temporal and occipital lobes)

bral cortex, six layers of cells are found (Fig. 16–7). The six layers, numbered from superficial to deep, are listed in Table 16–3 (on p. 323).

This list of cortical layers is a generalization; different areas of the cerebral cortex have distinctive arrangements of cells. For example, although layers II through V can be distinguished in visual cortex, stellate cells predominate in these four layers. In 1909, Brodmann published a map of the cortex that distinguished 52 histological areas (Fig. 16–8 on p. 323). **Brodmann's areas** are commonly used to designate cortical locations.

Mapping of Cerebral Cortex

People undergoing brain surgery have allowed neurosurgeons to stimulate and record from various areas of the cerebral cortex. During these surgeries, the patients were fully conscious. Some experiments consist of placing recording electrodes on the surface of the brain and then stimulating various parts of the body to determine if the cortical area being recorded responds to the stimulus. For example, when the surgeon touches the patient's fingertip, only a small, specific area of the cortex, located in a consistent position of the cortex among various people, responds.

Other experiments involve mild electrical stimulation of the cortex. In these experiments, the stimulation may elicit movements of a part of the body or cause the patient to recall a particular situation. Imaging techniques can also be used to investigate brain function without invasive procedures. For example, the brain activity when a person is having a conversation can be recorded and analyzed.

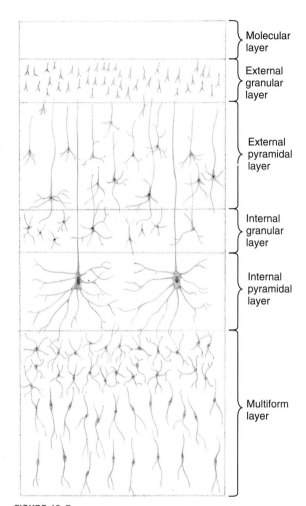

FIGURE 16–7
Layers of the cerebral cortex.

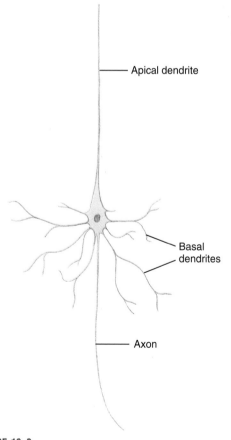

FIGURE 16–6
Pyramidal cell.

Localized Functions of Cerebral Cortex*

Different areas of the cerebral cortex are specialized to perform a variety of functions. Based on their functions, five categories of cortex have been identified:

- The primary sensory cortex discriminates among different intensities and qualities of sensory information.

---

*In neuroscience, the term *localization of function* is used to connote that an area contributes to the performance of a specific neural activity. Neural functions are achieved by networks of neurons, not by isolated centers.

TABLE 16–3	
LAYERS OF THE CEREBRAL CORTEX	
Name	**Description**
I	Molecular layer; mainly axons and dendrites; contains few cells
II	External granular layer; many small pyramidal and stellate cells
III	External pyramidal layer; pyramidal cells
IV	Internal granular layer; mainly stellate cells
V	Internal pyramidal layer; predominantly pyramidal cells, with stellate and other interneurons
VI	Multiform layer; primarily fusiform cells

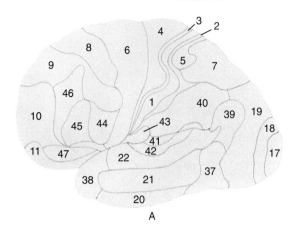

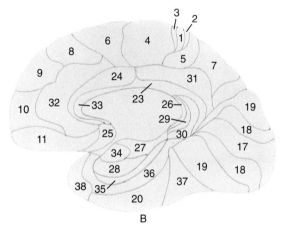

FIGURE 16–8

Brodmann's areas.

- The sensory association cortex performs more complex analysis of sensation.
- The motor planning areas organize movements.
- The primary motor cortex provides descending control of motor output.
- The association cortex controls behavior, interprets sensation, and processes emotions and memories.

Each type of cortex may play a role in response to a stimulus. For example, when one sees a bell, the primary visual cortex discriminates its shape and its brightness from the background. The visual association cortex analyzes the bell's color. The association cortex may recall the name of the object, what sound the bell makes, and specific memories associated with bells. The association cortex also participates in the decision of what to do with the bell. If the decision is to lift the bell, premotor areas plan the movement, and then the primary motor cortex sends commands to neurons in the spinal cord. The flow of cortical activity from the primary sensory cortex to cortical motor output is illustrated in Figure 16–9. Figure 16–9 is a simplified schematic that only applies to movement generated in response to an external stimulus. An equally plausible alternative would begin with a decision in the association cortex leading to movement.

PRIMARY SENSORY AREAS OF THE CEREBRAL CORTEX

Primary sensory areas receive sensory information directly from the ventral tier of thalamic nuclei. Each primary sensory area discriminates among different intensities and qualities of one type of input. Thus there are separate primary sensory areas for somatosensory, auditory, visual, and vestibular information. Most primary sensory areas are located within and adjacent to landmark cortical fissures (Fig. 16–10 on p. 325). The primary somatosensory cortex is located within the central sulcus and on the adjacent postcentral gyrus. The primary auditory cortex is located in the lateral fissure and on the adjacent superior temporal gyrus. The primary visual cortex is within the calcarine sulcus and on the adjacent gyri. Only the primary vestibular cortex is not close to a landmark fissure; instead, the primary vestibular cortex is posterior to the primary somatosensory cortex.

Primary Somatosensory Cortex The primary somatosensory cortex receives information from tactile

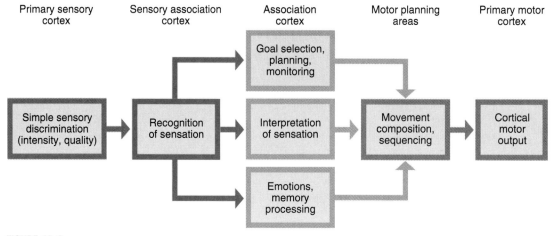

FIGURE 16–9

Flow of cortical information from the primary sensory cortex to motor output.

and proprioceptive receptors via a three-neuron pathway: peripheral afferent/dorsal column neuron, medial lemniscus neuron, and thalamocortical neuron. Although crude awareness of somatosensation occurs in the ventral posterolateral and the ventral posteromedial nuclei of the thalamus, neurons in the primary somatosensory cortex identify the location of stimuli and discriminate among various shapes, sizes, and textures of objects. The cortical termination of nociceptive and temperature pathways is more widespread than the discriminative tactile and proprioceptive information and thus is not limited to the primary somatosensory cortex.

Primary Auditory and Primary Vestibular Cortices
The primary auditory cortex receives information from the cochlea of both ears via a pathway that synapses in the inferior colliculus and medial geniculate body before reaching the cortex (see Chap. 15). The primary auditory cortex provides conscious awareness of the intensity of sounds. The primary vestibular cortex receives information regarding head movement and head position relative to gravity by a vestibulothalamocortical pathway (see Chap. 15).

Primary Visual Cortex Visual information travels to the cortex via a pathway from the retina to the lateral geniculate body of the thalamus, then to the primary visual cortex. Individual neurons in the primary visual cortex are specialized to distinguish between light and dark, various shapes, location of objects, and movement of objects.

SENSORY ASSOCIATION AREAS

Sensory association areas analyze sensory input from both the thalamus and the primary sensory cortex. Sensory association areas contribute to the analysis of one type of sensory information. For example, if one picks up a pen, the primary somatosensory cortex registers that the object is small, smooth, and cylindrical. The somatosensory association area recognizes the object as a pen, although a different area of the cortex is required to name the object. Somatosensory association areas integrate tactile and proprioceptive information from manipulating an object. Neurons in the somatosensory association area provide stereognosis by somehow comparing somatosensation from the current object with memories of other objects.

The visual association cortex analyzes colors and motion, and its output to the tectum directs visual fixation, the maintenance of an object in central vision. The auditory association cortex compares sounds with memories of other sounds and then categorizes the sounds as language, music, or noise. Sensory association areas are illustrated in Figure 16–11.

PRIMARY MOTOR CORTEX AND MOTOR PLANNING AREAS OF THE CEREBRAL CORTEX

The primary motor cortex is located in the precentral gyrus, anterior to the central sulcus. The primary motor cortex is the source of many cortical upper mo-

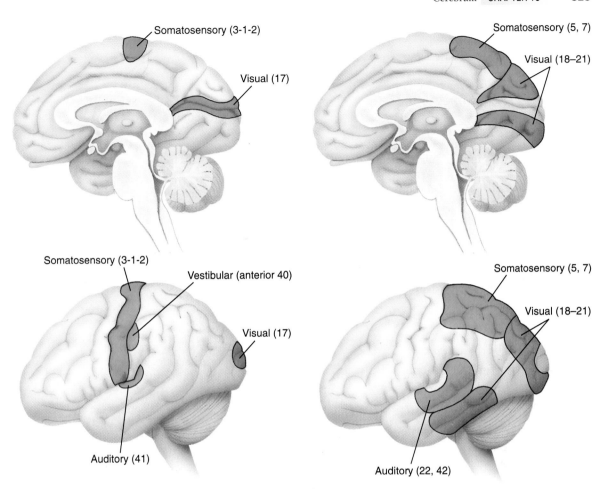

FIGURE 16–10

Primary sensory areas of the cerebral cortex. Corresponding Brodmann's areas are indicated in parentheses.

Cortical Area	Function
Primary somatosensory	Discriminates shape, texture, or size of objects
Primary auditory	Conscious discrimination of loudness and pitch of sounds
Primary visual	Distinguishes intensity of light, shape, size, and location of objects
Primary vestibular	Discriminates among head positions and head movements

FIGURE 16–11

Sensory association areas of the cerebral cortex. Corresponding Brodmann's areas are indicated in parentheses.

Cortical Area	Function
Somatosensory association	Stereognosis and memory of the tactile and spatial environment
Visual association	Analysis of motion, color; control of visual fixation
Auditory association	Classification of sounds

toneurons and controls contralateral voluntary movements, particularly the fine movements of the hand and face. Because the primary motor cortex is unique in providing precise control of hand and lower face movements, a much greater proportion of the total area of primary motor cortex is devoted to neurons that control these parts of the body than is devoted to the trunk and proximal limbs, where more gross motor activity is required. The hand, foot, and lower face representations in the motor cortex are entirely contralateral. In contrast, many muscles that tend to be active bilaterally simultaneously—muscles of the back, for example—are controlled by the primary motor cortex on both sides.

The cortical motor planning areas include the following:

- Supplementary motor area
- Premotor area
- Broca's area
- The area corresponding to Broca's area in the opposite hemisphere

Motor Planning Areas The cortex anterior to the primary motor cortex consists of three areas: **supplementary motor area, premotor area**, and **Broca's area** (or, on the contralateral side, the area corresponding to Broca's area). The supplementary motor cortex, located anterior to the lower body region of the primary motor cortex, is important for initiation of movement, orientation of the eyes and head, and planning bimanual and sequential movements. The premotor area, located anterior to the upper body region of the primary motor cortex, controls trunk and girdle muscles via the medial activation system.

Broca's area, inferior to the premotor area and anterior to the face and throat region of the primary motor cortex, is usually in the left hemisphere. Broca's area is responsible for planning movements of the mouth during speech and the grammatical aspects of speech. An area analogous to Broca's area, in the opposite hemisphere, plans nonverbal communication, including emotional gestures and adjusting the tone of voice. These areas will be considered further in a later section on communication.

Connections of the Motor Areas Premotor, supplementary motor, and Broca's areas receive information from sensory association areas. Both the primary motor cortex and motor planning areas receive information from the basal ganglia and cerebellum, relayed by the thalamus. The primary motor cortex receives somatosensory information relayed by the thalamus and from the primary somatosensory cortex and motor instructions from the motor planning areas. Cortical motor output, including the corticospinal tracts, corticobulbar tracts, corticopontine tracts, and cortical projections to the putamen, originates in the primary motor and primary somatosensory cortex and motor planning areas (Fig. 16–12).

ASSOCIATION AREAS OF THE CEREBRAL CORTEX

Areas of cortex not directly involved with sensation or movement are called the association cortex. Three areas of cortex are designated the association cortex:

- Prefrontal, the anterior part of the frontal lobe
- Parietotemporal association, at the junction of the parietal, occipital, and temporal lobes
- Limbic, in the anterior temporal lobe and inferior frontal lobe

Enormously complex abilities are localized in the association areas: personality, integration and interpretation of sensations, processing of memory, and generation of emotions. For example, damage to the orbitofrontal cortex can alter personality characteristics, while damage to other cortical areas has little effect on personality. Thus, although the neurophysiology of personality is not understood, personality is said to be localized to the orbitofrontal cortex. Similarly, intelligence, as measured by intelligence tests, and integration and interpretation of sensations are localized in the parietotemporal association areas. Conscious emotions are localized in the limbic association areas (Fig. 16–13 on p. 328).

The **prefrontal cortex** connects extensively with the sensory association areas in the parietal, occipital, and temporal lobes and with limbic areas. Prefrontal cortex functions include self-awareness and executive functions (also called goal-oriented behavior). Executive functions include the following:

- Deciding on a goal
- Planning how to accomplish the goal
- Executing the plan
- Monitoring the execution of the plan

Decisions ranging from the trivial to the momentous are made in the prefrontal area; what to wear, whether to move, and whether to have children are decided in and carried out by instructions from the prefrontal cortex.

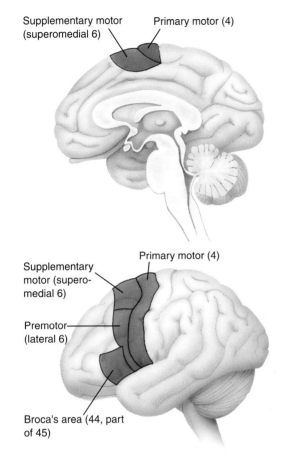

Motor Areas	Function
Primary motor cortex	Voluntarily controlled movements
Premotor area	Control of trunk and girdle muscles, anticipatory postural adjustments
Supplementary motor area	Initiation of movement, orientation planning, bimanual and sequential movements
Broca's area	Motor programming of speech (usually in left hemisphere only)
Area analogous to Broca's in opposite hemisphere	Planning nonverbal communication (emotional gestures, tone of voice; usually in the right hemisphere)

FIGURE 16–12

Motor areas of the cerebral cortex. Corresponding Brodmann's areas are indicated in parentheses.

Intelligence is primarily a function of the **parieto-temporal association areas,** in the posterior parietal and temporal cortices. Here, problem solving and comprehension of communication and of spatial relationships occur.

The third, and final, cortical association area is the **limbic association area,** located in the anterior temporal lobe and in the orbitofrontal cortex located above the eyes. The limbic association area connects with areas regulating mood (subjective feelings), affect (observable demeanor), and processing of some types of memory.

LIMBIC SYSTEM

The term *limbic* means "border" and refers to the border between the diencephalon and telencephalon. The term *border* could also be applied to the limbic system's activity as a border region between conscious and nonconscious areas of the brain.

Limbic structures form a ring around the thalamus (Fig. 16–14). Although a full consensus on which structures compose the limbic system has not been reached, most authorities include the following areas:

- Hypothalamus
- Anterior and medial nuclei of the thalamus
- Limbic cortex (cingulate gyrus, parahippocampal gyrus, uncus)
- Hippocampus
- Amygdala
- Basal forebrain: septal area, preoptic area, nucleus accumbens, and the nucleus basalis of Meynert

The hypothalamus and thalamus are described earlier in this chapter. **Limbic cortex** is a **C**-shaped region of cortex located on the medial hemisphere, consisting of the **cingulate gyrus, parahippocampal gyrus,** and **uncus** (a medial protrusion of the parahippocampal gyrus). The **hippocampus** is named for its fancied resemblance, in coronal section, to the shape of a seahorse. The hippocampus is formed by the gray and white matter of two gyri rolled together in the medial temporal lobe. The **amygdala** is an almond-shaped collection of nuclei deep to the uncus in the temporal lobe, at the end of the caudate tail.

In the basal forebrain, the **septal area** is a region of cortex and nuclei anterior to the anterior commissure, and the **preoptic area** is anterior to the septal area. The **nucleus accumbens** (also called ventral striatum) is the region where the caudate and putamen blend, and the **nucleus basalis of Meynert** is inferior to the preoptic area.

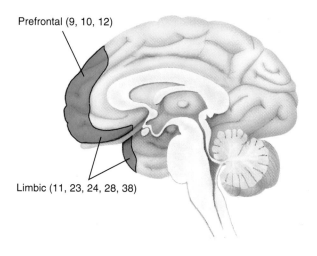

Prefrontal (9, 10, 12)

Limbic (11, 23, 24, 28, 38)

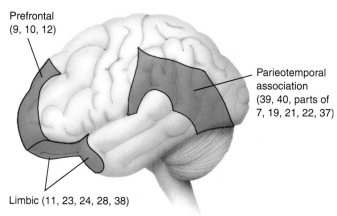

Prefrontal
(9, 10, 12)

Parieotemporal
association
(39, 40, parts of
7, 19, 21, 22, 37)

Limbic (11, 23, 24, 28, 38)

Association Cortex	Function
Prefrontal association	Goal-oriented behavior, self-awareness
Parietotemporal association	Sensory integration, problem solving, understanding language and spatial relationships
Limbic association	Emotion, motivation, processing of memory

FIGURE 16–13

Association areas of the cerebral cortex. Corresponding Brodmann's areas are indicated in parentheses.

Connections of the Limbic System

Connections within the limbic system are extensive. Two major fiber bundles are the **fornix** and **medial forebrain bundle** (Fig. 16–15). The fornix is an arch-shaped fiber bundle connecting the hippocampus with the mamillary body and anterior nucleus of the thalamus. The medial forebrain bundle connects an-terior structures (septal area, nucleus accumbens, amygdala, anterior cingulate gyrus), the hypothalamus, and the midbrain reticular formation. Because the connection between limbic areas and reticular areas of the midbrain are so important in behavior, combining the systems into a reticulolimbic system has been proposed. Additional connections of the limbic system are shown in Figure 16–15. Output

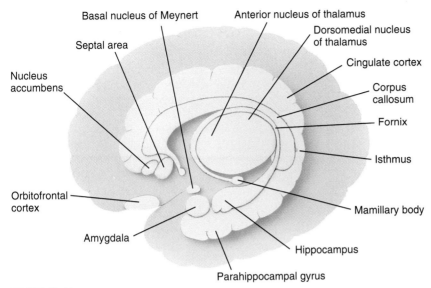

FIGURE 16–14
Limbic areas.

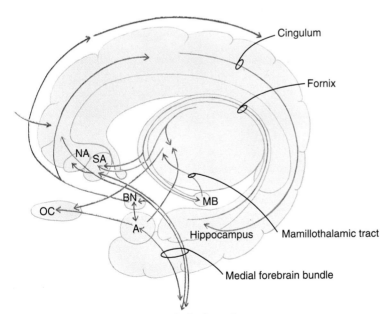

FIGURE 16–15
Connections of the limbic system. A, amygdala; OC, orbitofrontal cortex; MB, mamillary body of the thalamus. All other areas labeled with initials are parts of the basal forebrain: NA, nucleus accumbens; SA, septal area; BN, basal nucleus of Meynert.

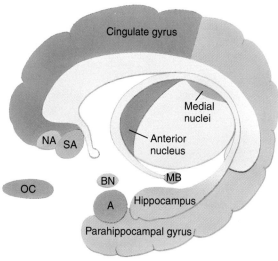

FIGURE 16–16

Sets of limbic structures: emotional and memory. Limbic structures involved in memory are shown in blue: amygdala, parts of hypothalamus, septal area, anterior nucleus of the thalamus, and anterior limbic cortex. Limbic structures involved in processing memory are shown in green: hippocampus, parahippocampal gyrus, medial thalamic nuclei, posterior limbic cortex, and basal forebrain. Septal area is involved in both emotional and memory functions.

from the limbic system travels via autonomic, somatic, reticular, and hormonal pathways.

Emotional and Memory Functions

Although the limbic system regulates feeding, drinking, defensive, and reproductive behaviors, in addition to visceral and hormonal functions, only two aspects of limbic function are considered in this text: emotional and memory functions. These functions use two fairly distinct subsets of limbic structures (Fig. 16–16). For emotions, the amygdala, areas in the hypothalamus, septal area, anterior nuclei of the thalamus, anterior limbic cortex, and limbic association area are required.* Unlike emotions, which are mediated within the limbic system, memory functions are widely distributed among limbic and nonlimbic areas of the brain. For processing some types of memory, the hippocampus, medial thalamic nuclei, posterior limbic cortex, and basal forebrain are essential.

*A highly influential hypothesis of the neural structures involved in emotions was proposed by Papez and became known as the circuit of Papez. Although the hypothesized connections have been confirmed, the circuit is not a major contributor to emotions. Instead, the circuit of Papez contributes to processing some types of memory.

EMOTIONS AND BEHAVIOR

Emotions are mediated within the limbic system, by the amygdala, areas in the hypothalamus, septal area, anterior nuclei of the thalamus, anterior cingulate cortex, and limbic association cortex. In humans, electrical stimulation of the amygdala elicits emotional experiences (Gloor, 1986). The amygdala receives information from all sensory systems and connects with the orbitofrontal cortex and anterior cingulate gyrus. Together the amygdala, orbitofrontal cortex, and anterior cingulate gyrus regulate emotional behaviors and motivation. The **amygdala** plays a vital role in social behavior, interpreting facial expressions and social signals (Young et al., 1995). Figure 16–17 illustrates the areas involved in the production of social behavior.

Emotions color our perceptions and influence our actions. For example, a person vexed by a difficult problem may misinterpret a question about progress in solving the problem as a threat and become angry. The person's facial expressions and abrupt, choppy movements indicating anger are easy to recognize. Immediate responses to a threat include somatic, autonomic, and hormonal changes, including increased muscle tension and heart rate, dilation of the pupils, and cessation of digestion. However, emotions also shape our lives in more subtle ways because emotions signal the nonconscious evaluation of a situation.

Conscious awareness of emotion occurs when information from the amygdala and from the autonomic system reaches the cortex. Emotion is intimately tied to decision making (Damasio et al., 1991; Bechara et al., 1994). Damasio et al. speculate that part of our decision-making process is imagining consequences and then attending to the resultant emotional signals from the visceral, muscular, and hormonal systems and neuromodulators. These emotional signals are based on prior experience and provide "gut feelings" about the actions being contemplated. The theory of Damasio et al. that emotions are crucial for sound judgment is called the somatic marker hypothesis. The emotional signals do not make decisions but are considered in the decision process.

PSYCHOLOGICAL AND SOMATIC INTERACTIONS

Clearly, the conceptual division between the sciences of immunology, endocrinology, and psychology/neuroscience is a historical artifact; the existence of a

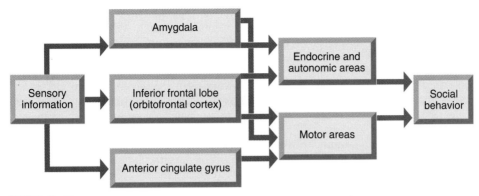

FIGURE 16-17

Social behavior: flow of information from sensory input to motor output.

communicating network of neuropeptides and their receptors provides a link among the body's cellular defense and repair mechanisms, glands, and brain.

—*Candace Pert, p. 1245*

Thoughts and emotions influence the functions of all organs. This occurs because of the bidirectional communication between the nervous system and the immune system (Fig. 16–18). Neurotransmitters and hormones regulated by the brain modulate immune system cells, and cytokines (chemicals secreted by white blood cells, including tumor necrosis factor and interleukins) regulate the neuroendocrine system. An individual's reaction to experiences can disrupt homeostasis; this is called a stress response. When an individual feels threatened, the stress response increases strength and energy to deal with the situation. Three systems create the stress response:

- Somatic nervous system: Motoneuron activity increases muscle tension.
- Autonomic nervous system: Sympathetic activity increases blood flow to muscles and decreases blood flow to the skin, kidneys, and digestive tract.
- Neuroendocrine system: Sympathetic nerve stimulation of the adrenal medulla causes the release of epinephrine into the bloodstream. Epinephrine increases cardiac rate and the strength of cardiac contraction, relaxes intestinal smooth muscle, and increases metabolic rate.

About 5 minutes after the initial response to stress, the hypothalamus stimulates the pituitary to secrete adrenocorticotropic hormone, causing the release of **cortisol** from adrenal glands. Cortisol mobilizes energy (glucose), suppresses immune responses, and

serves as an anti-inflammatory agent. These actions begin the return to homeostasis. Ideally, this would be the end of the story: stress response, followed by restoration of homeostasis. Unfortunately, often the stress response does not terminate because the stress is maintained by circumstances or by the individual's thinking patterns. For example, a social slight that would go unnoticed by one person may cause an-

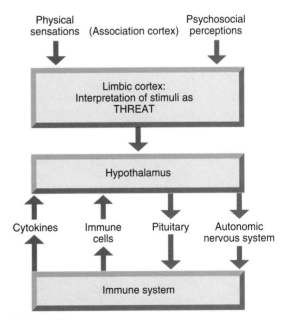

FIGURE 16-18

Chemical signaling between the nervous system and immune system in response to stress. Cytokines are nonantibody proteins that participate in the immune response (e.g., interferons and interleukins).

other person to extensively contemplate why they were snubbed and how they should respond.

Excessive amounts of cortisol are associated with stress-related diseases, including colitis, cardiovascular disorders, and adult-onset diabetes. Excessive cortisol also causes emotional instability and cognitive deficits (Murphy, 1991) and interferes with memory and attention in healthy elderly people (Lupien et al., 1994).

Short-term effects of negative behavior on immune function have been documented in a study of newly-wed couples (Kiecolt-Glaser et al., 1993). The participants were selected based on high marital satisfaction and good mental and physical health. Each couple discussed a marital problem for 30 minutes. Their immune responses and blood pressure were monitored before the discussion and at intervals following the discussion; final measurements were taken 24 hours later. Couples who used sarcasm, interruptions, and criticism during the discussion showed decreased immune responses and larger, more persistent increases in blood pressure than couples who displayed less negative behavior.

When the stress response is prolonged, persistent high levels of cortisol continue to suppress immune function. Immune suppression is advantageous for decreasing inflammation and regulating allergic reactions and autoimmune responses. However, stress-induced immune suppression may lead to unchecked growth of tumors. Research indicates that persons who experience severe problems at work occurring over a period of 10 years have a risk of colorectal cancer 5.5 times higher than the risk of adults without a history of work-related difficulties (Courtney et al., 1993). Thus the effects of the stress response can be beneficial or damaging, depending on the situation and whether the response is prolonged. Figure 16–19 illustrates the consequences of prolonged psychological stress. As noted in the figure, immune cells respond to neurotransmitters, neurohormones, and neuropeptides.

Researchers are also beginning to analyze ways that immune function can be improved. Given that an individual's interpretation of and reaction to a challenge are important in determining the physiological outcome of stressful situations, changing a person's attitudes may have an effect on immune function. Supportive group therapy has been demonstrated to double the survival time in patients with metastatic breast cancer, with the effect continuing after group therapy had ended (Spiegel et al., 1989). Short-term benefits of relaxation have been demonstrated in people practicing relaxation exercises. An average 48% increase in the level of an immune response protein (interleukin-1) was found when subjects' fingertips were warm and their breathing was relaxed (Keppel

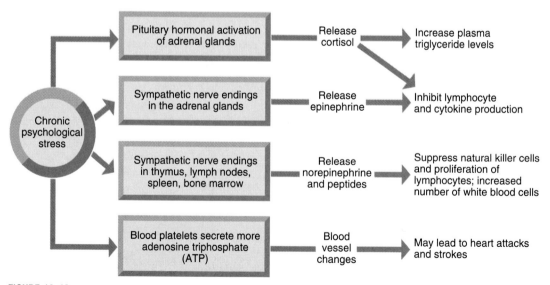

FIGURE 16–19

Effects of prolonged psychological stress on immune and blood-vascular system function.

et al., 1993). Thus cognition, emotions, and immune activity are intertwined. Another component in this complex is how records of new experiences are formed and used to guide subsequent activities. Memory functions involve many more areas of the brain than just limbic structures, as the following section illustrates.

MEMORY

Three memory systems serve distinct types of information. For example, each of the following types of memory is different:

- Remembering feeling elated
- Recalling what happened yesterday
- Knowing how to ride a bike

Each type of memory is dependent on different brain regions. Types of memory include emotional (feelings), declarative (facts, events, concepts, locations), and procedural (how-to) (Fig. 16–20).

Very little is known about the emotional memory system other than that memory for fear involves the amygdala and that damage to either of the other two memory systems does not affect the emotional memory system. Therefore the emotional memory system will not be considered further.

Declarative Memory

Declarative memory refers to recollections that can be easily verbalized. Declarative memory is also called conscious, explicit, or cognitive memory. Unlike emotional and procedural memory, declarative memory requires attention during recall. Declarative memory has three stages:

- Immediate memory (also called sensory register) lasts only 1 to 2 seconds. Information is processed through primary sensory and sensory association areas of cortex, but not by the limbic system.
- Short-term memory is brief storage of stimuli that have been recognized. Loss of information occurs within 1 minute unless the material is continually rehearsed.
- Long-term memory is relatively permanent storage of information that has been processed in short-term memory. The conversion of short-term to long-term storage is called consolidation.

A probable circuit of neural activity leading to the development of declarative memory is shown in Figure 16–21. In the frontal lobes, voluntary control is exerted over the processing of declarative memory, both in selecting information for storage and accessing stored information (Moscovitch, 1995). Electrical

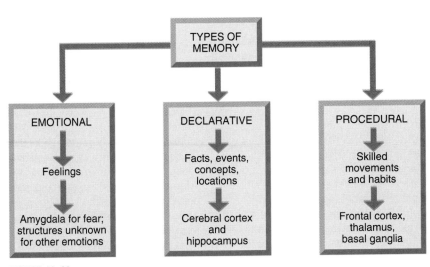

FIGURE 16–20

The three types of memory.

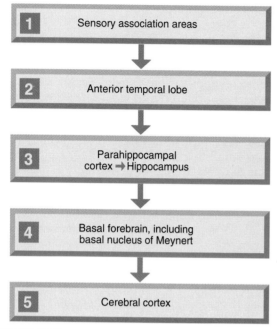

FIGURE 16–21

Formation of declarative memories. The green boxes indicate the contents that are part of the limbic system. Although parts of the limbic system are essential for converting short-term memory into long-term memory, memories are not stored in limbic structures.

stimulation of the anterior temporal lobe cortex causes people to report that it seems as if a past event or experience were occurring during the stimulation, despite their awareness of actually being in surgery (Penfield, 1958).

A famous case of unintended consequences of a surgery to relieve severe epilepsy contributed significantly to the understanding of memory. The patient, H.M., suffered severe, frequent seizures. Because his seizures originated in the medial temporal lobes, which contain the hippocampus, this area of his brain was removed bilaterally. The epilepsy improved, but his memory was permanently damaged. In the 40 years subsequent to the surgery, H.M. has been unable to remember any new information from 1 year prior to the surgery to the present. He cannot recall text he read minutes ago, nor can he remember people he has met repeatedly subsequent to the operation. Earlier memories are intact, and he is able to learn new skills. These outcomes indicate that the role of the hippocampus is processing memory from

short term to long term but that declarative memories are not stored in the hippocampus. Long-term storage is distributed among various cortical areas by the basal nucleus of Meynert.

The mechanism for converting short-term memories to long-term memories is not understood. Short-term memory is assumed to reflect temporary changes in cell membrane excitability. Long-term memory is believed to involve structural changes in neurons. **Long-term potentiation** (see Chap. 4) is a current proposal for explaining the cellular basis for memory, but more evidence is required to substantiate the hypothesis. Long-term potentiation consists of persistent enhancement of synaptic transmission following activation of specific receptors by high-frequency stimulation of presynaptic axons (Bliss and Collingridge, 1993).

Procedural Memory

Procedural memory refers to recall of skills and habits. This type of memory is also called skill, habit, nonconscious memory, or implicit memory. Practice is required to store procedural memories. Once the skill or habit is learned, attention is not needed while performing the task. For example, the initially difficult skill of driving a car in traffic becomes automatic with practice.

For learning motor skills, three learning stages have been identified:

- Cognitive
- Associative
- Autonomous

During the cognitive stage, the beginner is trying to understand the task and find out what works. Often beginners verbally guide their own movements, as seen in people who talk their way through descending stairs with crutches: "First the crutches, then the cast, then the right leg. . . ." During the associative stage, the person refines the movements selected as most effective. Movements are less variable and less dependent on cognition. During the autonomous stage, the movements are automatic, not requiring attention. When movements are automatic, attention can be devoted to having a conversation or other activities while the movements are being executed.

Although a decade ago the cerebellum was widely considered a likely candidate for mediating procedural memory, recent research has not supported cerebellar primacy. Instead, current evidence supports a frontothalamostriatal system for processing motor memory (Appollonio et al., 1993; Gabrieli, 1993). Evidence comes both from studies of people with specific lesions and from imaging studies of people with intact nervous systems performing movements. People with Huntington's disease who perform normally on a verbal memory task are impaired in their ability to learn a motor skill (Heindel and Salmon, 1989). The primary motor cortex rapidly reorganizes following practice (see review by Donoghue and Sanes [1994]).

Karni et al. (1995) demonstrated that some motor learning occurs in the primary sensory and motor cortices (together called sensorimotor). Participants practiced a specific sequence of finger movements. As performance improved, enlargement of the cortical area activated by the movement occurred. Moving the fingers in an unpracticed pattern activated a smaller area. Thus the sensorimotor area neurons appear to form networks delineating specific movement patterns.

Sietz and Roland (1992) also studied sequential finger movements. Initially, learning a pattern of sequential finger movements activated the contralateral sensorimotor hand area, premotor area, somatosensory association area, and ipsilateral cerebellum. As learning progressed, activity in the somatosensory association area declined as the participants reported no longer using internal verbal cues to guide the movements and as the movements became less dependent on somatosensory information. While activity in the somatosensory association area declined, activity of the globus pallidus and putamen increased. Sietz and Roland concluded that the basal ganglia are critical in the establishment of motor memories.

The abilities of H.M., the man with both hippocampi removed, illustrate the dissociation of declarative and procedural memories. He is able to learn new motor skills but cannot consciously remember that he has learned them. Thus his procedural memory is intact, despite his total loss of ability to consciously recall having practiced a task. H. M.'s communication abilities are intact because different brain areas are responsible for communication.

COMMUNICATION

People use both language and nonverbal methods to communicate. In approximately 95% of people, the cortical areas responsible for understanding language and producing speech are found in the left hemisphere. The distinction between language, a communication system based on symbols, and speech, the verbal output, is clinically important because different regions of the brain are responsible for each function.

Comprehension of spoken language occurs in **Wernicke's area,** a subregion of the left parietotemporal cortex. Broca's area, in the left frontal lobe, provides instructions for language output. These instructions consist of planning the movements to produce speech and providing grammatical function words, such as the articles *a, an,* and *the.* The contributions of the cortical and subcortical areas involved in normal conversation are shown in Figure 16–22.

In contrast to the auditory neural networks used during conversation, reading requires intact vision, visual association areas for visual recognition of the written symbols, plus connections with an intact Wernicke's area for interpreting the symbols. Writing requires motor control of the hand in addition to connections with Wernicke's and Broca's areas. Broca's area provides the grammatical relationship between words when writing, and Wernicke's area provides formulation of language.

Given that the right hemisphere typically does not process language, what do the contralateral areas corresponding to Wernicke's and Broca's areas contribute? In most people, activity in these areas of the right hemisphere is associated with nonverbal communication. Gestures, facial expressions, tone of voice, and posture convey meanings in addition to a verbal message. In the right hemisphere, the **area corresponding to Wernicke's area** is vital for interpreting the nonverbal signals from other people. The right hemisphere **area corresponding to Broca's area** provides instructions for producing nonverbal communication, including emotional gestures and intonation of speech.

Cerebral Dominance

Traditionally the hemisphere that manages language is called the dominant hemisphere, and the hemisphere with less language capacity is considered nondominant. The hemisphere that is dominant for language is also superior at logic and analytical tasks. This terminology may be misleading because the nondominant hemisphere is superior in understanding and producing nonverbal communication and comprehending spatial relationships.

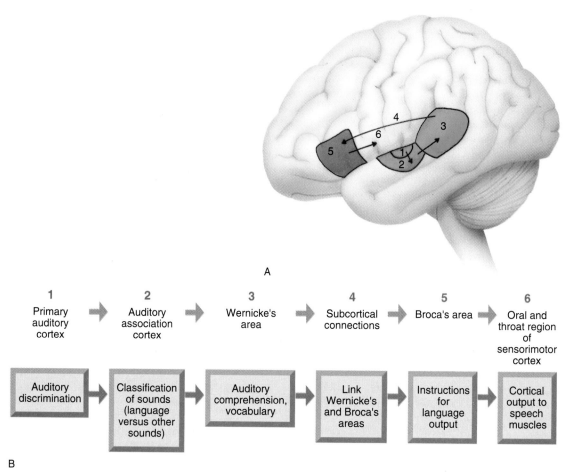

A

1	2	3	4	5	6
Primary auditory cortex	Auditory association cortex	Wernicke's area	Subcortical connections	Broca's area	Oral and throat region of sensorimotor cortex

| Auditory discrimination | → | Classification of sounds (language versus other sounds) | → | Auditory comprehension, vocabulary | → | Link Wernicke's and Broca's areas | → | Instructions for language output | → | Cortical output to speech muscles |

B

FIGURE 16–22

Flow of information during conversation, from hearing speech to replying.

COMPREHENSION OF SPATIAL RELATIONSHIPS

The area corresponding to Wernicke's area in the right hemisphere comprehends spatial relationships, providing schemas of the following:

- The body
- The body in relation to its surroundings
- The external world

The body schema, also known as the body image, is a mental representation of how the body is anatomically arranged (e.g., with the hand distal to the forearm). Schemas of the self in relation to the surroundings enable us to locate objects in space and to navigate accurately, finding our way within rooms and hallways and outside. Schemas of the external world provide the information necessary to plan a route from one site to another.

USE OF VISUAL INFORMATION

Visual information processed by the visual association cortex flows in two directions: dorsally, in an action stream to the posterior parietal cortex, and ventrally, in a perceptual stream to the occipitotemporal lobes (Fig. 16–23). Information in the action stream is used

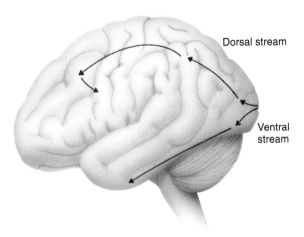

FIGURE 16–23

Use of visual information by the cerebral cortex: the action stream (dorsal) and the perceptual stream (ventral).

to adjust limb movements. For example, when a person reaches for a cup, the visual information in the dorsal stream is used to orient the hand and position the fingers appropriately during the reach. In contrast, information in the perceptual stream is used to identify objects, as in recognizing the cup. The two streams operate independently. As will be noted in Chapter 17, people with damage in the dorsal stream have problems with visually guided movements but no difficulty identifying objects, while people with damage in the ventral stream cannot identify objects by sight but are able to use visual information to adjust their movements.

CONSCIOUSNESS

Waking and sleeping, paying attention, and initiation of action are the province of the consciousness system. Various aspects of consciousness require differ-

ent subsystems. Aspects of consciousness include the following:

- Generalized arousal level
- Attention
- Selection of object of attention, based on goals
- Motivation and initiation for motor activity and cognition

Each of these aspects of consciousness is associated with activity of specific neuromodulators produced by brain stem neurons (Robbins and Everitt, 1995) and delivered to the cerebrum by the reticular activating system (see Chap. 14). The neuromodulators are serotonin, norepinephrine, acetylcholine, and dopamine. Serotonin is widely distributed throughout the cerebrum and modulates the general level of arousal. Norepinephrine contributes to attention and vigilance via locus ceruleus projections primarily to sensory areas. Acetylcholine activation of the anterior cingulate gyrus (Posner, 1995) contributes to voluntary direction of attention toward an object. Finally, dopamine contributes to the initiation of motor or cognitive actions, based on cognitive activity. Figure 16–24 summarizes the function and distribution of each brain stem modulator involved in consciousness.

Although the brain stem is the source of neuromodulators that regulate consciousness, consciousness also requires activity of the thalamus and cerebral cortex. Thus lesions of the brain stem, thalamus, and/or cerebral cortex may result in the alterations of consciousness listed in Chapter 14.

In addition to the general effects indicated above, specific anatomical locations have been proposed for the ability to attend to and orient to stimuli (Posner, 1995). Maintaining attention requires the right frontal and parietal lobes. Orientation uses the right parietal lobe to disengage attention from the present focus, the midbrain to direct attention to a new focus, and the pulvinar to focus attention by restricting unwanted inputs.

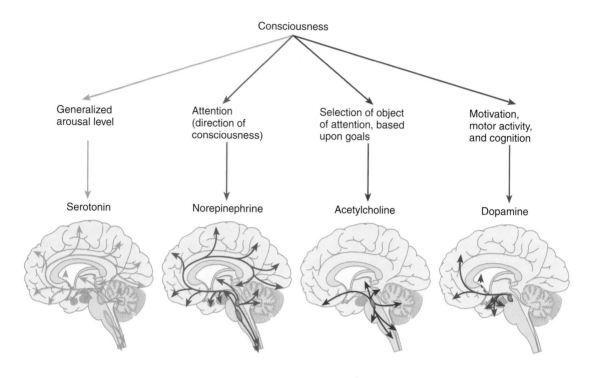

DISTRIBUTION OF TRANSMITTER

Transmitter		Serotonin	Norepinephrine	Acetylcholine	Dopamine
Origin		Raphe nuclei	Locus ceruleus and medial reticular zone	Pedunculopontine nucleus	Substantia nigra and ventral tegmental area
Limbic	Amygdala				
	Nucleus accumbens, septal area				
Basal forebrain					
Neocortex					
					Frontal only
Thalamus					
Striatum					
Cerebellar cortex					

FIGURE 16–24

The function and distribution of neurotransmitters involved in consciousness. The neurotransmitters are produced in the brain stem and delivered to the cerebrum by the reticular activating system. Colored boxes below each transmitter indicate that the transmitter is distributed to the indicated brain area.

Summary

Subcortical structures are involved in the nonconscious regulation of sensory, autonomic, and motor functions. Distinct areas of the cerebral cortex are devoted to analyzing sensation, planning and controlling movements, communication, behavioral control, and intellectual activity. Both cortical and subcortical structures are involved in consciousness, emotions, and memory.

REVIEW QUESTIONS

1. What neural connections would be lost with lesions of each of the thalamic relay nuclei?

2. Why is compression of or damage to the hypothalamus potentially life-threatening?

3. What signs would follow destruction of the genu region of the internal capsule?

4. What are the five functional categories of cerebral cortex?

5. Draw a flow chart of the cortical areas activated to comply with the request "Please pass the salt."

6. How is the stress response produced?

7. What are the effects of excessive, prolonged cortisol secretion?

8. What is the role of the hippocampus in memory?

9. Which structures are important for formation of motor memories?

10. How do we use visual information in the ventral stream?

11. Bob is reading intently when he hears someone call his name. He looks up and begins a conversation with a friend. What brain areas contribute to Bob's ability to maintain his attention while reading, then disengage from reading and shift his attention to his friend?

References

Anthoney, T. R. (1994). Neuroanatomy and the neurologic exam: A thesaurus of synonyms, similar-sounding non-synonyms, and terms of variable meaning. Boca Raton, FL: CRC Press.

Alexander, G. E., Crutcher, M. D., and DeLong, M. R. (1990). Basal ganglia–thalamocortical circuits: Parallel substrates for motor, oculomotor, "prefrontal" and "limbic" functions. Prog. Brain Res. 85:119–146.

Appollonio, I. M., Grafman, J., Schwartz, V., et al. (1993). Memory in patients with cerebellar degeneration. Neurology 43(8): 1536–1544.

Bechara, A., Damasio, A. R., Damasio, H., et al. (1994). Insensitivity to future consequences following damage to human prefrontal cortex. Cognition 50(1–3):7–15.

Bliss, T. V. P., and Collingridge, G. L. (1993). A synaptic model of memory: Long-term potentiation in the hippocampus. Nature 361:31–39.

Courtney, J. G., Longnecker, M. P., Theorrell, T., et al. (1993). Stressful life events and the risk of colorectal cancer. Epidemiology 4(5):407–414.

Damasio, A. R., Tranel, D., and Damasio, H. (1991). Somatic markers and the guidance of behavior: Theory and preliminary testing. In: Levin, H. S., Eisenberg, H. M., and Benton, A. L. (Eds.). Frontal lobe function and dysfunction. New York: Oxford University Press, pp. 217–229.

Donoghue, J. P., and Sanes, J. N. (1994). Motor areas of the cerebral cortex. J. Clin. Neurophys. 11(4):382–396.

Gabrieli, J. D. (1993). Disorders of memory in humans. Curr. Opin. Neurol. Neurosurg. 6(1):93–97.

Gloor, P. (1986). Role of the human limbic system in perception, memory, and affect: Lessons from temporal lobe epilepsy. In: Doane, B. K., and Livingston, K. E. (Eds.). The limbic system: Functional organization and clinical disorders. New York: Raven Press, pp. 159–169.

Heindel, W. C., and Salmon, D. P. (1989). Neuropsychological evidence for multiple implicit memory systems: A comparison of Alzheimer's, Huntington's, and Parkinson's disease patients. J. Neurosci. 9(2):582–587.

Karni, A., Meyer, G., Jezzard, P., et al. (1995). Functional MRI evidence for adult motor cortex plasticity during motor skill learning. Nature 377(6545):155–158.

Keppel, W. H., Regan, D. H., Heffeneider, S. H., et al. (1993). Effects of behavioral stimuli on plasma interleukin-1 activity in humans at rest. J. Clin. Psychol. 49(6):777–789.

Kiecolt-Glaser, J. K., Malarkey, W. B., Chee, M., et al. (1993). Negative behavior during marital conflict is associated with immunological down-regulation. Psychosom. Med. 55(5): 395–409.

Lupien S., Lecours, A. R., Lussier, I., et al. (1994). Basal cortisol levels and cognitive deficits in human aging. J. Neurosci. 14(5):2893–2903.

Moscovitch, M. (1995). Models of consciousness and memory. In: Gazzaniga, M. S. (Ed.). The cognitive neurosciences. Cambridge, Mass: MIT Press, pp. 1341–1356.

Murphy, B. E. P. (1991). Steroids and depression. J. Steroid Biochem. Mol. Biol. 38:537–559.

Penfield, W. (1958). Functional localization in temporal and deep Sylvian areas. Res. Publ. Assoc. Res. Nerv. Ment. Dis. 36: 210–226.

Pert, C. B., Ruff, M. R., Weber, R. J., et al. (1985). Neuropeptides and their receptors: A psychosomatic network. J. of Immunol. 135(2 Suppl):8205–8265.

Posner, M. I. (1995). Attention in cognitive neuroscience: An overview. In: Gazzaniga, M. S. (Ed.). The cognitive neurosciences. Cambridge, Mass: MIT Press, pp. 615–624.

Robbins T. W., and Everitt, B. J. (1995). Arousal systems and attention. In: Gazzaniga, M. S. (Ed.). The cognitive neurosciences. Cambridge, Mass: MIT Press, pp. 703–720.

Sietz, R. J., and Roland, P. E. (1992). Learning of sequential finger movements in man: A combined kinematic and positron emission tomography study. Eur. J. Neurosci. 4:154–165.

Spiegel, D., Kraemer, H., Bloom, J. R., et al. (1989). Effect of psychosocial treatment on survival of patients with metastatic breast cancer. Lancet 2(8868):888–891.

Young, A. W., Aggleton, J. P., Hellawell, D. J., et al. (1995). Face processing impairments after amygdalotomy. Brain 118:15–24.

17

Cerebrum: Clinical Applications

O n July 4th, almost 2 years ago, I had served a brunch for family and friends. Everything seemed fine. After our guests left, my husband found me collapsed on the floor. I don't remember anything about July that year. An aneurysm burst and took away some parts of my life. I had surgery to repair the aneurysm on July 5 and another surgery on August 3 to insert a shunt. I remember things since the second surgery. I had 3 weeks of rehabilitation in the hospital and then physical and occupational therapy twice a week for almost a year.

Now my movements are still too slow; everything takes me twice as long as before. I can't move my right foot, so I wear a brace to keep from turning my ankle or tripping. I used to bicycle long distances. Now I can't bicycle independently, so I ride a tandem bicycle. I can move my right arm from the wrist up but can't move my right hand. Writing is almost impossible because I was right-handed. I can type on the computer keyboard using my left hand only. Cooking takes me a long time, and I have trouble lifting things out of the oven. I enjoy traveling, but it's hard to get around in other countries. Many places don't have stair railings, and that makes it tough to go up or down stairs. I have minimal problems with language; my mouth works slower, and sometimes I forget parts of what I want to say.

I've made a lot of progress since the aneurysm burst. At first I could barely speak; trying to figure out the words was too difficult. I had to use a wheelchair because my balance was so bad. Now I can walk long distances, and I am completely independent.

In therapy we worked on walking, strengthening, balance, and stretching. I used an electrical stimulator to help contract the muscles that lift the front of the foot up, but that didn't seem to help. I haven't taken any medications for my condition.

—Jane Lebens

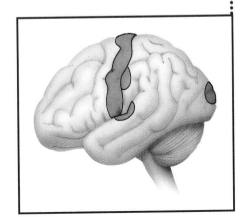

SIGNS OF DAMAGE TO CEREBRAL SYSTEMS

Thalamic Injury

Thalamic lesions involving the relay nuclei interrupt ascending pathways, severely compromising or eliminating contralateral sensation. Usually proprioception is most affected. Rarely, a thalamic pain syndrome ensues after damage to the thalamus, producing severe contralateral pain that may occur with or without provoking external stimuli.

Subcortical White Matter Lesions

Occlusion or hemorrhage of arteries supplying the **internal capsule** are common. Because the internal capsule is composed of many projection axons, even a small lesion may have severe consequences. For example, a lesion the size of a nickel could interrupt the posterior limb and adjacent gray matter. This would prevent messages in corticospinal, corticobulbar, corticopontine, corticoreticular, and thalamocortical fibers from reaching their destinations, resulting in the following:

- Contralateral loss of voluntary movement control
- Contralateral decreased automatic movement control
- Contralateral loss of conscious somatosensation

If the lesion extended more posteriorly, into the retrolenticular and sublenticular part of the capsule, conscious vision from the contralateral visual field would be lost because optic radiation fibers would be interrupted.

CALLOSOTOMY

Remarkable outcomes occur when the huge fiber bundle connecting the hemispheres, the **corpus callosum,** is surgically severed. The surgery (callosotomy) is performed in cases of intractable epilepsy when the excessive neuronal activity that characterizes epilepsy cannot be controlled by medication or surgical damage of a single cortical site. Callosotomy is usually successful in preventing excessive firing from spreading from one hemisphere to the other, thus limiting the seizure to one hemisphere. Although

people with callosotomies are rarely seen for rehabilitation, because callosotomies are performed infrequently and because recovery is usually spontaneous, the results of callosotomies illustrate the difference in function between the cerebral hemispheres.

Initially, after recovery from the surgery, many people with callosotomies report conflicts between their hands: the left hand will begin a task, and the right hand will interfere with the left hand's activity. A physical therapist working with person postcallosotomy reported to the doctor, "You should have seen Rocky yesterday—one hand was buttoning up his shirt and the other hand was coming along right behind it undoing the buttons!" (Bogen, 1993). Typically, these competitive hand movements resolve with time. Following recovery, compensation occurs, allowing the person with a "split brain" to interact normally in social situations and to perform normally on most traditional neurological examinations. Specialized tests designed to assess the performance of a single hemisphere are required to demonstrate abnormalities.

The most commonly used specialized tests involve assessment of vision and stereognosis. Results from right-handed people with callosotomies are summarized here. When words are presented briefly to the right visual field, people are able to read the words. However, when words are flashed in the left visual field, people are unable to read them and often report seeing nothing.

For somatosensory tests, people handle objects that are out of sight. For example, when handling a comb in the right hand, a person with a collosotomy is able to name and verbally describe the comb, yet unable to demonstrate using the comb. If the comb is handled by the left hand, the same person is able to demonstrate its use but unable to name it.

Why the great disparity in the abilities of the separated hemispheres? Information presented to the right visual field or right hand projects to the language-dominant left hemisphere, so the person is able to name and describe the word or object. Information from the left visual field or left hand is processed in the right cerebral hemisphere, which excels at comprehending space, manipulating objects, and perceiving shapes (Franco and Sperry, 1977). Thus the person is able to manipulate the object appropriately but cannot name or verbally describe the object, because, in most people, the right hemisphere does not process language.

Basal Ganglia Disorders

In contrast to the movement disorders associated with lenticular dysfunction (see Chap. 10), lesions or dysfunctions of the caudate rarely cause motor disorders but instead cause behavioral disturbances. The most common behavioral abnormality secondary to caudate damage is apathy, with loss of initiative, spontaneous thought, and emotional responses (Bhatia and Marsden, 1994). Conversely, excessive activity of the circuit connecting the caudate, thalamus, and lateral orbitofrontal cortex is correlated with obsessive-compulsive disorders (Swedo et al., 1989). People with obsessive-compulsive disorder have a tendency to perform certain acts repetitively, as in an irresistible urge to wash their hands hundreds of times per day.

Disorders of Specific Areas of the Cerebral Cortex

PRIMARY SENSORY AREAS: LOSS OF DISCRIMINATIVE SENSORY INFORMATION

Lesions of the primary sensory areas impair the ability to discriminate intensity and quality of stimuli, severely interfering with the capacity to use the sensations. Lesions of the primary somatosensory cortex interfere most with the localization of tactile stimuli and with proprioception. Crude awareness of touch and thermal stimuli is not affected in lesions of the primary somatosensory cortex, since crude awareness occurs in the thalamus. Also, localization of pain is not compromised by lesions confined to the primary somatosensory cortex. Pain information is believed to be processed in the sensory association cortex rather than the primary somatosensory cortex.

Because auditory information has extensive bilateral projections to the cortex, a lesion in the primary auditory cortex only interferes with the ability to localize sounds (see Chap. 15). Lesions in the primary vestibular cortex interfere with conscious awareness of head position and movement. Primary visual cortex lesions cause contralateral homonymous hemianopia (see Chap. 15). The consequences of lesions in primary sensory areas are illustrated in Figure 17–1.

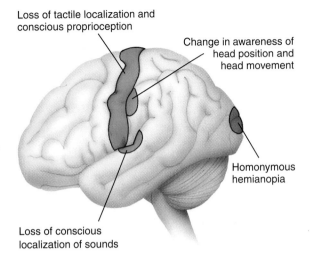

Functional Change	Cortical Area
Loss of tactile localization and conscious proprioception	Primary somatosensory
Loss of localization of sounds	Primary auditory
Homonymous hemianopia	Primary visual
Change in awareness of head position and movement	Primary vestibular

FIGURE 17–1

Results of lesions in primary sensory areas.

SENSORY ASSOCIATION AREAS: AGNOSIA

Agnosia is the general term for the inability to recognize objects when using a specific sense, even though discriminative ability with that sense is intact. The forms of agnosia are as follows:

- Astereognosis
- Visual agnosia
- Auditory agnosia

Astereognosis Astereognosis is the inability to identify objects by touch and manipulation despite intact discriminative somatosensation. A person with **astereognosis** would be able to describe an object being palpated but not recognize the object by touching and manipulating it. Astereognosis results from lesions in the somatosensory association area. A person with astereognosis affecting the information from one hand may avoid using that hand as a result of perceptual changes if information from the other hand is processed normally.

Visual Agnosia Similarly, lesions in the visual association area interfere with the ability to recognize objects in the contralateral visual field, although the capacity for visual discrimination remains intact. **Visual agnosia** is the term for inability to visually recognize objects despite intact vision. A person with visual agnosia can describe the shape and size of objects using vision but cannot identify the objects visually.

A highly specific type of visual agnosia is prosopagnosia. People with this rare condition are unable to visually identify people's faces, despite being able to correctly interpret emotional facial expressions and being able to visually recognize other items in the environment. Only visual recognition is defective; people can be identified by their voices or by mannerisms. Prosopagnosia is usually associated with bilateral damage to the inferior visual association areas (part of the ventral stream).

Auditory Agnosia Destruction of the auditory association cortex spares the ability to perceive sound but deprives the person of recognition of sounds. If the lesion destroys the left auditory association cortex, the person is unable to understand speech (see later section). Destruction of the right auditory association cortex interferes with the interpretation of noises. For example, a person cannot distinguish between the sound of a doorbell and the sound of footsteps. The areas of cortex involved in agnosias are illustrated in Figure 17–2.

Agnosia results from damage to sensory association areas.

MOTOR PLANNING AREAS: APRAXIA, MOTOR PERSEVERATION, AND BROCA'S APHASIA

Apraxia is the inability to perform a movement or sequence of movements despite intact sensation, automatic motor output, and understanding of the task. Thus persons with apraxia may be unable to touch their nose on request, but then easily scratch their

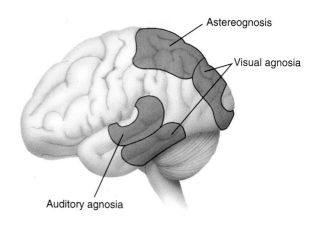

Functional Change	Cortical Area
Astereognosis	Somatosensory association
Visual agnosia	Visual association
Auditory agnosia	Auditory association

FIGURE 17–2

Results of lesions in sensory association areas.

nose if it itches. Apraxia occurs as a result of damage to the premotor or supplementary motor areas.

Motor perseveration is the uncontrollable repetition of a movement. For example, a person may continue to lock and unlock the brakes of a wheelchair despite intending to lock the brakes. Motor perseveration is associated with damage to the supplementary motor area (Gelmers, 1983).

Broca's aphasia is an inability to express oneself by language or symbols. A person with Broca's aphasia cannot communicate by speaking or writing. Broca's aphasia occurs with damage to Broca's area and will be discussed further in a later section.

PRIMARY MOTOR CORTEX: LOSS OF MOVEMENT FRACTIONATION AND DYSARTHRIA

Damage of the primary motor cortex is characterized by contralateral paresis and loss of fractionation of movement (see Chap. 10). The worst effects are distal: people with complete destruction of the primary motor cortex cannot voluntarily move their contralateral hand, lower face, and/or foot because these areas rely exclusively on contralateral primary motor cortex control.

Dysarthria is the term used to describe speech disorders resulting from paralysis, incoordination, or spasticity of the muscles used for speaking. Two types of dysarthria can be distinguished: spastic and flaccid. Damage to upper motoneurons causes **spastic dysarthria,** characterized by harsh, awkward speech. In contrast, damage of the lower motoneurons (cranial nerves IX, X, and/or XII) produces **flaccid dysarthria,** resulting in breathy, soft, and imprecise speech. In pure dysarthria, only the production of speech is impaired; language generation and comprehension are unaffected. The difficulty is with the mechanics of producing sounds accurately, not with finding words or with grammar. Lesions in areas of cortex that produce motor disorders are illustrated in Figure 17–3.

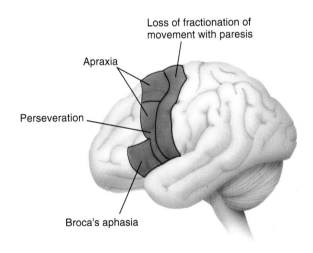

FIGURE 17–3

Results of lesions in motor areas of the cerebral cortex.

Functional Change	Motor Areas
Paresis, loss of fine motor control, spastic dysarthria	Primary motor cortex
Apraxia	Premotor area
Apraxia, perseveration	Supplementary motor area
Broca's aphasia or difficulty producing nonverbal communication	Broca's area in language-dominant hemisphere or analogous area in opposite hemisphere

PREFRONTAL ASSOCIATION CORTEX: LOSS OF EXECUTIVE FUNCTIONS AND DIVERGENT THINKING

Although physical therapy does not focus on remediation of association cortex deficits, these deficits may have a profound influence on compliance and outcomes. Apathy and lack of goal-directed behavior are typical of people with lesions in the prefrontal area. People with damage to this region have difficulty choosing goals, planning, executing plans, and monitoring the execution of a plan. The lack of initiative may interfere with the ability to live independently and to be employed. The behavior of people with prefrontal damage may be misinterpreted as uncooperative or unmotivated, when they have actually lost the neural capacity to initiate goal-directed action.

Perhaps surprisingly, lesions in the prefrontal cortex have little effect on intelligence as measured by conventional intelligence tests. People with prefrontal damage are able to perform paper-and-pencil problem-solving tasks nearly as well as they were able to prior to the damage. This may be because conventional intelligence tests assess convergent thinking, or the ability to choose one correct response from a list of choices. In people with prefrontal lesions, divergent thinking, the ability to conceive of a variety of possibilities, is impaired. For example, if asked to list possible uses of a stick, they perform much worse than people without brain damage. Despite the ability to perform normally on conventional intelligence tests, people with prefrontal lesions function poorly in daily life because of lack of goal orientation and lack of behavioral flexibility.

LIMBIC ASSOCIATION CORTEX: PERSONALITY AND EMOTIONAL CHANGES

Damage to the area of cortex above the eyes (orbitofrontal cortex) leads to inappropriate and risky behavior (Bechara et al., 1994). People with orbitofrontal lesions have intact intellectual abilities but use poor judgment and have difficulty conforming to social conventions. They have problems with inhibitory control, saying and doing things that are socially unacceptable. Minor frustrations may lead to outbursts of physical and verbal aggression. In some cases, the lack of tact, lack of concern for others, and poor judgment can have disastrous consequences, causing violent behavior, damaged relationships, and inability to be employed.

Damasio (1994) discovered that patients with damage to the orbitofrontal cortex were unable to make sound decisions in an experimental card game. Unlike people with intact nervous systems or patients with brain damage to other areas, patients with lesions in the orbitofrontal cortex showed no elevation of galvanic skin response prior to picking a card from a high-risk deck. Thus a possible explanation of the inappropriate behavior is lack of a sense of risk, that is, no emotional concern about outcomes (Damasio, 1994).

PARIETOTEMPORAL ASSOCIATION AREAS: PROBLEMS WITH COMMUNICATION, UNDERSTANDING SPACE, AND DIRECTING ATTENTION

Parietotemporal association areas are specialized for communication and for comprehending space. Damage to this area in the left hemisphere causes a lan-guage disturbance called Wernicke's aphasia; damage to the same area in the right hemisphere causes deficits in directing attention, comprehending space, and understanding nonverbal communication. Detailed discussion of these areas and disorders is deferred to a later section. The results of lesions in the association cortex are illustrated in Figure 17–4.

Disorders of Emotions

Understanding of the neural basis of emotions is in its infancy. Only the functions of the amygdala and orbitofrontal cortex and the neurochemistry of depression currently enjoy widespread consensus.

A woman with damage to both amygdalae has been studied extensively due to her inability to mediate emotions (Adolphs et al., 1994). She has difficulty recognizing the emotions conveyed by people's facial expressions, and she makes poor social and personal decisions. Despite these deficits, her memory for facts

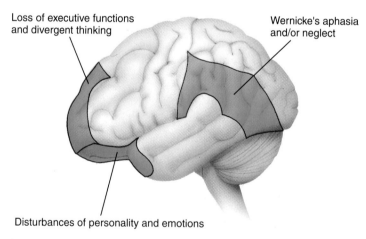

Loss of executive functions and divergent thinking

Wernicke's aphasia and/or neglect

Disturbances of personality and emotions

Functional Change	Association Cortex
Loss of executive functions and divergent thinking	Prefrontal association
Wernicke's aphasia or neglect	Parietotemporal association in dominant hemisphere or analogous area in opposite hemisphere
Disturbances of personality and emotions	Limbic association

FIGURE 17–4

Results of lesions in the association cortex.

and events is completely intact. Young et al. (1995) reported a similar case, concluding that the amygdala has a role in social learning and behavior associated with personal interactions.

Damasio (1994) reported that a man with damage to the orbitofrontal cortex was unable to choose between two dates for a return appointment. For nearly half an hour, the man considered the pros and cons of the dates without approaching a conclusion. When told to come on the second date, he quickly accepted the suggestion. According to Damasio, in the absence of emotional cues that some considerations were more important than others and without the sense that the decision was trivial, the man with orbitofrontal damage was unable to make decisions. In other circumstances, that is, driving on icy roads, the same man performed well because he remained calm even when witnessing accidents.

Depression, a syndrome of hopelessness and a sense of worthlessness, with aberrant thoughts and behavior, has been linked to neurotransmitter rather than structural abnormalities. People with depression have reduced levels of serotonin metabolites in their cerebrospinal fluid. Drugs that effectively treat depression all increase the effectiveness of serotonin transmission. Tricyclic antidepressants inhibit both norepinephrine and serotonin uptake. Newer antidepressants, such as fluoxetine (Prozac), selectively inhibit serotonin uptake, prolonging the availability of serotonin in synapses.

Memory Disorders

Amnesia is the loss of long-term memory. Most commonly amnesia is retrograde, involving the loss of memories for events that occurred prior to the trauma or disease that caused the condition. In H.M. (see Chap. 16), who had both hippocampi removed, past memories are intact, and he cannot remember events occurring after the surgery. This loss of memory for events following the event that caused the amnesia is called anterograde amnesia.

People with amnesia affecting only declarative memory retain the ability to form new preferences, despite lacking cognitive awareness of the preferences. A patient with postencephalitic amnesia was studied to determine whether he would learn to distinguish among different response patterns of staff members. Three staff members consistently provided different responses to his request for special foods: positive, negative, and neutral. When asked whom he

would ask for special foods, he indicated the staff member who responded positively to his requests, although he was totally unable to show familiarity with any of the staff members (Damasio et al., 1989). The patient's behavior demonstrated his ability to unconsciously recall the staff member who would provide him with special foods.

The dissociation of declarative and procedural memory is important clinically. People with severe declarative memory deficits following head trauma learn new motor skills, despite their inability to consciously recall having practiced the tasks (Ewert and Levin, 1989). Similarly, a 62-year-old woman with a significant declarative memory deficit secondary to a left cerebrovascular accident showed improved performance on a mirror learning task. In this task, the woman could see her hand and a maze in a mirror, but direct vision of her hand was blocked by a screen. Thus when she drew a line upward, in the mirror the direction of movement appeared to be downward. Learning was demonstrated by improvement in the time required to complete the maze (Cushman and Caplan, 1987). Learning of motor skills may proceed even when declarative memory fails.

Language Disorders

Disorders of language can affect spoken language (**aphasia**), comprehension of written language (**alexia**), and/or the ability to write (**agraphia**). Because aphasia has the most severe impact on communication during treatment, the following discussion focuses on aphasia. The common types of aphasia are Broca's, Wernicke's, conduction, and global.

Broca's aphasia is the inability to express oneself using language. The ability to understand language and to control the muscles used in speech for other purposes (swallowing, chewing) are not affected. People with Broca's aphasia may not produce any language output, or they may be able to generate habitual phrases, such as "Hello. How are you?" in the appropriate social setting, and may be able to produce emotional speech (obscenities, curses) when upset. People with Broca's aphasia are usually aware of their language difficulties and are frustrated by their inability to produce normal language. Usually writing is as impaired as speaking. The ability to understand spoken language and to read are spared. Motor, expressive, and nonfluent aphasia are synonymous with Broca's aphasia.

In **Wernicke's aphasia,** language comprehension is impaired. People with Wernicke's aphasia easily produce spoken sounds, but the output is meaningless. Listening to other people speak is equally meaningless, despite the ability to hear normally. Wernicke's aphasia also interferes with the ability to comprehend and produce symbolic movements, as in sign language. Because the ability to comprehend language is lost, people with Wernicke's aphasia have alexia (inability to read) and inability to write meaningful words. Unlike people with Broca's aphasia, people with Wernicke's aphasia often appear to be unaware of the disorder. In mild cases, word substitution, called paraphrasia, is common. For example, a person might say or write "captain of the school" instead of "principal." Synonyms for Wernicke's aphasia are receptive, sensory, or fluent aphasia, although language output is also abnormal.

Conduction aphasia results from damage to the neurons that connect Wernicke's and Broca's areas. In the most severe form, the speech and writing of people with conduction aphasia are meaningless. However, their ability to understand written and spoken language is normal. In mild cases, only paraphasias occur.

The most severe form of aphasia is **global aphasia,** an inability to use language in any form. People with global aphasia cannot produce understandable speech, comprehend spoken language, speak fluently, read, or write. Global aphasia is usually secondary to a large lesion damaging much of the lateral left cerebrum: Broca's area, Wernicke's area, intervening cortex, the adjacent white matter, caudate, and anterior thalamus. Common types of aphasia are summarized in Table 17–1.

Disorders of Nonverbal Communication

Damage to the right cortex in the area corresponding to Broca's area may cause the person to speak in a monotone, to be unable to effectively communicate nonverbally, and to lack emotional facial expressions and gestures. These consequences are sometimes referred to as flat affect. If the area corresponding to Wernicke's is damaged on the right side, the person has difficulty understanding nonverbal communication. Thus the person may be unable to distinguish between hearing "get out of here" spoken jokingly and "GET OUT OF HERE" spoken in anger. As noted

earlier, the area corresponding to Wernicke's area is also important for body image and for understanding the relationship between self and the environment. Damage to the area corresponding to Wernicke's area may cause neglect.

Neglect

The tendency to behave as if one side of the body and/or one side of space does not exist is called neglect. People with neglect fail to report or respond to stimuli present on the contralesional side. Neglect usually affects the left side of the body because the right parietal area is necessary for directing attention and the area analogous to Wernicke's in the right hemisphere comprehends spatial relationships. There are two types of neglect: personal and spatial. Aspects of personal neglect include the following:

- Unilateral lack of awareness of sensory stimuli
- Unilateral lack of personal hygiene and grooming
- Unilateral lack of movement of the limbs

Personal neglect results from a failure to direct attention, affecting awareness of one's own body parts. Therefore, personal neglect is also called hemi-inattention. Some people with personal neglect are able to localize light touch and to distinguish between sharp and dull if a stimulus is presented unilaterally but fail to respond to stimulation on one side when both sides of the body are stimulated concurrently. This phenomenon is called bilateral simultaneous extinction (see Chap. 7).

A form of denial, anosognosia, occurs in some people with severe hemiparesis and personal neglect. People with anosognosia deny their inability to use the paretic limbs, claiming they could clap their hands or climb a ladder. However, when asked what the experimenter would be able to do if he had exactly the same impairments as theirs, people with anosognosia who claimed they could perform the tasks reported that the experimenter would be impaired or unable to do the same task (Marcel and Tegner, 1993).

Spatial neglect is characterized by a unilateral lack of understanding of spatial relationships, resulting in a deranged internal representation of space. In an intriguing investigation of spatial neglect, Bisiach and Luzzatti (1978) asked two people with neglect to describe from memory what they would see looking at the main square in Milan from the steps of the cathedral and then describe the same scene looking across

TABLE 17-1
COMMUNICATION DISORDERS

Name	Synonyms	Characteristics	Comprehend Spoken Speech	Speak Fluently	Produce Meaningful Language	Normal Use of Grammatical Words	Read	Write	Structures Involved
Dysarthria	None	Lack motor control of speech muscles	Yes	No	Yes, although difficult to understand	Yes	Yes	Yes	Lower motoneurons or corticobulbar neurons
Broca's aphasia	Motor, expressive, or nonfluent aphasia	Grammatical omissions and errors, short phrases, effortful speech	Yes	No	Yes, although grammatical words missing	No	Yes	No	Broca's area, usually in left hemisphere
Wernicke's aphasia	Sensory, receptive, or fluent aphasia	Cannot comprehend language; speaks fluently but unintelligibly	No	Yes	No	No	No	No	Wernicke's area, usually in left hemisphere
Conduction aphasia	Disconnection aphasia	Understands language; language output unintelligible	Yes	Yes	No	No	Yes	No	Neurons connecting Wernicke's area with Broca's area
Global aphasia	Total aphasia	Cannot speak fluently; cannot communicate verbally	No	No	No	No	No	No	Wernicke's area, Broca's area, and the intervening cortical and subcortical areas

the square at the cathedral. Describing the view from the steps of the cathedral, both people consistently mentioned buildings on the right side of the visualized scene. When asked to mentally change their perspective, imagining looking at the cathedral, both described buildings on the right side and omitted buildings they had described moments earlier. Similarly, one of my patients who had been a successful artist painted the right half of a scene, leaving the left half of the canvas blank. She claimed the painting was finished and appeared perplexed when questioned about the missing parts of the boy in the painting. When I inverted the canvas to show her that the painting was incomplete, she began a new, different painting on the fresh canvas, oblivious to the image on the left side.

Some aspects of neglect are currently unexplained. Bisiach and Berti (1989) have shown that when a person with spatial neglect was asked to copy three figures, he completed both the right and left figures but only drew half of the central figure. Attentional theories of neglect would predict the person would omit the left figure, not part of the central figure.

People with neglect may have only one sign (e.g., lack of awareness of people or objects on their left) or any combination of signs. In most cases, neglect follows damage to the right cortex in the parietal lobe or in the area corresponding to Wernicke's area. However, similar syndromes occasionally occur with brain stem, thalamic, and cingulate cortex lesions. Rafal and Robertson (1995) summarize the variety of lesion locations that may contribute to the neglect syndrome. Various aspects of neglect may result from damage to the following brain areas:

- Parietal lobe
- Right temporal-parietal junction
- Midbrain reticular formation
- Intralaminar thalamic nuclei
- Premotor cortex
- Cingulate gyrus
- Prefrontal cortex

Thus neglect is a complex phenomenon, with different presentations and diverse etiologies.

Inability to Use Visual Information

Goodale and Milner (1992) reviewed several earlier reports to illustrate the independent use of visual information in the ventral and dorsal streams. A woman with damage to the ventral stream was profoundly unable to recognize the shape, orientation, or size of objects, yet she was able to pick up the unrecognized objects using normal approach and anticipatory positioning of her hand and fingers. Thus, despite visual agnosia, use of visual information for controlling movement was normal. The opposite impairment was noted in another woman with damage in the parietal lobe: she was unable to adjust her reach and hand orientation appropriately to the size and shape of objects, yet she was able to describe and identify the objects. Thus optic ataxia does not affect the ability to consciously perceive visual information.

DISEASES AND DISORDERS AFFECTING CEREBRAL FUNCTION

Loss of Consciousness

At any age, a blow to the head may cause a temporary loss of consciousness. The loss of consciousness results from movement of the cerebral hemispheres relative to the brain stem, causing torque of the brain stem, and from the abrupt increase in intracranial pressure. Consciousness may also be impaired by large, space-occupying lesions of the cerebrum, located in the diencephalon or exerting pressure on the brain stem.

Attention Deficit Disorder

Difficulty sustaining attention with onset during childhood is called attention deficit disorder (ADD). People with ADD display developmentally inappropriate inattention and impulsiveness. The etiology has not been established, but recent research implicates neurotransmitter abnormalities and possibly abnormal development of the frontal lobe and caudate nucleus (Giedd et al., 1994; Castellanos et al., 1994).

Epilepsy

Epilepsy is characterized by sudden attacks of excessive neuronal discharge interfering with brain function. Involuntary movements, disruption of autonomic regulation, illusions, and hallucinations may occur. Partial seizures affect only a restricted area of the cortex. Generalized seizures affect the entire cortex. The two main types of generalized seizures are absence seizures, identified by brief loss of consciousness without motor manifestations, and tonic-clonic seizures, which begin with tonic contraction of the skeletal muscles followed by alternating contraction and relaxation of muscles. Typically the tonic and clonic phases last about 1 minute each. After the seizure, the person is confused for several minutes and has no memory of the seizure.

Disorders of Intellect

Mental retardation, dementia, and dyslexia all reduce the capability for understanding and reasoning. Common causes of mental retardation are trisomy 21 and untreated phenylketonuria.

TRISOMY 21

Trisomy 21, also known as Down syndrome, is a genetic disorder due to an extra copy of chromosome 21. People with trisomy 21 have round heads, slanted eyes, a fold of skin extending from the nose to the medial end of the eyebrow, and simian creases on the palms of their hands. The weight of the brain and the relative size of the frontal lobes are both reduced compared to normal brains.

PHENYLKETONURIA

Phenylketonuria is an autosomal recessive defect in metabolism, resulting in retention of a common amino acid, phenylalanine. The accumulation of phenylalanine results in demyelination and, later, neuronal loss. If the condition is diagnosed in infancy (by blood and urine tests), nervous system damage may be prevented by a diet low in phenylalanine.

DEMENTIA

In contrast to mental retardation, dementia usually occurs late in life. Dementia is generalized mental deterioration, characterized by disorientation and impaired memory, judgment, and intellect. Many different etiologies lead to dementia. Among the most common causes of dementia are multiple infarcts and Alzheimer's disease. Multiple infarcts in the cerebral hemispheres result in focal neurological signs in addition to deterioration of intellectual functions.

Alzheimer's disease causes progressive mental deterioration, consisting of memory loss, confusion, and disorientation. Typically symptoms become apparent after age 60, and death follows in 5 to 10 years. Initially the disease presents with signs of forgetfulness, progressing to an inability to recall words and finally failure to produce and comprehend language. People with Alzheimer's disease may become lost easily, and in the late stage they neglect to dress, groom, or feed themselves. The cellular level changes in Alzheimer's disease include neurofibrillary tangles (tangled masses of neurofibrils), senile plaques (deposits of amyloid material), and severe atrophy of the cerebral cortex, amygdala, and hippocampus. Virtually all people with trisomy 21 develop similar cellular level changes by age 40, although in most cases behavioral changes are not obvious due to the prior low level of function. In a small percentage of cases, an abnormal gene on chromosome 21 has been linked to Alzheimer's disease, suggesting a similar genetic mechanism in trisomy 21 and some cases of Alzheimer's disease.

LEARNING DISABILITIES

In contrast to the generalized intellectual deficits of mental retardation and dementia, learning disabilities arise from a failure to develop specific types of intelligence. The most common learning disability is dyslexia, a condition of inability to read at a level commensurate with the person's overall intelligence. People with dyslexia have difficulty with reading, writing, and spelling words, yet their conversational and visual abilities are normal. They can interpret visual objects and illustrations without difficulty. Some cases of dyslexia have been traced to abnormalities of a gene on chromosome 6.

Traumatic Brain Injury

A majority of traumatic brain injuries occur in motor vehicle accidents. The impact tends to damage the orbitofrontal, anterior, and inferior temporal regions and to cause diffuse axonal injury. Axonal injury primarily affects the basal ganglia, superior cerebellar peduncle, corpus callosum, and midbrain. Because frontal, temporal, and limbic areas are typically damaged, people show poor judgment, decreased executive functions (planning, initiating, monitoring behavior), memory deficits, slow information processing, attentional disorders, and poor complex problem solving. The inability to use new information effectively results in concrete thinking, inability to appropriately apply rules, and trouble distinguishing relevant from irrelevant information. Due to impaired judgment, people with traumatic brain injury are at significant risk for problems with substance abuse, aggression, and inappropriate sexual behaviors. Even people who sustain relatively minor head injuries often have decreased frustration tolerance, leading to easily aroused anger, and require more time and direction to complete tasks than they needed prior to the injury.

Traumatic brain damage in infants is most frequently attributable to accidental falls, but brain damage consequent to most falls is relatively minor. More severe brain injury usually requires greater force than a typical fall, forces that sometimes are generated when an infant is violently shaken. The trauma from shaking is due to the impact of the brain's striking the skull repeatedly. Soon after the incident, cerebral edema may increase the infant's head circumference and cause bulging of the anterior fontanelle. Brain scans show hemorrhage and edema. Survivors may exhibit motor signs similar to cerebral palsy and have a pattern of cognitive deficits similar to those of adults with traumatic brain injury. (See the Traumatic brain injury due to blunt trauma without fracture box.)

Stroke

The neurological outcome of interruption of blood flow to the cerebrum depends on the etiology, location, and size of the infarct or hemorrhage. Infarcts occur when an embolus or thrombus lodges in a vessel, obstructing blood flow.

SIGNS AND SYMPTOMS OF STROKE

The signs and symptoms of stroke depend on the location and size of the lesion; a small insult to the cortex may produce no symptoms, while the same size or a smaller lesion in the brain stem could cause death. Edema secondary to large infarcts or large hemorrhages can cause death regardless of location by compressing vital structures. The following acute neurological deficits have each been reported to affect more than 25% of people surviving infarctions: hemiparesis, ataxia, hemianopsia, visual-perceptual deficits,

---

TRAUMATIC BRAIN INJURY DUE TO BLUNT TRAUMA WITHOUT FRACTURE

Pathology
Diffuse axonal injury; contusion, hemorrhage, swelling, and/or laceration

Etiology
Trauma

Speed of onset
Acute

Signs and symptoms

Personality
Decreased goal-directed behavior (executive functions) if dorsolateral prefrontal cortex involved; impulsiveness and other inappropriate behaviors if orbitofrontal cortex damaged; low tolerance for frustration

Consciousness
May be impaired temporarily or for a prolonged period; often have difficulty directing attention (distractibility)

Communication and memory
Communication usually normal; declarative memory impairments may be temporary or prolonged

Sensory
Usually normal

Autonomic
Normal

Motor
Perseveration of responses

Region affected
Most frequently affects the anterior frontal and temporal lobes

Demographics
For traumatic brain injury (including open head injury and closed injuries with and without fractures), the highest rates of brain injury occur between the ages of 15 and 25 years old, with males injured four times as often as females (Hardman, 1997)

Prognosis
Severity of injury and age at time of injury determine outcome; of survivors, approximately 15% will have severe physical and mental disability, and about 40% will have moderate disability (Taylor, 1992)

aphasia, dysarthria, sensory deficits, memory deficits, and problems with bladder control. Chapter 18 presents localization of deficits in the context of the vascular supply.

Although hemiplegia and hemisensory deficits resulting from stroke often appear to be unilateral, "uninvolved side" is usually a misnomer. Pai et al. (1994) report that subjects with right hemiparesis who were able to walk independently (some with assistive devices) were only able to successfully transfer and maintain their weight to the nonparetic side on 48% of trials and to the paretic side on 20% of trials. In patients with chronic hemiplegia who had suffered penetrating brain wounds, Smutok et al. (1989) found that motor function is impaired in the ipsilateral upper limb. Compared to normals, subjects with hemispheric lesions showed decreased grip or pinch strength and poorer finger tapping and pegboard performance ipsilateral to the lesion. See Chapter 10 for possible mechanisms of ipsilateral involvement in cerebrovascular accident.

RECOVERY FROM STROKE

Following a cerebrovascular accident, the earlier recovery begins, the better the prognosis. Typically, functional improvement is quicker during the first few months after the stroke than later. The speed of initial recovery is related to decreasing cerebral edema, improved blood supply, and removal of necrotic tissue. However, with therapy, functional gains can continue years later (Tangeman et al., 1990); nervous system plasticity accounts for these later gains.

How can recovery after a stroke be optimized? Currently, a definitive answer is not available. As noted in Chapter 10, forced use of the upper limb has been demonstrated to be an effective technique for improving functional recovery of the paretic limb (Taub et al., 1993; Wolf and Lecraw, 1989). In the forced use technique, the nonparetic upper limb is physically restrained, forcing the individual to use the paretic upper limb.

In physical and occupational therapy, an ongoing controversy is the long-term effectiveness of compensation, remediation, and motor control approaches to stroke rehabilitation. Compensation approaches emphasize performing tasks using either the paretic limb with an adapted approach or using the nonparetic limb to perform the task. The compensation approaches assume that damaged neural mechanisms cannot be restored, so external aids or environmental supports are used to assist patients in daily activities. For example, the ankle on the paretic side might be braced to allow early ambulation.

Remediation approaches attempt to decrease the severity of the neurological deficits. Here the assumption is that activation or stimulation of damaged processes will result in change at both the behavioral and neural level. Using the remediation approach, the therapist might use hands-on techniques to inhibit muscle tone and work on a sequence of activities from supine to upright prior to gait training. During gait training, the therapist might move the client's hips.

Motor control approaches emphasize task specificity, that is, practicing the desired task in a specific context. If the goal is independent walking outdoors, walking outdoors is practiced, rather than preparatory activities like standing balance or lateral weight transfers in standing (Richards et al., 1993; Winstein et al., 1989). Using the motor control approach, the client might be asked to pay attention to foot position during heel strike and to modify the movement at the paretic ankle to match the nonparetic side.

A nationwide panel of 18 specialists in poststroke rehabilitation and medicine concluded that research on the effectiveness of different types of poststroke physical therapy techniques has, as a whole, shown no differences in outcome between compensation and remediation approaches (Gresham et al., 1994). Physical and occupational therapy evaluation and treatment recommendations are also available in the document.

Schizophrenia

Schizophrenia is a group of disorders consisting of disordered thinking, delusions, hallucinations, and social withdrawal. The syndrome involves both anatomical and neurotransmitter abnormalities. The prefrontal and medial temporal cortices and the hippocampus are smaller in people with schizophrenia than in normals (Nopoulos, 1995). Drugs that block the reuptake of either dopamine or serotonin reduce symptoms in many people with schizophrenia. Thus abnormality of neurotransmitter regulation may contribute to the symptoms of schizophrenia.

Testing Cerebral Function

Therapists often briefly assess cerebral function. Part of this assessment may include evaluating the level of consciousness. Normal consciousness requires intact function of the ascending reticular activating system, thalamus, and thalamic projections to the cerebral cortex in addition to the cerebral cortex. Functions that are localized in the cerebral cortex include language, orientation, declarative memory, abstract thought, identification of objects, motor planning, and comprehension of spatial relationships (Table 17–2). Consciousness and language are assessed first because the other tests, except comprehension of spatial relationships, require that the person be alert and able to understand language. The message that the therapist wants the person to copy

a drawing, in order to assess comprehension of spatial relationships, may be conveyed by gestural cues.

Difficulties with certain tests of mental function indicate lesions in specific parts of the cerebral cortex. For example, Broca's aphasia indicates damage to Broca's area. Difficulty with other tests does not implicate any specific part of the cerebral cortex, but instead indicates more generalized dysfunction. The significance of difficulty with each test is listed in Table 17–2, in the "Interpretation" column.

The brief assessment indicated in Table 17–2 may be supplemented with neurological testing information from Nolan (1996) or Mancall (1981). For people with moderate to severe speech and/or language difficulties, consultation with a speech/language, pathologist is recommended.

TABLE 17–2
EVALUATION OF MENTAL FUNCTION

Function	Test	Interpretation
Consciousness level	Observe the person's interaction with the environment. Levels of consciousness are classified as: Alert attends to ordinary stimuli Lethargic tends to lose track of conversations and tasks, falls asleep if little stimulation provided Obtunded becomes alert briefly in response to strong stimuli, cannot answer questions meaningfully Stupor alert only during vigorous stimulation Coma little or no response to stimulation	Levels of consciousness depend upon neural activity in the ascending reticular activating system, thalamus, thalamic projections to the cerebral cortex, and the cerebral cortex.
Language and speech	Evaluate the spontaneous use of words, grammar, and fluency of speech.	Disorders may be due to Broca's or global aphasia or dysarthria. Brain areas involved may be Broca's area, both Wernicke's and Broca's areas and their connections, premotor and/or motor cortex, corticobulbar fibers, or motor cranial nerves.
	Comprehension Ask the person to answer a question similar to the following: "Is my brother's sister a man or a woman?"	Difficulty may be due to receptive aphasia or a hearing disorder. Brain areas involved may include Wernicke's area, or peripheral or central auditory structures.
	Naming Ask the person to identify objects: pencil, watch, paper clip; and body parts: nose, knee, eye.	If the person can produce automatic social speech (e.g., "Hello, how are you") but cannot name objects, the difficulty may be due to dysfunction of Wernicke's area (Wernicke's aphasia)
	Reading Ask the person to read a simple paragraph aloud. Then ask questions about the paragraph.	Assuming the person has intact speech, difficulty may be due to alexia, dyslexia, short term memory deficit, visual deficit, or illiteracy. Wernicke's area is the site of dysfunction in alexia.

(continued)

TABLE 17–2

EVALUATION OF MENTAL FUNCTION *Continued*

Function	Test	Interpretation
	Writing Ask the person to write answers to simple questions.	Difficulty may be due to agraphia, visual deficit, impaired motor control of the upper limb, or illiteracy. Wernicke's area is the site of dysfunction in agraphia.
	All of the following tests require language abilities. Some, as noted, also require intact speech.	

Orientation	Assess the person's orientation to person, place, and time. Questions similar to the following may be used.
	    Person
	        What is your name?
	        Where were you born?
	        Are you married?
	    Place
	        Where are we now?
	        What city and state are we in?
	    Time
	        What time is it?
	        What day of the week is this?
	        What year is this?
		Test assumes intact speech abilities. These questions assess declarative memory. The questions about the person assess long term memory and the questions about time and place assess short term memory. Difficulty with these questions may indicate dysfunction of the hippocampus, or a language disorder or speech disorder, or a generalized cortical processing disorder, as occurs with drug toxicity, psychosis, or extreme anxiety.

Declarative memory	Short term memory Tell the person that you are going to check their memory by asking them to remember three words for a few minutes. Give them three unrelated words and have the person repeat the words. Then converse about other topics, and after three minutes, ask them what the three words were. People with intact short term memory can recall all three words. Examples of words used include: clock, telephone, shoe. Recent memory Ask the person about activities in the past several days. Examples include: What did you have for breakfast? Who visited you yesterday? Long term memory Ask the person to name U.S. presidents, or about historical events, or about their school and work experience.	Test requires speech abilities. Declarative memory problems occur with damage to the hippocampus or with temporary disruptions of cerebral function, as may occur during psychosis, extreme anxiety, or following acute head trauma.
Interpretation of proverbs	Ask the person to explain what a proverb means. For example, "What does 'a rolling stone gathers no moss' mean?"	Test requires speech abilities. A concrete answer, such as "it means that a rock that keeps moving does not grow moss," indicates difficulty with abstract thinking.
Calculation	Serial 7s Ask the person to substract 7 from 100 and to keep subtracting 7s from each result. Ask the person simple addition, subtraction, multiplication, or division problems. For example, "What is 6 × 30?"	Test requires speech abilities. Difficulty may indicate problems with maintaining attention, or a problem with abstract thinking.
Stereognosis	Ask the person to close their eyes, then place a small object in the person's hand and ask them to identify the object. Objects may include a paper clip, a key, or a coin.	Test requires speech abilities. Asterognosis indicates damage to the somatosensory association area of the cerebral cortex.

(continued)

TABLE 17–2

EVALUATION OF MENTAL FUNCTION *Continued*

Function	Test	Interpretation
Visual identification	Show the person an object and ask them to identify it.	Test requires speech abilities. If the person cannot identify the object visually, but can identify the object by touch or another sense, the disorder is visual agnosia. Visual agnosia is caused by damage to the visual association areas in the cerebral cortex of the occipital lobe.
Motor planning	Ask the person to show hairbrushing, using a screwdriver, or buttoning a shirt.	Assuming intact sensation, understanding of the task, and motor control, inability to produce specific movements is apraxia. Apraxia usually occurs as a result of damage to the premotor or supplementary motor areas.
Comprehension of spatial relationships	Ask the person to copy a simple drawing, or to draw a person, a house, or a flower from memory.	Difficulty may indicate motor impairment, neglect, or a generalized decline of cerebral function. Neglect occurs with damage to the area that corresponds to Wernicke's area.

CLINICAL NOTES

CASE 1

A famous case in the right-to-die debate involved Karen Quinlan. After ingesting a tranquilizer, an analgesic, and alcohol, she suffered cardiopulmonary arrest that permanently damaged her brain. She became the focus of a conflict between doctors intent on keeping her alive and her parents, who requested she be allowed to die because no hope for recovery existed. A court ordered the doctors to remove her ventilator. However, she continued to breathe without the ventilator and survived in a vegetative state for 9 more years. She never regained consciousness. Although her brain damage was assumed to be in the cerebral cortex, subsequent analysis of her brain showed that the cortex was relatively intact, and the region with severe damage was the thalamus (Kinney, 1994).

Questions

1. Why does thalamic damage interfere with consciousness?

2. What other structures are required for consciousness?

CASE 2

H.A. is a 47-year-old woman postsurgery to remove a benign tumor in the optic chiasm region. One day postsurgery, the therapist arrives to assess the patient. The therapist notes that the patient is unconscious and has no bedcovers, the air conditioning is on full, and fans are placed to blow across the patient's body, yet the temperature of the patient's skin is unusually warm. When the therapist arrives the next day, the patient is warmly covered, the heater is on, and the room temperature is near 90°F, yet the patient's skin temperature is cool.

(continued)

Questions

1. What is the hospital staff trying to do by manipulating the room temperature?

2. What part of H.A.'s cerebrum is not functioning optimally?

CASE 3

K.L. is a 72-year-old man, transferred to rehabilitation 2 weeks after left cerebrovascular accident. He complains of weakness of his right limbs and of being unable to button his clothing or tie his shoes. Right hand movements are clumsy. On the right side of his body, K.L. is unable to localize tactile stimuli or to distinguish between passive flexion and extension of his joints. He is able to correctly report whether he was touched or not and whether a stimulus is sharp or dull.

Question

Where is the lesion?

CASE 4

R.B. is a 19-year-old boy who was rescued, unconscious, after falling from a 40-foot cliff. One day later he regained consciousness. Strength, position sense, touch localization, and two-point discrimination were normal on both sides. R.B. was easily able to identify unseen objects in his left hand but totally unable to recognize the same objects using his right hand. Although he was right-handed prior to the accident, after the accident he used his left hand whenever possible.

Questions

1. Name the deficit in ability to recognize an object by palpation.

2. Why does R.B. avoid using his right hand?

CASE 5

F.S., a 26-year-old schoolteacher, sustained head trauma and multiple femoral fractures in a traffic accident. She was comatose for 3 days. On regaining consciousness, sensation, movement, and her ability to communicate were intact. However, she seemed listless and did not initiate conversations or activities. F.S. was unable to learn a partial weight-bearing gait, flailing the crutches rather than bearing weight on them. She appeared totally apathetic, even about her situation and her family.

Question

Where is the lesion?

CASE 6

B.G., a 34-year-old stockbroker, suffered multiple fractures of the frontal skull in a motorcycle accident. After recovery from surgical repair, he was hemiparetic on the right side. Muscle strength on the right expressed as a percentage of strength on the left was as follows: girdle muscles, 80%; elbow/knee, 50%; and distal muscles, 0%. With an ankle-foot orthosis and a cane, he was able to walk with minimal assistance. However, he frequently attempted to walk independently without the cane or orthosis, and he fell each time. He began making tactless comments and became impulsive, frequently grabbing or pushing people and objects. After 3 days in rehabilitation, he left the hospital against medical advice. A friend drove him to work. Within an hour he was fired because of his behavior toward coworkers and readmitted to the hospital.

(continued)

Where is the lesion?

CASE 7

Critchley (1953) reported a patient who seemed normal but put a tea bag in the teapot, set the pot on the stove, and poured cold water into a cup. She lit a match, put the match to the gas burner, blew out the match, and turned on the gas.

Name the disorder.

CASE 8

H.M., a 66-year-old man, is 5 days poststroke. He is able to walk using a step-to-gait, with minimal assist for balance. He is unable to voluntarily move his right arm. The right arm is adducted at the shoulder, and the elbow, wrist, and fingers are flexed. His speech is strained, harsh, and slow, and some sounds are produced incorrectly, as in "Ow are oo? I am fime." His ability to produce and understand language and his writing are entirely normal.

1. Name the communication disorder.

2. Where is the lesion?

CASE 9

V.M., a 72-year-old woman, was admitted to the hospital with right hemisensory loss, hemiplegia, and communication problems. She greeted visitors with a halting, effortful, and garbled "Hello, how you?" Language output was marred by poor articulation and omission of grammatical function words. She appeared extremely frustrated by her inability to express herself. Her attempts at writing left-handed also showed omission of grammatical function words. She was able to easily follow simple verbal or written commands, indicating intact ability to comprehend language.

1. Where is the lesion?

2. Name the communication disorder.

CASE 10

P.D., an 86-year-old man, was referred to physical therapy following a total hip replacement 1 week ago. Two days ago, while his wife was visiting him in the hospital, he abruptly began speaking nonsense in a conversational tone as if he were speaking normally. His speech was a mixture of jargon and English. For example, he insisted, "I get creekons, tallings, and you must uffners." He became agitated when his wife didn't understand him, and he was unable to understand her questions. With hospital staff he continued to speak freely in his mixture of jargon and English. He showed no indication of comprehending spoken or written language, nor any awareness that his language output was defective. Communication was strictly limited to gestures. No other signs or symptoms are evident. In therapy today, he was cooperative if the therapist pantomimed the desired movements. If the therapist tried to instruct him verbally, he became withdrawn and uncooperative.

(continued)

Questions

1. Name the communication disorder.

2. Where is the lesion?

CASE 11

A.G., a 68-year-old man, is 2 weeks post–right cerebrovascular accident. In all situations, he ignores the left side of his body and objects and people on his left. He never looks toward the left, does not respond to touch or pinprick on his left side, does not eat food from the left side of a plate, does not move his left limbs, does not shave the left side of his face, nor dress his left side. Last week, two fingers on his left hand were lacerated while caught in his wheelchair spokes. Although the entrapped fingers prevented the wheelchair from moving, A.G. continued to try to move forward until stopped by the therapist. Gait requires maximal assistance because he does not bear full weight on the left leg and attempts to take steps using only the right leg, dragging the left leg behind. He becomes lost easily and his attempts to copy drawings are distorted because he omits features that should be included on the left side of the drawing.

Questions

1. Name the disorder.

2. What specific subtypes of the disorder does A.G. have?

CASE 12

A 32-year-old man was hit on the side of the head by a baseball 1 week ago. He complains of clumsiness in picking up objects, although he has no difficulty visually identifying objects. When he reaches for objects, he does not orient the position of his hand to the object; for example, when reaching for a pen held by the examiner, he uses a forearm pronated approach regardless of whether the pen is vertical or horizontal. In reaching for a cup, he does not adjust the opening between fingers and thumb to the size of the cup.

Questions

1. What is this condition called?

2. What area(s) of the brain is (are) damaged?

CASE 13

H.L. is a 17-year-old boy who sustained a closed head injury in an auto accident 1 month ago. H.L. was comatose for 2 weeks. During week 3, he became responsive to simple commands but was mute. Now H.L. talks, he believes he is at home, and he cannot report the correct year or month despite daily reminders of time and place. He does not initiate any activities unless prompted. H.L. is frequently verbally and physically aggressive. All limb movements are ataxic and dysmetric. Coming from sit to stand and gait require moderate assistance due to balance impairments, bilateral weakness, and poor coordination.

Question

What areas of the brain are impaired?

REVIEW QUESTIONS

1. What is the most likely location of a lesion in a person with loss of conscious somatosensation and voluntary movement on the left side of the body and face and loss of conscious vision from the left visual field?

2. Define each of the following terms and identify the area of cortex most commonly damaged with each sign: astereognosis, visual agnosia, apraxia, and spastic dysarthria.

3. A person unable to understand nonverbal communication and exhibiting signs of left neglect probably has a lesion where?

4. Is Broca's aphasia an upper motoneuron disorder?

5. What is the difference between dysarthria and aphasia?

6. A 64-year-old woman has right hemiparesis, hemisensory loss, Broca's aphasia, and intact vision. Where is the most likely site for the lesion?

7. When working with a person who has global aphasia, what type of communication is most effective?

8. J.H. is 2 weeks poststroke. Although she has been in the same hospital room for 10 days, she cannot find the bathroom or the hallway. What is the name for this problem?

References

Adolphs, R., Tranel, D., Damasio, H., et al. (1994). Impaired recognition of emotion in facial expressions following bilateral damage to the human amygdala. Nature 372(6507):669–672.

Bechara, A., Damasio, A. R., Damasio, H., et al. (1994). Insensitivity to future consequences following damage to human prefrontal cortex. Cognition 50(1–3):7–15.

Bhatia, K. P., and Marsden, C. D. (1994). The behavioural and motor consequences of focal lesions of the basal ganglia in man. Brain 117(4):859–876.

Bisiach, E., and Berti, A. (1989). Unilateral misrepresentation of distributed information: Paradoxes and puzzles. In: Brown, W. J. (Ed.). Neuropsychology of visual perception. Hillsdale, NJ: Lawrence Erlbaum Associates, pp. 145–161.

Bisiach, E., and Luzzatti, C. (1978). Unilateral neglect of representational space. Cortex 14:129–133.

Bogen, J. E. (1993). The callosal syndromes. In: Heilman, K. M., and Valenstein, E. (Eds.). Clinical neuropsychology. New York: Oxford University Press, pp. 337–407.

Castellanos, F. X., Giedd, J. N., Eckberg, P., et al. (1994). Quantitative morphology of the caudate nucleus in attention deficit hyperactivity disorder. Am. J. Psychiatry 151(12):1791–1796.

Critchley, M. (1953). The parietal lobes. London: E. Arnold.

Cushman, L., and Caplan, B. (1987). Multiple memory systems: evidence from stroke. Percept. Mot. Skills 64(2):571–577.

Damasio, A. R. (1994). Descartes' error: Emotion, reason, and the human brain. New York: G. P. Putnam's Sons.

Damasio, A. R., Tranel, D., and Damasio, H. (1989). Amnesia caused by herpes simplex encephalitis, infarctions in basal forebrain, Alzheimer's disease, and anoxia/ischemia. In: Boller, F., and Grafman, J. (Eds.). Handbook of neuropsychology, Volume 3. Amsterdam: Elsevier, pp. 149–166.

Ewert, J., and Levin, H. (1989). Procedural memory during posttraumatic amnesia in survivors of severe closed head injury: Implications for rehabilitation. Arch. Neurol. 46(8):911–916.

Franco, L., and Sperry, R. W. (1977). Hemisphere lateralization for cognitive processing of geometry. Neuropsychologia 15: 107–114.

Gelmers, H. J. (1983). Non-paralytic motor disturbances and speech disorders: The role of the supplementary motor area. J Neurol, Neurosurg & Psychiatry. 46(11):1052–1054.

Giedd, J. N., Castellanos, F. X., Kasey, B. J., et al. (1994). Quantitative morphology of the corpus callosum in attention deficit hyperactivity disorder. Am. J. Psychiatry 151(5):665–669.

Goodale, M. A., and Milner, A. D. (1992). Separate visual pathways for perception and action. Trends Neurosci. 15(1):20–25.

Gresham, G. E., Duncan, P. W., Stason, W. B., et al. (1994). Poststroke rehabilitation. Clin. Pract. Guidel. 16. Rockville, MD: Agency for Health Care Policy and Research, Public Health Service, U.S. Department of Health and Human Services.

Hardman, J. M. (1997). Cerebrospinal trauma. In: Davis, R. L., and Robertson, D. M. (Eds.). Textbook of neuropathology. (3rd ed.). Baltimore: Williams & Wilkins.

Kinney, H. C., Korein, J., Panigraphy, A., et al. (1994). Neuropathological findings in the brain of Karen Ann Quinlan: The role of the thalamus in the persistent vegetative state [see comments]. N. Eng. J. Med. 330(21):1469–1475.

Mancall, E. L. (1981). Alpers and Mancall's Essentials of the neurologic examination. (2nd ed.). Philadelphia: F.A. Davis.

Marcel, A. J., and Tegner, R. Conference presentation "Knowing one's plegia" at 11th European Workshop on Cognitive Neuropsychology. Cited in: Bisiach, E., and Berti, A. (1993). Consciousness in dyschiria. In: Gazzaniga, M. (Ed.). The cognitive neurosciences. Cambridge, Mass: MIT Press.

Nolan, M. F. (1996). Introduction to the neurologic examination. Philadelphia: F.A. Davis.

Nopoulos, P., Torres, I., Flaum, M., et al. (1995). Brain morphology in first-episode schizophrenia. Am. J. Psychiatry 152(12): 1721–1723.

Pai, Y.-C., Rogers, M., Hedman, L., and Hanke, T. (1994). Alterations in weight-transfer capabilities in adults with hemiparesis. Phys. Ther. 74:647–659.

Rafal, R., and Robertson, L. (1995). The neurology of visual attention. In: Gazzaniga, M. (Ed.). The cognitive neurosciences. Cambridge, Mass: MIT Press, p. 644.

Richards, C. L., Malouin, F., Wood-Dauphinee, S. et al. (1993). Task-specific physical therapy for optimization of gait recovery in acute stroke patients. Arch. Phys. Med. Rehabil. 74(4): 612–620.

Smutok, M. A., Grafman, J., Salazar, A. M., et al. (1989). Effects of unilateral brain damage on contralateral and ipsilateral upper extremity function in hemiplegia. Phys. Ther. 69: 195–203.

Swedo, S. E., Schapiro, M. B., Grady, C. L., et al. (1989). Cerebral glucose metabolism in childhood-onset obsessive-compulsive disorder. Arch. Gen. Psychiatry 46:518–523.

Tangeman, P. T., Banaitis, D. A., and Williams, A. K. (1990). Rehabilitation of chronic stroke patients: Changes in functional performance. Arch. Phys. Med. Rehabil. 71(11):876–880.

Taub, E., Miller, N. E., Novack, T. A., et al. (1993). Technique to improve chronic motor deficit after stroke. Arch. Phys. Med. Rehabil. 74:347–354.

Taylor, D. (1992). Outcomes and predictors of outcome. In: Long, C. H., and Leslie, K. (Eds.): Handbook of head trauma: Acute care to recovery. New York: Plenum Press.

Winstein, C. J., Gardner, E. R., McNeal, D. R., et al. (1989). Standing balance training effects on balance and locomotion in hemiparetic adults. Arch. Phys. Med. Rehabil. 70:755–762.

Wolf, S. L., and Lecraw, D. E. (1989). Forced use of hemiplegia extremities to reverse the effect of learned nonuse among chronic stroke and head-injured patients. Exp. Neurol. 104:125–132.

Young, A. W., Aggleton, J. P., Hellawell, D. J., et al. (1995). Face processing impairments after amygdalotomy. Brain 118:15–24.

18

Support Systems: Blood Supply and Cerebrospinal Fluid System

*I*am a 51-year-old professor of neuroanatomy. I teach physical and occupational therapy students, medical students, dental students, and undergraduates. My particular interest, in both teaching and research, is recovery of function. For 10 years or so, I was doing research on recovery from spinal cord injury using rats, but the money dried up, and I haven't done research in several years.

I have had two strokes. The first stroke was when I was 3 years old. But it was misdiagnosed at the time (they thought I had polio), and I didn't know until my 20s that I had had a stroke. People certainly recover a lot better when they are young. The second stroke occurred when I was 41.

The first signs of this stroke were a very severe headache and (so I am told, since my memory of this time was wiped out) a collapse on my left side due to left hemiparesis. My wife asked me to move my left arm. I said, "My left arm is gone. All I've got is a big hole there." This was the first sign of left side neglect. I had no transient ischemic attacks or other warning signs of impending stroke.

When I was taken into a nearby emergency room, they immediately did a CAT scan, which showed a serious right hemisphere hemorrhage. I think I was also given a lumbar puncture and an angiogram. The physician threw up his hands at the results of the scans and placed me under a no-code order that night.

I had four major effects of the stroke. One, I had a loss of proprioception, which was particularly noticeable. I could never tell where my left arm was without looking. I also had a patchy loss of touch and pain sensation (when starting dialysis, I would feel one needle going in but not the other), but I never got it mapped out. Two, I have left-side hemiparesis. I walk with a quad cane, and my left fingers are tonically flexed so that my left arm is not usable. Three, left-side neglect.

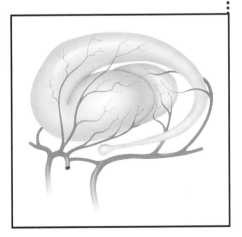

At first I would bump into drinking fountains that I just didn't see. This was worst immediately after the stroke, when I missed the first word of every line I read. The neglect has gotten much better over time and is no longer a real problem. Four, I have short-term memory loss. For some reason, the short-term memory loss is worst with food. I can't remember what I eat each day, but otherwise the memory loss doesn't cause me much problem.

All of these problems have improved over time, so that they are no longer the problems they were. This is probably partly because I have learned how to get around them.

I received lots of physical therapy! Intensive PT during recovery right after the stroke (9 weeks inpatient, several months outpatient). Learning how to stand and transfer, as well as how to walk, was the most important. I have also received PT after two fractures—one of the pelvis, one of the hip. The therapy helped me get going again.

All of the physical therapy was very effective; I couldn't function without it! Working on my own, the best exercise I get is walking as much as possible. I do some other exercises, but not too often.

I take phenobarbital, 400 ml/day, to prevent seizures but occasionally they happen and I have to increase the dosage. I had one grand mal (generalized tonic-clonic) seizure about 2 years after the stroke, but no subsequent grand mal seizures after being on this medication. I have had a number of minor atonic seizures, most of which caused no problems. My atonic seizures ("drop attacks") hit without any warning—I go along, minding my business, and suddenly find myself on the ground. I am never aware of falling, and I don't know if I lose consciousness, but probably very briefly if so. Most of the time I fall like a rag doll (no muscle tone) and don't hurt myself. As soon as I am aware of being down, I have to figure out how to get up again, which I can't do myself. Fortunately, there has always been someone around to help me up. Only twice have I had serious problems. Once, it hit me as I was getting in the shower, and I fell into the shower door, discovering on the way down that it was not shatterproof glass. I came to, lying in a sea of shards and bleeding profusely. I was lucky my wife was home, or I may not have made it . . . The other time was last February when I collapsed while walking home from the bus stop one night and fractured my hip. That was nasty, requiring 4 months of hospitalization.

The biggest change due to the stroke was not a physical one but a mental one. I felt very positive, despite the stroke, and felt that life was really good! In addition I discovered new social skills that I never had before and also had wonderfully creative thoughts drop in on me. These changes are described in my book *Life at a Snail's Pace,* published by Peanut Butter Publishing in Seattle, Washington, 1995.

—Dr. Roger Harris

INTRODUCTION

Two fluid systems support the neurons and glial cells of the nervous system: the cerebrospinal fluid (CSF) system and the vascular system. The CSF system includes the ventricles, the meninges, and the CSF. The vascular system includes the arterial supply, veins and venous sinuses, and mechanisms to regulate blood flow.

CEREBROSPINAL FLUID SYSTEM

CSF system regulates the extracellular milieu and protects the central nervous system. The CSF is formed primarily in the ventricles and then circulates through the ventricles and into the subarachnoid space (between the arachnoid and pia mater) prior to being absorbed into the venous circulation. The CSF supplies water, certain amino acids, and specific ions to the extracellular fluid and probably removes metabolites from the brain. The CSF and extracellular fluid freely communicate in the brain. The meninges and the buoyancy of the fluid provide protection to the brain by absorbing some of the impact when the head is struck.

Ventricles

The CSF-filled spaces inside the brain form a system of four ventricles (Fig. 18–1). The **lateral ventricles** are paired, one in each cerebral hemisphere. The **C**-shaped lateral ventricles consist of a body, an atrium, and anterior, posterior, and inferior horns. The spaces extend into each lobe of the hemispheres. Much of the outside wall of the lateral ventricle is formed by the caudate nucleus, and the tail of the caudate is above the inferior horn. Below the body of the lateral ventricle is the thalamus; above is the corpus callosum. The lateral ventricles connect to each other and to the third ventricle by the interventricular foramina (foramina of Monro).

The **third ventricle** is a narrow slit in the midline of the diencephalon; thus its walls are the thalamus and hypothalamus. An interthalamic adhesion often crosses the center of the third ventricle. A canal through the midbrain, the cerebral aqueduct (aqueduct of Sylvius) connects the third and fourth ventricles.

The **fourth ventricle** is a space between the pons and medulla anteriorly and cerebellum posteriorly.

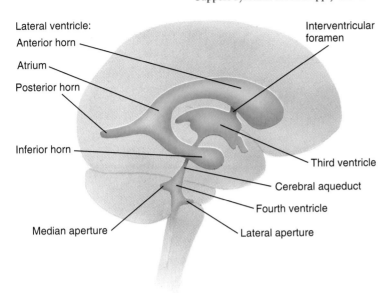

Lateral ventricle:
Anterior horn
Atrium
Posterior horn
Inferior horn
Median aperture

Interventricular
foramen
Third ventricle
Cerebral aqueduct
Fourth ventricle
Lateral aperture

A

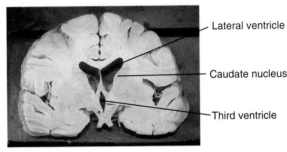

Lateral ventricle
Caudate nucleus
Third ventricle

B

FIGURE 18–1

Ventricles. *A,* Lateral view of the ventricles.
B, Coronal section of the brain showing the
lateral and third ventricles.

Inferiorly the fourth ventricle is continuous with the central canal of the spinal cord. The fourth ventricle drains into the subarachnoid space via three small openings: the two lateral foramina (foramina of Luschka) and a midline opening (foramen of Magendie).

Meninges

Three layers of **meninges** cover the brain and spinal cord. From external to internal, the layers are the dura mater, arachnoid, and pia mater. The **dura mater** surrounds the brain and consists of an outer layer firmly bound to the inside of the skull and an inner layer. The inner layer attaches to the arachnoid. The two layers are fused except at the dural sinuses, which are spaces for the collection of venous blood and CSF. The inner layer of dura has two projections:

the falx cerebri, separating the cerebral hemispheres, and the tentorium cerebelli, separating the cerebellum from the cerebral hemispheres. Spinal dura is continuous with the inner layer of brain dura.

The **arachnoid** is a delicate membrane loosely attached to the dura. Projections of arachnoid form arachnoid villi, which pierce the dura and protrude into the venous sinuses. The arachnoid villi allow CSF to flow into the sinuses. Clusters of arachnoid villi form arachnoid granulations (Fig. 18–2).

Pia mater, the innermost layer, is tightly apposed to the surfaces of the brain and spinal cord. Arachnoid trabeculae (collagen fibers) connect the arachnoid and pia mater, serving to suspend the brain in the meninges. The subarachnoid space, between the pia and arachnoid, is filled with CSF. Extensions of the pia, the denticulate ligaments, anchor the spinal cord to the dura mater.

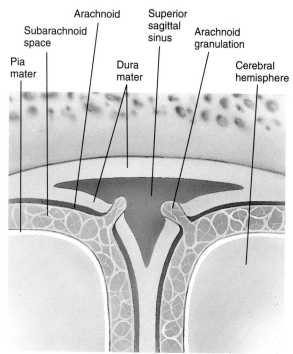

FIGURE 18–2

A coronal section through the skull, meninges, and cerebral hemispheres. The section shows the midline structures near the top of the skull. The three layers of meninges, the superior sagittal sinus, and arachnoid granulations are indicated.

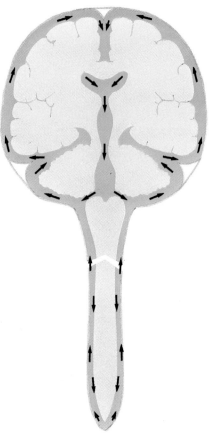

FIGURE 18–3

The flow of cerebrospinal fluid from the lateral ventricles, third ventricle, and fourth ventricle into the subarachnoid space surrounding the brain and spinal cord. The cerebrospinal fluid is reabsorbed into the venous sinuses.

Formation and Circulation of Cerebrospinal Fluid

Although some CSF is formed by extracellular fluid leaking into the ventricles, most CSF is secreted by **choroid plexuses** in the ventricles. Choroid plexus is a network of capillaries embedded in connective tissue and epithelial cells. Through three layers of cells (capillary wall, connective tissue, and epithelium), CSF is formed from blood by filtration, active transport, and facilitated transport of certain substances. These processes result in formation of a fluid similar to plasma.

The CSF flows from the lateral ventricles into the third ventricle via the interventricular foramina and from the third ventricle into the fourth via the cerebral aqueduct (Fig. 18–3). The CSF exits the fourth ventricle through the lateral and medial foramina, entering the subarachnoid space. Within the subarachnoid space, the fluid flows around the spinal cord and

brain. Finally the CSF is absorbed through the arachnoid villi, which project through the dura and into the venous sinuses. In the unidirectional flow of CSF into venous blood, all contents of the CSF (proteins, microorganisms) are included.

Clinical Disorders of the Cerebrospinal Fluid System

Common disorders of the CSF system include epidural and subdural hematoma and hydrocephalus. The hematomas are usually a consequence of trauma. Normally only potential spaces exist between the dura and skull and between the dura and arachnoid.

Bleeding into either of these potential spaces can cause separation of the layers, resulting in an epidural or subdural hematoma. **Epidural hematoma** results from arterial bleeding between the skull and dura mater. Most often an epidural hematoma occurs when the middle meningeal artery is torn by a fracture of the temporal or parietal bone. Because arteries bleed rapidly, signs and symptoms develop swiftly. After a blow to the head, the person may have a few hours of normal function and then develop a worsening headache, vomiting, decreasing consciousness, hemiparesis, and Babinski's sign. In contrast, signs and symptoms of **subdural hematoma** gradually worsen over a prolonged period (days to months). Bleeding is slow in subdural hematoma because the hematoma is produced by venous bleeding, where the blood pressure is less than in arteries. The signs and symptoms are similar to epidural hematoma, with confusion being more prominent. Both types of hematoma are potentially life-threatening because neural tissue is compressed and displaced.

If CSF circulation is blocked, pressure builds in the ventricles, causing **hydrocephalus** (Fig. 18–4). In infants, the cranial bones have not yet fused, so the pressure causes the ventricles, hemispheres, and cranium to expand. Signs of hydrocephalus include disproportionately large head size for age, poor feeding, inactivity, and downward gaze of the eyes (compression of oculomotor nerve center). Common causes of congenital hydrocephalus include failure of the fourth ventricle foramina to open, blockage of the cerebral aqueduct, cysts in the fourth ventricle (Dandy-Walker cysts), and the Arnold-Chiari malformation (see Chap. 5). Rarely, hydrocephalus may result from excessive production or inadequate reabsorption of CSF. In older children or adults, because the cranium cannot expand, excessive pressure in the ventricles compresses the nervous tissue, particularly the white matter. This commonly results in gait and balance impairments, incontinence and headache. Frequently, frontal lobe functions are also involved (i.e., some features of emotions, planning, memory, and intellect). Language, spatial awareness, and declarative memory are spared. In progressive hydrocephalus, a shunt is implanted, usually draining a ventricle into the peritoneum (Fig. 18–5). Generally the shunts are not removed.

A technique alleged to evaluate and treat the CSF system is craniosacral therapy. Advocates of this therapy claim that CSF production is periodic, with each period of secretion followed by a period during which no CSF is produced. The fluid pressure changes purportedly produce a rhythmical movement of the dura that can be palpated (Upledger and Vredevoogd, 1983). Currently no evidence exists for the existence

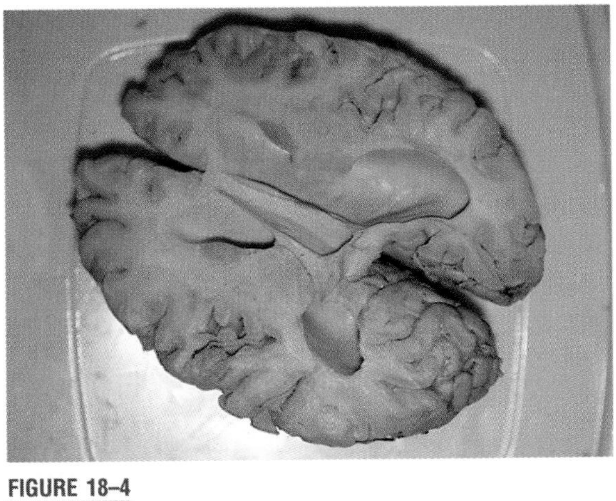

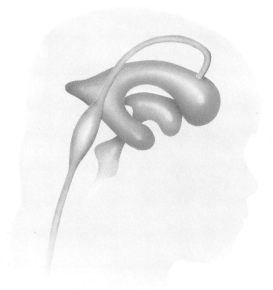

FIGURE 18–4

Horizontal section showing enlarged ventricles characteristic of hydrocephalus. Note the displacement of white matter by the excessive cerebrospinal fluid pressure.

FIGURE 18–5

The placement of a shunt into the lateral ventricle to drain excessive cerebrospinal fluid. The swelling in the shunt is the location of a valve that prevents reverse flow of fluid in the shunt.

of craniosacral rhythm (pulselike movement of CSF transmitted to dural mater and to body fascia), and physical therapist's attempts to assess the rhythm have been demonstrated to be unreliable (Wirth-Pattullo and Hayes, 1994).

VASCULAR SUPPLY

This section is presented regionally, beginning with blood supply to peripheral nerves, then spinal cord, followed by vasculature of the brain. Peripheral nerves are accompanied by blood vessels. Branches from the blood vessels pierce the epineurium surrounding the peripheral nerves. Arterioles and venules travel parallel to fascicles of neurons (Fig. 18–6) to provide ionic exchange and nourishment.

To supply the spinal cord, branches of the vertebral, cervical, thoracic, and lumbar arteries enter the vertebral canal and split into anterior and posterior radicular arteries (Fig. 18–7). These radicular arteries form three spinal arteries running longitudinally along the cord: one is in the anterior midline and two

are posterior, on either side of midline but medial to the dorsal roots. The anterior spinal arteries supply the anterior two-thirds of the cord. The posterior spinal arteries supply the posterior third of the cord. The vertebral arteries that supply the upper spinal cord enter the skull through the foramen magnum to supply the brain.

Four arteries provide blood to the brain. The paired vertebral arteries supply the brain stem, cerebellum, and posteroinferior cerebrum. The paired internal carotid arteries supply the anterior, superior, and lateral cerebral hemispheres. The arterial branches discussed in the following sections are only the major arteries. Each artery has many branches, an elaborate capillary bed, and multiple arteriovenous junctions.

Vascular Supply to Brain Stem and Cerebellum

The brain stem and cerebellum are supplied by branches of the vertebral arteries and branches of the basilar artery (Fig. 18–8). The basilar artery is formed by the union of the vertebral arteries. Each **vertebral artery** has three main branches: the anterior and posterior spinal arteries and the posterior inferior cerebellar artery. The medulla receives blood from all three branches of the vertebral arteries. The posterior inferior cerebellar artery also supplies the inferior cerebellum.

Near the pontomedullary junction, the vertebral arteries join to form the **basilar artery.** The basilar artery and its branches (anterior inferior cerebellar, superior cerebellar) supply the pons and most of the cerebellum. At the junction of the pons and midbrain, the basilar artery divides to become the **posterior cerebral arteries.** The posterior cerebral artery is the primary blood supply to the midbrain.

Vascular Supply to Cerebral Hemispheres

INTERNAL CAROTID AND POSTERIOR CEREBRAL ARTERIES

The cerebrum is entirely supplied by the internal carotid and posterior cerebral arteries. The internal carotid arteries enter the skull through the temporal

FIGURE 18–6

Blood supply to a fascicle within a peripheral nerve.

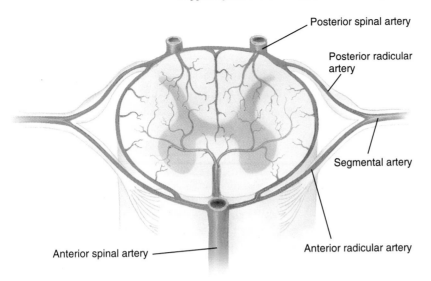

FIGURE 18–7
Blood supply of the spinal cord.

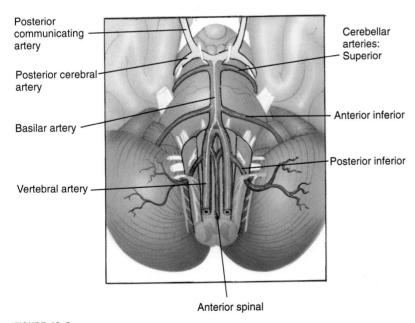

FIGURE 18–8
Blood supply of the brain stem and cerebellum.

bones; small branches from each internal carotid become posterior communicating arteries that join the internal carotid with the posterior cerebral artery. Near the optic chiasm, the internal carotid divides into anterior and middle cerebral arteries (see following sections). Together, the posterior cerebral arteries and branches of the internal carotid arteries form the circle of Willis to supply the cerebrum.

CIRCLE OF WILLIS

The circle of Willis is an anastomotic ring of nine arteries, supplying all of the blood to the cerebral hemispheres (Fig. 18–9). Six large arteries anastomose via three small communicating arteries. The large arteries are the anterior cerebral artery (a branch of the internal carotid), the internal carotid, and the posterior cerebral (branches of the basilar). The anterior communicating artery (unpaired) joins the anterior cerebral arteries together, and the posterior communicating artery links the internal carotid with the posterior cerebral artery.

CEREBRAL ARTERIES

Each of the three major cerebral arteries (anterior, middle, posterior) has both cortical branches (supplying the cortex and outer white matter) and deep branches (to central gray matter and adjacent white matter).

From its origin, the **anterior cerebral artery** moves medially and anteriorly, into the longitudinal fissure. The artery sweeps up and back above the corpus callosum, its branches supplying the medial surface of the frontal and parietal lobes and the anterior head of the caudate (Fig. 18–10). The **middle cerebral artery** passes through the lateral sulcus, and then its branches fan out to supply most of the lateral hemisphere. The **posterior cerebral artery** wraps around and supplies the midbrain and then supplies the occipital lobe and parts of the medial and inferior temporal lobes. The major cerebral arteries connect at their beginning (via the circle of Willis) and at their ends (watershed area [see Fig. 18–10]). The **watershed area** is an area of marginal blood flow on the

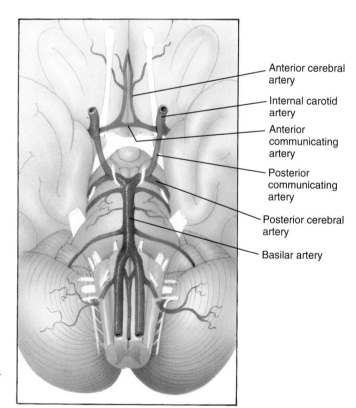

Anterior cerebral artery

Internal carotid artery

Anterior communicating artery

Posterior communicating artery

Posterior cerebral artery

Basilar artery

FIGURE 18–9

From anterior to posterior, the arteries that form the circle of Willis are the anterior communicating, two anterior cerebrals, two internal carotids, two posterior communicating, and two posterior cerebral arteries.

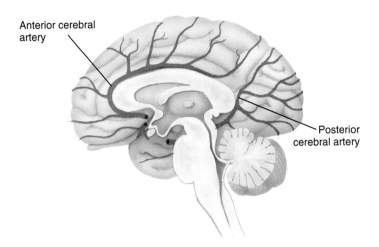

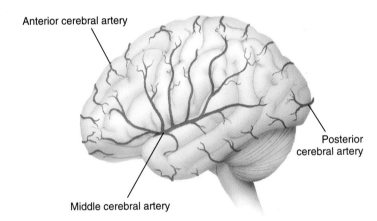

FIGURE 18–10

The large cerebral arteries: anterior, middle, and posterior. In the lower panel, the region supplied by small anastomoses at the ends of the arteries is the watershed area.

surface of the lateral hemispheres, where small anastomoses link the ends of the cerebral arteries. If the anterior cerebral arteries were deprived of blood flow, what functional effects would result? What would happen with occlusion of the middle or posterior cerebral artery?

> The anterior cerebral artery supplies the anterosuperior parts of the medial cerebral hemisphere. The middle cerebral artery supplies the lateral cerebral hemisphere. The posterior cerebral artery supplies the midbrain, occipital lobe, and parts of the medial and inferior temporal lobe.

In addition to the branches of major cerebral arteries supplying deep structures, two other arteries supply only deep structures: the **anterior and posterior choroidal arteries.** The anterior choroidal (a branch of the internal carotid) supplies the optic tract, choroid plexus in the lateral ventricles, and parts of the optic radiations, putamen, thalamus, internal capsule, and hippocampus. The posterior choroidal (a branch of the posterior cerebral artery) supplies the choroid plexus of the third ventricle and parts of the thalamus and hippocampus (Fig. 18–11).

Disorders of Vascular Supply

Interrupting the blood flow to a part of the brain usually produces a focal loss of function, except in cases of subarachnoid hemorrhage. The effects of blood

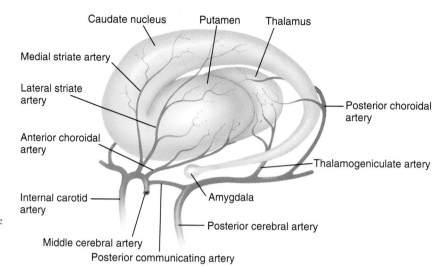

FIGURE 18–11

Branches of the internal carotid artery supply the caudate and putamen. The supply to the putamen is via the anterior choroidal artery. The posterior choroidal artery, a branch of the posterior cerebral artery, supplies part of the thalamus.

flow interruption range from a brief loss of function followed by complete recovery to permanently life-altering impairments and disabilities to death. Episodes of focal functional loss following vascular incidents are classified according to both the pattern of progression and etiology. The patterns of progression from the time of onset include the following:

- **Transient ischemic attack:** A brief, focal loss of brain function, with full recovery from neurological deficits within 24 hours. Transient ischemic attacks are believed to be due to ischemia.
- **Completed stroke:** Neurological deficits from vascular disorders that persist more than 1 day and are stable (not progressing or improving).
- **Progressive strokes:** Some people with ischemic stroke have deficits that increase intermittently over time. These are believed to be due to repeated emboli (blood clots that formed elsewhere and were transmitted by the blood to a new location) or continued formation of a thrombus (blood clot that stays where it formed).

TYPES OF STROKE

The term *cerebrovascular accident* is synonymous with stroke. Currently some members of the medical community are advocating "brain attack" as a lay term to replace *stroke,* to emphasize that prompt treatment may benefit some people who have strokes, just as prompt treatment is effective for some heart attacks.

Brain Infarction Infarcts occur when an embolus or thrombus lodges in a vessel, obstructing blood flow. Typically, an **embolus** abruptly deprives an area of blood, resulting in almost immediate onset of deficits. Sometimes the embolus breaks into fragments and is dislodged, resulting in quick resolution of deficits. More commonly, residual brain damage is permanent, resulting in prolonged and incomplete functional recovery.

Onset of signs from thrombic ischemia may be abrupt or may worsen over several days. Recovery from a **thrombus** is usually slow, and significant residual disability is common.

Obstructions of blood flow in small, deep arteries result in **lacunar infarcts.** Lacunae are small cavities that remain after the necrotic tissue is cleared away (Fig. 18–12). Lacunar infarcts occur most often in the basal ganglia, internal capsule, thalamus, and brain stem. Signs of lacunar infarcts develop slowly and are often either purely motor or purely sensory, and good recovery is the norm.

The slow occlusion of an artery has a very different outcome from an abrupt occlusion. For example, if one internal carotid artery is slowly occluded, the anastomotic connections and collateral circulation among the unaffected arteries may be adequate to maintain brain function. Less frequently, an abrupt internal carotid occlusion is fatal due to infarction of the anterior two-thirds of the cerebral hemisphere. The difference in outcome is explained by the time course of occlusion, the location of the occlusion,

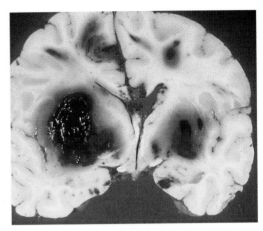

FIGURE 18–13
Multiple hemorrhages within the brain, secondary to head trauma. (Courtesy of Dr. Melvin J. Ball.)

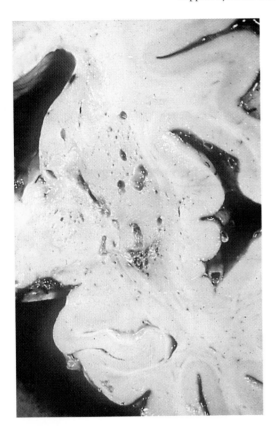

FIGURE 18–12
Coronal section of a cerebral hemisphere, with a lateral ventricle appearing near the top left corner. The small cavities in the basal ganglia are lacunar infarcts, produced by occlusions of small, deep arteries. (Courtesy of Dr. Melvin J. Ball.)

blood pressure at the time of the occlusion, and individual variation in collateral connections. Gradual occlusion may allow development of increased collateral circulation. Low blood pressure during the occlusion makes adequate perfusion of the brain less likely.

Hemorrhage Hemorrhage deprives the downstream vessels of blood, and the extravascular blood exerts pressure on the surrounding brain. Generally, hemorrhagic strokes present with the worst deficits within hours of onset, and then improvement occurs as edema decreases and extravascular blood is removed. Severe hemorrhage within the brain tissue is shown in Figure 18–13.

Subarachnoid Hemorrhage Bleeding into the subarachnoid space usually causes sudden, excruciating

headache with a brief (a few minutes) loss of consciousness. Unlike other hemorrhages, the initial findings often are not focal. Deficits from subarachnoid hemorrhage are progressive because of continued bleeding or secondary hydrocephalus. Vasospasm and infarction are common sequelae of subarachnoid hemorrhage. Figure 18–14 shows a subarachnoid hemorrhage.

STROKE SIGNS AND SYMPTOMS BY ARTERIAL LOCATION

Pathology may involve the main arteries, smaller branches, the capillary network, or arteriovenous formations.

Vertebral and Basilar Arteries Disruptions of the blood supply in the brain stem cerebellar region occur fairly frequently. Because the vertebral arteries are subject to shear forces at the atlantoaxial joint, abrupt neck rotation or hyperextension can cause brain stem ischemia. Strokes attributable to chiropractic manipulation and to spontaneous head movement are discussed by Frisoni and Anzola (1991).

A complete occlusion of the basilar artery causes death due to ischemia of brain stem nuclei and tracts that control vital functions. Partial occlusions of the basilar artery can cause tetraplegia (descending motor tracts), loss of sensation (ascending sensory tracts), coma (reticular activating system), and cranial nerve signs. The most common symptoms in acute cerebel-

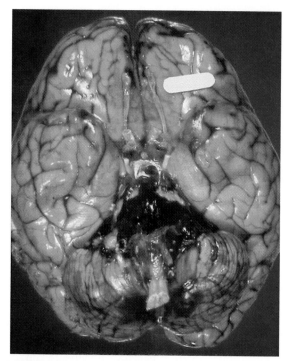

FIGURE 18–14

Subarachnoid hemorrhage is visible as dark areas, most prominent in the brain stem region. (Courtesy of Dr. Melvin J. Ball.)

lar infarction are dizziness and/or vertigo, lack of balance, nausea and vomiting, dysarthria, and headache (Macdonell et al., 1987).

Cerebral Arteries Occlusion of the cortical branches of the **anterior cerebral artery** results in personality changes (frontal lobe) with contralateral hemiplegia and hemisensory loss. The hemiplegia and hemisensory loss are more severe in the lower limb than in the face and upper limb because the medial sensorimotor cortex and adjacent white matter are affected. Lack of blood supply to the deep branches of the anterior cerebral artery results in motor dysfunction due to damage of the anterior putamen and of frontopontine axons in the internal capsule.

Occlusion of the cortical branches of the **middle cerebral artery** deprives the optic radiation and the lateral parts of the sensorimotor cortex and adjacent white matter of blood. This produces homonymous hemianopia combined with contralateral hemiplegia and hemisensory loss involving the upper limb and face more than the lower limb because the neurons regulating movement and processing conscious sensation of the upper body are located in the lateral cerebral cortex. Language impairment is common if the lesion is in the language-dominant hemisphere (usually the left hemisphere). Difficulty understanding spatial relationships, neglect, and impairment of nonverbal communication often occur with lesions in the hemisphere that is nondominant for language (usually the right hemisphere). Deep branches of the middle cerebral artery (**striate arteries**) supply the striatum and the genu and limbs of the internal capsule. Loss of blood supply to the deep branches deprive axons passing through the internal capsule, with the consequence of involving the upper and lower extremities and the face equally. Most ischemic strokes occlude the middle cerebral artery, often producing a stereotyped posture: a pattern of increased muscle tone in the hemiplegic limbs, with characteristic adduction at the shoulder, flexion at the elbow, and extension throughout the lower limb. Occlusion of the anterior choroidal artery, a branch off the internal carotid, produces contralateral hemiplegia and hemisensory loss with homonymous hemianopia by depriving axons in the posterior internal capsule of blood.

Occlusion of the midbrain branches of the **posterior cerebral artery** can result in contralateral hemiparesis (cerebral peduncle) and eye movement paresis or paralysis, not including lateral or inferomedial eye movements (oculomotor nerve and its controlling nuclei or descending neurons). Occlusion of the branches to the calcarine cortex result in cortical blindness affecting information from the contralateral visual field (see Chap. 15). Deep branches of the posterior cerebral artery supply much of the diencephalon and hippocampus. Lack of blood flow to the thalamus can cause thalamic syndrome, characterized by severe pain, contralateral hemisensory loss, and flaccid hemiparesis. Vascular compromise of the hippocampus interferes with declarative memory (see Chap. 17). Occlusion of the posterior choroidal branch prevents blood from reaching part of the thalamus and hippocampus (Fig. 18–15).

The watershed area where the distal branches of the cerebral arteries anastomose is vulnerable to ischemia (Fig. 18–16). Lack of blood to the watershed region often causes upper limb paresis and paresthesias. Hypotension may result in decreased blood flow in the watershed area, thereby decreasing the effectiveness of the anastomoses.

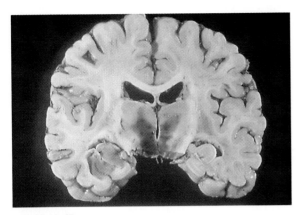

FIGURE 18–15

Occlusion of the posterior choroidal artery, producing necrosis in part of the thalamus. (Courtesy of Dr. Melvin J. Ball.)

intracerebral hemorrhage, or both, depending on the location of the malformation.

An **aneurysm** is a dilation of the wall of an artery or vein. These swellings have thin walls that are prone to rupture. Saccular aneurysms are most common, affecting only side of the vessel wall. A berry aneurysm, a type of saccular aneurysm, is a small sac that protrudes from a cerebral artery and has a thin connection with the artery (Fig. 18–17). The hemorrhage resulting from aneurysm rupture may be massive, causing sudden death, or causing a wide variety of signs and symptoms depending on the location and extent of the bleeding. The bleeding is into the subarachnoid space, producing subdural hematoma.

The effects of a stroke depend on the etiology, the severity, and the location of the stroke.

Disorders of Vascular Formation

Arteriovenous malformations are developmental abnormalities with arteries connected to veins by abnormal, thin-walled vessels larger than capillaries. The malformations usually do not cause signs or symptoms until they rupture; then the bleeding causes dysfunction due to lack of blood to the area the arteries normally supply and due to pressure exerted by the extravascular blood. Rupture of an arteriovenous malformation can cause subdural hematoma,

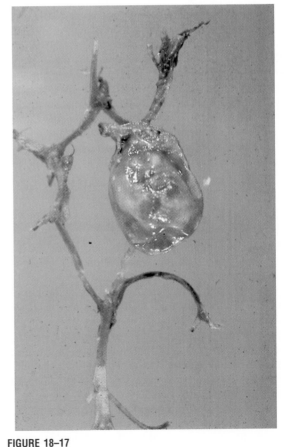

FIGURE 18–17

A large berry aneurysm at the end of the right internal carotid artery. (Courtesy of Dr. Melvin J. Ball.)

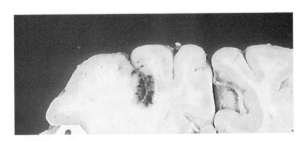

FIGURE 18–16

A coronal section near the top of the skull, showing an infarction in the watershed area. (Courtesy of Dr. Melvin J. Ball.)

Blood-Brain Barrier

The blood-brain barrier is a specialized permeability barrier between the capillary endothelium of the central nervous system and the extracellular space. The barrier is formed by tight junctions between the capillary endothelial cells that exclude large molecules (free fatty acids, proteins, specific amino acids). This exclusion is useful for preventing many pathogens from entering the central nervous system; however, the barrier also prevents access of certain drugs and protein antibodies to the brain. For example, in the early stages of Parkinson's disease, dopamine delivered to the brain can ameliorate the signs and symptoms. However, dopamine cannot cross the blood-brain barrier. Therefore a metabolic precursor of dopamine, called L-dopa, is given to people with Parkinson's disease. L-dopa can cross the blood-brain barrier. Once L-dopa is in the brain, it is converted into dopamine. Currently, the intentional disruption of the blood-brain barrier is an experimental method of delivering some medications to the central nervous system.

The blood-brain barrier is absent in areas of the brain that directly sample the contents of the blood or secrete into the bloodstream. These regions include parts of the hypothalamus and other specialized areas around the third and fourth ventricles. Specialized ependymal cells (tanycytes) separate the leaky regions from the rest of the brain; these special cells may prevent proteins, viruses, and some drugs from entering the brain via the leaky regions.

Cerebral Blood Flow

Because the brain cannot store glucose or oxygen effectively, a consistent blood supply is essential. Oxygen consumption increases from brain stem to cerebral cortex, leaving the cerebral cortex more vulnerable to hypoxia than vital centers in the lower brain stem. This differential oxygen requirement explains some incidents of persistent vegetative state. In some cases of persistent vegetative state, severe head trauma or anoxia destroys the cerebral and cerebellar cortices, yet the person survives because the brain stem and spinal cord functions continue.

Cerebral arteries **autoregulate** local blood flow, depending primarily on two factors: blood pressure and metabolites. The arteries dilate if blood pressure, oxygen, or pH levels are inadequate, or if carbon dioxide or lactic acid are excessive. Conversely, when blood pressure, oxygen, or pH levels are excessive, or car-

bon dioxide or lactic acid levels are below functional levels, the arteries constrict. A minor role in regulating arterial diameter is played by autonomic and other neuron systems within the brain; these mechanisms are currently not well understood. Autoregulation is vitally important to ensure adequate blood flow and to prevent brain edema.

CEREBRAL EDEMA

Cerebral edema is the accumulation of excess tissue fluid in the brain. Concussion frequently causes cerebral edema because trauma allows fluid to leak from the damaged capillaries. Edema is often progressive because the fluid pressure results in ischemia, causing arterioles to dilate, increasing the capillary pressure, and producing more edema. Also, lack of oxygen to a region of the brain makes the capillaries more permeable, and thus more fluid escapes into the extracellular compartment. Edema can be alleviated by shunts or medications (e.g., dexamethasone, mannitol).

EFFECTS OF SPACE-OCCUPYING LESIONS IN THE BRAIN

Cerebral edema, hydrocephalus, tumors, and other lesions that occupy space in the brain can cause an increase in intracranial pressure. Symptoms include vomiting and nausea (pressure on vagus nerve), headache (increased capillary pressure), drowsiness, and visual and eye movement problems (pressure on optic and oculomotor nerves).

Space-occupying lesions may produce herniation (protrusion) of part of the brain. Pressure from a hemorrhage, edema, or a tumor can cause displacements of brain structures that have grave consequences.

Uncal Herniation Uncal herniation occurs when a space-occupying lesion in the temporal lobe displaces the uncus medially, forcing the uncus into the opening of the tentorium cerebelli. In turn, this compresses the midbrain, interfering with the function of the oculomotor nerve and consciousness (effect on ascending reticular activating system). Figure 18–18 shows an infarct secondary to uncal herniation.

Central Herniation Central herniation occurs when a space-occupying lesion in the cerebrum exerts pressure on the diencephalon, moving the diencephalon, midbrain, and pons inferiorly. This movement stretches the branches of the basilar artery, causing

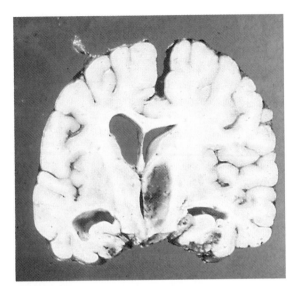

FIGURE 18–18

A large subdural hematoma displaced the right cerebral hemisphere, causing uncal herniation. Note the distortion of the shape of the lateral ventricles and that both lateral ventricles and the third ventricle are to the left of the midline. An infarct of the posterior cerebral artery occurred secondary to the uncal herniation. (Courtesy of Dr. Melvin J. Ball.)

brain stem ischemia and edema. Bilateral paralysis ensues (due to damage of upper motoneurons), and consciousness and oculomotor control are impaired.

Tonsillar Herniation Pressure from an uncal herniation, a tumor in the brain stem/cerebellar region, hemorrhage, or edema may force the cerebellar tonsils (small lobes forming part of the inferior surface of the cerebellum) through the foramen magnum. Tonsillar herniation compresses the brain stem, interfering with vital signs, consciousness, and flow of cerebrospinal fluid.

LABORATORY EVALUATION OF CEREBRAL BLOOD FLOW

Blood flow to the brain can be evaluated by **positron emission tomography** (PET scan) or **angiography**. A PET scan is a computer-generated image based on metabolism of injected radioactively labeled substances (Fig. 18–19). A PET scan records local variations in blood flow, reflecting neural activity. Angiography consists of a radiopaque dye injected into a carotid or vertebral artery followed by a sequence of x-rays (Fig. 18–20). Typically the end of a plastic catheter inserted into the femoral artery is moved to

the origin of the vessel to be visualized, and then the dye is injected. In the first series of x-rays, the arteries are visible; later, as the dye circulates, the veins are seen. Angiography is particularly useful for visualizing aneurysms, occlusions, and malformations of the arteriovenous system; however, thrombosis and embolization are risks with this invasive procedure.

Venous System

The spinal cord and lower medulla drain into small veins that run longitudinally. These veins drain into radicular veins, which then empty into the epidural venous plexus.

The major venous system of the brain consists of cerebral veins. These veins drain into **dural sinuses** (Fig. 18–21) and eventually into the internal jugular vein (Fig. 18–22). Cerebral veins interconnect extensively. Two sets of veins drain the cerebrum: superficial and deep. The superficial veins drain cortex and the adjacent white matter, then empty into the superior sagittal sinus or one of the sinuses around the

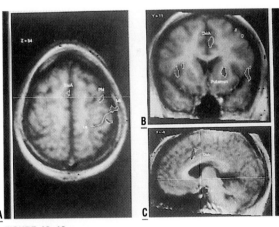

FIGURE 18–19

Positron emission tomography scan. Shows the cortical areas that have significantly more regional blood flow during self-paced finger flexions than during visually triggered finger flexions or during rest. The colors indicate the level of metabolic activity: red is highest, orange is high, yellow is moderate, green is low, and blue is lowest. SMA, supplemental motor area; PM, premotor area; M1, primary motor cortex; and CMA, cingulate motor cortex. The CMA is a region that has not been studied extensively. A, Horizontal section. Posterior to the central sulcus (CS), increased activation of the primary somatosensory cortex is visible. B, Coronal section. C, Midsaggital section. (Reproduced with permission of Rapid Science Publishers Ltd. from Larsson, J., Bulyas, B., and Roland, P. E. [1996]. Cortical representation of self-paced finger movement. NeuroReport 7[2]:466.)

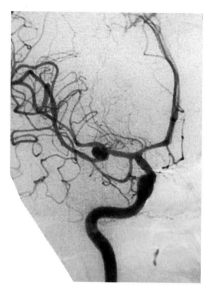

FIGURE 18–20

Angiogram showing an aneurysm arising from the middle cerebral artery.

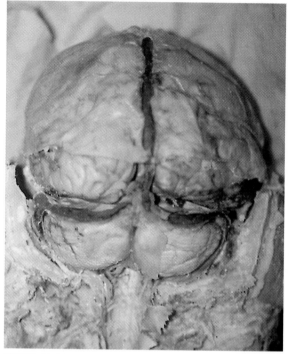

FIGURE 18–21

A posterior view of the dura mater covering the brain, with the dural (venous) sinuses exposed. The superior sagittal sinus, between the superior parts of the cerebral hemispheres, and the transverse sinuses, between the cerebral and cerebellar hemispheres, are visible.

inferior cerebrum. The deep cerebral veins drain the basal ganglia, diencephalon, and nearby white matter, then empty into the straight sinus. The superior sagittal and straight sinuses join at the confluence of the sinuses. The transverse sinuses arise from the confluence and connect with the internal jugular veins.

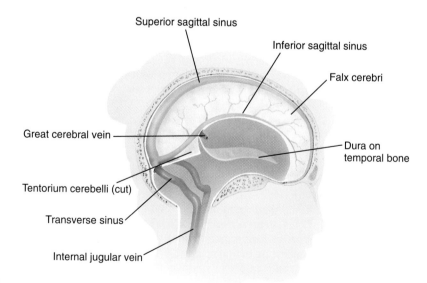

FIGURE 18–22

The venous system of the brain. The venous sinuses eventually drain into the internal jugular vein.

CLINICAL NOTES

For each of the following cases, answer the following questions:

1. Which vertical systems are involved?

2. Where is the lesion?

3. What is the likely etiology?

CASE 1

A 54-year-old man reports sudden and complete loss of ability to move his legs. No trauma occurred.

- Sensory examination reveals bilateral loss of pain and temperature sensations below T9, intact above T9.

- Localized touch, vibration, and position senses are intact throughout the entire body.

- Motor examination reveals bilateral paralysis below T9, motor system normal above T9.

CASE 2

A 72-year-old woman awoke 3 days ago with severe weakness and loss of sensation on the left side of her body and lower face. All sensory and motor functions on the right side are within normal limits.

- Sensory testing reveals responsiveness only to deep pinch on the left side.

- She cannot move any joint on the left side independent of the movement of other joints. When she attempts to reach forward, no flexion occurs at the shoulder; instead, her shoulder elevates, and elbow flexion increases.

- In sitting or assisted standing, she does not bear weight on the left. She cannot walk, even with assistance. She steps forward with the right lower limb, then lurches forward and attempts to continue stepping with the right lower limb, dragging the left lower limb.

- Her gaze tends to be directed toward the right. Even with cuing or loud noises on the left side, she does not turn her head or eyes to the left of midline. She seems unaware of the left side of her body.

- Her ability to converse is normal.

CASE 3

K.F., a 9-month-old boy, has an enlarged cranium and is being assessed for possible developmental delay.

- Sensation is normal.

- K.F. cannot sit unsupported. In supported sitting, he is unable to hold his head in neutral for more than 10 seconds. He moves very little. In supine position, his limbs tend to flop out to the sides, and he does not turn from back to side.

- His gaze is directed downward.

(Note: Healthy children achieve unsupported sitting between the ages of 4 and 8 months; turning from back to side is usually achieved by 7 months.)

REVIEW QUESTIONS

1. What are the functions of the cerebrospinal fluid?

2. Where is cerebrospinal fluid located?

3. Why is there a difference in pattern of progression between an epidural hematoma and a subdural hematoma?

4. In an infant, an abnormally large head size, inactivity, insufficient feeding, and downward gaze of both eyes may indicate what disorder?

5. Can hydrocephalus occur in adults?

6. Which blood vessels supply the pons?

7. What blood vessel supplies the midbrain?

8. List the arteries that form the circle of Willis.

9. What is the watershed area?

10. What is transient ischemic attack?

11. What is a lacuna?

12. A partial occlusion of which artery can result in tetraplegia, loss of sensation, coma, and cranial nerve signs?

13. Hemiplegia and hemisensory loss that are more severe in the lower limb than in the upper limb and face indicate that what part of the brain is affected? What artery supplies this region?

14. Neglect, poor understanding of spatial relationships, and impairment of nonverbal communication are signs of damage to what part of the brain? Which artery supplies this region?

15. If hemiplegia and hemisensory loss affect the upper and lower limbs and the face equally, where is the lesion? Branches of what major artery supply this region?

16. Eye movement paresis with sparing of lateral and inferomedial eye movements combined with contralateral hemiplegia indicates a lesion located where? Which artery supplies this region?

17. What arteries supply the watershed area?

18. What is an arteriovenous malformation?

19. What is an aneurysm?

20. What is an uncal herniation?

21. What is a PET scan?

References

Frisoni, G. B., and G. P. Anzola. (1991). Vertebrobasilar ischemia after neck motion. Stroke 22:1452–1460.

Macdonell, R. A., Kalnins, R. M., and Donnan, G. A. (1987). Cerebellar infarction: Natural history, prognosis, and pathology. Stroke 18(5):849–855.

Upledger, J. E., and Vredevoogd, J. D. (1983). Craniosacral therapy. Seattle: Eastland Press.

Wirth-Pattullo, V., and Hayes, K. W. (1994). Interrater reliability of craniosacral rate measurements and their relationship with subjects' and examiners' heart and respiratory rate measurements. Phys. Ther. 74:908–920.

Suggested Readings

Amarenco, P., and Hauw, J.-J. (1990). Cerebellar infarction in the territory of the anterior and inferior cerebellar artery: A clinicopathological study of 20 cases. Brain 113:139–155.

Bogousslavsky, J., and Regli, F. (1990). Anterior cerebral artery territory infarction in the Lausanne Stroke Registry: Clinical and etiologic patterns. Arch. Neurol. 47:144–150.

Broderick, J., Phillips, S., Whisnant, J., et al. (1989). Incidence rates of stroke in the eighties: The end of the decline in stroke? Stroke 20:577–582.

Cutler, R., and Spertell, R. (1982). Cerebrospinal fluid: A selective review. Ann. Neurol. 11:1–10.

Decroix, J., Graveleau, R., Masson, M., et al. (1986). Infarction in the territory of the anterior choroidal artery. Brain 109: 1071–1085.

Dennis, M., Fritz, C., Netley, C., et al. (1981). The intelligence of hydrocephalic children. Arch. Neurol. 38:607–615.

Haaxma-Reiche, H., Piers, D., and Beekhuis, H. (1989). Normal cerebrospinal fluid dynamics: A study with intraventricular injection of 111In-DPTA in leukemia and lymphoma without meningeal involvement. Arch. Neurol. 46:997–999.

Leech, R., and Brumback, R. (1990). Hydrocephalus: Current clinical concepts. St. Louis: Mosby–Year Book.

Nolte, J. (1993). Ventricles and cerebrospinal fluid; Blood supply of the brain. In: The human brain: An introduction to its functional anatomy (3rd ed.). St. Louis: Mosby–Year Book, pp. 48–92.

Romero-Sierra, C. (1986). Vascular supply. In: Neuroanatomy: A conceptual approach. New York: Churchill Livingstone.

Toole, J. (1990). Cerebrovascular disorders. New York: Raven Press.

Toole, J., Yoson, C., and Janeway, R. (1978). Transient ischemic attacks: A study of 225 patients. Neurology 28:746–753.

Appendix:
Neurotransmitters and Neuromodulators

INTRODUCTION

Neurotransmitters and neuromodulators are chemical compounds that convey information among neurons. Classically, a neurotransmitter has been defined as a chemical released by a presynaptic neuron that causes excitation or inhibition of the postsynaptic membrane. Their effect on the postsynaptic membrane occurs less than a millisecond after release and is terminated within a tenth of a second. Neuromodulators alter neural function by activating G proteins, which in turn activate membrane channels or intracellular enzymes (called second messengers). This leads to prolonged opening of membrane ion channels, activation of genes, and/or adjustment of calcium levels inside the postsynaptic cell. Neuromodulators require seconds before their effects are manifest; the effects last from minutes to days.

Rigid distinctions cannot be made between neurotransmitters and neuromodulators because a single compound may act as a neurotransmitter at some sites in the nervous system and as a neuromodulator at other sites. For example, substance P acts as a neurotransmitter between the first- and second-order neurons in the nociceptive pathway but as a neuromodulator in the hypothalamus.

NEUROTRANSMITTERS

Generally, acetylcholine and the amino acids act as neurotransmitters. The amino acid transmitters include glycine, GABA, and glutamate. The actions of the amino acid transmitters on postsynaptic membranes are predictable: GABA and glycine are always inhibitory, and glutamate is always excitatory.

Acetylcholine

Acetylcholine plays the major role in transmitting information in the peripheral nervous system. Acetylcholine is the transmitter released by motoneurons, parasympathetic neurons, and preganglionic sympathetic neurons. In the central nervous system, acetylcholine is involved with the selection of objects of attention and with autonomic regulation. Sources of acetylcholine in the central nervous system include the pedunculopontine nucleus, the basal nucleus of Meynert, and the basal forebrain. Receptors for acetylcholine are nicotinic (brief opening of ion channels) or muscarinic (slow-acting G-protein mediated effects). An autoimmune disease, myasthenia gravis, causes the destruction of acetylcholine receptors on skeletal muscle membranes.

GABA

GABA is a major inhibitory neurotransmitter. Two types of GABA receptors exist: $GABA_A$ and $GABA_B$. When GABA binds with $GABA_A$ receptors, Cl^- channels open, producing hyperpolarization of the postsynaptic membrane. Benzodiazepines (anti-anxiety and anticonvulsant drugs) and barbiturates (tranquilizing drugs) also activate $GABA_A$ receptors and thus hyperpolarize postsynaptic membranes. $GABA_B$ receptors are linked to ion channels via second messenger systems. Baclofen, a muscle relaxant used to treat spasticity, increases the presynaptic release of GABA that activates the $GABA_B$ receptors in the spinal cord (Cooper, 1996).

Glycine

Glycine inhibits postsynaptic membranes, primarily in the brain stem and spinal cord. Glycine also prevents desensitization of the N-methyl-D-aspartate (NMDA) receptor.

Glutamate

Glutamate is the excitatory neurotransmitter that activates the NMDA receptor. The NMDA receptor is an excitatory amino acid receptor with six different binding sites. The NMDA receptor has been implicated in long-term potentiation, a possible mechanism of neural plasticity during development and learning. Overactivity of NMDA receptor channels may produce epileptic seizures. Excitotoxicity, the death of neurons from overexcitation, is due to persistent opening of many NMDA receptor channels. Other non–NMDA glutamate receptors are both the direct action ion-channel type and the G-protein-mediated-receptor type.

NEUROMODULATORS

The monoamines and the peptides usually act as neuromodulators. The monoamines include the catecholamines dopamine, norepinephrine, and epinephrine, and the indolamine, serotonin. Dopamine, norepinephrine, and serotonin usually have inhibitory effects on postsynaptic membranes, but sometimes are excitatory. The peptides include substance P and enkephalins.

Dopamine

Dopamine has effects on motor activity, motivation, and cognition. Major sources of dopamine are the substantia nigra and the ventral tegmental area. Loss of dopamine from the substantia nigra is the primary deficit in Parkinson's disease. The involvement of dopamine in certain aspects of psychosis is demonstrated by the action of some antipsychotic medications that prevent the binding of dopamine to certain receptor sites. These drugs decrease hallucinations, delusions, and disorganized thinking. However, because these drugs prevent the binding of dopamine, a side effect of many of these medications is tardive dyskinesia. Tardive dyskinesia is a hyperkinetic disorder, characterized by involuntary muscle contractions. Clozapine is an antipsychotic drug that binds only to one type of dopamine receptor and does not produce tardive dyskinesia. The motivational aspects of dopamine are evident in addiction to certain drugs.

The action of dopamine is potentiated by cocaine, because cocaine interferes with a protein that removes dopamine from its binding site. Amphetamines increase the release of dopamine and block dopamine reuptake. Finally, although some people with schizophrenia have an excess of a subtype of dopamine receptor, current evidence does not rule out effects of drug treatment as the possible cause of the abnormal concentration of dopamine receptors.

Serotonin

Serotonin adjusts the general arousal level and suppresses sensory information. For example, serotonin plays a role as part of the descending pain control system. Highest levels of serotonin are coincident with alertness, levels are low in non–REM sleep, and lowest during REM sleep. Low levels of serotonin are associated with depression and suicidal behavior. The antidepressant Prozac (fluoxetine) is a selective blocker of serotonin reuptake.

Norepinephrine

Norepinephrine plays a vital role in active surveillance of surroundings by increasing attention to sensory information. Highest levels of norepinephrine are associated with vigilance and lowest levels of norepinephrine occur during sleep. Norepinephrine binds to alpha and beta receptors. Norepinephrine is essential in producing the "fight or flight" reaction to stress.

Overactivity of the norepinephrine system produces fear, and in extreme cases, panic, by action on cortical and limbic regions. Panic disorder is the abrupt onset of intense terror, a sense of loss of personal identity, the perception that familiar things are strange or unreal, combined with the signs of increased sympathetic nervous system activity. Panic disorder is produced by excessive levels of norepinephrine. Andrenergic antagonists, such as propranolol, prevent the activation of beta receptors. This action prevents the sweating, rapid heart beat, and other signs of sympathetic activation that may otherwise occur in stressful situations. Musicians and actors often take propranolol before a performance.

Posttraumatic stress disorder also involves excessive norepinephrine activity. This has been demonstrated by intravenous administration of a drug, yohimbine, that stimulates norepinephrine activity. Veterans with posttraumatic stress disorder experience flashbacks to the traumatic event, panic, grief, intrusive thoughts about the traumatic event, and emotional numbness when given yohimbine. Control subjects experience little effect of yohimbine.

Drugs Used to Treat Depression

Drugs effective in treating depression include monoamine oxidase inhibitors, tricyclic antidepressants, and serotonin-selective reuptake blockers. Monoamine oxidase (MAO) degrades catecholamines, so norepinephrine levels are lowered. The main effect of tricyclic antidepressants seems to be increased activity of serotonin and alpha$_1$ (norepinephrine) receptors, and decreased activity of central beta-receptors (norepinephrine). The serotonin-selective uptake blockers include Prozac (fluoxetine).

Substance P

Substance P is found in the dorsal horn of the spinal cord, substantia nigra, amygdala, hypothalamus, and cerebral cortex (Cooper, 1996). Within the spinal cord, substance P acts as a neurotransmitter in the nociceptive pathway. At the other sites, substance P acts as a neuromodulator, usually producing long-duration excitation of postsynaptic membranes.

Endorphins

Endorphins are found in areas with opiate receptors, including the substantia gelatinosa, hypothalamus, periventricular gray, and periaqueductal gray. Their primary action is the inhibition of slow nociceptive information.

The receptors for the various neurotransmitters are summarized in Table A–1.

TABLE A–1
NEUROTRANSMITTERS AND THEIR RECEPTORS

Transmitter	Receptors
Acetylcholine	nicotinic, muscarinic
GABA	GABA$_A$, GABA$_B$
Glycine	glycine
Glutamate	NMDA, non–NMDA
Dopamine	D1, D2, D3
Serotonin	5-HT1, 5-HT2, 5-HT3
Norepinephrine	$\alpha_1, \alpha_2, \beta_1, \beta_2$
Substance P	NK1 (neurokinin 1)
Endorphins	$\mu_1, \mu_2, \delta, \kappa_1, \kappa_2$ (opiate)

Reference

Cooper, J. R., Bloom, F. E., and Roth, R. H. (1996). The biochemical basis of neuropharmacology. (7th ed.). New York: Oxford University Press.

Suggested Readings

Bremner, J. D., Innis, R. B., Ng, C. K., et al. (1997). Positron emission tomography measurement of cerebral metabolic correlates of yohimbine administration in combat-related posttraumatic stress disorder. Archives of General Psychiatry. 54(3):246–254.

Dowling, J. E. (1992). The chemistry of synaptic transmission. In: Neurons and networks: An introduction to neuroscience. Cambridge, MA: The Belknap Press of Harvard University Press, pp. 125–150.

Koslow, S. H., Meinecke, D. L., Lederhendler, I. I., et al. (Eds.) (1995). The neuroscience of mental health II. Rockville, MD: National Institutes of Health/National Institute of Mental Health.

Robbins, T. W., and Everitt, B. J. (1995). Arousal systems and attention. In: Gazzaniga, M. S., Bizzi, E., Black, I. B., et al., (Eds.). The cognitive neurosciences. Cambridge, MA: MIT Press, pp. 703–720.

Answers

CHAPTER 2

Case 1

1. The loss of myelin in the peripheral nervous system involves destruction of Schwann cells.
2. The loss of myelin lowers the membrane resistance (Rm) and allows the leakage of electrical current. The loss of myelin does not affect the Ra.
3. The propagation of action potentials is impaired because of the lower membrane resistance. Receptor potentials are not impaired, since the nerve endings of the sensory neurons are not damaged.

Case 2

1. Destruction of oligodendrocytes causes demyelination of axons within the central nervous system.
2. Increases in body temperature may alter the activity of the voltage-gated Na^+ channels, preventing the generation of an action potential.

Review Questions

1. Glial cells have no dendrites or axons, and glial cells cannot conduct an electrical potential.
2. Oligodendrocytes and Schwann cells form the myelin sheath around axons to promote the propagation of an action potential.
3. Dendritic projections are input units for the neuron.
4. The sensory neurons that convey information from the body to the spinal cord are pseudounipolar cells. They appear to have one process, however, one axon connects the periphery to the cell body and a second axon connects the cell body to the spinal cord.
5. Multipolar cells are specialized to receive and accommodate huge amounts of synaptic input to their many dendrites.
6. Sodium (Na^+), potassium (K^+), and chloride (Cl^-) contribute to the resting potential of the cell membrane.
7. Depolarization occurs when the membrane potential becomes less negative with respect to the resting membrane potential. Hyperpolarization occurs when the membrane potential becomes more negative with respect to the resting membrane potential.
8. A membrane channel that opens when it is bound by a neurotransmitter is a ligand-gated channel.
9. The term *graded* means that both the amplitude and duration of the electrical potential can vary depending on the stimulus and that it is not an all-or-none event.
10. Temporal summation and spatial summation of local potentials can bring the membrane to the threshold level.
11. The generation of an action potential requires the influx of Na^+. This influx is mediated by a voltage-gated channel.
12. Large-diameter axons promote faster conduction velocity of an action potential.
13. Nodes of Ranvier have a high density of voltage-gated Na^+ channels, which promote generation of an action potential.
14. Networks composed of interneuronal convergence and divergence are found throughout the central nervous system.

CHAPTER 3

Case 1

1. The influx of Ca^{++} into the axon terminal promotes the release of neurotransmitters. If there are fewer than normal numbers of Ca^{++} channels, then intracellular levels of Ca^{++} cannot be elevated sufficiently to cause release of the transmitter ACh.
2. Physical therapy will not be beneficial for increasing strength, since the antibodies are preventing the release of neurotransmitter necessary for muscle contraction.

Case 2

1. Small amounts of botulinum toxin A reduce involuntary muscle activity by decreasing the amount of ACh released at the neuromuscular junction.
2. No, the action of ACh on a nicotinic, ligand-gated receptor is different from the action on the muscarinic, G-protein mediated receptor. An EPSP is initiated via direct opening of an ion channel associated with the ligand-gated receptor, whereas either an EPSP or an IPSP may be initiated by the action of an intracellular G protein associated with the G-protein mediated receptor.

Review Questions

1. Postsynaptic inhibition involves the flux of K$^+$ and/or Cl$^-$ through the membrane of a receiving, postsynaptic cell. This prevents the cell from generating an action potential. Presynaptic inhibition involves the inhibition of an axon terminal by preventing an influx of Ca^{++}. With decreases in Ca^{++} influx associated with presynaptic inhibition, there is a decreased release of neurotransmitter from the terminal, resulting in decreased stimulation of the postsynaptic cell.
2. The release of neurotransmitter is dependent on Ca^{++}.
3. Direct activation of a membrane channel by a neurotransmitter results in the faster generation of a synaptic potential.
4. The binding of the neurotransmitter causes a change in the shape of the membrane ion channel, resulting in its opening.
5. G-proteins function as shuttles that move through the cytoplasm of the neuron to activate an effector molecule that causes cellular events.
6. The effects of a neurotransmitter depend on the action of the receptor.

CHAPTER 4

Case 1

1. Yes, it is possible for peripheral nerve axons to regenerate and recover following injury.
2. Yes, most likely with recovery of the nerve, the normal sensation should return.

Case 2

1. Damage is often not confined to the oxygen-deprived neurons. As these neurons die, they may cause death of adjacent neurons via the processes of excitotoxicity.
2. Glutamate is the principal excitatory neurotransmitter involved in excitotoxicity.

Review Questions

1. Long-term potentiation (LTP) requires cooperativity, associativity, and specificity.
2. Wallerian degeneration is the process by which an axon, isolated from the cell body, undergoes a process of degeneration followed by death of the entire distal segment.
3. Yes, cortical motor and sensory maps can change even in the adult mammal.

4. Excitotoxicity is the process by which overexcitation of a neuron leads to cell death.
5. Lactic acid is an end product of glycolysis that contributes to cell death.
6. Excessive levels of intracellular calcium promote cell death by activating calcium-dependent proteases and by activating pathways that produce oxygen free radicals.
7. Yes, some brain damage can potentially be reduced with the administration of pharmaceutical agents.

CHAPTER 5

Case 1

1. All three systems are involved. Lack of response to stimulation indicates involvement of the sensory system, inability to voluntarily control the bladder and bowels is an autonomic nervous system problem, and inability to voluntarily move his legs is a motor deficit.
2. The lesion is in the spinal region of the nervous system. The surgery was to move nervous tissue protruding through the bony defect (meningomyelocele) into the vertebral column as much as possible to protect the nervous tissue from infection and injury. Spinal lesions interrupt ascending and descending pathways within the cord, preventing sensory information from being conveyed to the brain and motor information from traveling from the brain to motoneurons. Peripheral lesions are usually not bilateral. Unimpaired sensation and movement of the upper limbs and torso indicate that the brain connections with the upper spinal cord are functionally and anatomically intact.

Review Questions

1. During the embryo stage, from the second to the end of the eighth week, the organs are formed.
2. a. A thickening of the ectoderm becomes the neural plate.
   b. The edges of the plate move toward each other, forming the neural groove.
   c. The folds touch in the dorsal midline, first in the future cervical region.
   d. The folds fuse together, forming a tube that detaches from the neural crest cells and from the ectoderm that will become skin.
3. In early development, a myotome is the part of the somite that will become muscle. After the embryo stage, a myotome is a group of muscles innervated by a segmental spinal nerve.
4. Neural crest cells become peripheral sensory neurons,

myelin cells, autonomic neurons, and endocrine organs (adrenal medulla and pancreatic islets).

5. After the superior neuropore closes, the region of the neural tube that will become the brain expands to form three enlargements: the hindbrain, midbrain, and forebrain. Next, the hindbrain divides into the myelencephalon and metencephalon, while the forebrain divides into the diencephalon and telencephalon. The myelencephalon and metencephalon further differentiate to become the medulla, pons, midbrain, and cerebellum. The telencephalon becomes the cerebral hemispheres.

6. The progressive developmental processes are cellular proliferation, migration, and growth; extension of axons to target cells; formation of synapses; and myelination of axons.

7. The regressive developmental processes are neuronal death and axon retraction.

8. Growing into deficit is the appearance of signs of nervous system damage during infancy and childhood due to nervous system damage that occurred earlier. The signs are not evident until the infant or child reaches the age when the damaged system(s) would normally have become functional.

9. Anencephaly is the development of only a rudimentary brain stem, without cerebral and cerebellar hemispheres. The Arnold-Chiari deformity is a developmental malformation of the hindbrain, with an elongated inferior cerebellum and medulla that protrude into the vertebral canal. The next three are all due to incomplete closure of the caudal neural tube: in spina bifida occulta, neural tissue does not protrude through the bony defect; in meningocele, the meninges protrude through the bony defect; and in meningomyelocele, neural tissue and meninges protrude outside the body.

10. Severe mental retardation is associated with defects in the structure of dendrites and dendritic spines.

11. Cerebral palsy is a disorder of movement and postural control due to permanent, nonprogressive damage of a developing brain. The major types of cerebral palsy are spastic, athetoid, and mixed.

12. Critical periods are the time when neuronal projections compete for synaptic sites. Normal function of neural systems is dependent on appropriate experience during the critical period.

CHAPTER 6

Review Questions

1. The three types of somatosensory receptors are mechanical, chemical, and temperature.

2. Nociceptors are receptors that respond to stimuli that damage or threaten to damage tissue.

3. Primary endings respond to stretch of muscle and to the rate of muscle stretch. Secondary endings only respond to stretch.

4. Firing of the gamma motoneurons causes contraction of the ends of the intrafusal fibers. The contraction maintains the stretch of the central region of the intrafusal fibers so that the sensory endings are able to respond to stretch of the muscle.

5. Muscle stretch information from primary endings in muscle spindles, tension in tendons from Golgi tendon organs, and tension in ligaments from ligament receptors is transmitted by large-diameter Ia and Ib axons.

6. Aδ and C axons convey nociceptive and temperature information.

7. The pathways that convey information to the brain are conscious relay, divergent, and unconscious relay.

8. High-fidelity, somatotopically arranged information is conveyed to the primary sensory cortex, located in the postcentral gyrus.

9. Neural signals that are interpreted as dull, aching pain travel in the paleospinothalamic tract; spinomesencephalic and spinoreticular tract information does not reach conscious awareness.

10. The unconscious relay pathways end in the cerebellum.

11. Synapses between neurons conveying discriminative touch information occur in the left nucleus gracilis in the medulla, right VPL nucleus in the thalamus, and primary sensory cortex.

12. Synapses between neurons conveying discriminative pain information occur in the left dorsal horn of spinal cord, right VPL nucleus in the thalamus, and primary sensory cortex.

13. Posterior spinocerebellar and cuneocerebellar pathways convey unconscious proprioceptive information. The anterior spinocerebellar and rostrospinocerebellar convey information about activity in spinal interneurons and about descending motor commands.

CHAPTER 7

Case 1

The left lower extremity has lost discriminative touch and conscious proprioception. The pathways for these ascend in the ipsilateral cord. The right lower-extremity loss is fast pain and discriminative temperature sense. The pathways for these sensations ascend in the contralateral cord. This pattern of sensory loss plus paralysis on the same side as loss of the dorsal column information indicates a hemisection of the cord. The left half of the cord is interrupted at about L2; the right half of the cord is intact. These signs together (paralysis and dorsal column signs on one side, spinothalamic signs on the opposite side) are called Brown-Séquard syndrome.

Case 2

The sensory and motor signs are both found on only one side of the body, there are no vertical tract signs, and reflexes are intact. Proximal strength and sensation are within normal limits. All these factors indicate that the spinal cord is not involved. Because there are no brain stem or cerebral signs, the most likely level for the lesion is peripheral. The pattern of sensory and motor loss corresponds to the median nerve distribution in the hand, not to a dermatomal distribution. The most likely etiology is carpal tunnel syndrome.

Case 3

Motor and sensory deficits are entirely on the left side of the body. The lower half of the face, trunk, and both limbs are involved, indicating damage to vertical tract neurons. The facial signs indicate a lesion above the lower midbrain because a spinal cord lesion would not affect the face and a lesion in most areas of the brain stem would have facial signs contralateral to the limb signs. The most likely location is cerebral. The abrupt onset indicates a vascular etiology.

Case 4

The 2 weeks when she was pain free and the recurrence of symptoms with high levels of stress indicate that there may be no current physical lesion. Instead, postural changes secondary to avoiding painful positions after the accident and anxiety may be contributing to muscle guarding, abnormal movement patterns, and disuse.

Review Questions

1. For a quick screening, proprioception and vibration are tested in the fingers and toes, and pinprick sensation is tested in the limbs, trunk, and face.
2. Complaints of abnormal sensations or sensory loss, skin lesions that are not painful, and/or localized weakness or atrophy indicate that thorough sensory testing is required.
3. The subject must be prevented from seeing the stimuli during testing, the subject must understand the testing and its purpose, prediction about stimuli must be prevented by irregular timing of stimuli and by varying the stimuli (for example, randomly presenting dull versus sharp stimuli).
4. The distal distribution of a peripheral nerve is stimulated by an electrical current delivered to a surface electrode. The electrical potential evoked in the nerve is recorded by surface electrodes placed along the course of the peripheral nerve.
5. Conscious proprioception and vibratory sense are typically most impaired by demyelination, because the information travels in large diameter axons that require heavy myelination.
6. A left hemisection of the spinal cord causes loss of voluntary motor control combined with loss of position, vibration, and discriminative touch sensation below the level of the lesion on the left side of the body. Pain and temperature information is lost from the right side of the body, one or two dermatomes below the level of the lesion, because collaterals of the first order proximal axons ascend and descend a few levels in the dorsolateral column (zone of Lissauer). This combination of signs is called Brown-Séquard syndrome.
7. Varicella zoster is a viral infection of the dorsal root ganglion that causes inflammation of the sensory nerves and painful eruptions on the skin. Usually limited to a single dermatome.
8. A lesion in the left posterolateral lower pons would cause loss of pain and temperature information from the left face and the right side of the body. This pattern arises because the nociceptive information from the left face travels in the left posterolateral pons, while the nociceptive information from the body has crossed midline in the spinal cord.
9. Sensory extinction is a lack of awareness of the stimulus presented on one side of the body when stimuli are provided simultaneously to both sides of the body. If a stimulus is presented only on the affected side, the person is aware of the stimulus.
10. According to the counterirritant theory, pressure on an injured finger stimulates mechanoreceptor afferents that facilitate enkephalin interneurons. The enkephalin released by the interneurons activates receptors on primary nociceptive afferents and interneurons that decrease the release of substance P by the primary afferents and hyperpolarize interneurons in the nociceptive pathway. Both of these actions decrease the transmission of nociceptive information.
11. The three supraspinal analgesic systems originate in the raphe nuclei, periaqueductal gray, and locus ceruleus.

12. Narcotics activate opiate receptors in the raphe nuclei, periaqueductal gray, and dorsal horn of the spinal cord. This activation inhibits nociceptive information by direct and interneuronal inhibition of wide dynamic range tract neurons, decreasing transmission of nociceptive information.

13. The levels of pain inhibition are:
    I. Peripheral
    II. Dorsal horn
    III. Neuronal descending
    IV. Hormonal
    V. Cortical

14. Referred pain is pain perceived as arising in a site different from the actual site of origin.

15. Nociceptive chronic pain is due to stimulation of nociceptive receptors. Neuropathic chronic pain is due to abnormal activity within the nervous system.

16. Examples of neuropathic chronic pain include: nerve compression, deafferentation pain, phantom pain, and sympathetically maintained pain.

17. Paresthesia is an abnormal, nonpainful sensation. Dysesthesia is an abnormal painful sensation.

18. Ectopic foci are sites along an injured nerve that are abnormally sensitive to mechanical stimulation.

19. Phantom limb pain is pain that seems to originate from a missing limb.

20. Waddell et al. postulate that frequently chronic low back pain results from muscle guarding, abnormal movements, and disuse syndrome. The effectiveness of exercise in treating chronic low back pain syndrome supports Waddell's hypothesis.

21. Research evidence does not support psychological factors as a cause of chronic pain syndromes; instead, limitations in daily activities secondary to chronic pain syndromes cause psychological difficulties.

CHAPTER 8

Case 1

1. The history of minor trauma, constant burning sensation, sharp pain, trophic changes in the fourth and fifth fingers, and unwillingness to move the limb indicate sympathetically maintained pain as a likely diagnosis.

2. The syndrome that affected C.M.'s pupil, eyelid, and sweating on one side of his face and neck was Horner's syndrome.

3. The diagnosis was confirmed when a pharmacologic block of sympathetic transmission eliminated the symptoms.

Case 2

R.D. fainted. The most likely cause was excitement, so the diagnosis was vasodepressor syncope.

Case 3

1. Activation of β-adrenergic receptors accelerates heart rate, increases heart contractility, and vasodilates arteries in the heart and in skeletal muscle. Blocking the β-adrenergic receptors benefits the heart by decreasing myocardial oxygen demand, allowing B.H. to accomplish more physiological work without compromising cardiac function.

2. Becaue β-blockers decrease heart rate, age-adjusted normal values for heart rate cannot be used as guidelines for exercise prescription. The following guidelines can be used to establish target heart rate. An exercise stress test, monitoring heart rate and blood pressure during treadmill or stationary bicycle exercise, must be administered to determine the heart rate at which symptoms occur, and a percentage (60% to 90%) of that heart rate can be used as a guideline for aerobic exercise (American College of Sports Medicine, 1990). If an exercise stress test cannot be performed, a prescription of an exercise heart rate 20 beats per minute above resting heart rate, with instructions to decrease or stop exercise if symptoms occur, is appropriate (Pollock and Wilmore, 1990).

Review Questions

1.

Receptor(s)	Respond to
Pressure receptors in the heart and carotid arteries	Increased blood pressure secondary to increased heart rate, stroke volume, and the increased venous return by the muscle pump
Stretch receptors in the lungs	Dilation of the bronchi and bronchioles
Chemoreceptors in the carotid and aortic bodies	Concentration of blood oxygen
Chemoreceptors in the medulla	Blood levels of H^+ and carbon dioxide
Chemoreceptors in the hypothalamus	Blood glucose levels and blood osmolality
Hypothalamic thermoreceptors	Increased blood temperature secondary to an increase in metabolic rate

2. Visceral afferents convey information from the internal organs and blood vessels into the central nervous system.

3. Neurons in the medulla and pons, and hormones from the pituitary directly control autonomic functions. Parts of the hypothalamus, thalamus, and limbic system modulate the brain stem control.

4. Autonomic regulation is primarily unconscious and can be orchestrated by hormones, functions of the organs regulated can be adjusted by local factors independent of the central nervous system, and the autonomic efferent pathways typically use two neurons as opposed to the single neuron of the somatic efferent pathways.

5. Sympathetic trunks are a series of interconnected paravertebral ganglia. Postganglionic neurons leaving the paravertebral ganglia join either the ventral or dorsal ramus and travel in a peripheral nerve to reach the vasculature in skeletal muscles or skin.

6. Splanchnic nerves are composed of preganglionic axons that innervate abdominal and pelvic viscera.

7. The sympathetic system optimizes blood supply according to the requirements of various organs. Sympathetic activation can elicit vasodilation and, at other times, vasoconstriction because different effects are elicited by activation of subtypes of receptors on the postsynaptic membrane.

8. Capacitance vessels are veins and venules that can hold large quantities of blood when their walls are relaxed. Vasoconstriction of their walls prevents fainting when a person is standing. Blood flow in skeletal muscle arterioles is controlled by α- and β_2-adrenergic receptors, muscarinic cholinergic receptors, and local blood chemistry.

9. Interference with the sympathetic innervation to the muscle that elevates the eyelid and to the pupil of the eye would result in a drooping eyelid and a constricted pupil.

10. The sympathetic system regulates body temperature; blood flow in internal organs, skeletal muscle, and skin; and metabolism.

11. Parasympathetic activity tends to promote energy conservation and storage. In addition, parasympathetic neurons innervate the lacrimal gland, the pupil and lens of the eye, and the bowels, bladder, and external genitalia.

CHAPTER 9

Review Questions

1. A lower motoneuron directly innervates muscle. Whenever a lower motoneuron fires, the muscle fibers it innervates contract. The cell body of a lower motoneuron is in the spinal cord or brain stem, and its axon is in a peripheral nerve. Upper motoneurons synapse with lower motoneurons and thus affect the activity of lower motoneurons. Upper motoneurons are entirely within the central nervous system. Cell bodies of upper motoneurons are in the cerebral cortex or brain stem. The axons of upper motoneurons form the descending motor pathways; examples include the corticospinal, corticobulbar, and reticulospinal tracts.

2. The general function of control circuits is to adjust the activity in descending motor pathways.

3. Slow twitch muscle fibers are usually activated before fast twitch muscle fibers because the alpha motoneurons innervating slow twitch muscle fibers have smaller cell bodies that depolarize earlier than the larger cell bodies of the alpha motoneurons innervating fast twitch muscle fibers.

4. The activity of a motor unit is determined by sensory information from peripheral receptors, by spinal connections, and by activity in descending pathways.

5. Alpha-gamma coactivation is the simultaneous firing of alpha and gamma motoneurons to a muscle so that the extrafusal and intrafusal muscle fibers contract simultaneously.

6. The phasic stretch reflex is elicited by a quick stretch, the afferent is a Ia fiber, and the connection in the spinal cord between the afferent and efferent is monosynaptic. The tonic stretch reflex is elicited by slow or sustained stretch, it uses both Ia and II afferent fibers, and the connections within the spinal cord are multisynaptic.

7. Information from the Golgi tendon organ normally adjusts muscle activity, in concert with information from the muscle and upper motoneurons. Rarely, information from the Golgi tendon organ may prevent injury by inhibiting excessive muscle contraction.

8. Changing a person's arousal level alters the response to a quadriceps tendon tap by changing the amount of descending motor pathway input to the lower motoneurons.

9. An H-reflex is produced by electrically stimulating the Ia afferents in a peripheral nerve. This produces signals that are transmitted into the spinal cord, where alpha motoneurons are activated monosynaptically. Then the alpha motoneurons transmit signals to the skeletal muscles to elicit muscle contraction. The depolarization of the muscle membrane is recorded by electrodes on the skin over the muscle. The H-reflex quantifies the level of excitation or inhibition of alpha motoneurons.

10. Reciprocal inhibition prevents antagonist opposition to movements.

11. To motor control researchers, the term *synergy* refers to coordinated activity of muscles that are often activated together by a normal nervous system. Clinicians often use the term *synergy* to indicate pathological synergies, as when a person with an upper motoneuron lesion cannot flex the knee without simultaneous, obligatory flexion of the hip.

12. A central pattern generator is a flexible network of interneurons that produce purposeful movements.

13. Response reversal demonstrates that somatosensory input can adjust central pattern generator output to environmental conditions so that the goal of the movement is achieved.

14. The medial activating pathways and their functions are as follows:

Pathway	Function
Medial corticospinal	Assist in control of neck, shoulder, and trunk muscles
Medial reticulospinal	Facilitate lower motoneurons to ipsilateral postural muscles and limb extensors
Medial vestibulospinal	Assist in control of neck and upper back muscles
Lateral vestibulospinal	Facilitate lower motoneurons to extensor muscles and inhibit flexor muscles
Tectospinal	Facilitate lower motoneurons that control neck muscles

15. The lateral activating pathways and their functions are as follows:

Pathway	Function
Lateral corticospinal	Influence lower motoneurons that innervate muscles of the hand; fractionation of movement
Corticobulbar	Influence lower motoneurons that innervate facial, tongue, laryngeal, and pharyngeal muscles
Rubrospinal	Influence lower motoneurons that innervate upper limb muscles
Lateral reticulospinal	Facilitate lower motoneurons to flexor muscles and inhibit lower motoneurons to extensor muscles (this action may be reversed in some circumstances)

16. The nonspecific activating pathways and their functions are as follows:

Pathway	Function
Ceruleospinal	Facilitate the activity of interneurons and motoneurons in the spinal cord
Raphespinal	Facilitate the activity of interneurons and motoneurons in the spinal cord

17. One route for information from the basal ganglia output nuclei to the lower motoneurons is output nuclei → motor thalamus → motor areas of the cerebral cortex → corticospinal and corticobulbar neurons → lower motoneurons. A second route is from the pedunculopontine nucleus → reticulospinal and vestibulospinal tracts → lower motoneurons.

18. The end result of dopamine binding in the basal ganglia direct and indirect pathways is increased activity in the motor cortex.

19. The basal ganglia compare proprioceptive information and movement commands, assist in the sequencing of movements, and adjust muscle tone and muscle force.

20. The major function of the cerebellum is comparing actual to intended motor activity.

21. The major sources of input to the cerebellum are the cerebral cortex, via connections in the pons; internal feedback about spinal cord interneuron activity; and proprioceptor and skin mechanoreceptors.

22. The cerebrocerebellum coordinates finger movements, the spinocerebellum coordinates gross limb movements, and the vestibulocerebellum coordinates postural adjustments.

CHAPTER 10

Case 1

The corticospinal and corticobulbar fibers have been interrupted, interfering with voluntary control of the left face and body. The sensory information from the left body and face is not reaching consciousness. Therefore, the lesion is in the right internal capsule. The lesion is focal and nonprogressive. The abrupt onset indicates a vascular etiology.

Case 2

The resting tremor, difficulty with initiating movements, and rigidity indicate involvement of the basal ganglia. The lesion involves loss of cell bodies in the substantia nigra and pedunculopontine nucleus. The pathology is progressive. This is Parkinson's disease; etiology is currently unknown.

Case 3

The loss of voluntary movement and sensation is unilateral and in the ulnar nerve distribution. The lesion is not spinal, because the deficits are not in a myotomal/dermatomal distribution and because no longitudinal signs are present. The lack of longitudinal signs also excludes the cerebral hemispheres and brainstem as locations of lesion. The lesion is focal and nonprogressive. Therefore the damage is to a peripheral nerve, probably secondary to the humeral fracture.

Case 4

Bilateral loss of voluntary movement and sensation indicate a central nervous system lesion. Given that her motor and sensory functions are intact above the L1 spinal cord level (corresponding the T10 vertebra), the injury is to the spinal cord. The lesion is focal, nonprogressive, and traumatic.

Case 5

The only system affected is the motor system. The bilateral Babinski's signs and the velocity-dependent hypertonia indicate upper motoneuron involvement. Muscle fibrillation is a sign of motor system pathology, but does not indicate whether upper or lower motor neurons are involved. Because no common disorder only affects upper motor neurons bilaterally, the most likely etiology is amyotrophic lateral sclerosis, which damages both upper and lower motoneurons. The lower motoneuron involvement could be confirmed by motor nerve conduction velocity studies.

Case 6

Normal active range of motion in the right upper limb and normal cutaneous sensation combined with muscle cramping during a specific motor activity indicate dystonia. The lesion is in the basal ganglia.

Review Questions

1. Hemiplegia is weakness or paralysis affecting one side of the body.
2. Fibrillations and abnormal movements always indicate pathology. Benign muscle spasms, cramps, and fasciculations may occur following excessive exercise.
3. Hypertonia is unusually strong resistance to passive movement. The differences between the two types of hypertonia, spasticity and rigidity, are that the resistance in spasticity is velocity dependent and the resistance in rigidity is velocity independent. This means that if spastic muscles are being passively stretched, faster stretch will elicit more resistance than slower stretch. If rigid muscles are being passively stretched, the resistance will remain the same regardless of the speed of stretch. Spasticity is due to intrinsic changes in muscle following upper motoneuron lesions. Rigidity is due to a direct effect on alpha motoneurons. Rigidity, in contrast to spasticity, is not associated with hyperreflexia, clonus, or clasp-knife response.

4. Spinal shock is a temporary condition following injury to the spinal cord, during which stretch reflexes cannot be elicited and muscles are hypotonic due to lack of descending facilitation by upper motoneurons.

5. Loss of reflexes, muscle atrophy, flaccid paralysis, and fibrillations indicate a lower motoneuron lesion. In contrast, upper motoneuron lesions produce hyperactive reflexes and spastic paralysis or paresis. The muscle atrophy in upper motoneuron lesions is less severe. Fibrillations may occur in upper motoneuron lesions.

6. Babinski's sign in an adult indicates damage to the corticospinal tract(s).

7. A lesion that produces abnormal cutaneous reflexes, abnormal timing of muscle activation, paresis, and spasticity interrupts upper motoneurons.

8. Clonus is repetitive stretch reflexes elicited by maintained passive dorsiflexion of the ankle or wrist.

9. Hyperactive stretch reflexes do not produce spasticity. Spasticity is due to intrinsic changes in muscles and is not reflex based. In the paretic lower limb of adults with chronic hemiplegia, lower motoneurons are less active than in the nonparetic lower limb. If hyperactive stretch reflexes produced spasticity, lower motoneurons to spastic muscles would be more active than lower motoneurons to nonspastic muscles.

10. Surface EMG can be used to quantify paresis by comparing EMG amplitude in a muscle on the hemiplegic side with the homologous muscle on the nonhemiplegic side. If the muscle on the hemiplegic side generates only 50% or less of the amplitude of the homologous muscle, paresis is a problem. Hypertonia is decreased passive range of motion without increased EMG output. Cocontraction is temporal overlap of EMG activity in antagonist muscles. Hyperreflexia is defined as EMG activity during muscle stretch, with a positive correlation between EMG amplitude and velocity of muscle stretch.

11. In a complete spinal cord injury, there is a total absence of upper motoneuron influence on lower mo-

toneurons. In the most typical cerebrovascular accident, upper motoneuron influences are abnormal because the corticospinal tracts are interrupted and the activity of the other descending activating pathways is altered.

12. In spastic cerebral palsy, reflex irradiation and abnormal cocontraction are observed. These motor signs do not occur in adult-onset upper motoneuron disorders.

13. Learned nonuse is the preference for using a nonparetic limb when using the paretic limb is difficult. Over time the person may learn to only use the nonparetic limb, unless intervention forces the use of the paretic limb.

14. Amyotrophic lateral sclerosis destroys both upper and lower motoneurons only. Amyotrophic lateral sclerosis spares sensory, autonomic, cognitive, language, and all other nonmotor nervous system functions.

15. Parkinson's disease is characterized by rigidity, hypokinesia, resting tremor, and visuoperceptive impairments. In Parkinson's disease, neurons in the substantia nigra compacta and pedunculopontine nucleus die off.

16. Huntington's disease, dystonia, and choreoathetotic cerebral palsy produce hyperkinesia.

17. Involuntary abnormal postures or repetitive twisting movements are signs of dystonia.

18. Slurred, poorly articulated speech and ataxic gait may indicate damage of the spinocerebellum (vermis and paravermis).

19. The signs of cerebrocerebellar lesions include dysdiadochokinesia, the inability to rapidly alternate movements; dysmetria, the inability to accurately move an intended distance; and action tremor, shaking of the limb during voluntary movement.

20. The long loop response is the second response of a contracting muscle to stretch. The long loop response is visible on EMG. In the biceps brachii, the long loop response occurs 50 to 80 milliseconds after the muscle stretch.

21. The asymmetrical tonic neck reflex is the extension of the limbs on the nose side and flexion of the limbs on the skull side when the head is turned toward the right or left. This reflex is frequently observed in normal infants. However, obligatory assumption of this position every time the head is turned to the side indicates that the nervous system has been damaged. An asymmetrical tonic neck reflex in an older child or an adult is a sign of nervous system abnormality.

22. Posturography can reveal the type of sensory information a person typically relies on and the pattern of muscle activity the person uses to maintain equilibrium.

23. Preparatory postural adjustments occur before stepping: the weight is shifted forward and also onto the stance limb.

24. Two identical stimuli may elicit different responses depending on the instructions a person is given. For example, if a person is asked to maintain 90° of flexion at the elbow when a weight is placed in the hand, three responses occur in the biceps: stretch reflex, long loop response, and a voluntary response. If the person is asked to let the hand drop when the weight is placed in the hand, only the stretch reflex occurs.

25. For a person with hemiplegia who is having difficulty initiating gait, placing the foot of the swing limb 10 to 15 cm behind the stance foot may facilitate gait initiation.

26. The two phases of reaching are fast approach and homing in.

27. Diagnostic EMG is used to distinguish between denervated muscle and myopathy.

28. Motor nerve conduction velocity tests are used to differentiate among dysfunctions in nerve, neuromuscular junction, and muscle. Surface EMG for analysis of movement is used to determine which of the following factors contribute to impaired movement: paresis, hypertonia, cocontraction, and/or hyperreflexia.

CHAPTER 11
Case 1

Rapid onset, motor system more involved than sensory, cranial nerve involvement, and respiratory paresis indicate Guillain-Barré syndrome. Nerve conduction velocity results indicate that the lesions affect the peripheral nerves, specifically the myelin. See Chapter 2 for a review of this disorder.

Case 2

1. Sensory and motor signs with the specific history of trauma indicate crushing injury (class II) of the ulnar nerve.
2. Nerve conduction velocity tests can isolate the location of the lesion prior to wallerian degeneration of the axon distal to the injury, and the probable rate of recovery of function can be calculated by assuming regrowth of the distal axon of about 1 mm/day.

Case 3

Lack of sensory and coordination deficits, combined with weakness and normal nerve conduction velocity, indicates a primary disease of muscle. The small-amplitude potentials recorded from muscle confirm the diagnosis of myopathy.

Review Questions

1. Because cutaneous branches of peripheral nerves contain sympathetic efferent axons to sweat glands and arterioles.
2. Complete severance of a peripheral nerve causes loss of sensation, abnormal sensations, lack of sweating and control of smooth muscle in arterial walls, and paralysis of muscles in the area supplied by the portion of the nerve distal to the lesion.
3. The trophic changes that occur in denervated tissues include muscle atrophy, shininess of the skin, brittle nails, and thickening of subcutaneous tissues.
4. Carpal tunnel syndrome is an example of a Class I mononeuropathy. In Class I mononeuropathy, the myelin is damaged.
5. Class II mononeuropathy is typically the result of crushing a nerve, which disrupts all sizes of axons. Reflexes are decreased or absent. Following the injury, wallerian degeneration and then muscle atrophy occur. Prognosis is good because regenerating axons are confined to intact connective tissue and myelin sheaths and are able to reinnervate appropriate targets.
6. Because the axons and connective tissues are completely severed, regenerating axons are not guided by intact sheaths to the appropriate target organ. The lack of guidance and the presence of tissues that physically interfere with the optimal direction of regrowth lead to inappropriate innervation and neuroma formation.
7. Multiple mononeuropathy is focal damage randomly affecting more than one nerve. The presentation of signs and symptoms is asymmetrical.
8. Polyneuropathy is most frequently the result of diabetes, nutritional deficiencies secondary to alcohol abuse, and autoimmune disorders.
9. Polyneuropathy usually presents with distal involvement first because the longest axons are most susceptible to inadequate axonal transport and to the random process of demyelination.
10. Myelinopathy can be distinguished from axonopathy by nerve conduction velocity studies. Myelinopathies produce severe slowing of NCV, and axonopathies primarily produce decreased amplitude of the evoked potential.
11. Electromyography can be used to distinguish neuropathy from myopathy. Myopathy will have small-amplitude compound muscle action potentials (see Chap. 10).

12. Neuronal hyperactivity produces muscle spasms, muscle fasciculations, and cyanosis of the skin. Neuronal hypoactivity produces absent reflexes, lack of sweating, and muscle paralysis.
13. The signs and symptoms listed are characteristic of central nervous system lesions.

CHAPTER 12
Case 1

1. The lesion is in the spinal cord and not in a root because the loss of sensation and motor control throughout the medial upper limb and the entire trunk and lower limb indicates that vertical tracts must be interrupted.
2. The signs include loss of sensation below the C6 dermatome and complete loss of motor control below C6. These signs indicate a complete lesion of the spinal cord at the C6 level.

Case 2

1. B.D. has anterior cord syndrome at C6. Anterior cord syndrome interrupts descending motor pathways, lower motoneurons, and the spinothalamic tracts. Because some neurons in the spinothalamic tract are intact, pain and temperature sensation is partially preserved. In this case, the incomplete injury slightly damaged anteriorly located tracts on the right side and moderately damaged the same tracts on the left side, leaving parts of the vertical tracts intact.
2. This patient's outcome is very different from the complete lesion of the cord in Case 1 because many axons in vertical tracts were spared.

Case 3

1. Vertical tracts conveying pain and temperature information are not impaired because the spinothalamic tracts are anterior to the lesion.
2. The dermatomal and myotomal patterns of loss indicate a lesion in the right posterolateral cord at the L2 spinal segment. Although the lesion does interfere with transmission of pain information from the dermatome that projects to the right L2 segment, no deficit is evident clinically because of the overlap of adjacent sensory fields in the periphery.
3. The earlier incident of weakness in the left leg, with full resolution, plus the current signs, indicate multiple sclerosis as a possible diagnosis.

Case 4

1. Complete loss of sensation begins at the S1 level on the left. Because the L5 level contributes to ankle kinesthesia, ankle position sense is partially retained. Preserved sensation at the left S4 and S5 levels and intact sensation in the right lower limb indicate that vertical tracts are not damaged. Thus the lesion is likely to be outside the spinal cord.

2. Given the S1 to S3 dermatomal distribution of symptoms, a lesion compressing the dorsal roots can be suspected. In this case, the lesion is a small tumor compressing the left S1 to S3 dorsal roots.

3. Touch and proprioceptive sensations are more affected than pain and temperature sensations because compression affects large axons more than small axons. A lesion in the dorsolateral cord would have interfered with ascending dorsal column information from the left lower limb and descending lateral corticospinal information to the left lower limb.

Review Questions

1. A spinal nerve is the brief union of the dorsal and ventral roots of a single spinal segment within the intervertebral foramen. The spinal nerve divides into the anterior and posterior rami.

2. A ventral root is a collection of motor and autonomic efferent axons from one segment of the spinal cord. The ventral primary ramus is a branch of the spinal nerve and provides innervation to the anterior and lateral trunk and the limbs.

3. A spinal segment is a section of the spinal cord that is connected with a specific dermatome, myotome, and sclerotome by a spinal nerve, its roots, and its rootlets.

4. The dorsal horn processes sensory information.

5. Lamina II is also known as the substantia gelatinosa.

6. Reflexes and voluntary motor control are not separate because afferent and descending voluntary information converges on the same spinal interneurons. This convergence allows descending voluntary signals to modify reflexive actions and allows afferent input to adjust movements elicited by descending commands.

7. Nonreciprocal inhibition coordinates muscular activity throughout a limb.

8. Voluntary voiding of urine requires that information regarding fullness of the bladder is conveyed into the sacral spinal cord by afferents, then to the cerebral cortex, where a decision is made. Then the brain initiates voiding by corticospinal inhibition of lower motoneurons that innervate the external sphincter and by brain stem pathways to the autonomic efferents that stimulate contraction of the bladder wall.

9. The differences between segmental and vertical tract signs are as follows: Segmental signs are limited to a dermatomal and/or myotomal distribution. The sensory signs include lost or abnormal sensation, and the motor signs include lower motoneuron signs: flaccid weakness, atrophy, fibrillations, and fasciculations. Vertical tract signs occur at all levels below the lesion and include decreased or lost sensation, decreased or lost voluntary control of pelvic organs, and upper motoneuron signs: muscle hypertonia, paresis, phasic stretch hyperreflexia, and Babinski's sign. Signs of interruption of the sympathetic tracts above the T6 level include autonomic dysreflexia, poor thermoregulation, and orthostatic hypotension.

10. The four spinal cord syndromes are central cord, anterior cord, Brown-Séquard, and cauda equina syndrome. See Figure 12–12 for illustrations of the location of the lesion in each syndrome.

11. Spinal cord functions are depressed or lost immediately below the lesion after a spinal cord injury because descending tracts that supply tonic facilitation to the spinal cord neurons are interrupted by the lesion.

12. Some people with spinal cord injuries have exaggerated withdrawal reflexes because the normal descending reticulospinal inhibition on the neurons within the withdrawal reflex circuit has been removed.

13. An incomplete spinal cord injury is damage to the spinal cord in which sensory and/or motor function is preserved in the lowest sacral segment. Anterior cord, Brown-Séquard, or central cord syndrome can produce incomplete spinal cord injury. Cauda equina syndrome is damage to the cauda equina, not the cord, so cauda equina damage does not produce incomplete spinal cord injury.

14. Autonomic dysreflexia, poor thermoregulation, and orthostatic hypotension arise when the spinal cord below the T6 level is deprived of descending sympathetic innervation.

CHAPTER 13

Case 1

The most likely location of the lesion is the trochlear nerve, damaged or severed by the skull fracture, because the trochlear nerve innervates the superior oblique muscle that moves the eye to look down and in. The intact pupillary and accommodation reflexes and the normal movements of the eyes up, down, in, and up and in indicate that the oculomotor nerve is undamaged. The ability to look laterally indicates that the abducens nerve is also intact.

Case 2

1. This is not an upper motoneuron lesion because an upper motoneuron lesion would spare the pupillary reflexes. In an upper motoneuron lesion, the afferent limb by the optic nerve would be intact, the connections in the brain stem would be intact, and the efferent limb via the oculomotor nerve would be intact.
2. The absence of any cognitive or consciousness disorders and the absence of vertical tract signs, such as paresis elsewhere in the body and/or a loss of sensation, indicate that the lesion is probably in the peripheral nervous system. The oculomotor supplies the extraocular muscles that move the pupil up, down, in, and up and in, as well as the papillary reflexes. The two muscles innervated by the trochlear and abducens nerves are intact. Therefore the lesion involves the oculomotor nerve.

Case 3

There are no signs of vertical tract involvement because the absence of signs below the face indicates that the connections between the brain and the spinal cord are normal. The lesion spares facial sensation, the muscles of mastication, and the tongue, so the trigeminal nerve and hypoglossal nerves are intact. The facial nerve controls the muscles of facial expression, including orbicularis oculi, so the lesion involves the facial nerve. The disorder affects either the nucleus of cranial nerve VII or its axons. If the axons are affected, the disorder is Bell's palsy.

Review Questions

1. Contraction of the lateral rectus muscle, innervated by the abducens nerve, moves the right eye toward the right. Next, contraction of the superior rectus muscle, innervated by the oculomotor nerve, moves the right eye upward. Contraction of the medial rectus muscle, innervated by the oculomotor nerve, moves the left eye toward the right. Contraction of the inferior oblique muscle, innervated by the oculomotor nerve, moves the left eye upward.
2. The oculomotor nerve provides the efferents for the pupillary reflex.
3. The hypoglossal nerve provides efferents to the tongue muscles.

4. The glossopharyngeal nerve provides afferents for the gag reflex.
5. The facial nerve provides control of the muscles of facial expression.
6. The trigeminal nerve provides somatosensation from the face.
7. See Figure 13–6 for an illustration of the accommodation reflex.
8. During rotational acceleration of the head, flow of fluid in the semicircular canals lags behind due to inertia. The fluid causes the cupula to bend, which also bends the ends of hair cells. Bending of the hairs produces excitation or inhibition of the vestibular nerve endings.
9. Head position is sensed by receptors called macula in the saccule and utricle. The macula consists of a gelatinous mass with embedded hair cells and topped by otoconia. In response to gravity, the weight of the otoconia causes displacement of the gelatinous mass and bending of the hairs. The bending of the hairs stimulates the hair cells, which in turn stimulate the vestibular nerve endings.
10. The organ of Corti converts mechanical displacement information into neural signals.
11. Different brain areas produce authentic and inauthentic smiles. The limbic system is the source of authentic smiles, and voluntary signals from the cerebral cortex produce inauthentic smiles.
12. Cranial nerves V, VII, IX, X, and XII are all involved in swallowing.
13. Double vision can result from lesions of cranial nerves III, IV, or VI, or their nuclei, or lesions of the medial longitudinal fasciculus.

CHAPTER 14

Case 1

1.

Functional losses	Structures involved
Lack of pain and temperature sensation from the right side of the body	Spinothalamic tract
Lack of somatosensation from the left side of the face	Spinal tract and nucleus of cranial nerve V
Ataxia on the left side of the body	Middle cerebellar peduncle
Paralysis of muscles of facial expression on the left	Facial nerve

Functional losses	Structures involved
Nystagmus, vertigo, oscil-lopsia, nausea, vomiting	Vestibular nuclei
Loss of corneal reflex on the left side	Nucleus of cranial nerve V

2. The cranial nerve signs indicate that the lesion must be in the brain stem. The lesion affects the left side of the brain stem because the loss of pain and temperature sensation is from the right side of the body (the axons conveying pain and temperature information from the body cross the midline in the spinal cord) and the cranial nerve signs involve the left side of the head. All of the structures involved are located in the lateral pons; thus the lesion is in the lateral pons.

Case 2

Normal voluntary control of both sides of the upper face, including the orbicularis oculi, plus the movement of the right lower face when she frowns in response to frustration, indicates that the facial nerve, cranial nerve VII, is intact. Thus the lesion is an upper motoneuron lesion that prevents corticobulbar information from the left cerebral cortex from reaching the right facial nerve nucleus, resulting in paralysis of voluntary movements of the right lower face. The lesion is in the corticobulbar tract that influences cranial nerve VII.

Case 3

1. M.Z.'s ability to respond appropriately to requests for eye movements indicates he is not in a vegetative state. In the vegetative state, the person is not conscious. M.Z.'s responses to the requests demonstrate that he is conscious.
2. Because he is unable to move any part of his body other than his eyes, the condition is locked-in syndrome.

Case 4

1.

Functional losses	Structures involved
Decreased control of trunk and proximal limb muscles and paralysis of distal limb muscles on the right side of the body	Upper motoneurons, including corticospinal tracts
Unable to voluntarily move his right lower face	Corticobulbar tracts

Functional losses	Structures involved
Inability to move his left eye medially, downward, or upward; drooping of left upper eyelid; dilated left pupil that is nonresponsive to light	Oculomotor nerve

2. The combination of cranial nerve signs on the left side (oculomotor nerve) and loss of voluntary motor control of the right lower face and right body indicate a brain stem lesion. To interrupt the left oculomotor nerve and the corticospinal tracts and the corticobulbar tract to the facial nucleus, the lesion is in the anterior midbrain on the left side. The lesion is on the left because the corticospinal and corticobulbar tracts cross the midline below the midbrain, and the functional loss from a cranial nerve lesion is ipsilateral to the lesion.

Review Questions

1. The dorsal column/medial lemniscus, spinocerebellar, some parasympathetic, corticobulbar, corticopontine, and corticoreticular tracts are modified in the brain stem.
2. The lateral zone of the reticular formation integrates sensory and cortical input and produces generalized arousal. The medial zone regulates vital functions and somatic motor activity. The midline zone adjusts the transmission of pain information, somatic motor activity, and consciousness levels.
3. The major reticular nuclei and the neuromodulators they produce are the ventral tegmental area: dopamine; pedunculopontine nucleus: acetylcholine; raphe nuclei: serotonin; and locus ceruleus and medial reticular zone: norepinephrine.
4. Dopamine from the ventral tegmental area is involved in activating cerebral areas essential for motivation and decision making.
5. The pedunculopontine nucleus affects movement by its connections with the globus pallidus, subthalamic nucleus, and reticular formation areas that are the source of the reticulospinal tracts.
6. The raphe nuclei in the medulla are part of the descending pain control system. Other structures that are part of the descending pain control system are the locus ceruleus and the periaqueductal gray, located in the pons and midbrain respectively.
7. Ascending fibers from the locus ceruleus direct attention.

8. For each function, the nucleus required and its brain stem location are as follows:

Function	Nucleus	Brain stem location
Control of voluntary muscles in the pharynx and larynx	Ambiguus	Upper medulla
Integration and transmission of pain information from the face	Spinal nucleus of the trigeminal nerve	Lower medulla
Control of tongue muscles	Hypoglossal	Upper medulla
Processing of information about sounds	Cochlear	Junction of medulla and pons
Signals the cerebellum when movement errors occur	Inferior olivary nucleus	Upper medulla
Control of muscles of mastication	Trigeminal motor nucleus	Pons
Contraction of the pupillary sphincter and change in curvature of the lens to focus on near objects	Parasympathetic oculomotor	Midbrain

9. The substantia nigra and pedunculopontine nucleus in the midbrain are part of the basal ganglia circuit.
10. The functions of the cerebellum include coordination of movements, motor planning, and cognitive functions, including rapid shifts of attention.
11. Because the second-order neuron in the dorsal column/medial lemniscus pathway crosses the midline in the inferior medulla.
12. The periaqueductal gray coordinates somatic and autonomic reactions to pain, threats, and emotions.
13. Corticobulbar tracts convey motor signals from the cerebral cortex to cranial nerve motor nuclei.
14. A complete facial nerve lesion produces paralysis of the ipsilateral muscles of facial expression, and the person cannot close the ipsilateral eye. Eye closure is intact with a corticobulbar lesion because cortical control of muscles in the upper face is bilateral.
15. A person with a complete left facial nerve lesion would not be able to smile on the left side of the face because there is no neural connection between the brain stem and the muscles of facial expression on the left side.
16. In the brain stem, damage to the reticular formation or the ascending reticular activating system can cause disorders of consciousness.
17. A person with complete loss of consciousness and normal vital functions is in a vegetative state.
18. Because the brain stem is tightly confined within the bone and dura, space-occupying lesions compress the brain stem and thus interfere with its function.

CHAPTER 15

Case 1

1. A.J. has deficient input from both the cochlear and vestibular branches of cranial nerve VIII on the right side.
2. The nerve damage secondary to the fracture interferes with hearing and disrupts the reciprocal relationship between signals from the semicircular canals. Alteration in the pattern of vestibular signals is misinterpreted as signifying movement, resulting in illusions of changes in head position and movement. The nervous system attempts to maintain equilibrium by responding to the illusory changes, resulting in inappropriate muscle activity and nystagmus. Over time, with movement and practice, the central nervous system will adapt to the altered signals, and balance will improve. However, if A.J. continues to restrict his movements to prevent provoking the dizziness and nausea, the signs and symptoms will remain unchanged or worsen.

Case 2

1. The cranial nerve signs—loss of sensation and movement of the face, inability to move the right eye toward the right, and deafness—are all on the right side. Ataxia of the limbs, a cerebellar sign, is also on the right side, and cerebellar signs are ipsilateral (see Chap. 10). The vertigo, nystagmus, oscillopsia, and vomiting indicate a lesion involving the vestibular system. B.F. also has loss of pain and temperature sensations on the left side of the body. The mix of cranial nerve signs on the right side of the head with vertical tract signs on the left side of the body indicate a brain stem lesion. The cranial nerve nuclei involved are trigeminal, abducens, facial, and vestibulocochlear. All of these cranial nerve nuclei are in the lateral pons,

and their axons supply ipsilateral structures. Thus the lesion is in the lateral pons on the right side.

2. The localized loss of function and the slow onset indicate a neoplastic etiology.

Case 3

1. Conscious awareness of visual information occurs only in the cerebral region, so the lesion must be in the cerebrum. The visual loss is restricted to the right visual field of both eyes. This means the information from the left half of each retina is not being conveyed to or processed in the left visual cortex. To interrupt the fibers from the left half of the retina of both eyes, the lesion must be posterior to the optic chiasm. So the lesion could be affecting the left optic tract, left lateral geniculate, left optic radiation, or left visual cortex.

2. Neural connections of the left motor and somatosensory cortex have been interrupted. The axons connecting motor and somatosensory cortex with subcortical areas travel through the internal capsule, and the initial section of the optic radiations travels in the internal capsule, so the internal capsule is the most likely location for the lesion.

Review Questions

1. The superior colliculus is important in orienting the eyes and head toward sounds.

2. The vestibular system contributes to gaze stabilization via the vestibulo-ocular reflex. Reflexive connections of the vestibular nuclei with the extraocular nuclei via the medial longitudinal fasciculus stabilize gaze.

3. "Each pair of semicircular canals produces reciprocal signals" means that increased frequency of signals from one canal occurs simultaneously with decreased signals from its partner.

4. The tracts that use vestibular information to control posture include the medial corticospinal, reticulospinal, tectospinal, and vestibulospinal.

5. Information from the left visual field reaches the right visual cortex via the following pathways: Light from the left visual field strikes the right half of each retina. From the left eye, this information travels in the left optic nerve, crosses the midline in the optic chiasm, travels in the optic tract to the right lateral geniculate body, and from the geniculate body to the right optic cortex. Visual information from the right eye travels in the right optic nerve, stays ipsilateral in the optic chi-

asm, and then follows the same route as visual information from the nasal half of the left eye.

6. The medial longitudinal fasciculus coordinates eye and head movements.

7. Nystagmus is involuntary back and forth movements of the eyes. Physiological nystagmus is a normal response of an intact nervous system to moving visual objects, head rotation, caloric stimulation of the semicircular canals, or extreme positions of the eyes. Pathological nystagmus is abnormal oscillating eye movements that are caused by a nervous system disorder.

8. Eye movements are directed by information from the vestibular, visual, proprioceptive, limbic, and voluntary movement systems.

9. BPPV is benign paroxysmal positional vertigo, a syndrome of a brief (less than 2 minutes) sensation of whirling movement evoked by a rapid change in head position.

10. Oscillopsia is the visual sensation of movement of nonmoving objects.

CHAPTER 16

Review Questions

1. Damage to the ventral anterior nucleus would interrupt fibers from the globus pallidus to the premotor areas, damage to the ventral lateral nucleus would interrupt circuits from the dentate nucleus to primary motor cortex and premotor areas, ventral posterolateral nucleus damage would prevent relay of somatic information from the body to the somatosensory cortex, damage to the ventral posteromedial nucleus would stop somatic sensation from the face from reaching the somatosensory cortex, and lesions of the medial and lateral geniculates would interrupt axons transmitting auditory and visual information to the cerebral cortex.

2. The hypothalamic regulation of body temperature, metabolic rate, and blood pressure are essential for survival.

3. A lesion of the genu of the internal capsule would cause contralateral loss of corticobulbar control, resulting in inability to voluntarily control the cranial nerves that receive cortical input: oculomotor, trochlear, trigeminal, abducens, facial, glossopharyngeal, accessory, and hypoglossal. Thus the person would be unable to voluntarily control the muscles that move the eyes, chew, form facial expressions, swallow, produce speech, elevate the shoulders, and turn the head. Cor-

tical control of reticular activity would be decreased because corticoreticular fibers travel through the genu and then project bilaterally to the reticular formation.

4. The functional categories of cerebral cortex are primary sensory, sensory association, association, motor planning, and primary motor.

5.

Primary auditory cortex → Auditory association cortex → Parietotemporal association areas → Visual cortex

↓

Sensorimotor cortex ← Premotor area ← Visual cortex ← Cortical eye fields

↓

Spinal cord

6. The stress response is produced by activity of the voluntary, autonomic, and neuroendocrine systems.

7. Excessive, prolonged cortisol secretion may contribute to the development of stress-related diseases, including colitis, adult-onset diabetes, cardiovascular disorders, and emotional and cognitive disorders.

8. The hippocampus consolidates declarative short-term memory into long-term memory, although the hippocampus does not store the long-term memories.

9. Sensorimotor cortex, thalamus, and basal ganglia are involved in formation of motor memories.

10. Visual information in the ventral stream is used to identify objects and people.

11. The right frontal and parietal lobes contribute to the maintenance of attention, the right parietal lobe disengages attention, and the midbrain shifts attention to a new focus.

CHAPTER 17

Case 1

1. Thalamic activity determines the level of cortical activity, so severe bilateral thalamic damage prevents consciousness.

2. The reticular formation of the brain stem, the reticular activating system, the thalamus, thalamic projections to the cerebral cortex, and the cerebral cortex are all required for consciousness.

Case 2

1. External temperature control is being substituted for the patient's impaired ability to adjust her own body temperature.

2. Pressure from edema following surgery is compressing

the hypothalamus, compromising blood flow and thus hypothalamic function. Despite lack of direct damage, the hypothalamus cannot currently regulate temperature. As the edema is reabsorbed, the hypothalamus will gradually regain its ability to regulate body temperature.

Case 3

The primary motor and somatosensory cortices are damaged. Primary motor cortex neurons control fractionation of movement, particularly of distal muscles, so the lesion interferes with fine motor control of the hand. The primary somatosensory cortex is necessary for localization of tactile stimuli and conscious proprioception. The lesion does not affect the somatosensory association area because the ability to distinguish sharp from dull is intact.

Case 4

1. R.B. demonstrated a form of agnosia called **astereognosis,** the inability to identify objects by touch and manipulation despite intact discriminative somatosensation. Astereognosis results from lesions in the somatosensory association area.

2. The disuse of a hand with normal strength may occur as a result of perceptual changes if information from the other hand is processed normally.

Case 5

This patient's apathy and lack of goal-directed behavior are typical of patients with lesions in the prefrontal area. Patients with damage to this region have difficulty choosing goals, planning, executing plans, and monitoring the execution of a plan.

Case 6

Despite intact intellectual ability, B.G. demonstrated the poor judgment and inability to conform to social conventions often seen in patients with orbitofrontal head injury. In addition, his left premotor and primary motor areas were also damaged, resulting in decreased voluntary movement on the right side.

Case 7

The disorder is apraxia, the inability to correctly perform purposeful movements despite intact sensation, comprehension, and physical ability to perform the movement. In this case, the errors were entirely sequencing errors, and discrete movements within the sequence were performed correctly.

Case 8

1. **Spastic dysarthria**, characterized by harsh, awkward speech, occurs when upper motoneurons are damaged.
2. In this case, the upper motoneurons could be damaged at their origin in the primary motor cortex, adjacent white matter, or in the internal capsule. The lesion also damaged neurons that control the arm and hand and somewhat affected control of the trunk and lower limb; again, the lesion is in the primary motor cortex, adjacent white matter, or internal capsule, disrupting corticospinal control of lower motoneurons in the cord.

Case 9

1. The lesion is in the left hemisphere because the sensory and motor losses are contralateral to cerebral lesions.
2. The language dysfunction is **Broca's aphasia**, characterized by grammatical omissions and errors, short phrases, and effortful speech. Interference with the programming of language output is consistent with a left hemisphere lesion because Broca's area usually is in the left hemisphere.

Case 10

1. The communication disorder is Wernicke's aphasia.
2. The lesion is probably in the left hemisphere, in Wernicke's area, disrupting the ability to comprehend language.

Case 11

1. The disorder is neglect.
2. A.G. shows signs of both personal and spatial neglect because he seems unaware of his own left side, he becomes lost easily, and his drawings omit features that should be included on the left side.

Case 12

1. The condition is optic ataxia.
2. The damage is to both parietal lobes.

Case 13

H.L. has focal contusions of the inferior frontal and anterior temporal and medial lobes, impairing judgment, self-control, and declarative memory. H.L. also has diffuse axonal injury to the superior cerebellar peduncle, which impairs cerebellar projections to the motor areas of the cerebral cortex via the ventrolateral nucleus of the thalamus.

Review Questions

1. The most probable location of the lesion is the posterior limb of the internal capsule.
2. Astereognosis: loss of the ability to recognize an object by touch and manipulation; lesion in somatosensory association area. Visual agnosia: lack of ability to recognize objects by vision, despite intact vision; lesion in visual association area. Apraxia: inability to perform voluntary movement, in spite of preserved sensation, muscular power, and coordination; lesion in premotor area. Spastic dysarthria: speech disorder due to upper motoneuron damage. The upper motoneuron damage can cause paralysis, spasticity, and/or uncoordinated activity of the speech muscles; lesion in primary motor and primary somatosensory cortex.
3. The lesion is in the area analogous to Wernicke's area in the right hemisphere.
4. Broca's aphasia is not an upper motoneuron disorder. Broca's area provides grammatical function words and planning of speech movements. Information from Broca's area projects to the adjacent sensorimotor cortex, which is the source of most upper motoneurons that control cranial nerves involved in producing speech.
5. Dysarthria interferes with the motor production of sounds. Thus dysarthria can be a lower or upper motoneuron disorder. The lower motoneuron form of dysarthria produces flaccidity of the muscles of speech, causing soft, imprecise speech. The upper motoneuron form of dysarthria is characterized by harsh, awkward speech. Unlike dysarthria, a disorder of speech, aphasia interferes with language. Broca's aphasia interferes with the language output, whereas Wernicke's aphasia interferes with understanding language.
6. The left inferior frontal cortex and adjacent parietal cortex are the most probable sites of the lesion.
7. Nonverbal communication, using gestures and demonstration, is most effective in conveying information to people who cannot understand language.
8. J.H. has spatial neglect.

CHAPTER 18

Case 1

Because the anterior spinal artery supplies the anterior two-thirds of the cord, occluding this artery can result in the deficits seen. The posterior spinal arteries supply the dorsal columns, so vibration and position senses are unaffected.

Case 2

Vertical systems: all sensations and voluntary movement are affected on one side. Lack of attention to the left side of the body (neglect of the left limbs, lack of response to stimuli on the left side) indicates a cortical lesion. Because at the cortical level all vertical systems project contralaterally, the lesion is in the right cortex. The sudden onset suggests a vascular etiology. The artery supplying the lateral part of the sensorimotor cortex is the middle cerebral. Because the lower limb is affected, the lesion involves the deep branches of the middle cerebral artery.

Case 3

The vertical sign is the motor deficits; the sensory system is unaffected. In conjunction with the enlarged cranium and eye position, the motor signs indicate hydrocephalus.

Review Questions

1. The cerebrospinal fluid provides water, some amino acids, and ions to the extracellular fluid and protects the central nervous system by absorbing some of the impact when the head receives a blow. Cerebrospinal fluid may also remove metabolites from the extracellular fluid.
2. Cerebrospinal fluid is located in the ventricles and the subarachnoid space.
3. An epidural hematoma arises from arterial bleeding. Because arteries bleed quickly, the signs and symptoms develop rapidly. In contrast, a subdural hematoma is produced by venous bleeding. Veins bleed slowly, causing slow progression of the signs and symptoms.
4. An enlarged head, difficult feeding, downward looking eyes, and inactivity in an infant may indicate hydrocephalus.
5. Hydrocephalus can occur in adults, but the head does not enlarge because bone growth has stopped, and thus the signs and symptoms differ from hydrocephalus in infants.

6. The basilar artery and its branches, the anterior inferior cerebellar and superior cerebellar arteries, supply the pons.
7. The posterior cerebral artery and its branches supply the midbrain.
8. The anterior communicating, anterior cerebral, internal carotid, posterior communicating, and posterior cerebral arteries form the circle of Willis.
9. The watershed area is the region on the surface of the lateral cerebral hemisphere that receives blood from small anastomoses linking the ends of the cerebral arteries. The watershed area is susceptible to insufficient blood flow.
10. A transient ischemic attack is a brief, focal loss of brain function that lasts less than 24 hours. The neurological deficits must be completely resolved within 24 hours.
11. A lacuna is a small cavity that remains after the necrotic tissue is cleared following an infarct in a small artery.
12. A partial occlusion of the basilar artery may result in tetraplegia, loss of sensation, coma, and cranial nerve signs because the basilar artery supplies both the left and right sides of the medulla and pons; thus interference with its blood flow deprives the descending upper motoneuron axons and the ascending sensory axons, as well as the reticular formation, cranial nerve nuclei, and cranial nerves of adequate perfusion.
13. More severe hemiplegia and hemisensory loss in the lower limb indicate that the lesion is in medial sensorimotor cortex and adjacent subcortical white matter. This region is supplied by the anterior cerebral artery.
14. Neglect, problems comprehending space, and impaired nonverbal communication occur with damage to the area analogous to Wernicke's area in the hemisphere that is nondominant for language (usually the right hemisphere). This area is supplied by the middle cerebral artery.
15. Hemisensory loss and hemiplegia affecting the limbs and face equally indicate a lesion in the internal capsule, supplied by deep branches of the middle cerebral artery.
16. The loss of eye movements combined with contralateral hemiplegia indicates a lesion of the anterior midbrain, supplied by branches of the posterior cerebral artery.
17. Distal branches of the anterior, middle, and posterior cerebral arteries supply the watershed area.
18. An arteriovenous malformation is an abnormal connection between arteries and veins, with thin-walled vessels larger than capillaries connecting the vessels. Arteriovenous malformations are developmental defects.

19. An aneurysm is a dilation of the wall of an artery or vein. Aneurysms are prone to rupture because their walls are thinner than normal vessel walls.

20. An uncal herniation is displacement of the uncus medially, so that the uncus protrudes through the tentorium cerebelli and compresses the midbrain.

21. A PET scan, positron emission tomography, is a computer-generated image reflecting the metabolic activity throughout biological tissues. PET scans of the central nervous system record the relative activity and inactivity of areas in the brain or spinal cord.

Glossary

accommodation adjustments of the eyes to view a near object: the pupils constrict, the eyes converge (adduct), and the lens becomes more convex. The optic nerve is the afferent (sensory) limb of the reflex, while the oculomotor nerve provides the efferent (motor) limb.

acetylcholine a neurotransmitter released by axons from the pedunculopontine nucleus and nucleus basalis of Meynert, by lower motoneurons, by preganglionic autonomic axons, by postganglionic parasympathetic axons, and by postganglionic sympathetic axons that innervate sweat glands. Binds with nicotinic or muscarinic receptors. Action on postsynaptic membranes is usually excitatory.

adrenergic 1. referring to neurons that secrete norepinephrine or epinephrine. 2. referring to drugs that bind with and activate the same receptors as norepinephrine or epinephrine. 3. referring to receptors that bind norepinephrine, epinephrine, or agonist drugs.

agnosia general term for the inability to recognize objects when using a specific sense, even though discriminative ability with that sense is intact. Specific types of agnosia include astereognosis, visual agnosia, and auditory agnosia.

agnosia, visual inability to visually recognize objects despite intact vision.

agraphia diminished or lost ability to produce written language.

alexia diminished or lost ability to comprehend written language.

allodynia sensation of pain in response to normally nonpainful stimuli.

all-or-none term applied to the generation of an action potential, indicating that every time even minimally sufficient stimuli are provided to generate an action potential, an action potential will be produced. Stimuli that are stronger than the minimally sufficient stimuli produce action potentials of the same voltage and duration as the minimally sufficient stimuli.

amygdala nuclei that interpret facial expressions and social signals. Together the amygdala, orbitofrontal cortex, and anterior cingulate gyrus regulate emotional behaviors and motivation. The amygdala consists of an almond-shaped collection of nuclei deep to the uncus in the temporal lobe.

analgesia, stress-induced absent or reduced perception of pain due to activation of the pain inhibition systems during an emergency or in competitive situations.

anencephaly a developmental defect characterized by development of a rudimentary brain stem without cerebral and cerebellar hemispheres.

aneurysm saclike dilation of the wall of an artery or vein. These swellings have thin walls that are prone to rupture.

angiography radiopaque dye injected into a carotid or vertebral artery followed by a sequence of x-rays.

aphasia disorder of language expression or comprehension. Deficit in the ability to produce understandable speech and writing, or the ability to understand written and spoken language.

aphasia, Broca's inability to express oneself by language or symbols. A person with Broca's aphasia cannot communicate by speaking or writing. Syn: motor aphasia, expressive aphasia.

aphasia, conduction a language disorder resulting from damage to the neurons that connect Wernicke's and Broca's areas. The ability to understand written and spoken language is normal. In mild cases, only paraphrasias occur. In the most severe form, speech and writing produced are meaningless.

aphasia, global inability to use language in any form. People with global aphasia cannot produce understandable speech, comprehend spoken language, speak fluently, read, or write.

aphasia, Wernicke's impairment of language comprehension. People with Wernicke's aphasia easily produce spoken sounds, but the output is often meaningless. Listening to other people speak is equally meaningless, despite the ability to hear normally. Syn: receptive aphasia, sensory aphasia.

apparatus, vestibular the part of the inner ear that detects position and movement of the head. Consists of the semicircular canals, the saccule, and the utricle.

apraxia inability to perform a movement or sequence of movements despite intact sensation, automatic motor output, and understanding of the task.

area, Broca's region of cortex that provides instructions for language output, including planning the movements to produce speech and providing grammatical function words, such as the articles *a, an,* and *the.* Located inferior to the premotor area and anterior to the face and throat region of the primary motor cortex, usually in the left hemisphere.

area, corresponding to Broca's area region of the cerebral cortex inferior to the premotor area and anterior to the face and throat region of the primary motor cortex. Usually in the right hemisphere. Plans nonverbal communication, including emotional gestures and adjusting the tone of voice.

area, corresponding to Wernicke's area subregion of the parietotemporal cortex where interpretation of the nonverbal signals from other people and understanding of spatial relationships occurs. Usually located on the right side.

area, lateral premotor a region of the cerebral cortex involved in preparing for movement and controlling trunk and girdle muscles via the medial activation system. Located anterior to the upper body region of the primary motor cortex, on the lateral surface of the hemisphere.

area, limbic association part of the cerebral cortex involved in regulating mood (subjective feelings), affect (observable demeanor), and processing of some types of memory. Located in the anterior temporal lobe and in the orbitofrontal cortex (above the eyes).

area, preoptic part of the limbic system, located anterior to the septal area.

area, septal part of the limbic system in the basal forebrain region, anterior to the anterior commissure.

area, somatosensory association a region of cerebral cortex that analyzes information from the primary somatosensory cortex and from the thalamus. Provides stereognosis and memory of the tactile and spatial environment. Located posterior to the primary somatosensory cortex.

area, supplementary motor a region of the cerebral cortex involved in preparing for movement, orientation of the eyes and head, and planning bimanual and sequential movements. Located anterior to the lower body region of the primary motor cortex, on the superior and medial surface of the hemisphere.

area, ventral tegmental a region in the midbrain that provides dopamine to cerebral areas important in motivation and in decision making.

area, watershed area of marginal blood flow on the surface of the lateral hemispheres, where small anastomoses link the ends of the cerebral arteries.

area, Wernicke's subregion of the parietotemporal cortex where comprehension of language occurs. Usually located on the left side.

areas, association regions of the cerebral cortex that are not directly involved with sensation or movement. Involved with personality, integration and interpretation of sensations, processing of memory, and generation of emotions. The three association areas are prefrontal, parietotemporal, and limbic.

areas, Brodmann's histologic regions of the cerebral cortex mapped by Brodmann. Often used to designate functional areas.

areas, motor planning regions of the cerebral cortex involved in organizing movement. Motor planning areas include supplementary motor area, premotor area, Broca's area, and the area corresponding to Broca's area.

areas, parietotemporal association part of the cerebral cortex devoted to intelligence, problem solving, and comprehension of communication and spatial relationships. Located at the junction of the parietal, occipital, and temporal lobes.

areas, primary sensory areas of the cerebral cortex that receive sensory information directly from the ventral tier of thalamic nuclei. Each primary sensory area discriminates among different intensities and qualities of one type of sensory input. Separate primary sensory areas are devoted to somatosensory, auditory, visual, and vestibular information.

areas, sensory association areas of the cerebral cortex that analyze sensory input from both the thalamus and from primary sensory cortex. Sensory association areas contribute to the analysis of one type of sensory information.

artery, anterior cerebral vessel that provides blood to the medial surface of the frontal and parietal lobes and the anterior head of the caudate. A branch of the internal carotid artery.

artery, anterior choroidal a branch of the internal carotid artery that provides blood to the optic tract, choroid plexus in the lateral ventricles, and parts of the optic radiations, putamen, thalamus, internal capsule, and hippocampus.

artery, basilar vessel that provides blood to the pons and most of the cerebellum. Formed near the pontomedullary junction by the union of the vertebral arteries. Divides to become the posterior cerebral arteries.

artery, internal carotid vessel that provides blood to the anterior, superior, and lateral cerebral hemispheres via its branches: the anterior and middle cerebral and the anterior choroidal arteries.

artery, middle cerebral vessel whose branches fan out to provide blood to most of the lateral hemisphere. A branch of the internal carotid artery.

artery, posterior cerebral vessel that provides blood to the midbrain, occipital lobe, and parts of the medial and inferior temporal lobes. A branch of the basilar artery.

artery, posterior choroidal a branch of the posterior cerebral artery that provides blood to the choroid plexus of the third ventricle and parts of the thalamus and hippocampus.

artery, striate any of several arteries arising from the proximal part of either the anterior or middle cerebral arteries to supply the basal ganglia and parts of the thalamus and internal capsule.

artery, vertebral vessel that provides blood to the brain stem, cerebellum, and the posteroinferior cerebrum. Branch of the subclavian artery.

astereognosis inability to identify objects by touch and manipulation despite intact discriminative somatosensation.

astrocytes macroglia that play a critical role in nutritive and cleanup functions within the central nervous system.

ataxia abnormal voluntary movements that are of normal strength but jerky and inaccurate.

athetosis slow, writhing, purposeless movements.

atrophy, disuse loss of muscle bulk resulting from lack of use.

atrophy, neurogenic loss of muscle bulk resulting from damage to the nervous system.

attack, transient ischemic a brief, focal loss of brain function, with full recovery from neurological deficits within 24 hours. Transient ischemic attacks are believed to be due to inadequate blood supply.

autoregulation adjustment of local blood flow to the demands of the surrounding tissues.

axon a process that extends from the cell body of a neuron. Most axons conduct signals away from the cell body. The only axons that conduct information toward the cell body are the distal axons of primary afferent neurons, which conduct signals to the dorsal root ganglion or cranial nerve ganglion.

axon hillock specialized region of a neuron cell body that gives rise to the axon. The axon hillock is densely populated with voltage-gated Na^+ channels.

axons, myelinated axons that are completely enveloped by a myelin sheath.

axons, unmyelinated axons that are only partially enveloped by myelin.

basal ganglia interconnected group of nuclei involved in comparing proprioceptive information and movement commands, sequencing movements, and regulating muscle tone and muscle force. May select and inhibit muscle synergies. Consist of the caudate, putamen, globus pallidus, subthalamic nucleus, substantia nigra, and pedunculopontine nucleus.

bundle, medial forebrain axons connecting anterior structures (septal area, nucleus accumbens, amygdala, anterior cingulate gyrus), the hypothalamus, and the midbrain reticular formation.

canals, semicircular three hollow rings in the inner ear, oriented at right angles to each other. Each canal has an enlargement, the ampulla, that contains the receptor mechanism for detecting rotational acceleration or deceleration of the head.

capacitance, membrane (Cm) the storage of electrical charge across a cell membrane.

cell body the metabolic center of any cell that contains the nucleus and the energy-producing/storing apparatus. The cell body of neurons also includes neurotransmitter-synthesizing mechanisms.

cells, bipolar neurons having two primary processes, a dendritic root and an axon, that extend from the cell body.

cells, multipolar neurons having multiple dendrites arising from many regions of the cell body and possessing a single axon.

cells, pseudounipolar neurons that have one projection from the cell body that later divides into two axonal roots. A pseudounipolar cell has no true dendrites.

cells, Renshaw interneurons that produce recurrent inhibition in the spinal cord. Act to focus motor activity.

cells, Schwann macroglia that form myelin sheaths enveloping only a single neuron's axon or partially surrounding several axons. Found in the peripheral nervous system.

cerebellum a part of the brain posterior to the brain stem. Involved in coordination of movement and postural control, motor planning, rapid shifting of attention, and regulation of muscle tone.

cerebral palsy, spastic motor disorder that develops in utero or during infancy. Characterized by muscle hypertonia.

cerebrocerebellum part of the cerebellum that coordinates voluntary movements via influence on corticofugal pathways, plans movements, judges time intervals, and produces accurate rhythms. Located in the lateral cerebellar hemispheres.

channels, G-protein mediated ion neuronal membrane channels that open in response to the activation of a G-protein or its second messenger.

channels, ligand-gated neuronal membrane ion channels that open in response to the binding of a chemical neurotransmitter.

channels, modality-gated membrane ion channels, specific to sensory neurons, that open in response to mechanical forces (i.e., stretch, touch, pressure) or thermal or chemical changes.

channels, voltage-gated membrane ion channels that open in response to changes in electrical potential across a neuron's cell membrane.

chemoreceptor a receptor that responds to chemical change. Found in the carotid body, brain stem respiratory centers, specialized sensory cells for taste and smell, and the skin, muscles, and viscera.

chiasm, optic site where the optic nerve fibers from the nasal half of the retina cross the midline.

cholinergic 1. referring to a neuron that secretes acetylcholine. 2. referring to drugs that bind with and activate the same receptors as acetylcholine. 3. referring to receptors that bind acetylcholine or agonist drugs.

chorea involuntary, jerky, rapid movements.

choreoathetosis a combination of involuntary, jerky, rapid movements and slow, writhing, purposeless movements.

choroid plexus a network of capillaries embedded in connective tissue and epithelial cells that produce cerebrospinal fluid.

circle of Willis anastomotic ring of nine arteries, supplying all of the blood to the cerebral hemispheres. Consists of two anterior cerebral arteries, two internal carotids, two posterior cerebral arteries, one anterior communicating artery, and two posterior communicating arteries.

circuits, control neural connections that adjust activity in the descending tracts, resulting in excitation or inhibition of the lower motoneurons. Consist of the basal ganglia and cerebellum.

cleft, synaptic the space between the presynaptic and postsynaptic nerve terminals.

clonus repetitive stretch reflexes elicited by passive dorsiflexion of the foot or passive extension of the wrist. Occurs in upper motoneuron lesions, secondary to the lack of descending motor control.

coactivation, alpha-gamma simultaneous firing of alpha and gamma motoneurons. Ensures that the muscle spindle maintains its sensitivity even when the extrafusal fibers surrounding the spindle contract.

cochlea snail shell–shaped organ, formed by a spiraling, fluid-filled tube. The cochlea contains a mechanism, the organ of Corti, that converts mechanical vibrations into the neural impulses that produce hearing.

cocontraction simultaneous contraction of agonist and antagonist muscles. May occur in an intact nervous system when learning a new movement, or may be a sign of neural dysfunction.

colliculus, superior part of the tectum of the midbrain, the superior colliculus integrates various sensory inputs and influences eye movement, head and body orientation, and postural adjustments.

columns, anterolateral white matter in the anterior and lateral spinal cord that contains spinothalamic and motor axons.

coma condition of being unarousable; no response to strong stimuli such as strong pinching of the Achilles tendon.

conduction, saltatory rapid propagation of an action potential by jumping from one node of Ranvier to the next along a myelinated axon.

convergence multiple inputs from a variety of different cells terminating on a single neuron.

corpus callosum large fiber bundle connecting the right and left cerebral cortices.

cortex, cerebral gray matter covering the cerebral hemispheres.

cortex, limbic part of the limbic system. A C-shaped region of cortex located on the medial hemisphere, consisting of the cingulate gyrus, parahippocampal gyrus, and uncus (a medial protrusion of the parahippocampal gyrus).

cortex, prefrontal anterior part of the frontal cortex, responsible for self-awareness and executive functions (also called goal-oriented behavior). Executive functions include deciding on a goal, planning how to accomplish the goal, executing the plan, and monitoring the outcome of the action.

cortex, primary motor part of the cerebral cortex. Origin of many cortical upper motoneurons that influence contralateral voluntary movements, particularly the fine, fractionated movements of the hand and face. Located in the precentral gyrus, anterior to the central sulcus.

cortex, primary sensory (primary somatosensory) cerebral cortex that receives somatosensory information from the body and face. Located posterior to the central sulcus.

cortisol a steroid hormone that mobilizes energy (glucose), suppresses immune responses, and serves as an anti-inflammatory agent. Secreted by the adrenal glands. Syn: hydrocortisone.

cramp severe and painful muscle spasm, associated with fatigue or local ionic imbalances.

crest, neural during development, the part of the ectoderm that will become the peripheral sensory neurons, myelin cells, autonomic neurons, and endocrine organs (adrenal medulla and pancreatic islets).

deafness, conductive hearing defect due to inability to transmit vibrations in the outer or middle ear.

deafness, sensorineural hearing defect due to damage of the receptor cells or the cochlear nerve.

deformity, Arnold-Chiari developmental malformation of the hindbrain, with elongation of the inferior cerebellum and medulla. The inferior cerebellum and medulla protrude into the vertebral canal.

degeneration, wallerian degeneration and death of the distal segment of a severed axon.

delirium reduced attention, orientation, and perception, associated with confused ideas and agitation.

dendrite process that extends from the cell body of a neuron. Dendrites conduct information toward the cell body.

depolarization the process whereby a neuron's cell membrane potential becomes less negative than its resting potential.

depolarized the electrical state of a neuron's cell membrane when the membrane potential becomes less negative than the resting potential.

dermatome the part of the somite that becomes dermis, or after the embryo stage, the dermis innervated by a single spinal nerve.

diencephalon centrally located part of the cerebrum, consisting of the thalamus, hypothalamus, epithalamus, and subthalamus.

disease, Huntington's autosomal dominant hereditary disorder that causes degeneration in many areas of the brain, primarily in the striatum and cerebral cortex. Characterized by hyperkinesia.

disease, Meniere's sensation of fullness in the ear, tinnitus (ringing in the ear), severe acute vertigo, nausea, vomiting, and hearing loss. Cause is unknown.

disease, Parkinson's the most common disorder of the basal ganglia, resulting from death of dopamine-producing cells in the substantia nigra compacta and acetylcholine-producing cells in the pedunculopontine nucleus. Characterized by muscular rigidity, slowness of movement, shuffling gait, droopy posture, resting tremors, diminished facial expression, and visuoperceptive impairments.

disorder, attention deficit difficulty sustaining attention with onset during childhood.

divergence the branching of a single neuronal axon to synapse with a multitude of neurons.

dopamine a neurotransmitter released by axons from the substantia nigra and the ventral tegmental area. Action on postsynaptic membranes is usually inhibitory.

dorsal root the afferent (sensory) root of a spinal nerve.

dysarthria speech disorder resulting from paralysis, incoordination, or spasticity of muscles used for speaking. Due to upper or lower motoneuron lesions or muscle dysfunction. Comprehension of spoken language, writing, and reading are not affected by dysarthria. Two types of dysarthria may be distinguished: spastic, due to damage of upper motoneurons, and flaccid, resulting from damage to lower motoneurons.

dysdiadochokinesis inability to rapidly alternate movements; for example, inability to rapidly pronate and supinate the forearm, or inability to rapidly alternate toe tapping.

dysesthesia painful abnormal sensation, including burning and aching sensations.

dysmetria inability to accurately move an intended distance.

dysreflexia, autonomic excessive activity of the sympathetic nervous system, usually elicited by noxious stimuli below the level of a spinal cord lesion.

dystonia hereditary movement disorder, usually nonprogressive, characterized by involuntary sustained muscle contractions causing abnormal postures or twisting, repetitive movements.

edema, cerebral accumulation of excess tissue fluid in the brain.

effectiveness, synaptic functional activation of postsynaptic receptors in response to the release of neurotransmitter from a presynaptic terminal.

electromyography recording of electrical activity produced by muscle fibers.

embolus a blood clot that formed elsewhere and has been transported to a new location before occluding a vessel.

ending, primary sensory ending of a type Ia axon that responds phasically to stretch of the central region of intrafusal fibers in the muscle spindle.

ending, secondary sensory ending of a type II axon that responds tonically to stretch of the central region of intrafusal fibers (primarily nuclear chain fibers) in the muscle spindle.

endorphins endogenous, or naturally occurring, substances that activate analgesic mechanisms. Endorphins include enkephalins, dynorphin, and β-endorphin.

enkephalin a neurotransmitter that, when bound to receptor sites, depresses the release of substance P and hyperpolarizes interneurons in the nociceptive pathway, thus inhibiting the transmission of nociceptive signals.

epilepsy sudden attacks of excessive neuronal discharge interfering with brain function.

epithalamus the major structure of the epithalamus is the pineal gland, an endocrine gland innervated by sympathetic fibers. The pineal gland is believed to help regulate circadian (daily) rhythms and influence the secretions of the pituitary gland, adrenals, and parathyroids.

equation, Nernst the mathematical equation used to determine the equilibrium potential for a diffusible ion.

excitotoxicity overexcitation of a neuron, leading to cell death.

extinction, sensory a form of unilateral neglect. Loss of sensation is only evident when symmetrical body parts are tested bilaterally.

facilitation, presynaptic at an axoaxonic synapse, the excitatory process by which transmitter released by one axon terminal causes the second axon terminal to release a greater than normal amount of neurotransmitter.

fasciculation   a quick twitch of muscle fibers in a single motor unit, which is visible on the surface of the skin.

fasciculus cuneatus   axons that transmit discriminative touch and conscious proprioceptive information from the upper half of the body to the brain. Located in the lateral section of the dorsal column of the spinal cord.

fasciculus gracilis   axons that transmit discriminative touch and conscious proprioceptive information from the lower half of the body to the brain. Located in the medial section of the dorsal column of the spinal cord.

fasciculus, medial longitudinal (MLF)   a brain stem tract that coordinates head and eye movements by providing bilateral connections among vestibular, oculomotor, and accessory nerve nuclei and the superior colliculus.

feedback   information resulting from a movement. For example, when a person flexes the elbow, feedback consists of information from sensory receptors in muscles, tendons, and skin.

feed-forward   neural preparation for anticipated movement, based on instruction, previous experience, and the ability to predict the movement requirements and/or outcome.

fiber, intrafusal   specialized muscle fiber inside the muscle spindle.

fibers, association   axons connecting cortical regions within one hemisphere.

fibers, commissural   axons connecting homologous areas of the nervous system.

fibers, corticobulbar   axons that influence the activity of lower motoneurons innervating the muscles of the face, tongue, pharynx, and larynx. Corticobulbar fibers arise in

motor planning areas of the cerebral cortex and the primary motor cortex, then project to cranial nerve nuclei in the brain stem.

fibers, extrafusal   contractile skeletal muscle fibers outside of the muscle spindle.

fibers, projection   axons connecting subcortical structures to the cerebral cortex, and axons connecting the cerebral cortex with the subcortical structures.

fibrillation   brief contraction of a single muscle fiber, not visible on the surface of the skin.

fibromyalgia   tenderness of muscles and adjacent soft tissues, stiffness of muscles, and aching pain. The painful area shows a regional rather than dermatomal or peripheral nerve distribution.

foci, ectopic   site on neural membrane that is abnormally sensitive to mechanical stimulation.

formation, reticular   complex neural network in the brain stem, including the reticular nuclei, their connections, and ascending and descending reticular tracts.

fornix   arch-shaped fiber bundle connecting the hippocampus with the mamillary body and anterior nucleus of the thalamus.

fractionation   ability to activate individual muscles independently of other muscles.

GABA (γ-aminobutyric acid)   a neurotransmitter released by the caudate nucleus and putamen and by cerebellar Purkinje cells and spinal interneurons. Action on postsynaptic membranes is inhibitory.

ganglion, dorsal root   collection of primary sensory neuron cell bodies located in the dorsal root.

gate theory of pain   theory that transmission of pain information can be blocked in the dorsal horn by stimulation of large-fiber primary afferent neurons.

generator, central pattern   a flexible

network of interneurons that activate repetitive, rhythmical, purposeful movements (e.g., walking or chewing movements).

geniculate, lateral   site of synapse between axons from the retina and neurons that project to the visual cortex. Part of the thalamus, located inferiorly and posteriorly.

geniculate, medial   site of synapse in the auditory pathway. Part of the thalamus, located inferiorly and posteriorly.

genu   most medial part of the internal capsule, containing cortical fibers that project to cranial nerve motor nuclei and to the reticular formation.

glia   the support cells of the nervous system, including oligodendrocytes, Schwann cells, astrocytes, and microglia.

glutamate   an excitatory amino acid neurotransmitter. Excessive amounts can be toxic to neurons.

glycine   a neurotransmitter released by axons from spinal cord interneurons.

gray, periaqueductal   area around the cerebral aqueduct in the midbrain. Involved in somatic and autonomic reactions to pain, threats, and emotions. Activity of the periaqueductal gray results in the fight-or-flight reaction and in vocalization during laughing and crying.

groove, neural   during development, the depression formed by the infolding of the neural plate; becomes the neural tube.

growing into deficit   signs and symptoms of nervous system damage that do not become evident until the systems damaged would have become functional.

gyrus, cingulate   gyrus on the medial cerebral hemisphere, superior to the corpus callosum.

gyrus, parahippocampal   most medial gyrus of the inferior temporal lobe.

habituation   a form of short-term

plasticity. Repeated stimuli result in a decreased response due to a decrease in the amount of neurotransmitter released from the presynaptic terminal of a sensory neuron.

hematoma, epidural   collection of blood between the skull and dura mater.

hematoma, subdural   collection of blood between the dura mater and arachnoid.

hemianopsia, bitemporal   loss of information from both temporal visual fields. Produced by damage to fibers in the center of the optic chiasm, interrupting the axons from the nasal half of each retina.

hemianopsia, homonymous   loss of visual information from one hemifield. A complete lesion of the visual pathway anywhere posterior to the optic chiasm, in the optic tract, lateral geniculate, or optic radiations, results in loss of information from the contralateral visual field.

hemiplegia   weakness or paralysis affecting one side of the body.

hemisphere, lateral   part of the cerebellar hemisphere lateral to the paravermis. Involved in coordination of voluntary movements, planning of movements, and the ability to judge time intervals and produce accurate rhythms.

herniation, central   movement of the diencephalon, midbrain, and pons inferiorly, caused by a lesion in the cerebrum exerting pressure on the diencephalon. This movement stretches the branches of the basilar artery, causing brain stem ischemia and edema.

herniation, tonsillar   protrusion of the cerebellar tonsils (small lobes forming part of the inferior surface of the cerebellum) through the foramen magnum.

herniation, uncal   protrusion of the uncus into the opening of the tentorium cerebelli, causing compression of the midbrain.

hippocampus   part of the limbic system. Important in processing, but not storage of, declarative memories. Formed by the gray and white matter of two gyri rolled together in the medial temporal lobe.

homunculus   a figure representing the parts of the body controlled by or transmitting sensory information to a specific part of the cerebral cortex.

horn, dorsal   the posterior section of gray matter in the spinal cord. Primarily sensory in function, the dorsal horn contains endings and collaterals of first-order sensory neurons, interneurons, and dendrites and somas of tract cells.

horn, lateral   the lateral section of gray matter in the spinal cord. Contains the cell bodies of preganglionic sympathetic neurons.

horn, ventral   the anterior section of gray matter in the spinal cord. Contains endings of upper motoneurons, interneurons, and dendrites and cell bodies of lower motoneurons.

hydrocephalus   accumulation of an excessive amount of cerebrospinal fluid in the ventricles.

hypereffectiveness, synaptic   increased response to a neurotransmitter because damage to some branches of a presynaptic axon results in larger than normal amounts of transmitter being released by the remaining axons onto postsynaptic receptors.

hyperkinetic   characterized by abnormal involuntary movements. Includes dystonic, choreic, athetotic, and choreoathetotic movements.

hyperpolarization   the process whereby a neuron's cell membrane potential becomes more negative than its resting potential.

hyperpolarized   the electrical state of a neuron's cell membrane when the membrane potential becomes more negative than its resting potential.

hypersensitivity, denervation   increased response to a neurotransmitter because new receptor sites have developed on the postsynaptic membrane.

hypertonia   abnormally strong resistance to passive stretch. Occurs in chronic upper motoneuron disorders and in some basal ganglia disorders. Two types: 1. spastic (resistance is dependent on velocity of stretch) and 2. rigid (resistance is independent of velocity of muscle stretch).

hypotension, orthostatic   an extreme fall in blood pressure when moving from prone or supine to sitting or standing.

hypothalamus   the ventromedial part of the diencephalon. Plays a major role in regulation of the autonomic and endocrine systems, and contributes to emotional and motivational states.

hypotonia   abnormally low muscular resistance to passive stretch. Occurs in cerebellar, lower motoneuron, and primary afferent neuron disorders. Also occurs temporarily following upper motoneuron lesions due to a period of neural shock (electrical silence) postinjury.

infarct, lacunar   obstruction of blood flow in a small, deep artery. Lacunae are small cavities that remain after the necrotic tissue is cleared away.

inhibition, autogenic   inhibition of the lower motoneurons innervating a muscle in response to stretch of the muscle's Golgi tendon organs.

inhibition, nonreciprocal   selective inhibition of agonist, synergist, and antagonist muscles. Coordinates the actions of various muscles during limb movements.

inhibition, presynaptic   at an axoaxonic synapse, the inhibitory process by which transmitter released by one axon terminal causes the second terminal to release a lower than normal amount of neurotransmitter.

inhibition, reciprocal decreased activity in an antagonist when an agonist is active.

inhibition, recurrent inhibition of agonists and synergists, combined with disinhibition of antagonists.

internal capsule axons connecting the cerebral cortex with subcortical structures. The internal capsule is white matter bordered by the thalamus medially and the caudate and lenticular nucleus laterally. The internal capsule has three parts: anterior limb, genu, and posterior limb. Anterior limb: located lateral to the head of the caudate, contains corticopontine fibers and fibers interconnecting thalamic and cortical limbic areas. Genu: most medial part of the internal capsule, containing cortical fibers that project to cranial nerve motor nuclei and to the reticular formation. Posterior limb: located between the thalamus and lenticular nucleus, with additional fibers traveling posterior and inferior to the lenticular nucleus (retrolenticular and sublenticular fibers). The posterior limb contains corticopontine, corticospinal, and thalamocortical projections.

interneurons neurons that either process information locally or convey information short distances from one site in the nervous system to another.

junction, neuromuscular synapse between a nerve terminal and the membrane of a muscle fiber. Acetylcholine is the neurotransmitter released at the neuromuscular junction.

labyrinth the inner ear, consisting of the cochlea and the vestibular apparatus.

laminae, Rexed's histological divisions of the spinal cord gray matter.

layer, mantle during development, the inner wall of the neural tube.

layer, marginal during development, the outer wall of the neural tube.

lemniscus, medial axons of second-order neurons conveying sensory information related to discriminative touch and conscious proprioception from the body to the cerebral cortex. Begins in the nucleus cuneatus and nucleus gracilis and ends in the ventral posterolateral nucleus of the thalamus.

level, neurological describing spinal cord injury, neurological level is the most caudal level with normal sensory and motor function bilaterally.

limb, anterior part of the internal capsule located lateral to the head of the caudate, contains corticopontine fibers and fibers interconnecting thalamic and cortical limbic areas.

limb, posterior part of the internal capsule located between the thalamus and lenticular nucleus, with additional fibers traveling posterior and inferior to the lenticular nucleus (retrolenticular and sublenticular fibers). Contains corticopontine, corticospinal, and thalamocortical projections.

locus ceruleus nucleus in the upper pons involved in direction of attention, nonspecific activation of interneurons and lower motoneurons in the spinal cord, and inhibition of pain information in the dorsal horn. Transmitter produced is norepinephrine.

M1 the shortest latency response after stretch of a muscle, produced by the monosynaptic phasic stretch reflex.

M2 the second response after stretch of a muscle, probably involving neural circuits in the brain stem. Also called the long loop response.

macroglia the large support cells of the nervous system, including oligodendrocytes, Schwann cells, and astroctyes.

malformations, arteriovenous developmental abnormalities with arteries connected to veins by abnormal, thin-walled vessels larger than capillaries. Arteriovenous malformations usually do not cause signs or symptoms unless they rupture.

marginal layer most dorsal part of the spinal gray matter. Involved in processing nociceptive information. Syn: lamina I.

mechanoreceptor a receptor that responds to mechanical stimulation (e.g., stretch or pressure). Examples include receptors in the muscle spindle, touch receptors in skin, and stretch receptors in viscera.

medulla inferior part of the brain stem. Contributes to the control of eye and head movements, coordinates swallowing, and helps regulate cardiovascular, respiratory, and visceral activity.

memory, declarative recollections that can be easily verbalized. Declarative memory is also called conscious, explicit, or cognitive memory.

memory, procedural recall of skills and habits. This type of memory is also called skill, habit, or nonconscious or implicit memory.

meninges membranes that enclose the brain and spinal cord. Include the dura mater, arachnoid, and pia mater.

meningomyelocele developmental defect; inferior part of the neural tube remains open.

messenger, second a molecule that diffuses through the intracellular environment of a neuron and initiates cellular events including opening or closing of membrane ion channels, activation of genes, or modulation of calcium concentrations inside the cell.

microglia the small support cells of the nervous system.

modulation long-lasting changes in the electrical potential of a neuron's cell membrane that alters the flow of ions across the cell membrane.

mononeuropathy dysfunction of a single peripheral nerve.

mononeuropathy, multiple dysfunction of several separate peripheral nerves. Signs and symptoms show an asymmetrical distribution.

motoneurons, alpha lower motoneurons that innervate extrafusal fibers in skeletal muscle. When these neurons fire, skeletal muscle fibers contract.

motoneurons, gamma lower motoneurons that innervate intrafusal fibers in skeletal muscle. When these neurons fire, the ends of intrafusal fibers contract, stretching the central region of muscle fibers within the muscle spindle.

motoneurons, lower neurons with their cell bodies in the spinal cord or brain stem whose axons directly innervate skeletal muscle fibers. Two types: alpha motoneurons innervate extrafusal muscle fibers, and gamma motoneurons innervate intrafusal muscle fibers.

motoneurons, upper neurons whose axons are located in a descending pathway. Upper motoneurons transmit information from the brain to lower motoneurons and movement-related interneurons in the spinal cord or brain stem. Although upper motoneurons do not directly innervate skeletal muscle, they contribute to the control of movement by influencing the activity of lower motoneurons.

myelin the sheath of proteins and fats formed by oligodendrocytes and Schwann cells to envelop the axons of nerve cells. Provide physical support and insulation for conduction of electrical signals by neurons.

myelination the process of acquiring a myelin sheath.

myopathy abnormality or disease intrinsic to muscle tissue.

myotome during development, the part of a somite that becomes muscle, or after the embryo stage, a group of muscles innervated by a segmental spinal nerve.

neglect tendency to behave as if one side of the body and/or one side of space does not exist.

nerve, abducens cranial nerve VI. Controls the lateral rectus muscle that moves the eye laterally.

nerve, accessory cranial nerve XI. Motor nerve innervating the trapezius and sternocleidomastoid muscles.

nerve, facial cranial nerve VII. Mixed nerve containing both sensory and motor fibers. The sensory fibers transmit touch, pain, and pressure information from the tongue and pharynx and information from taste buds of the anterior tongue to the solitary nucleus. Motor innervation by the facial nerve includes the muscles that close the eyes, move the lips, and produce facial expressions. The facial nerve provides the efferent limb of the corneal reflex and also innervates salivary, nasal, and lacrimal (tear-producing) glands.

nerve, glossopharyngeal cranial nerve IX. Mixed nerve containing both sensory and motor fibers. The sensory fibers transmit somatosensation from the soft palate and pharynx and taste information from the posterior tongue. The motor component innervates a pharyngeal muscle and the parotid salivary gland.

nerve, hypoglossal cranial nerve XII. Motor nerve, providing innervation to the intrinsic and extrinsic muscles of the ipsilateral tongue.

nerve, oculomotor cranial nerve III. Controls the superior, inferior, and medial rectus, the inferior oblique, and the levator palpebrae superioris muscles. These muscles move the eye upward, downward, and medially; rotate the eye around the axis of the pupil; and assist in elevating the upper eyelid. Parasympathetic efferent fibers in the oculomotor nerve innervate the ciliary muscle and the sphincter pupillae, controlling reflexive constriction of the pupil and the thickness of the lens of the eye.

nerve, olfactory cranial nerve I. Transmits information about odors.

nerve, optic cranial nerve II. Transmits visual information from the retina to the lateral geniculate body of the thalamus and to nuclei in the midbrain.

nerve, spinal a nerve located in the intervertebral foramen, formed by the dorsal and ventral roots, that contains both afferent and efferent axons. Spinal nerves branch to form dorsal and ventral rami.

nerve, trigeminal cranial nerve V. Mixed nerve containing both sensory and motor fibers. The sensory fibers transmit information from the face and temporomandibular joint. The motor fibers innervate the muscles of mastication. Has three branches: ophthalmic, maxillary, and mandibular.

nerve, trochlear cranial nerve IV. Controls the superior oblique muscle, which rotates the eye or, if the eyes is adducted, depresses the eye.

nerve, vagus cranial nerve X. Provides sensory and motor innervation of the larynx, pharynx, and viscera.

nerve, vestibulocochlear cranial nerve VIII. Sensory nerve with two distinct branches. The vestibular branch transmits information related to head position and head movement. The cochlear branch transmits information related to hearing.

neuralgia, postherpetic severe pain that persists more than 1 month after an infection with varicella-zoster virus. Occurs along the distribution of a peripheral nerve or branch of a peripheral nerve.

neuralgia, trigeminal dysfunction of the trigeminal nerve, producing severe, sharp, stabbing pain in the distribution of one or more branches of the trigeminal nerve.

neuritis, vestibular inflammation of the vestibular nerve, usually caused by a virus. Dysequilibrium, spontaneous nystagmus, nausea, and severe vertigo persist up to 3 days.

neuron the electrically excitable nerve cell of the nervous system.

neuron, afferent neuron that brings information into the central nervous system.

neuron, efferent neuron that relays commands from the central nervous system to the smooth and skeletal muscles and glands of the body.

neuron, postganglionic autonomic neuron with its cell body in an autonomic ganglion and its termination in an effector organ.

neuron, preganglionic autonomic neuron with its cell body in the brain stem or spinal cord and its termination in an autonomic ganglion.

neuron, wide dynamic range second-order neuron in the nociceptive pathway that responds to information from cutaneous, musculoskeletal, and visceral receptors.

neurotransmitters chemicals contained in the presynaptic terminal that are released into the synaptic cleft to transmit information between neurons.

nociceptive able to receive or transmit information about stimuli that damage or threaten to damage tissue.

nociceptor free nerve ending that responds to stimuli that damage or threaten to damage tissue.

nodes of Ranvier interruptions in the myelin sheath that leave small patches of axon unmyelinated. These unmyelinated patches contain a high density of voltage-gated Na^+ channels that contribute to the generation of action potentials.

norepinephrine a neurotransmitter released by axons from the locus ceruleus and medial reticular zone and by postganglionic sympathetic axons except those innervating sweat glands. Binds with α- and β-adrenergic receptors.

nuclei, association thalamic nuclei that connect reciprocally with large areas of cerebral cortex. Association nuclei are found in the anterior thalamus, medial thalamus, and dorsal tier of the lateral thalamus.

nuclei, cochlear site of synapse between first- and second-order neurons involved in hearing. Located laterally at the pontomedullary junction.

nuclei, nonspecific thalamic nuclei that receive multiple types of input and project to widespread areas of cortex. This functional group includes the reticular, midline, and intralaminar nuclei, important in consciousness and arousal.

nuclei, raphe brain stem nuclei that modulate activity throughout the central nervous system. Major source of serotonin. The midbrain raphe nuclei are important in mood regulation and onset of sleep. Pontine raphe nuclei modulate activity in the brain stem and cerebellum. Medullary raphe nuclei modulate activity in the spinal cord via raphespinal tracts. Projections to the spinal cord inhibit transmission of nociceptive information, adjust levels of interneuron activity, and produce nonspecific activation of lower motoneurons.

nuclei, relay thalamic nuclei that receive specific information and serve as relay stations by sending the information directly to localized areas of cerebral cortex. All relay nuclei are found in the ventral tier of the lateral nuclear group.

nuclei, vestibular site of synapse between first- and second-order neu-

rons involved in detecting head movement and head position. Located laterally at the pontomedullary junction.

nucleus collection of nerve cell bodies in the central nervous system.

nucleus accumbens part of the limbic system in the basal forebrain region, where the caudate and putamen blend together.

nucleus basalis of Meynert part of the limbic system in the basal forebrain region, inferior to the preoptic area.

nucleus, Clarke's site of synapse between first- and second-order neurons that convey unconscious proprioceptive information to the cerebellum. The second-order axon is in the posterior spinocerebellar tract. Clarke's nucleus is located in the medial dorsal horn of the spinal cord, from T1 to L2 spinal segments. Syn: nucleus dorsalis.

nucleus cuneatus site of synapse between fasciculus gracilis and medial lemniscus neurons. Relays discriminative touch and conscious proprioceptive information. Located in the dorsal part of the lower medulla.

nucleus dorsalis site of synapse between first- and second-order neurons that convey unconscious proprioceptive information to the cerebellum. The second-order axon is in the posterior spinocerebellar tract. Clarke's nucleus is located in the medial dorsal horn of the spinal cord, from T1 to L2 spinal segments. Syn: Clarke's nucleus.

nucleus gracilis site of synapse between fasciculus gracilis and medial lemniscus neurons. Relays discriminative touch and conscious proprioceptive information. Located in the dorsal part of the lower medulla.

nucleus, inferior olivary nucleus in the upper medulla that receives input from most motor areas of the brain and spinal cord. May alert the cerebellum to errors in movement. Axons from the inferior olivary nucleus project to the contralateral cerebellar hemisphere.

nucleus, lateral cuneate nucleus that receives proprioceptive information from the upper body. Relays unconscious proprioceptive information to the cerebellum, via the cuneocerebellar tract. Located in the dorsolateral medulla.

nucleus, main sensory of trigeminal site of synapse between first- and second-order discriminative touch neurons in the trigeminothalamic pathway.

nucleus, mesencephalic of the trigeminal nerve location of cell bodies of primary afferents conveying proprioceptive information from the muscles of mastication and extraocular muscles.

nucleus, pedunculopontine a nucleus within the caudal midbrain that influences movement via connections with the globus pallidus, subthalamic nucleus, and reticular areas. The neurons produce acetylcholine.

nucleus proprius part of the dorsal gray matter in the spinal cord. Processes proprioceptive and two-point discrimination information. Syn: laminae III and IV.

nucleus, red sphere of gray matter that receives information from the cerebellum and cerebral cortex and projects to the cerebellum, spinal cord (via rubrospinal tract), and reticular formation. Activity in the rubrospinal tract contributes to upper limb flexion.

nucleus, solitary main visceral sensory nucleus. Receives information from the oral cavity and thoracic and abdominal viscera via the vagus, glossopharyngeal, and facial nerves. Involved in regulation of visceral function. Located in the dorsal medulla.

nucleus, spinal trigeminal site of synapse between first- and second-order neurons conveying nociceptive information from the face. Located in the lower pons and medulla.

nucleus, trigeminal main sensory nucleus that receives touch information from the face. The information is transmitted to the ventral posteromedial nucleus of the thalamus, then to the cerebral cortex.

nucleus, ventral posterolateral of the thalamus site of synapse between neurons that convey somatosensory information from the body to the cerebral cortex. The neospinothalamic and medial lemniscus axons end in this nucleus.

nucleus, ventral posteromedial site of synapse between neurons that convey somatosensory information from the face to the cerebral cortex. Located in the thalamus.

nystagmus involuntary back and forth movements of the eyes. Physiological nystagmus is a normal response that can be elicited in an intact nervous system by rotational or temperature stimulation of the semicircular canals or by moving the eyes to the extreme horizontal position. Pathological nystagmus, a sign of nervous system abnormality, is abnormal oscillating eye movements that occur with or without external stimulation.

nystagmus, pathological abnormal oscillating eye movements that occur with or without external stimulation.

obtunded sleeping more than awake, drowsy and confused when awake.

oculomotor complex oculomotor nucleus, supplying efferent somatic fibers to the extraocular muscles innervated by the oculomotor nerve, plus the oculomotor parasympathetic (Edinger-Westphal) nucleus, supplying parasympathetic control

of the pupillary sphincter and the ciliary muscle (adjusts thickness of the lens in the eye).

oligodendrocytes macroglia that form myelin sheaths, enveloping several axons from several neurons. Found within the central nervous system.

olive small oval lump on the anterolateral medulla that lies external to the inferior olivary nucleus.

organ of Corti the organ of hearing, located within the cochlea.

organs, otolithic the utricle and saccule, part of the inner ear. Contain receptors that respond to head position relative to gravity and to linear acceleration and deceleration of the head.

oscillopsia lack of visual stabilization. The world appears to bounce up and down due to failure of the vestibulo-ocular reflex.

outflow, craniosacral parasympathetic nervous system.

outflow, thoracolumbar the sympathetic nervous system.

pain, chronic persistent pain. There are three major types: (1) continuing tissue damage, (2) neuropathic, and (3) chronic pain syndrome. Continuing tissue damage arises from rheumatoid arthritis, reflex sympathetic dystrophy, and other physically identifiable causes. Chronic neuropathic pain is due to abnormal neural activity within the central nervous system. Chronic pain syndrome is pain that persists more than 6 months after normal healing would have been expected. Disuse syndrome may be a contributing factor.

pain, fast discriminative information about stimuli that damage or threaten to damage tissue. Conveyed to cerebral cortex.

pain, myofascial a controversial diagnosis: pressure on sensitive points (called trigger points) reproduces the person's pattern of referred pain. Advocates contend that diagnosis is confirmed when stretch or injecting local anesthetic into the trigger points eliminates the pain.

pain, neospinothalamic discriminative information about stimuli that damage or threaten to damage tissue. Conveyed to cerebral cortex. Syn: fast pain.

pain, neuropathic chronic persistent pain due to abnormal neural activity in various locations in the nervous system.

pain, nociceptive chronic persistent pain due to stimulation of nociceptive receptors.

pain, paleospinothalamic nonlocalized information about stimuli that damage or threaten to damage tissue. Conveyed to cerebral cortex.

pain, referred pain that is perceived as arising in a site different from the actual site producing the nociceptive information.

pain, slow nonlocalized information about stimuli that damage or threaten to damage tissue. Conveyed by divergent pathways to areas in the midbrain and reticular formation and to the medial and intralaminar nuclei of the thalamus. This information reaches widespread areas of the cerebral cortex.

pain, sympathetically maintained syndrome of pain, vascular changes, and atrophy. An aberrant response of the sympathetic nervous system to trauma produces the syndrome.

palsy, Bell's paralysis or paresis of the muscles of facial expression on one side of the face, caused by a lesion of the facial nerve.

palsy, cerebral movement and postural disorder resulting from permanent, nonprogressive damage to the developing brain.

paralysis inability to voluntarily contract muscle(s). Reflexive contraction may be intact if the paralysis is due to an upper motoneuron

lesion. Reflexive contraction is absent if paralysis is due to a complete lower motoneuron lesion.

paralysis, flaccid loss of voluntary movement and muscle tone.

paraphrasia word substitution. Syn: paraphasia.

paraplegia paresis or paralysis of both lower limbs. May also involve part of the trunk.

paravermis part of the cerebellar hemisphere adjacent to the vermis; influences the activity of the lateral activation pathways.

paresis muscle weakness.

paresthesia nonpainful abnormal sensation, often described as pricking and tingling.

parkinsonism a general term for basal ganglia disorders with signs and symptoms characteristic of Parkinson's disease. Includes Parkinson's disease and similar disorders due to drugs, infection, or trauma.

pathway, conscious relay three-neuron series that transmits somatosensory information about location and type of stimulation to the cerebral cortex.

pathway, divergent series of neurons that transmit somatosensory information to the brain stem and cerebrum.

pathway, paleospinothalamic axons that convey nonlocalized nociceptive information to the medial and intralaminar nuclei of the thalamus. The information is then transmitted to limbic and other areas of the cerebral cortex. Involved in arousal, withdrawal, autonomic, and affective responses to pain.

pathway, retinogeniculocalcarine neural connections that convey visual information from the retina to the visual cortex.

pathway, spinomesencephalic series of neurons that transmit nociceptive information to the superior colliculus and to the periaqueductal

gray in the midbrain. Activates parts of the descending pain control system.

pathway, spinoreticular series of neurons that convey nonlocalized nociceptive information to the reticular formation. The information is transmitted to the medial and intralaminar nuclei of the thalamus.

pathway, trigeminoreticulothalamic pathway that conveys slow pain information from the face to the reticular formation, then to the thalamic intralaminar nuclei. The information is finally transmitted to widespread areas of the cerebral cortex.

pathway, unconscious relay neurons that transmit unconscious proprioceptive and other movement-related information to the cerebellum.

pathways, descending motor axons that convey movement-related information from the brain to lower motoneurons in the spinal cord or brain stem.

pathways, fine movement axons involved in the descending control of skilled, voluntary movements.

pathways, nonspecific activating upper motoneurons that influence the general level of activity in lower motoneurons.

pathways, postural/gross movement control automatic skeletal muscle activity.

pathways, unconscious relay neurons that convey proprioceptive information from the spinal cord or information from spinal interneurons to the cerebellum. The information does not reach consciousness and is used to adjust movements.

peduncles, cerebellar bundles of axons that connect the cerebellum with the brain stem. The superior peduncle connects with the midbrain, the middle peduncle with the pons, and the inferior peduncle with the medulla.

peduncles, cerebral the most anterior part of the midbrain, formed by axons descending from the cerebrum to the pons, medulla, and spinal cord. Specifically the corticospinal, corticobulbar, and corticopontine tracts.

period, absolute refractory the time period during an action potential when no stimulus, no matter how strong, will elicit another action potential.

period, critical time that neuronal projections are competing for synaptic sites.

period, relative refractory the time period soon after the peak of an action potential when only a stronger than normal stimulus can elicit another action potential.

perseveration, motor uncontrollable repetition of a movement, usually associated with damage to the supplementary motor area.

plaques patches of demyelination.

plasmapheresis replacement of blood plasma with a plasma substitute, to remove circulating antibodies.

plate, association dorsal section of the neural tube that becomes the dorsal horn in the spinal cord.

plate, neural during development, the thickened ectoderm on the surface of an embryo; becomes the neural tube.

polyneuropathy generalized disorder of peripheral nerves that typically presents distally and symmetrically.

posturography recording of force plate information and electromyograms from postural muscles during postural tests.

potential, action a large change in the electrical potential of a neuron's cell membrane, resulting in the rapid spread of an electrical signal along the cell membrane.

potential, equilibrium the electrical membrane potential at which any diffusible ion is electrically and chemically distributed equally on the two sides of the membrane.

potential, excitatory postsynaptic (EPSP) an electrical depolarization of a neuron's cell membrane. Initiated by the binding of a neurotransmitter to membrane receptors and produced by the instantaneous flow of Na^+, K^+, or Ca^{++} into the cell.

potential, inhibitory postsynaptic (IPSP) an electrical hyperpolarization of a cell membrane. Initiated by the binding of a neurotransmitter to membrane receptors and produced by the instantaneous flow of Cl^- into the cell and/or K^+ out of the cell.

potential, local a small change in the electrical potential of a neuron's cell membrane that is graded in both amplitude and duration.

potential, resting membrane the difference in electrical potential across the cell membrane of a neuron when the neuron is neither receiving nor transmitting information, i.e., the electrical state of a neuron's cell membrane when the cell is at rest (neither electrically excited nor inhibited).

potentials, auditory evoked a method of testing brain stem function by auditory stimulation combined with recording electrical potentials from the scalp.

potentials, receptor local potentials generated at the receptor of a sensory neuron.

potentials, synaptic local potentials generated at a postsynaptic membrane.

potentiation, long-term (LTP) a cellular mechanism for memory that results from the synthesis and activation of new proteins and the growth of new synaptic connections.

proprioception, conscious awareness of the movements and relative position of body parts.

propriospinal within the spinal cord. Usually refers to neurons that are located entirely within the spinal cord.

pyramids ridges on the anteroinferior medulla, formed by the lateral corticospinal tracts.

radiculopathy lesion of a dorsal or ventral nerve root. Clinical use of the term may refer to a spinal nerve lesion.

ramus, dorsal primary branch of a spinal nerve that innervates the paravertebral muscles, posterior parts of the vertebrae, and overlying cutaneous areas.

ramus, ventral primary branch of a spinal nerve that innervates the skeletal, muscular, and cutaneous areas of the limbs and/or of the anterior and lateral trunk.

receptor, muscarinic receptor on an organ innervated by postganglionic parasympathetic neuron. Acetylcholine binding to muscarinic receptors initiates a G-protein mediated response.

receptor, phasic a sensory nerve ending that adapts to a constant stimulus and stops responding.

receptor, tonic a sensory nerve ending that responds as long as a stimulus is present.

receptors, adrenergic receptors in the sympathetic nervous system that respond to norepinephrine or epinephrine or to adrenergic drugs. Subtypes are α- and β-adrenergic receptors.

receptors, cholinergic receptors that respond to acetylcholine. Found in the autonomic and central nervous systems and on the motor end plate in skeletal muscle membranes. Subtypes include nicotinic and muscarinic.

receptors, nicotinic receptors on postsynaptic neurons in autonomic ganglia. Acetylcholine binding to nicotinic receptors causes a fast excitatory postsynaptic potential in the postsynaptic membrane. Also found on the motor end plate of skeletal muscle.

reflex   an involuntary response to an external stimulus.

reflex, consensual   constriction of the pupil in the opposite eye when a bright light is shined into one eye. The optic nerve is the afferent (sensory) limb of the reflex, while the oculomotor nerve provides the efferent (motor) limb.

reflex, crossed extension   extension of the opposite lower limb when one lower limb is moved away from a stimulus.

reflex, H-   reflexive muscle contraction elicited by electrically stimulating the skin over a peripheral nerve. Used to assess the degree of excitation of alpha motoneurons.

reflex, phasic stretch   muscle contraction in response to quick stretch. Syn: myotatic reflex, muscle stretch reflex, deep tendon reflex.

reflex, pupillary   pupil constriction in the eye directly stimulated by a bright light. The optic nerve is the afferent (sensory) limb of the reflex, while the oculomotor nerve provides the efferent (motor) limb.

reflex, tendon organ   pressure on Golgi tendon organs activates neurons that inhibit the alpha motoneurons to the muscle associated with the stretched tendon.

reflex, vestibulo-ocular   automatic movements of the eyes that stabilize visual images during head and body movements.

reflex, withdrawal   movement of a limb away from a stimulus.

resistance, axoplasmic (Ra)   the ability of the cytoplasm inside a neuron to resist the movement of ionic charges.

resistance, membrane (Rm)   the ability of the neuron's cell membrane to resist the movement of ionic charges.

response, clasp-knife   when a spastic muscle is slowly and passively stretched, resistance to stretch is suddenly inhibited at a specific point in the range of motion.

response, long loop   the second response after stretch of a muscle, probably involving neural circuits in the brain stem. Also called M2.

response reversal   modification of ongoing motor activity to adapt the movement to environmental conditions. For example, if one catches a foot under an object while walking, the foot is moved to clear the object rather than continue to collide with the object.

rhizotomy, dorsal   surgical severance of selected dorsal roots. Purpose is to decrease pain or to decrease hyperflexia.

rigidity   velocity-independent muscle hypertonia.

saccade   high-speed eye movement.

saccule   part of the inner ear that contains receptors that respond to head position relative to gravity and to linear acceleration and deceleration of the head.

schizophrenia   group of disorders consisting of disordered thinking, delusions, hallucinations, and social withdrawal.

sclerosis, amyotrophic lateral   a disease that destroys only the lateral activating pathways and anterior horn cells in the spinal cord, thus producing upper and lower motoneuron signs.

sclerosis, multiple   a disease characterized by random, multifocal demyelination limited to the central nervous system. Signs and symptoms include numbness, paresthesias, Lhermitte's sign, asymmetrical weakness, and/or ataxia.

sclerotome   during development, the part of a somite that becomes the vertebrae and skull.

section, basilar   anterior part of the brain stem, containing predominantly motor system structures.

sensitize   to make neurons fire with less stimulation than is usually required.

serotonin   a neurotransmitter released by axons from the raphe nuclei. See *nuclei, raphe* for a summary of functions.

sheath, myelin   a covering of fat and protein that surrounds axons.

sign, Babinski's   reflexive extension of the great toe, often accompanied by fanning of the other toes. The sign is elicited by firm stroking of the lateral sole of the foot, from the heel to the ball of the foot, then across the ball of the foot.

sign, Lhermitte's   radiation of a sensation like electrical shock down the back or limbs, elicited by neck flexion.

sign, Tinel's   a sensation of pain or tingling in the distal distribution of a peripheral nerve, elicited by tapping on the skin over an injured nerve.

sinus, dural   spaces between layers of dura mater that collect venous blood.

soma   cell body, the metabolic center of a cell.

somite   during development, the part of the mesoderm that will become dermis, bone, and muscle.

spasm, muscle   sudden, involuntary contraction of muscle fibers.

spasticity   velocity-dependent muscle hypertonia.

spina bifida   a developmental defect resulting from failure of the inferior part of the neural tube to close.

spinal cord injury, complete   lack of sensory and motor function in the lowest sacral segment (American Spinal Cord Injury Association definition).

spinal cord injury, incomplete   preservation of sensory and/or motor function in the lowest sacral segment (American Spinal Cord Injury Association definition).

spindle, muscle   sensory organ embedded in muscle that responds to stretch of the muscle.

spinocerebellum functional name for the vermis and paravermal region of the cerebellum. Controls ongoing movements.

spondylosis, cervical degeneration of the cervical vertebrae and disks that produces narrowing of the vertebral canal and intervertebral foramina.

sprouting the regrowth of damaged axons.

sprouting, collateral reinnervation of a denervated target by branches of intact axons.

sprouting, regenerative injured axon sends out side sprouts to a new target.

stage, embryonic developmental stage lasting from the second to the end of the eighth week in utero; during this time, the organs are formed.

stage, fetal developmental stage lasting from the end of the eighth week in utero until birth; nervous system continues to develop, and myelination begins.

stage, pre-embryonic developmental stage lasting from conception to the second week in utero.

state, vegetative complete loss of consciousness, without alteration of vital functions.

stereognosis ability to use touch and proprioceptive information to identify an object.

stream, action stream of visual information that flows dorsally and is used to direct movements.

stream, perception stream of visual information that flows ventrally and is used to recognize visual objects.

stroke sudden onset of neurological deficits due to disruption of the blood supply in the brain. Syn: cerebrovascular accident (CVA), brain attack.

stroke, completed neurological deficits resulting from vascular disorders affecting the brain that persist more than one day and are stable (not progressing or improving).

stroke, progressive neurological deficits, resulting from vascular disorders, that increase intermittently over time. Progressive strokes are believed to be due to repeated emboli or continued formation of a thrombus in the brain.

stupor condition of being arousable only by strong stimuli, such as strong pinching of the Achilles tendon.

substance P neurotransmitter produced by primary nociceptive neurons. Also produced in other areas of the central nervous system.

substantia gelatinosa part of the dorsal gray matter in the spinal cord. Involved in processing nociceptive information. Syn: lamina II.

substantia nigra one of the nuclei in the basal ganglia circuit, located in the midbrain. The compacta part provides dopamine to the caudate nucleus and putamen. The reticularis part serves as one of the output nuclei for the basal ganglia circuit.

subthalamus part of the basal ganglia circuit, involved in regulating movement. The subthalamus facilitates the basal ganglia output nuclei. The subthalamus is located superior to the substantia nigra of the midbrain.

summation, spatial the cumulative effect of receptor or synaptic potentials occurring simultaneously in different regions of the neuron.

summation, temporal the cumulative effect of a series of either receptor potentials or synaptic potentials that occur within milliseconds of each other.

syncope fainting. Loss of consciousness due to an abrupt decrease in blood pressure that deprives the brain of adequate blood supply.

syndrome, anterior cord signs and symptoms produced by interruption of ascending spinothalamic tracts, descending motor tracts, and damage to the somas of lower motoneurons. This spinal cord syndrome interferes with pain and temperature sensation and with motor control.

syndrome, Brown-Séquard's signs and symptoms produced by a hemisection of the spinal cord. Segmental losses are ipsilateral and include loss of lower motoneurons and all sensations. Below the level of the lesion, voluntary motor control, conscious proprioception, and discriminative touch are lost ipsilaterally, and temperature and nociceptive information are lost contralaterally.

syndrome, cauda equina signs and symptoms produced by damage to the lumbar and/or sacral nerve roots, causing sensory impairment and flaccid paralysis of lower limb muscles, bladder, and bowels.

syndrome, central cord signs and symptoms produced by interruption of spinothalamic fibers crossing the midline, producing loss of pain and temperature sensation at the involved segments. Larger lesions also impair upper limb motor function because the lateral corticospinal tracts to the upper limb are located in the medial part of the white matter and because the lesion typically occurs in the cervical region.

syndrome, chronic pain physiological impairment consisting of muscle guarding, abnormal movements, and disuse syndrome.

syndrome, Horner's drooping of the upper eyelid, constriction of the pupil, and vasodilation with absence of sweating on the ipsilateral face and neck. Due to lesions of the cervical sympathetic chain or its central pathways.

syndrome, locked-in complete inability to move, despite intact consciousness. Due to damage to descending activating pathways.

syringomyelia rare, progressive disorder. A syrinx, or fluid-filled cavity, develops in the spinal cord, almost always in the cervical region. Segmental signs occur in the upper limbs, including loss of sensitivity to pain and temperature stimuli. Upper motoneuron signs in lower limbs include paresis, muscle hypertonicity, and phasic stretch hyperreflexia. Often loss of bowel and bladder control also occurs.

system, consciousness neural connections governing alertness, sleep, and attention. Includes the reticular formation, the ascending reticular activating system, the basal forebrain (anterior to the hypothalamus), the thalamus, and the cerebral cortex.

system, dorsal column/medial lemniscus pathway that transmits information about discriminative touch and conscious proprioception to the cerebral cortex.

system, lateral activation upper motoneurons that influence the activity of lower motoneurons innervating limb muscles. Includes the lateral corticospinal, rubrospinal, and lateral reticulospinal tracts.

system, limbic group of structures involved in emotions, processing of declarative memories, and autonomic control. Includes parts of the hypothalamus, thalamus, limbic cortex (cingulate gyrus, parahippocampal gyrus, uncus), hippocampus, amygdala, and the basal forebrain (septal area, preoptic area, nucleus accumbens, and the nucleus basalis of Meynert).

system, medial activation upper motoneurons that influence the activity of lower motoneurons innervating postural and girdle muscles.

tectum part of the midbrain posterior to the cerebral aqueduct, consisting of the pretectal area and the superior and inferior colliculi. In-

volved in reflexive movements of the eyes and head.

tegmentum posterior part of the brain stem, including sensory nuclei and tracts, reticular formation, cranial nerve nuclei, and the medial longitudinal fasciculus.

terminal, postsynaptic the membrane region of a cell containing receptor sites for neurotransmitter.

terminal, presynaptic the end projection of an axon, specialized for releasing neurotransmitter into the synaptic cleft.

tetraplegia impairment of arm, trunk, lower limb, and pelvic organ function, usually due to damage involving the cervical spinal cord.

theory, counterirritant theory that inhibition of nociceptive signals by stimulation of non-nociceptive receptors occurs in the dorsal horn of the spinal cord.

thermoreceptor a receptor that responds to changes in temperature.

threshold 1. the least amount of stimulation that can be perceived when testing sensation. 2. the minimum stimulus necessary to produce action potentials in an axon.

thrombus a blood clot within the vascular system.

tinnitus the sensation of ringing in the ear.

tomography, positron emission (PET scan) computer-generated image based on metabolism of injected radioactively labeled substances. The PET scan records local variations in blood flow, reflecting neural activity.

tone, muscle amount of tension or electrical activity in resting muscle.

touch, discriminative localization of touch and vibration, and the ability to discriminate between two closely spaced points touching the skin.

tract, anterior spinocerebellar axons that transmit information about the activity of spinal interneurons and of descending motor signals

from the cerebral cortex and brain stem. The neurons arise in the thoracolumbar spinal cord and end in the cerebellar cortex. The information does not reach consciousness and is used to adjust movements.

tract, ceruleospinal axons originating in the locus ceruleus that enhance activity in spinal interneurons and motoneurons. The effects of ceruleospinal activity are generalized, not related to specific movements. Other ceruleospinal neurons inhibit the nociceptive pathway neurons in the dorsal horn.

tract, cuneocerebellar axons that transmit high-fidelity, somatotopically arranged tactile and proprioceptive information from the upper half of the body to the cerebellar cortex. The information does not reach consciousness and is used to adjust movements.

tract, dorsolateral white matter dorsal to the dorsal horn in the spinal cord. Axons of first-order nociceptive neurons ascend or descend in this tract before synapsing in the dorsal horn lamina. Syn: zone of Lissauer.

tract, geniculocalcarine axons conveying visual information from the lateral geniculate body of the thalamus to the visual cortex.

tract, internal feedback axons of neurons that monitor the activity of spinal interneurons and of descending motor signals from the cerebral cortex and brain stem. The information is transmitted to the cerebellum. The information does not reach consciousness and is used to adjust movements.

tract, lateral corticospinal axons that arise in motor planning areas of the cerebral cortex and the primary motor cortex and synapse with lower motoneurons that innervate limb muscles. Essential for fractionated hand movements.

tract, lateral (medullary) reticulospinal axons originating in the medullary reticular formation that descend bilaterally to facilitate flexor muscle motoneurons and to inhibit extensor muscle motoneurons.

tract, lateral vestibulospinal axons arising in the lateral vestibular nucleus that project ipsilaterally to facilitate lower motoneurons to extensor muscles and simultaneously inhibit lower motoneurons to flexor muscles via interneurons.

tract, medial corticospinal axons that convey information from motor areas of the cerebral cortex to the spinal cord. The axons end in the cervical and thoracic cord and influence the activity of lower motoneurons that innervate neck, shoulder, and trunk muscles.

tract, medial reticulospinal axons that project from the pontine reticular formation to the spinal cord. Activation of this tract facilitates ipsilateral lower motoneurons innervating postural muscles and limb extensors.

tract, medial vestibulospinal axons arising in the medial vestibular nucleus that project bilaterally to the cervical and thoracic spinal cord. Affect the activity of lower motoneurons controlling neck and upper back muscles.

tract, neospinothalamic axons of second-order nociceptive neurons that convey localized pain information from the spinal cord to the ventral posterolateral nucleus of the thalamus. Part of the discriminative pain and temperature conscious relay pathway to the cerebral cortex.

tract, optic axons conveying visual information from the optic chiasm to the lateral geniculate body of the thalamus.

tract, posterior (dorsal) spinocerebellar transmits high-fidelity, somatotopically arranged tactile and proprioceptive information from Clarke's nucleus (information from the lower half of the body) to the cerebellar cortex. The information does not reach consciousness and is used to adjust movements.

tract, raphespinal 1. axons originating in the raphe nuclei that enhance activity in spinal interneurons and motoneurons. The effects of raphespinal activity are generalized, not related to specific movements. 2. axons originating in the raphe nuclei that inhibit the transmission of nociceptive information in the spinal cord.

tract, rostrospinocerebellar axons that transmit information about the activity of spinal interneurons and of descending motor signals from the cerebral cortex and brain stem. The neurons arise in the cervical spinal cord and end in the cerebellar cortex. The information does not reach consciousness and is used to adjust movements.

tract, rubrospinal axons that originate in the red nucleus of the midbrain, cross to the opposite side, then descend to synapse with lower motoneurons primarily innervating upper limb flexor muscles.

tract, tectospinal axons that project from the superior colliculus to synapse with lower motoneurons in the cervical spinal cord. Involved in reflexive movements of the head toward stimuli.

tracts, corticobulbar axons that convey motor signals from the cerebral cortex to cranial nerve nuclei in the brain stem.

tracts, high-fidelity two groups of axons, the posterior spinocerebellar and cuneocerebellar, that relay accurate, detailed, somatotopically arranged tactile and proprioceptive information from the spinal cord to the cerebellar cortex. The information does not reach consciousness and is used to adjust movements.

tracts, spinocerebellar groups of axons that convey proprioceptive information or information from spinal interneurons to the cerebellum. The information does not reach consciousness and is used to adjust movements.

transmission, ephaptic cross-excitation of axons, due to loss of myelin. Excitation of one axon induces activity in a parallel axon.

transport, anterograde movement of proteins and neurotransmitters from the soma to the axon.

transport, retrograde movement of some substances from the axon back to the soma for recycling.

tremor, action shaking of a limb during voluntary movement.

tremor, resting repetitive alternating contraction of the extensor and flexor muscles of the distal extremities during inactivity. The tremor diminishes during voluntary movement. A classic resting tremor is the movement of the hands as if using the thumb to roll a pill along the fingertips (pill-rolling tremor); characteristic of parkinsonism.

uncus most medial part of the parahippocampal gyrus.

unit, motor alpha motoneuron and the muscle fibers it innervates.

unmasking of silent synapses the disinhibition or reactivation of functional synapses that are unused unless injury to other pathways necessitates their activation.

utricle part of the inner ear that contains receptors that respond to head position relative to gravity and to linear acceleration and deceleration of the head.

ventral root the efferent (motor) root of a spinal nerve.

ventricle a space in the brain that contains cerebrospinal fluid. The lateral ventricles are within the cerebral hemispheres, the third ventricle is in the midline of the diencephalon, and the fourth ventricle is located between the pons and medulla anteriorly and the cerebellum posteriorly.

vermis   the midline part of the cerebellum, involved in controlling ongoing movements and posture via the brain stem descending pathways.

vertigo   an illusion of motion, common in vestibular disorders.

vertigo, benign paroxysmal positional   acute onset of vertigo provoked by change of head position that quickly subsides even if the provoking head position is maintained.

vessels, capacitance   vessel whose relaxed walls expand to contain more blood. Blood pools in these vessels.

vestibulocerebellum   functional name for the flocculonodular lobe of the cerebellum. Influences the activity of eye movements and postural muscles.

zoster, varicella   infection of a dorsal root ganglion or cranial nerve ganglion with varicella-zoster virus.

Index

Note: Page numbers in *italics* refer to illustrations; page numbers followed by t refer to tables.

A

Abducens nerve (VI), 10t, *250, 251–255, 251t, 254*
 disorders of, 268
 testing of, 272t
Abducens nucleus, *254*
Absolute refractory period, 33
Accessory nerve (XI), 10t, *250,* 251t, *266, 266*
 disorders of, 270
 testing of, 273t
Accommodation, ocular, 252t, 255, *256*
Acetylcholine, 135–136, *136,* 145t, 383
 in consciousness, *338*
 pedunculopontine nucleus production of, *282, 283*
Acetylcholine receptors, antibodies to, 54
Action potential, 29, *30,* 31t, 33–34, *34, 35*
 all-or-none nature of, 33
 propagation of, 35–37, *36*
 saltatory conduction of, 36–37, *36*
 threshold stimulus intensity for, 33
Action tremor, 192
Adenosine monophosphate, cyclic (cAMP), as second messenger, *52, 52*
Adrenal medulla, autonomic nervous system regulation of, 144t
 sympathetic efferents to, 137, *138*
Adrenergic receptors, 136, *136,* 140, *140,* 144t, 145t
Afferent, definition of, 26
Agnosia, 343, *344*
Agonist, 53, 140
Agraphia, 347
Alexia, 347
Allodynia, 123
Alpha–gamma coactivation, 152
Alzheimer's disease, 351
Ambulation, 196
γ-Aminobutyric acid (GABA), 383
Amnesia, 347
Ampullae, *262*
Amputation, phantom limb sensation in, 124
Amygdala, *13,* 327, *329*
 in social behavior, 330, *331*
 lesions of, 346–347

Amyotrophic lateral sclerosis, 185, *185,* 186
Analgesia, endogenous, 119–120, *119*
 stress-induced, 120
Anencephaly, stepping reflex in, 159
Aneurysm, berry, 375, *375*
Angiography, in cerebral blood flow evaluation, 377, *377*
Anosognosia, 348
Antagonist, definition of, 53
Anterior cord syndrome, 235, *236*
Anterolateral column, of conscious relay pathway, 97–100, *99, 101*
 vs. dorsal column, 100
Antibodies, acetylcholine receptor, 54
 calcium channel, 54
Aphasia, 347
 Broca's, 344, *345,* 347, 349t
 conduction, 348, 349t
 global, 348, 349t
 Wernicke's, 348, 349t
Appendicitis, autonomic nervous system in, 134, *134*
Apraxia, 344
Arachidonic acid, as second messenger, 52, *52*
Arachnoid, 14, 365, *366*
Arnold-Chiari malformation, 78–79, *78,* 83t
Arteriovenous malformation, 375
Association areas, of cerebral cortex, 326–327, *328*
Association fibers, of subcortical white matter, 319, 321t
Association nuclei, of thalamus, 317, *317,* 318t
Association plate, 73, *73*
Astereognosis, 343–344, *344*
Astrocyte, 20–21, *22*
Ataxia, 191
Attention deficit disorder, 350
Auditory agnosia, 344, *344*
Auditory association area, 300, 324, *325*
Auditory cortex, 300, *300,* 324, *325*
 lesions of, 309, 343, *343*
Auditory evoked potentials, 294
Auditory system, 300, *300, 301,* 324, *325*
 disorders of, 308–309, 343, *343*
Autogenic inhibition, 154, *155*
Autonomic dysreflexia, in spinal cord injury, 241, *242*
Autonomic nervous system, 131–147, *132*
 adrenergic neurons of, 136, *136*
 adrenergic receptors of, 136, *136*

Autonomic nervous system (*continued*)
  afferent pathways of, 132, *133*
  cholinergic neurons of, 135–136, *136*
  cholinergic receptors of, 135–136, *136*
  efferent pathways of, 135–136, *136*
  in brain stem injury, 143
  in cerebral injury, 144
  in pain syndrome, 144–145
  in peripheral injury, 143
  in spinal cord injury, 143, 241, *242*
  in syncope, 145–146
  in visceral regulation, 132–135, *134*
  muscarinic receptors of, 136, *136*
  neurotransmitters of, 135–136, *136*
  nicotinic receptors of, 136, *136*
  parasympathetic, 141–143. See also *Parasympathetic nervous system.*
  sensory receptors of, 132
  sympathetic, 137–141, *137, 138, 140.* See also *Sympathetic nervous system.*
  tests of, 146
  vs. somatic nervous system, 135
Autonomic system, 4
Axon(s), 23, *24.* See also *Neuron(s).*
  action potential propagation along, *30,* 35–37, *36*
  anterograde transport of, 25, *26*
  axoplasmic transport of, 25, *26*
  diameter of, 87, 92, 93t
  injury to, predevelopmental, 78
    recovery from, 59–64, *60–63*
  myelination of, 20, 78
  of peripheral nervous system, 87, *87,* 209, 209t
  propriospinal, 225
  retraction of, 77–78
  retrograde transport of, 25, *26*
  sprouting of, 60, *61*
  unmyelinated, 20
Axon hillock, 23, *24*
Axoplasmic resistance, in action potential propagation, 35

B
Babinski's sign, 178–179, *179*
Baclofen, in upper motoneuron lesion treatment, 184
Basal ganglia, 11–12, 164–166, *165,* 165t, *166,* 319–320
  disorders of, 185–189, *186–189,* 343
  function of, 166
  functional connections of, 165, *166*
  in motor learning, 335
  neurotransmitters of, 166
Basilar artery, 368, *369*
  occlusion of, 373–374

Basilar membrane, 261, *263*
Basis pedunculi, *287,* 288
Behavior, emotion and, 330, *331*
  immune system and, 330–333, *331, 332*
Behavioral neuroscience, 3
Bell's palsy, 268, 269
Benign paroxysmal positional vertigo, 309, 309t
Berry aneurysm, 375, *375*
Bitemporal hemianopsia, 310–311, *311*
Bladder, autonomic nervous system regulation of, 144t
  dysfunction of, 237–238, *237*
  function of, 231–232, *233*
Blindness, 267
  cortical, 311
Blood flow, cerebral, 368, 370–371, *370–372,* 376–377, *377*
  disorders of, 371–374, *373–375.* See also *Stroke.*
  laboratory evaluation of, 377, *377, 378*
  in skeletal muscle, 139–140
Blood pressure, measurement of, 146
Blood supply, 368–378
  disorders of, 371–374
  to brain stem, 368, *369*
  to cerebellum, 368, *369*
  to cerebral hemispheres, 368, 370–371, *370–372*
  to fascicle, 368, *368*
  to peripheral nerves, 368, *368*
  to spinal cord, 368, *369*
Blood vessels, stretch of, 132, *133*
  sympathetic regulation of, 144t
Blood-brain barrier, 376
Body schema, 336
Body temperature, dysregulation of, in spinal cord injury, 241, *242*
  regulation of, 139
Botulinum toxin, in upper motoneuron lesion treatment, 184
Botulinum toxin A (Botox), 53
Botulism, 217
Bowel, autonomic nervous system regulation of, 144t
  dysfunction of, 238
  spinal control of, 232
Brachial plexus, 209, *210*
Brain. See also *Cerebral cortex.*
  blood flow of, 376–377, *377, 378*
  blood supply to, 368, 370–371, *370–372*
    disorders of, 371–374, *373–375.* See also *Stroke.*
  conscious relay pathways to, 92, 93t, 94–100, *95, 97–99, 101*
  divergent pathways to, 93t, 94, 101–102
  edema of, 376

Brain (continued)
    formation of, 74–76, 75–77
    hemorrhage within, 373, 373
    ischemia of. See Stroke.
    lacunar infarcts of, 372, 373
    space-occupying lesions of, 376–377, 377
    traumatic injury to, 351–352
    unconscious relay pathways to, 93t, 94, 102–104, 103
    venous system of, 377–378, 378
Brain stem, 4, 4, 278–295
    anatomy of, 8–10, 8, 9, 10t, 278–280, 278–280, 279t
    basilar section of, 278, 280
    blood supply to, 368, 369
        disorders of, 373–374
    compression of, 294
    cranial nerve nuclei in, 279
    disorders of, 289–294, 290–294, 294t
    in consciousness, 284, 290, 294, 294, 294t
    in intrinsic analgesia, 119–120, 119
    lesions of, 117–118, 117, 143
    longitudinal sections of, 278, 280
    medulla of, 284–286, 285. See also Medulla.
    midbrain of, 288. See also Midbrain.
    neurotransmitter production in, 282–284, 283, 338
    pons of, 286–288, 287. See also Pons.
    reticular formation of, 280–281, 281, 300, 300
    reticular nuclei of, 282–284, 283
    tectum of, 280, 280, 288
    tegmentum of, 280, 287, 288
    vertical tracts of, 278, 279t, 280
        disorders of, 289–290, 290–293
Broca's aphasia, 344, 345, 347, 349t
Broca's area, 164, 326, 326, 335, 336
Brodmann's areas, 321, 323
Brown-Séquard syndrome, 235, 236

C
C fibers, sensitization of, 100
Calcium, in neuronal excitotoxicity, 64–65, 64
    inositol triphosphate–stimulated release of, 52–53
Calcium channel, antibodies to, 54
Calculation, evaluation of, 355t
Callostomy, 342
Cancer, Lambert-Eaton syndrome in, 54
Capacitance vessels, 139–140
Carotid artery (arteries), 14, 15
    internal, 368, 370
Carpal tunnel syndrome, 213, 215
Cats, motor system in, 157–158, 157, 158
Cauda equina syndrome, 235, 236
Caudate nucleus, 165, 165

Causalgia, 145
Cellular neuroscience, 2–3
Central cord syndrome, 235, 236
Central nervous system, 4–5, 4. See also Brain; Spinal cord.
Cerebellum, 4, 4, 10, 10, 75, 166–170, 288–289
    anatomy of, 10, 10, 166–168, 167, 168
    blood supply to, 368, 369
    formation of, 74–76, 75
    functions of, 168–170, 169, 170t, 171, 288–289
    high-fidelity pathways to, 102–103, 103
    internal feedback pathways to, 103–104, 103
    lesions of, 191–192, 192t
    lobes of, 167, 168
    unconscious relay pathways to, 102–104, 103
    vertical sections of, 167–168, 168
Cerebral artery (arteries), 14, 15, 368, 369, 370–371, 371
    anterior, 371
        occlusion of, 374
    middle, 370, 371
        occlusion of, 374
    occlusion of, 374
    posterior, 370, 371
        internal, 368, 370
        occlusion of, 374
Cerebral blood flow, 368, 370–371, 370–372, 376–377, 377, 378
    disorders of, 371–374, 373–375. See also Stroke.
    laboratory evaluation of, 377, 377, 378
Cerebral cortex, 11, 320–327, 322. See also Brain.
    association areas of, 326–327, 328
    Brodmann's areas of, 321, 323
    communication and, 335, 336, 347–348, 349t
    functional reorganization of, 61–63, 63
    functions of, 322–327, 324–326
    in consciousness, 337, 338, 350
    layers of, 320–321, 322, 323t
    lesions of, 343–346, 343–346
    mapping of, 321–322
    memory and, 333, 333–335, 334
    motor planning areas of, 326, 326, 344
    prefrontal, 345
    primary motor areas of, 324–326, 326, 344–345, 345
    primary sensory areas of, 323–324, 325, 343, 343
    pyramidal cells of, 320, 322
    sensory association areas of, 324, 325, 343–344, 344
    spatial relationships and, 336, 350
    visual association cortex of, 336–337, 337, 350
Cerebral dominance, 335
Cerebral hemispheres. See also Cerebral cortex.
    anatomy of, 11–12, 12
    formation of, 76, 77

Cerebral palsy, 81–82, 83, 83t
  ataxic, 81
  athetoid, 81
  choreoathetotic, 191
  spastic, 81, 183, *183,* 243
Cerebral peduncle, *8*
Cerebral veins, 377–378, *378*
Cerebrocerebellum, 169–170, *169, 170t, 171*
Cerebrospinal fluid, 5, 14
  circulation of, 366, *366*
  formation of, 366
  shunt drainage of, 367, *367*
Cerebrospinal fluid system, 364–368
  craniosacral therapy for, 367–368
  disorders of, 366–368, *367*
  ventricles of, 364–366, *365*
Cerebrovascular accident. See *Stroke.*
Cerebrum, 4–5, *4,* 316–338, *316*
  anatomy of, 10–12, *11, 12*
  association fibers of, 319, *321,* 321t
  basal ganglia of, 319–320. See also *Basal ganglia.*
  blood supply to, 368, 370–371, *370–372*
  cerebral cortex of, 320–327. See also *Cerebral cortex.*
  commissural fibers of, 319, *321,* 321t
  diencephalon of, 316–319, *316, 317,* 318t, *319*
  disorders of, 350–356. See also *Stroke.*
    evaluation of, 354t–355t, 356
    signs of, 342–350, *343–346,* 349t
  edema of, 376
  epithalamus of, 318
  evaluation of, 354t–355t, 356
  gray matter of, 6, 7
  hypothalamus of, 317–318. See also *Hypothalamus.*
  limbic system of, 327–330, *329.* See also *Limbic system.*
  projection fibers of, 319, *320, 321,* 321t
  subcortical structures of, 319–320, *320, 321,* 321t
    lesions of, 342
  thalamus of, 316–317, *317,* 318t. See also *Thalamus.*
  watershed area of, 370–371
    ischemia in, 374, *375*
  white matter of, 6, 7
Ceruleospinal tract, 164, 228t
Cervical ganglia, 137–139, *138*
Cervical spondylosis, 244
Cervicothoracic ganglion (stellate ganglion), 137, *138*
  lesions of, 143
Chemoreceptors, 86, 132
Cholinergic receptors, 135–136, *136*
Choroidal arteries, 371, *372*
  occlusion of, 374, *375*
Chromatolysis, central, 60

Chronic pain syndrome, 124–126, *125*
  fibromyalgia in, 125
  myofascial pain in, 125
  psychological factors in, 125–126
Cingulate cortex, *13*
Cingulate gyrus, 327, *329*
Circle of Willis, 370, *370*
Clasp-knife response, 180
Clonus, 180
Clozapine, 384
Cocaine, in utero exposure to, 80, 83t
Cochlea, 261, *261, 263*
Cochlear duct, 261, *263*
Cochlear nucleus, 286
Cocontraction, electromyographic definition of, 181
Cognitive neuroscience, 3
Colliculus, inferior, 288, 300, *300*
  superior, 160, 288, 307
Colorectal cancer, stress response and, 332
Coma, 294t
Commissural fibers, 11, *12,* 319, *321,* 321t
  lesions of, 342
  surgical severance of, 342
Communication, 335, *336*
  disorders of, 347–348, 349t
Computerized axial tomography (CAT), 2
Conduction aphasia, 348, 349t
Conductive deafness, 308
Conscious proprioception, 94–96, *95, 97*
Conscious relay somatosensory pathway, 92, 93t, 94–100, *95, 97–99, 101*
Consciousness, 284, 337, *338*
  disorders of, 290, 294, *294,* 294t, 350
  evaluation of, 354t
Consensual reflex, 254–255, *255*
Convergence, neuronal, 37, *37*
Corneal reflex, 252t, 259
Coronal plane, 5, *6*
Corpus callosum, 11, 319, *321,* 321t
  surgical severance of, 342
Corticobulbar fibers, 161, *163*
Corticobulbar pathways, *163*
Corticobulbar tract, 279t, *280*
  disorders of, 289, *290*
Corticopontine tract, 279t, *280*
Corticoreticular tract, 279t, *280*
Corticospinal tract, 279t, *280*
  lateral, 161, *162,* 228t
  lesions of, *291, 293*
  medial, *160,* 161, 228t
Cortisol, in disease, 332

Counterirritant theory of pain, 119, *119*
Cramps, 176
Cranial nerve(s), 9–10, *9,* 10t, 249–273, *250*
   I (olfactory), 10t, *250,* 251, 251t
      disorders of, 267
      testing of, 271t
   II (optic), 10t, *250,* 251, 251t, *253, 254t*
      disorders of, 267
      testing of, 271t
   III (oculomotor), 10t, *142, 250,* 251–255, 251t, *254*
      disorders of, 267–268, *293*
      parasympathetic fibers of, 253–254
      testing of, 271t
   IV (trochlear), 10t, *250,* 251–255, 251t, *254*
      disorders of, 268
      testing of, 271t
   V (trigeminal), 10t, *250,* 251t, 255–259, *257, 258*
      branches of, 255, *257*
      disorders of, 268
      testing of, 272t
   VI (abducens), 10t, *250,* 251–255, 251t, *254*
      disorders of, 268
      testing of, 272t
   VII (facial), 10t, *142, 250,* 251t, 259, *260*
      disorders of, 268, 289, *290, 292*
      testing of, 272t
   VIII (vestibulocochlear), 10t, *250,* 251t, 259–261, *261–263*
      disorders of, 269–270
      testing of, 272t–273t
   IX (glossopharyngeal), 10t, *142, 250,* 251t, 261, *263, 265*
      disorders of, 270
      testing of, 273t
   X (vagus), 10t, *142, 250,* 251t, 265, *265*
      disorders of, 270, *292*
      testing of, 273t
   XI (accessory), 10t, *250,* 251t, 266, *266*
      disorders of, 270
      testing of, 273t
   XII (hypoglossal), 10t, *250,* 266, *266*
      disorders of, 270
      testing of, 273t
   brain stem locations of, *279*
   functions of, 250
   infection of, 116, *116*
   testing of, 270, 271t–273t
Craniosacral therapy, 367–368
Critical period, 82
Crossed extension reflex, 229, *229*
Cuneate nucleus, *95,* 96, 103

Cuneocerebellar pathway, 103, *103*
Cupula, *262*

D
Deafferentation, experimental, 108
Deafness, 308–309
Decision making, emotion and, 330
Declarative memory, 333–334, *333, 334,* 347
   evaluation of, 355t
Deep tendon reflex, 153–154, *153*
Defecation, dysfunction of, 238
   spinal control of, 232
Delirium, 294t
Dementia, 351
Demyelination, in Guillain-Barré syndrome, *37, 38*
   in multiple sclerosis, 39, *39,* 41, 244
Dendritic spines, defects in, 81
Depression, 347, 385
Dermatomes, 72, *73,* 74, 88, *89*
Diabetic polyneuropathy, 215–217, *215, 216*
Diagnosis, in neurological evaluation, *16,* 17
Diencephalon, *75, 76,* 316–319, *316, 317,* 318t, *319*
   anatomy of, 10–11, *11*
Diffuse lesion, definition of, 15
Divergence, neuronal, 37, *37*
Divergent somatosensory pathway, 93t, 94, 101–102
Dizziness, 310
Dopamine, 384
   in consciousness, *338*
   of basal ganglia, 166
   ventral tegmental area production of, 282, *283*
Dorsal column, 94–96, *95, 97,* 228t, 279t, *280*
   lesions of, *97*
   somatotopic arrangement of, 96
   vs. anterolateral column, 100
   vs. medial lemniscus, 100
Dorsal horn, of spinal gray matter, 226
Dorsal root, 224, *225*
   lesions of, 243–244
Dorsal root ganglion, 224
   infection of, 116
Dorsolateral tract (Lissauer's marginal zone), 98
Dura mater, 14, *14*
Dural sinuses, 14
Dysarthria, 191, 345, *345,* 349t
   disorders of, 270
Dysdiadochokinesia, 192
Dyslexia, 351
Dysmetria, 192
Dysreflexia, autonomic, in spinal cord injury, 241, *242*
Dystonia, 190–191

E

Ectoderm, *71*

Edema, cerebral, 376

Efferent, definition of, 26

Electromyography, 199t, 200–201, 217

   in upper motoneuron lesions, 181

Embolus, in brain infarction, 372

Emotion, behavior and, 330, *331*

   decision making and, 330

   disorders of, 346–347

   limbic system and, 328, 330, *330*

End-feet, of astrocyte, *22*

Endoderm, *71*

Endoneurium, 208, *208*

Endoplasmic reticulum, of neuron, 25t

Endorphins, 119, 385

Enkephalin, 119

Epidural hematoma, 367

Epilepsy, 350–351

Epinephrine, 136, *136*, 145t

Epineurium, 208, *208*

Epithalamus, 10–11, *11*, 318

Equilibrium, 259–261, *262*

   disorders of, 309–310, 309t

Equilibrium membrane potential, 27

Excitatory postsynaptic potentials, 46, *47*

Excitotoxicity, of neuron, 64–65, *64*

Exercise, in peripheral neuropathy, 220

Extrafusal fibers, of muscle spindle, 88–90, *90*

Extraocular muscles, *253*

   innervation of, *254*

Eye(s), accommodation of, 252t, 255, *256*

   autonomic nervous system regulation of, 144t

   disorders of, 267–268, 269–270

   examination of, 271t, 272t

   movement of, 252–253, *254*, 305–308, *307*

      cortical control of, 308, *308*

      disorders of, 311

      visually guided, 307

   sympathetic regulation of, 140

F

Face, discriminative touch information from, *95*, 96, 256, *258*

   disorders of, 268, 289, *290*

   examination of, 272t

   fast pain information from, *99*, 100, *258*

   sensation from, *95*, 96, 256, *258*

   slow pain information from, 102, *258*, 259

Facial nerve (VII), 10t, 132, *133*, *142*, 250, 251t, 259, *260*

   disorders of, 268, 289, *290*, 292

   testing of, 272t

Fasciculations, 176

Fasciculus cuneatus, 94, *95*

Fasciculus gracilis, 94, *95*

Fear, physiological response to, 139

Fetal alcohol syndrome, 79, 83t

Fibrillations, 176

Fibromyalgia, in chronic pain syndrome, 125

Flaccid dysarthria, 345, *345*

Focal lesion, definition of, 15

Forebrain, 74, *75*, 77t

   malformation of, 79

Forebrain bundle, medial, 327, *329*

Fornix, 327, *329*

Fourth ventricle, 364–365, *365*

Frontal lobe, 11, *12*

Frontopontine tract, lesions of, *293*

Functional electrical stimulation, in spinal cord injury, 243

   in upper motoneuron lesion treatment, 185

G

G proteins, 49–51, *50*

   subunits of, 49

GABA (γ-aminobutyric acid), 383

Gag reflex, 252t

Ganglion (ganglia), basal, 11–12, 164–166, *165*, 165t, *166*, 319–320

   disorders of, 185–189, *186–189*, 343

   function of, 166

   functional connections of, 165, *166*

   in motor learning, 335

   neurotransmitters of, 166

   cervical, 137–139, *138*

   paravertebral, 137–139, *138*

   stellate (cervicothoracic ganglion), 137, *138*

   lesions of, 143

Gate theory of pain, 118–119

Gaze, stabilization of, 302

Geniculate body, medial, of auditory system, 300, *300*

Genitalia, autonomic nervous system regulation of, 144t

Glia, 3, 20–22, *21*, *22*

Global aphasia, 348, 349t

Globus pallidus, 11–12, 165, *165*, 165t

Glossopharyngeal nerve (IX), 10t, 132, *133*, *142*, 250, 251t, 261, 263, *265*

   disorders of, 270

   testing of, 273t

Glutamate, 384

   in neuronal excitotoxicity, 64–65, *64*

   of basal ganglia, 166

Glycine, 383

Golgi apparatus, of neuron, 25t, *26*

Golgi tendon organ, 91, *91*

Golgi tendon organ reflex, 154, *155*
Graphesthesia, 110t
Grasping movement, 161–164, *162–164*, 196–197, *196, 197*
Gray matter, of cerebrum, 6, *7*
    of spinal cord, 7, *8*, 225–227, *226, 227*
Group therapy, cancer survival and, 332–334
Growth cone, of neuron, 77
α-Guanosine triphosphate (GTP), in G-protein–gated ion channel, *50, 51*
Guillain-Barré syndrome, demyelination in, *37*, 38

H
Habituation, 58
Hair cells, of semicircular canal, *262*
Hand(s), function of, on sensory testing, 113
    strength training of, 185
    vasomotor test of, 146
Head, trauma to, 309, 309t
Hearing, 261, *264, 301*. See also *Auditory system.*
    examination of, 273t
    loss of, 308–309
Heart, autonomic regulation of, 143, *143,* 144t
Hematoma, 367
Hemianopsia, 310–311, *311*
Hemiplegia, 176
Hemorrhage, brain, 373, *373, 374*
    subarachnoid, 373, *374*
Henneman's size principle, 152
Herniation, central, 376–377
    tonsillar, 377
    uncal, 376, *377*
Hindbrain, 74, *75,* 77t
    Arnold-Chiari deformity of, 78–79, *78*
Hippocampus, *13*
    in memory, 334
History, in neurological evaluation, 17
Holoprosencephaly, 79, 83t
Homonymous hemianopsia, 310–311, *311*
Homunculus, 96, *98*
Horizontal plane, 5, *6*
Horner's syndrome, 143
H-reflex, 156, *156*
Huntington's disease, 190, *190,* 192t
Hydrocephalus, 367, *367*
Hyperkinetic disorders, 189–191, *190*
Hyperreflexia, electromyographic definition of, 181
    vs. hypertonia, 180–181
Hypertonia, 177
    electromyographic definition of, 181
    treatment of, 184
    vs. hyperreflexia, 180–181

Hypoglossal nerve (XII), 10t, *250, 266, 266*
    disorders of, 270
    testing of, 273t
Hypotension, orthostatic, in spinal cord injury, 241, *242*
Hypothalamus, *8,* 10–11, *11,* 76, 317–318, *319*
    functions of, 317–318, *319*
    in autonomic nervous system function, 134
    lesions of, 144
Hypotonia, 177

I
Immune system, behavior and, 330–333, *331, 332*
Infant, stepping reflex of, 158–159
    traumatic brain injury in, 352
Inhibitory postsynaptic potentials, 46–48, *47*
Inositol triphosphate, as second messenger, 52–53, *52*
Insula, *75, 76*
Insular lobe, 11, *12*
Intellect, disorders of, 351
Internal capsule, 11, *12,* 319, *321,* 321t
    lesions of, 342
Interneurons, *25, 26*
    of motor system, 157–159, *157, 158*
    spinal, 227, 228–231, *229–232*
        in spinal cord injury, 238–239
Intrafusal fibers, of muscle spindle, 88–90, *90*
Intralaminar nucleus, 102
Intraocular muscles, innervation of, *254*
Ion channels, 27–37
    cAMP-dependent, 52, *52*
    G-protein second messenger modulation of, 51, *51*
    G-protein–gated, 49–51, *50*
    ligand-gated, 27, 29, *30,* 31, 31t, *32,* 49, *50*
    modality-gated, 27
    voltage-gated, 27, 28–34, *29, 30,* 31t, 33, *34–35*

J
Joint(s), injury to, pain of, 120–121
    receptors of, *91, 92*
    somatosensory testing of, 111t

K
Kinesthesia, directional, 110t

L
Labyrinth, 259, *261*
Lacrimal glands, autonomic nervous system regulation of, 144t
Lambert-Eaton syndrome, 54
Language, disorders of, 347–348, 349t
    evaluation of, 354t

Lateral activation system, of descending motor pathways, 161–164, *162–164*

Lateral horn, of spinal gray matter, 226

Learning, 58–59

  motor, *333, 334–335*

Learning disability, 351

Lhermitte's sign, in multiple sclerosis, 244

Ligament receptors, *91,* 92

Limbic association area, 327, *328*

Limbic association cortex, lesions of, 345–346

Limbic cortex, 327, *329*

Limbic lobe, 11, *12*

Limbic system, 12, *13,* 327–330, *329*

  connections of, 327–328, *329*

  emotion and, 328, 330, *330*

  immune system and, 330–333, *331, 332*

  in autonomic nervous system function, 134

  memory and, 328, 330, *330*

Liver, autonomic nervous system regulation of, 144t

Locked-in syndrome, 294, *294*

Locus ceruleus, *283,* 284

  in intrinsic analgesia, *119,* 120

Longitudinal fasciculus, medial, 253, 302, *304*

Long-term potentiation, 58–59, *59*

Low back pain, 124–126, *125*

Lumbar plexus, 209, *211*

Lungs, autonomic nervous system regulation of, 144t

M

Macroglia, 20–21, *21*

Magnetic resonance imaging, 2

Mamillary bodies, *8*

Mantle layer, *71,* 72

Marginal layer, *71,* 72

Masseter reflex, 252t

Mechanoreceptors, 86, 132

Medial activation system, of descending motor pathways, 159–161, *160*

Medial lemniscus, 94–96, *95,* 228t

  vs. dorsal column, 100

Median nerve, compression of, 213, 215

  conduction velocity study of, *114, 200,* 217, *218, 219*

Medulla, *8, 9, 75,* 284–286

  anatomy of, *278,* 284

  cranial nerve nuclei of, 286

  functions of, 286

  in autonomic nervous system function, 134

  in locked-in syndrome, 294, *294*

  inferior, 284–285, *285*

  lesions of, 117–118, *117,* 205, *291–292*

Medulla (*continued*)

  upper, *285,* 286

    lesions of, 290, *291–292*

Meissner's corpuscles, 87, *89,* 93t

Membrane capacitance, in action potential propagation, 35

Membrane channels, 27–37

  cAMP-dependent, 52, *52*

  G-protein second messenger modulation of, 51, *51*

  G-protein–gated, 49–51, *50*

  ligand-gated, 27, 29, *30,* 31, 31t, *32,* 49, *50*

  modality-gated, 27

  voltage-gated, 27, 28–34, *29, 30,* 31t, *34–35*

Membrane potential. See also *Action potential.*

  equilibrium, 27

  local, 29, 31, 31t, *32*

  modulation of, 28–29

  receptor, 31

  resting, 27–28, *28*

  synaptic, 31

Membrane resistance, in action potential propagation, 35

Memory, 58–59, 333–335

  declarative, 333–334, *333, 334,* 347

    evaluation of, 355t

  limbic system and, 328, 330, *330*

  procedural, *333, 334–335*

Meniere's disease, 309, 309t

Meninges, 5, 14, 365, *366*

  spinal, 227

Meningocele, spina bifida with, *80*

Meningomyelocele, 243

  spina bifida with, *80*

Mental function, evaluation of, 354t–355t, 356

Mental retardation, 81, 83t

Merkel's disks, 87, *89,* 93t

Mesencephalic nucleus, 256

Mesoderm, *71*

Metabolism, parasympathetic regulation of, 141

  sympathetic regulation of, 141

Metencephalon, 74, *75*

Methylprednisolone, in spinal cord trauma, 241

Microglia, 20, 21–22

  in disease, 22

Midbrain, *8, 9,* 74, *75,* 76, 77t, *287, 288*

  cranial nerve nuclei of, *287, 288*

  lesions of, *293*

Midsagittal plane, 5, *6*

Mitochondria, of neuron, 25t

Molecular neuroscience, 2

Mononeuropathy, 212, 213–214, 213t, *214*

  evaluation of, 217, 219, 219t, 220t

  multiple, 212, 213t, 214

Motoneurons, alpha, 151, *151,* 151t, 152
    gamma, *90,* 91, 151, *151,* 151t, 152
    lower, 150, *151,* 151–152, 151t
        disorders of, 177–178, *177, 178,* 192t, 234
    peripheral sensory input to, 152
    spinal pools of, 159, *159*
    upper (descending motor pathways), 159–164, *160, 162–164*
        disorders of, 178–185, *179,* 180t, *183, 184,* 192t, 234–235
        lateral, 161–164, *162–164*
        medial, 159–161, *160*
Motor cortex, 324–326, *326*
    lesions of, 344–345, *345*
Motor perseveration, 344, *345*
Motor planning, evaluation of, 355t
Motor planning area, 326, *326*
    lesions of, 344
Motor plate, 73, *73*
Motor skills, learning of, *333,* 334–335
Motor system, 150–172, *150, 172*
    basal ganglia of, 164–166, *165,* 165t, *166*
    central pattern generators of, 157–159, *157, 158*
    cerebellum of, 166–170, *167–169,* 170t
    control circuits of, 151, 164–170, *165,* 165t, *166–169*
    descending pathways (upper motoneurons) of, 150–151, 159–164, *160, 162–164*
        disorders of, 178–185, *179,* 180t, *183, 184,* 192t, 234–235
    disorders of, 176–201. See also specific disorders.
        basal ganglia lesions and, 185–189, *186–189,* 192t
        cerebellar lesions and, 191–192, 192t
        hyperkinetic, 189–191, *190,* 192t
        lower motoneuron lesions and, 177–178, *177, 178,* 192t
        types of, 176–177
        upper motoneuron lesions and, 178–185, *179,* 180t, *183,* 192t
    electrodiagnostic studies of, 198, 199t, 200–201
    in cats, 157–158, *157, 158*
    inhibitor circuits of, 229–231, *230–232*
    interneurons of, 157–159, *157, 158*
    lateral activation system of, 161–164, *162–164*
    lower motoneurons of, 150, *151,* 151–152, 151t
        disorders of, 177–178, *177, 178,* 192t, 234
    medial activation system of, 159–161, *160*
    motor unit of, 151–152, *151*
    nerve conduction studies of, 198, 200, *200*
    nonreciprocal inhibition in, 230–231, *231*
    reciprocal inhibition in, 230, *230*
    recurrent inhibition in, 230, *230*

Motor system *(continued)*
    reflexes of, 152–156, *153–156,* 229, *229.* See also *Reflex(es).*
    spinal region coordination of, 157–159, *157, 158, 229–232,* 230–231
    testing of, 198–201, 198t–199t, *200*
Motor units, 151–152, *151*
    number of, 152
Mouth, examination of, 273t
Movement, 150–151, *150, 172.* See also *Motor system.*
    ambulatory, 196
    disorders of, 176–201. See also specific disorders.
        basal ganglia lesions and, 185–189, *186–189,* 192t
        cerebellar lesions and, 191–192, 192t
        hyperkinetic, 189–191, *190,* 192t
        lower motoneuron lesions and, 177–178, *177, 178,* 192t
        types of, 176–177
        upper motoneuron lesions and, 178–185, *179,* 180t, *183,* 192t
    eye, 252–253, *254,* 305–308, *307, 308,* 311
    feedback in, 196–197, *197*
    feedforward in, 196–197, *197*
    fine, 161–164, *162–164,* 196–197, *196, 197*
    fractionation of, 161
        loss of, 344–345, *345*
    grasping, 196–197, *196, 197*
    gross, 159–161, *160*
    inhibition of, 229–231, *230–232*
    planning for, 196–197, *196, 197*
    postural, 159–161, *160,* 193–196, *193–195*
        adjustment to, 302, *303*
        disorders of, 309–310, 309t
        testing of, 199t
    posturography of, 194–195, *195*
    reaching, 196–197, *196, 197*
    reflex, 152–156, *153–156,* 229, *229.* See also *Reflex(es).*
    somatosensation in, 107. See also *Somatosensation.*
Multifocal lesion, definition of, 15
Multiple sclerosis, 244
    demyelination in, 39, *39,* 41
Muscarinic receptors, 53, 136, *136,* 144t, 145t
Muscle(s), 88–92, *90, 91,* 93t. See also *Motor system; Movement; Reflex(es).*
    atrophy of, 176
    autogenic inhibition of, 154, *155*
    blood flow in, 139–140
    delayed activation of, 179
    fast twitch, 152
        neuronal sculpting of, 78
    H-reflex of, 156, *156*

Muscle(s) (*continued*)
   injury to, pain of, 120–121
   innervation of, 88–92, *90, 91*
   involuntary contraction of, 176–177
   neuronal sculpting of, 78
   nonreciprocal inhibition of, 230–231, *231*
   paresis of, 176, 179, 181
   reciprocal inhibition of, 154, *154,* 230, *230*
   recurrent inhibition of, 230, *230*
   slow twitch, 152
      neuronal sculpting of, 78
   spasticity of, 177, 179–180, 180t, 181
   strength training of, 185
   stretch of, 88, 90–91, *90*
   stretch reflex of, 153–154, *153*
   synergy of, 157
   testing of, 198t–199t
   tone of, 176–177, 180–181, 180t, 184
   tonic stretch reflex of, 154
   withdrawal reflex of, 154, *155*
Muscle fibers, 78, 88–89, 152
Muscle spindles, 88–91, *90,* 153–154, *153, 154*
Myasthenia gravis, 54–55, 217
Myelencephalon, 74, *75*
Myelin sheath, 78
Myelination, 78. See also *Demyelination.*
   in action potential propagation, 35
Myelopathy, cervical, 244
Myeloschisis, spina bifida with, *80*
Myofascial pain, in chronic pain syndrome, 125
Myopathy, 217. See also *Peripheral neuropathy.*
   electromyographic features of, 200–201
Myotatic reflex, 153–154, *153*
Myotome, 72, 73, *73*

N
Narcotics, 120
Neck reflexes, 194, *194*
Neglect, 348, 350
Neospinothalamic tract, 97, 98–100, *99,* 279t
Nernst equation, 27
Nerve(s), cranial, 9–10, *9,* 10t, 249–273, *250.* See also *Cranial nerve(s).*
   peripheral, 207–220. See also *Peripheral nervous system.*
   spinal, 224–225, *225*
Nerve conduction studies, 198, 200, *200*
Nerve conduction velocity, 217, *218, 219*
   testing of, 113, *114*
Nerve roots, 224, *225,* 243–244
Nervous system, 4–5, *4.* See also specific components.
   anatomy of, 5–14. See also *Neuroanatomy.*

Nervous system (*continued*)
   development of, 70–83, 77t
      at day 28, 74–76, *75, 76*
      at days 18–26, *71, 72*
      cellular level of, 76–78
      disorders of, 78–82, *78, 80,* 83t
      embryonic stage of, 70–76, *71–77*
      fetal stage of, *71,* 76, 77t
      postnatal, 82–83
      preembryonic stage of, 70, *70*
Neural crest, *71, 72,* 73–74
Neural groove, *71, 72*
Neural plate, *71, 72*
Neural tube, *71, 72*
   defects of, 78–79, *78, 80,* 83t
Neuralgia, postherpetic, 116
Neuritis, vestibular, 309, 309t
Neuroanatomy, 5–14, 5t, *6*
   cellular level of, 5–6, *7*
   of brain stem, 8–10, *8, 9,* 10t, 278–280, *278–280,* 279t
   of cerebellum, 166–168, *167, 168*
   of cerebrospinal fluid system, *13, 14*
   of cerebrum, 10–12, *11, 12*
   of peripheral nervous system, 7, 208–211, *208*
   of spinal region, 7–8, *7, 8,* 224–227, *225–227*
   of vascular system, 14, *15*
   planes of, 5, *6*
   terms for, 5t
Neurogenic shock, 145
Neurological evaluation, 15–17, *16.* See also *Sensory examination.*
   diagnosis in, *16,* 17
   flow chart for, *16*
   history in, 17
Neuromodulators, 384–385
Neuromuscular electrical stimulation, in upper motoneuron lesion treatment, 185
Neuromuscular junction, 151, 212
   dysfunction of, 217
Neuron(s), 3, 20, 23–25
   action potential of, 27, 28–34, *29, 30,* 31t, *34–36*
      all-or-none nature of, 33
      propagation of, 35–37, *36*
      saltatory conduction of, 36–37, *36*
      threshold stimulus intensity for, 33
   adrenergic, 136, *136*
   afferent, 26
   axon hillock of, *23,* 24
   axon of, *23,* 24. See also *Axon(s).*
   bipolar, 24, *25*
   cell body (soma) of, 23, *23*

Neuron(s) (continued)
  cellular components of, 24–25, 25t, 26
  cholinergic, 135–136, 136
  convergence of, 37, 37
  death of, 64–65, 64, 77
  dendrites of, 23, 23
  depolarization of, 28, 30
    threshold for, 29
  development of, 76–77
  divergence of, 37, 37
  efferent, 26
  electrical potentials of, 27–37, 30, 31t
    absolute refractory period of, 33, 35
    modulation of, 28–29
    relative refractory period of, 33, 35
    resting, 27–29, 28, 29
    spatial summation of, 31, 32
    temporal summation of, 31, 32
    threshold stimulus intensity of, 33, 35
  ephaptic transmission (cross-excitation) of, 124
  equilibrium potential of, 27
  excitotoxicity of, 64–65, 64
  first, 100
  growth cone of, 77
  hyperpolarization of, 28
  information transmission in, 26, 27–37, 28–30, 31t, 32, 34–37
  interaction between, 37, 37
  local potential of, 29, 31, 31t, 32
  membrane channels of, 27–37. See also Membrane channels.
  multipolar, 24, 25
  peripheral, somatosensory, 87
  postganglionic, 135
  preganglionic, 135
  presynaptic terminals of, 23, 24
  projection, 92
    ascending, 100–101
  pseudounipolar, 24, 25
  receptor potential of, 31, 31t
  resting membrane potential of, 27–28, 28
    changes from, 28–29, 29
  spinoreticular, 102
  structure of, 23–25, 23, 25, 25t
  synapse of, 23, 24, 44–46, 45. See also Synapse.
  synaptic potential of, 31, 31t
  types of, 24, 25
Neuropathy, 115. See also Peripheral neuropathy.
Neuroplasticity, 58–67
  cortical reorganization in, 61–63, 63
  experimental manipulation of, 66–67

Neuroplasticity (continued)
  in axonal injury, 60, 60, 61
  in habituation, 58
  in learning, 58–59
  in memory, 58–59, 59
  in neurological disorder treatment, 66–67
  neurotransmitter release in, 63–64
  rehabilitation effects on, 65–66
  synaptic changes in, 60–61, 62
Neuropore, 71
Neurotransmitters, 24, 44–46, 44, 45, 53–55, 53t, 383–384
  activity-related release of, 63–64
  agonists of, 53–54
  antagonists of, 53–54
  brain stem production of, 282–284, 283, 338
  G-protein–mediated receptor binding of, 49–53, 50–52
  ligand-gated channel binding of, 49, 50
  of autonomic nervous system, 135–136, 136
  receptors for, 49–53, 50–52, 385t
Nicotinic receptors, 53, 136, 136, 145t
Nociceptors, 86, 88, 132. See also Pain.
Nodes of Ranvier, in action potential propagation, 35–36, 36
Nonreciprocal inhibition, 230–231, 231
Nonverbal communication, disorders of, 348
Norepinephrine, 136, 136, 145t, 384–385
  formation of, 283, 284
  in consciousness, 338
Nucleus, of neuron, 25t
Nucleus (nuclei), abducens, 254
  caudate, 165, 165
  cochlear, 286
  cuneate, 95, 96, 103
  gracile, 95, 96
  intralaminar, 102
  mesencephalic, 256
  oculomotor, 254, 254
  of brain stem, 279, 282–284, 283
  of medulla, 286
  of midbrain, 287, 288
  of pons, 165, 282, 283, 287, 287–288
  olivary, 285, 286
  pedunculopontine, 165, 282, 283, 287, 288
  raphe, 282, 283, 284
    in intrinsic analgesia, 119–120, 119
  red, 165, 165, 280, 287, 288
    lesions of, 293
  reticular, 282–284, 283
  solitary, 132, 134
  subthalamic, 165, 165, 165t
  thalamic, 11, 95, 96, 118, 316–317, 317, 318t
    in slow pain reception, 102

Nucleus (nuclei), abducens (*continued*)
trigeminal, *95, 96,* 259
trochlear, *254*
ventral posterolateral, *95, 96*
lesions of, 118
ventral posteromedial, *95, 96*
lesions of, 118
vestibular, 286, 302–303, *304*
lesions of, *292*
Nucleus accumbens, 327
Nucleus ambiguus, lesions of, *292*
Nucleus basalis of Meynert, 327, *329*
Nucleus dorsalis (Clark's nucleus), 102–103
Nucleus gracilis, *95, 96*
Nystagmus, 307
pathological, 311

O
Obtundation, 294t
Occipital lobe, 11, *12*
Oculomotor complex, *287, 288*
Oculomotor nerve (III), 10t, *142, 250,* 251–255, 251t, *254*
disorders of, 267–268, *293*
parasympathetic fibers of, 253–254
testing of, 271t
Oculomotor nucleus, 254, *254*
Olfactory nerve (I), 10t, *250,* 251, 251t
disorders of, 267
testing of, 271t
Oligodendrocyte, 20, *21*
Olivary nucleus, *285, 286*
Olive, *8, 278, 284*
Optic nerve (II), 10t, *250,* 251, 251t, *253,* 254t
disorders of, 267
testing of, 271t
Optic tract, *8*
Orbitofrontal cortex, lesions of, 345–346, *347*
Organ of Corti, 261, *263*
Orientation, evaluation of, 354t
Orthostatic hypotension, in spinal cord injury, 241, *242*
Oscillopsia, 310
Ossicles, 261
Otoconia, 260–261, *262*
Otoliths, 260–261, *262*
Otolithic organs, 260, *261, 262,* 301–302
Oval window, *261*

P
Pacinian corpuscles, of skin, 87, *89,* 93t
Paciniform corpuscles, of joint, *91, 92*

Pain, 118–126
acute, 123t
analgesic systems and, 119–120, *119*
chronic, 123–126, 123t, *125*
central nervous system dysfunction in, 124
ectopic foci in, 123–124
ephaptic transmission in, 124
neuropathic, 123–124
nociceptive, 119, *120,* 123
psychological explanations of, 125–126
surgical treatment of, 124
sympathetic nervous system dysfunction in, 124
with amputations, 124
counterirritant theory of, 119, *119*
discriminative, 98–100, *99, 101*
fast, 98–100, *99*
testing of, 111t
gate theory of, 118–119
in muscle injury, 120–121
inhibition of, 119, *119,* 120
intensification of, 120
receptors for, 86, 88, *88,* 132
referred, 121–122, *122*
slow, 100–102
sympathetic nervous system in, 144–145, *146*
transmission of, 120, *120*
Paleospinothalamic tract, 98, *99,* 102
Pancreas, autonomic nervous system regulation of, 144t
Panic disorder, 384
Parahippocampal gyrus, *13,* 327, *329*
Paralysis, 176
botulinum toxin A–induced, 53
Paraplegia, 176, 238
prognosis for, 241–243, 242t
Parasympathetic nervous system (craniosacral outflow), 141–143, *142, 143*
functions of, 141
vs. sympathetic nervous system, 141, 143, *143,* 144t, 145t
Paravertebral ganglia, 137–139, *138*
Paresis, 176
electromyographic definition of, 181
in upper motoneuron syndrome, 179
Paresthesias, 115
in chronic pain, 123
Parietal lobe, 11, *12*
lesions of, 197
Parietotemporal association area, 327, *328*
lesions of, 346, *346*
Parkinsonism, 189

Parkinson's disease, 185–188, *186–189,* 192t
  treatment of, 187–188
Pedunculopontine nucleus, 165, 282, *283, 287,* 288
Periaqueductal gray, *287,* 288
  in intrinsic analgesia, *119,* 120
Perineurium, 208, *208*
Peripheral nerves. See also *Cranial nerve(s).*
  blood supply to, 368, *368*
  compression of, 115, 213, *214*
  crush injury of, 213–214
  cutaneous, 87–88, *87, 89*
  entrapment of, 213, 215
  severance of, 115, 143, *214*
Peripheral nervous system, 4, *4,* 208–220
  anatomy of, 7, 208–211, *208*
  axons of, 87, *87,* 209, 209t
  dysfunction of, 212–217, 213t. See also *Peripheral neu-
    ropathy.*
  neuromuscular junction of, 212, 217
  plexuses of, 209, *210, 211*
Peripheral neuropathy, 212–217, 213t
  electrodiagnostic studies in, 217, *218, 219*
  evaluation of, 217, 219, 219t, 220t
  exercise for, 220
  multiple-nerve, 213t, 214–217, *215, 216*
  single-nerve, 213–214, 213t, *214,* 219t
  treatment of, 219–220
  vs. spinal region lesion, 220t, 235, 235t
Personal neglect, 348, *350*
Personality, disorders of, 345–346, *346*
Phantom limb sensation, 124
Phasic stretch reflex, 153–154, *153*
Phenylketonuria, 351
Pia mater, 365, *366*
Plaques, in multiple sclerosis, 39, *39*
Plasmapheresis, in Guillain-Barré syndrome, 38
Plasticity, 65–66
Poliomyelitis, *177,* 177–178, *178*
Polyneuropathy, 212, 213t, 214–217, *215, 216*
Pons, *8, 9, 75,* 286–288, *287*
  in autonomic nervous system function, 134
  lesions of, 118, *291–292*
  nuclei of, 165, 282, *283, 287,* 287–288
Positron emission tomography, in cerebral blood flow eval-
    uation, 377, *377*
Postherpetic neuralgia, 116
Postpolio syndrome, 177–178
Postsynaptic potentials, 46–48, *47, 48*
  excitatory, 46, *47*
  inhibitory, 46–48, *47*
Postsynaptic terminal, 44, *44, 45*

Posttraumatic stress disorder, 384–385
Posture, adjustment to, 302, *303*
  control of, 159–161, *160,* 193–196, *193–195*
  disorders of, 309–310, 309t
  testing of, 199t
Posturography, 194–195, *195*
Prefrontal association cortex, lesions of, 345
Prefrontal cortex, 327, *328*
Premotor area, 326, *326*
  lateral, 161–162, *164*
Preoptic area, 327
Presynaptic facilitation, 48–49, *48*
Presynaptic inhibition, 48–49, *48*
Presynaptic terminal, *23, 24,* 44, *44, 45*
Pretectal area, 288
Projection fibers, of subcortical white matter, 319, *320,
    321,* 321t
Propranolol, 384
Proprioception, conscious, 94–96, *95, 97*
Proverbs, interpretation of, 355t
Psychological factors, in chronic pain syndrome, 125–126
Pupil, reflexes of, 252t, 254–255, *255*
Putamen, 11–12, 165, *165,* 165t
Pyramid, *8,* 284
Pyramidal cells, of cerebral cortex, 320, *322*

Q
Quadriplegia, 238

R
Radiculopathy, 243–244
Raphe nuclei, 282, *283,* 284
  in intrinsic analgesia, 119–120, *119*
Raphespinal tract, 164, 228t
Reaching movement, 161–164, *162–164,* 196–197,
    196, 197
Reading, 335, *336*
Receptive field, of peripheral sensory neurons, 87, *87*
Receptor(s), adrenergic, 136, *136,* 140, *140,* 144t, 145t
  agonists of, 140
  blockers of, 140
  cholinergic, 135–136, *136*
  G protein–mediated, 49–51, *50*
  joint, *91, 92*
  ligament, *91, 92*
  muscarinic, 53, 136, *136,* 144t, 145t
  nicotinic, 53, 136, *136,* 145t
  pain, 86, 88, *88,* 132
  phasic, 86
  sensory, 86, 87–88, *89,* 93t
  synaptic, 49–53, *50–52*

Receptor(s), adrenergic (*continued*)
  thermal, 86, 88, *88*
  tonic, 86
  touch, 87–88, *89,* 93t
Reciprocal inhibition, 154, *154,* 230, *230*
Recurrent inhibition, 230, *230*
Red nucleus, 165, *165, 280, 287, 288*
  lesions of, *293*
Referred pain, 121–122, *122*
Reflex(es), 152–156, 229, *229*
  consensual, 254–255, *255*
  corneal, 252t, 259
  cranial nerve, 252t
  crossed extension, 229, *229*
  cutaneous, 154, *155*
    abnormalities of, 178–179, *179*
  H, 156, *156*
  masseter, 252t
  movement and, 154, 156
  neck, 194, *194*
  phasic stretch, 153–154, *153*
    hyperreflexia of, 180–181
  proprioceptive, 153–154, *153*
  pupillary, 252t, 254–255, *255*
  stepping, 158–159
  swallow, 267t
  tendon organ, 154, *155*
  testing of, 198t
  tonic stretch, 154
  vestibulo-ocular, 194, 306–307, *307*
  withdrawal, 154, *155,* 229, *229*
Reflex sympathetic dystrophy, 145
Refractory period, absolute, 33
  relative, 33, 35
Relative refractory period, 33, 35
Relay nuclei, of thalamus, 316–317, 318t
Resting membrane potential, 27–28, *28*
Reticular formation, 280–282, *281*
  in auditory system, 300, *300*
  lateral zone of, 281, *281*
  medial zone of, 281, *281, 284*
  midline zone of, 281–282, *281*
Reticular nuclei, 282–284, *283*
Reticulospinal tract, lateral (medullary), *162,* 228t
  medial, 160–161, 228t
  pontine (medial), *160*
Retina, 303–304, *305, 306*
Retinogeniculocalcarine pathway, 303–304, *306*
Rexed's laminae, 226, *227*
Rhizotomy, dorsal, in chronic pain, 124
    in spastic cerebral palsy, 243
Ribosomes, of neuron, 25t

Rigidity, 177
Romberg's test, 199t
Rostrospinocerebellar pathway, *103,* 104
Rubrospinal tract, 162, *162,* 228t
Ruffini's endings, of joint, *91, 92*
  of skin, 87–88, *89,* 93t

S
Saccule, 260, *261, 262*
Sacral plexus, 209, *211*
Sagittal plane, 5
Salivary glands, autonomic nervous system regulation of, 144t
  sympathetic regulation of, 140
Saltatory conduction, in action potential propagation, 35–36, *36*
Scalp, sympathetic regulation of, 140
Schizophrenia, 356
Schwann cell, 20, *21*
Sclerotome, 72, *73*
Second messengers, 51–53, *51, 52*
Semicircular canals, 259–261, *261, 262,* 301, *302*
Sensation. See *Somatosensation; Somatosensory system.*
Sensorineural deafness, 308
Sensory association areas, 96, 324, *325*
  lesions of, 343–344, *344*
Sensory cortex, 96
  conscious relay pathway to, 92, 93t, 94–100, *95, 97, 99, 101*
  somatotopic arrangement of, 96, *98*
Sensory examination, 108–113, *109,* 110t–111t, *112*
  brain stem lesions on, 117–118, *117*
  cortical lesions on, 118
  electrodiagnostic studies in, 113–115, *114*
  interpretation of, 109, 113
  nerve conduction velocity testing in, 113, *114*
  peripheral nerve lesions on, 115
  quick, 109
  reporting of, *112*
  somatosensory evoked potentials in, 113, 115
  spinal lesions on, 115–116, *116*
  thalamic lesions on, 118
Sensory extinction, 118
Sensory inattention, 118
Sensory receptors, 86, 87–88, *89,* 93t
Septal area, 327
Serotonin, 384
  in consciousness, *338*
  raphe nuclei production of, 282, *283,* 284
Sexual function, dysfunction of, 238
  spinal control of, 232
Shingles, 116, *116*

Shock, neurogenic, 145
Sight, 303–305, *305, 306*
  blind, 311
  vestibulo-ocular reflexes in, 306–307, *307*
Skin. See also *Somatosensation; Somatosensory system.*
  innervation of, 87–88, *87, 89*
  pain receptors of, 88, *88*
  receptive fields of, 87, *87*
  sensations from, 87–88, *88, 89*
  thermal receptors of, 88, *88*
  touch receptors of, 87–88, *89*
Sleep, function of, 284
Smell, disorders of, 267
  examination of, 271t
Smile, 267
Sodium channel, 33
Sodium ion channel, 28, *29*
Sodium-potassium (Na+-K+) pump, in action potential, 33, *34*
  in resting membrane potential, 27–28, *28*
Solitary nucleus, 132, *134*
Somatic motor system, 4
Somatosensation, 107–128, *108–113*
  examination of, 108–113, *109,* 110t–111t, *112*
    electrodiagnostic studies in, 113–115, *114*
    nerve conduction velocity testing in, 113, *114*
    reporting of, *112*
    somatosensory evoked potentials in, 113, *115*
  in movement, 107
  loss of, 108
    map of, 109, 113
    self-injury with, 108
Somatosensory association areas, 96, 324, *325*
  lesions of, 324, *325,* 343–344, *344*
Somatosensory cortex, 323–324, *325*
  lesions of, 118, 343, *343*
Somatosensory evoked potentials, 113, *115*
Somatosensory system, 4, 85–104
  axon diameters in, 92, 93t
  conscious relay pathways of, 92, 93t, 94–100, *95, 97–99, 101*
  cutaneous innervation of, 87–88, *87–89*
  divergent pathways of, 93t, 94, 101–102
  examination of, 108–113. See also *Sensory examination.*
  Golgi tendon organs of, 91, *91*
  joint receptors of, *91,* 92
  musculoskeletal innervation of, 88–92, *90*
  neurons of, 87, *87*
  receptive fields of, 87, *87*
  receptors of, 86, 87–92, *88, 91,* 93t
  unconscious relay pathways of, 93t, 94, 102–104, *103*
  vs. autonomic nervous system, 135

Somatotopy, of sensory cortex, 92, 96, *98*
Somite, *71, 72, 72*
Spasms, 176
Spastic dysarthria, 345, *345*
Spasticity, 177, 180, 180t, 181
  in upper motoneuron syndrome, 179–180, 180t
  treatment of, 184
Spatial neglect, 348, 350
Spatial relationships, 336
  comprehension of, evaluation of, 355t
Spatial summation, of local membrane potential, 31, *32*
Speech, 266, 267, 335, *336*
  development of, critical period for, 82–83
  disorders of, 270
Spina bifida, 79, *80,* 81, 83t
Spina bifida cystica, *80*
Spinal cord, anatomy of, 7–8, *7, 8,* 224–227, *225, 226*
  anterolateral tracts of, 94, 97–100, *99, 101*
  blood supply to, 368, *369*
  chronic injury of, 238–239, *240*
  compression of, 244
  cross section of, *7, 8*
  development of, 74, *74*
    disorders of, 243
  dorsal columns of, 94–96, *95, 96*
  dorsal horn of, 226
  functions of, 227–232, 228t, *229–233*
  gray matter of, *7, 8,* 225–227, *226, 227*
  hemisection of, 115–116
  internal structure of, 225–227, *226*
  interneurons of, 228–229
  lateral horn of, 226
  meninges of, 227
  motor coordination by, 157–159, *157, 158,* 229–231, *230–232.* See also *Reflex(es).*
  Rexed's laminae of, *226, 227*
  sacral, 231–232, *233*
    lesions of, 237–238, *237*
  segments of, 224, *225*
  syrinx of, 244
  transection of, 115, 143, *189*
  traumatic injury to, 143, 182, 238–243
    autonomic dysfunction in, 241, *242*
    autonomic dysreflexia in, 241, *242*
    classification of, 239–240, *240*
    complete, 239
    incomplete, 239
    interneuron abnormalities in, 238–239
    neurologic level of, 239–240, 239t, *240*
    orthostatic hypotension in, 241, *242*
    prognosis for, 241–243, 242t
    sensory syndrome with, 115–116

Spinal cord (continued)
  traumatic injury to (continued)
    thermoregulatory dysfunction in, 241, 242
    treatment of, 241–243, 242t
  ventral horn of, 226
  white matter of, 7, 8, 225, 226
Spinal nerves, 224–225
  rami of, 225
Spinal region, 4, 4, 224–244. See also Spinal cord.
  anatomy of, 224–227, 225–227
  compression in, 244
  developmental disorders of, 243
  dorsal root of, 224, 225
    lesions of, 243–244
  interneurons of, 227, 228–229
  lesions of, 232–243. See also specific disorders.
    segmental dysfunction and, 232, 233–234, 234
    vertical tract dysfunction and, 232–233, 234, 234–235
    vs. peripheral region lesions, 220t, 235, 235t
  meninges of, 227, 229–231, 229–232
  nerves of, 224–225, 225
  tumors of, 244
  ventral root of, 224
    lesions of, 243–244
Spinal shock, 238
Spinocerebellar tract, 102–104, 228t, 279t, 280
  anterior, 103, 103–104
  posterior, 102–103, 103
Spinocerebellum, 169, 169, 170t, 171
Spinomesencephalic tract, 99, 101–102
Spinoreticular tract, 99
Spinothalamic tract, 228t, 280
  lesions of, 292, 293
Splanchnic nerves, 139
Stellate ganglion (cervicothoracic ganglion), 137, 138
  lesions of, 143
Stepping reflex, of infant, 158–159
Stereognosis, conscious, 94
  evaluation of, 355t
  testing of, 111t
Stress response, 331–332
Striate arteries, occlusion of, 374
Stroke, 182, 183, 188, 352–353, 356, 372–373, 373
  metabolic effects of, 64, 64–65
  paresis after, 179
  recovery from, 353, 356
  signs and symptoms of, 353, 373–374, 375
  types of, 372–373, 373
Strychnine, 56
Stupor, 294t
Subarachnoid hemorrhage, 373, 374

Subdural hematoma, 367
Substance P, 100, 385
Substantia nigra, 165, 165, 165t, 287, 288
Subthalamic nucleus, 165, 165, 165t
Subthalamus, 319
Sudeck's atrophy, 145
Summation, of local membrane potential, 31, 32
Supplementary motor area, 161–162, 164, 326, 326
Sural nerve, biopsy of, 216
Swallowing, 252t, 266, 267t
  disorders of, 270
Sweat glands, autonomic nervous system regulation of, 144t
Sweat test, 146
Sympathetic nervous system (thoracolumbar outflow), 137–141, 137, 138, 140
  adrenal medulla of, 137, 138
  efferent neurons of, 137–139, 137, 138
    to abdominal organs, 138, 139
    to periphery, 137–139, 138
  functions of, 139–141
    in blood flow regulation, 139–140
    in body temperature regulation, 139
    in brain stem injury, 143
    in cerebral injury, 144
    in eye regulation, 140
    in heart regulation, 140, 140–141
    in metabolism, 141
    in pain maintenance, 144–145, 146
    in pain syndromes, 144–145
    in peripheral injury, 143
    in spinal injury, 143
    in syncope, 145–146
    in visceral regulation, 140, 140–141
  paravertebral ganglia of, 137–139, 138
  tests of, 146
  vs. parasympathetic nervous system, 141, 143, 143, 144t, 145t
Synapse, 23, 24, 44–46, 45
  axo-axonic, 44
  axo-dendritic, 44
  axo-somatic, 44
  denervation hypersensitivity of, 61, 62
  electrical potentials at, 46–49, 47, 48
  G-protein mediated receptors of, 49–51, 50
  hypereffectiveness of, 61, 62
  injury-related changes in, 60–61, 62
  ligand-gated ion channels of, 49, 50
  neurotransmitter action at, 24, 44–46, 44, 45, 49–53. See also Neurotransmitters.
  postsynaptic membrane of, 44, 44, 45, 46–48, 47

Synapse (continued)
  presynaptic terminal of, 23, 24, 44, *44, 45*, 48–49, *48*
  receptors of, 49–53, *50–52*
  second-messenger systems of, 51–53, *51, 52*
  silent, unmasking of, 61
  structure of, 44, *44*
Synaptic cleft, 24, 44
Syncope (fainting), 139–140, 145–146, 294t
Synergy, muscle, 157
Syringomyelia, 244
Systems neuroscience, 3

T

Tardive dyskinesia, 384
Tectospinal tract, 160, *160*, 228t
Tectum, 280, *280*, 288
Tegmentum, 280, 287, 288
Telencephalon, *75*, 76
Temperature, body, dysregulation of, in spinal cord injury, 241, 242
    regulation of, 139
  perception of, testing of, 111t
  receptors for, 86, 88, *88*
  sensation of, 97–98, *101*, 102
Temporal lobe, 11, *12*
Temporal summation, of local membrane potential, 31, *32*
Tendon organ reflex, 154, *155*
Tetanus toxin, 56
Tetraplegia, 176, 238
Tetrodotoxin, 56
Thalamus, 10–11, *11, 13*, 76
  in autonomic nervous system function, 134
  lesions of, 118, 342
  nuclei of, 316–317, *317*, 318t
    in slow pain reception, 102
Thermoreceptors, 86, 88, 132
Third ventricle, 364, *365*
Thrombus, in brain infarction, 372
Tic douloureux, 268, 269
Tiltboard, 199t
Tinel's sign, in chronic pain, 124
Tinetti balance scale, 199t
Tongue, disorders of, 270
  examination of, 273t
Tonic stretch reflex, 154
Tonsillar herniation, 377
Touch, 87–88, *89*, 93t
  discriminative, 94–96, *95, 97*
  tests of, 110t
  two-point discrimination test of, 109, *109*, 110t
Tract cells, 225

Transcutaneous electrical nerve stimulation, 118–119
  in upper motoneuron lesion treatment, 184–185
Transient ischemic attack, 353, 372
Trigeminal nerve, in nociception, *99*, 100
  in touch reception, *95*, 96
  varicella zoster infection of, 116, *116*
Trigeminal nerve (V), 10t, *250*, 251t, 255–259, *257, 258*
  branches of, 255, *257*
  disorders of, 268, 269
  testing of, 272t
Trigeminal neuralgia, 268, 269
Trigeminal nucleus, *95, 96*, 259
Trigeminoreticulothalamic pathway, *99*, 102, 259
Trigeminothalamic tract, *99*
  lesions of, 293
Trisomy 21 (Down syndrome), 83t, 351
Trochlear nerve (IV), 10t, *250*, 251–255, 251t, *254*
  disorders of, 268
  testing of, 271t
Trochlear nucleus, 254
Tumors, spinal, 244
  stress response and, 332
  survival with, 332–333
Two-point discrimination, 109, *109*, 110t
Tympanic membrane, 261, *264*

U

Unconscious relay somatosensory pathway, 93t, 94, 102–104, *103*
Uncus, 327
  herniation of, 376, 377
Upper motoneuron syndrome, 178–181, *179*, 180t
Urination, dysfunction of, 237, 237–238
  spinal control of, 231–232, *233*
Utricle, 260, *261, 262*

V

Vagus nerve (X), 10t, 132, *133, 142, 250*, 251t, 265, *265*
  disorders of, 270, 292
  testing of, 273t
Varicella zoster, cranial nerve infection with, 116, *116*
Vasodepressor syncope, 145
Vasomotor test, 146
Vasovagal attack, 145
Vegetative state, 294t
Veins, blood pooling in, 139–140
  of brain, 377–378, *378*
Ventral horn, of spinal gray matter, 227
Ventral posterolateral nucleus, *95, 96*
  lesions of, 118

Ventral posteromedial nucleus, in touch sensation, *95, 96*
  lesions of, 118
Ventral root, 224
  lesions of, 243–244
Ventral tegmental area, 282, *283*
Ventricle(s), *13,* 14, 364–366, *365*
  formation of, 76, *76*
  fourth, *13*
  lateral, *13*
  third, *13*
Vertebral artery (arteries), 14, *15,* 368, *369*
  occlusion of, 373
Vertigo, 309, 309t
Vestibular apparatus, 259–261, *261, 262*
Vestibular cortex, 324, *325*
Vestibular neuritis, 309, 309t
Vestibular nuclei, 286, 302–303, *304*
  lesions of, *292*
Vestibular reflexes, 194
Vestibular system, 301–303
  disorders of, 309t, 309–310
  evaluation of, 272t–273t
  in motor control, 302, *303*
  otolithic organs of, 301–302
  physical therapy for, 310
  receptors of, 301
  semicircular canals of, 301, *302*
Vestibulocerebellum, 169, *169,* 170t, *171*
Vestibulocochlear nerve, *261*
Vestibulocochlear nerve (VIII), 10t, *250,* 251t, 259–261,
    261–263
  disorders of, 269–270
  testing of, 272t–273t

Vestibulo-ocular reflexes, 306–307, *307*
Vestibulospinal tract, lateral, *160,* 161, 228t
  medial, *160,* 161, 228t
Vibration, perception of, testing of, 111t
Viscera, autonomic nervous system regulation of, 144t
  central regulation of, 132–135, *133, 134*
  sympathetic regulation of, 140–141, *140*
Vision, 303–305, *305, 306*
  vestibulo-ocular reflexes in, 306–307, *307*
Visual agnosia, 344, *344*
Visual association area, 305, 324, *325*
Visual cortex, 304–305, *306,* 324, *325*
  development of, critical period for, 82
  lesions of, 311
Visual system, 303–308, 336–337, *337*
  disorders of, 310–311, *311, 350*
  for eye movement, 305–308, *307, 308*
  for sight, 303–305, *305, 306*
Voiding, dysfunction of, 237–238, *237*
  spinal control of, 231–232, *233*

W
Walking, development of, critical period for, 82–83
Wallerian degeneration, 60, *60*
Wernicke's aphasia, 348, 349t
Wernicke's area, 335, *336*
  of auditory system, 300
White matter, of cerebrum, 6, *7*
  of spinal cord, 7, *8,* 225, *226*
Withdrawal reflex, 154, *155,* 229, *229*

Y
Yohimbine, 385